UNDERSTANDING CHEMISTRY

Milwaukee

THE WILEY SENIOR CHEMISTRY PROGRAM

First Year Chemistry Course

UNDERSTANDING CHEMISTRY
UNDERSTANDING CHEMISTRY LABORATORY MANUAL
UNDERSTANDING CHEMISTRY TEACHER'S RESOURCE GUIDE (for textbook and Laboratory Manual)

Second Year Chemistry Course

FUNDAMENTALS OF CHEMISTRY, THIRD EDITION
FUNDAMENTALS OF CHEMISTRY TEACHER'S MANUAL
FUNDAMENTALS OF CHEMISTRY LABORATORY MANUAL
FUNDAMENTALS OF CHEMISTRY TEACHER'S MANUAL FOR LABORATORY MANUAL

SI METRIC

Why do fireworks such as those shown on our cover produce different colours? The manufacturers of the fireworks have used their knowledge of the characteristics of metals to produce the different effects that you see. As the fireworks are lighted, the different metals in compounds of the fireworks produce different colours. For example, red colours in fireworks result from the addition of strontium compounds and yellow colours from the addition of sodium compounds. In Chapter Two – *The Nature of Matter,* you will learn more about this characteristic of metals. Each metal, you will find, has a different colour when held in a flame.

Cover Photo: Mike Dobel / Masterfile

UNDERSTANDING CHEMISTRY

Herman J. Bruckman
Sandwich Secondary School

Allan Cruickshanks
W.F. Herman Secondary School

Find the symbols for ~~these~~ the elements. You
will find a hot message.

Flurione (9), Uranium (92), Carbon (6), Potassium (19)
F   U   C   K

Yttrium (39), Oxygen (8), Uranium (92)
Y   O   U

Bismuth (83), Technetium (43), Hydrogen (1)
Bi   Tc   H

I was here but now I'm <u>not</u>,
I went out to smoke a <u>pot</u>,
I left this here to prove a <u>point</u>,
Life is nothing without a <u>joint</u>,

F

John Wiley & Sons
Toronto   New York   Chichester   Brisbane   Singapore

Communications Branch, Consumer and Corporate Affairs Canada has granted permission for the use of the National Symbol for Metric Conversion.

Canadian Cataloguing in Publication Data

Cruickshanks, Allan
  Understanding chemistry

For use in high schools.
Includes index.
ISBN 0-471-79684-0

1. Chemistry   I. Bruckman, Herman.   II. Title.

QD33.C78 1988   540   C87-095087-8

Design: Michael van Elsen Design Inc.
Illustration: James Loates *Illustrating*
Typesetter: Q-Composition (1984) Inc.
Assembly: Compeer Typographic Services Ltd.
Printer: The Bryant Press Limited

Printed and bound in Canada.

10   9   8   7   6   5   4   3   2

TABLE OF CONTENTS

APPENDICES

SPECIAL TOPICS IN THIS TEXTBOOK

Acknowledgements

The authors would like to thank the staff of John Wiley & Sons Canada Limited for their help during the past three years. Special thanks are extended to Trudy Rising and Doug Macnamara for their early support and initiative, and to Noreen Rankin for her help with photo research. Most of all, we must thank Jonathan Bocknek for his encouragement, patience, ideas, enthusiasm, and guidance.

We must also acknowledge the contributions of our reviewers, many of whose ideas have been incorporated into the textbook. However, we alone are responsible for the final content of the book.

Doug Abe, Sir Wilfrid Laurier Collegiate Institute
Carolyn Anco, Don Bosco Separate School
Peter Au, Smith Falls District Collegiate Institute
Jack Elgood, Perth & District Collegiate Institute
Reginald Friesen, Department of Chemistry, University of
   Waterloo
David Hopkins, Ashbury College
David Knight, General Brock High School
Robert Richardson, Nelson High School
Paul Tamblyn, Acton District High School
Pash Ummat, Port Dover Composite School
Doug Wrigglesworth, Sir Sandford Fleming Secondary School
Harold Wright, Oakwood Collegiate Institute

For his role in writing and coordinating much of the Laboratory Manual to accompany this textbook, we would like to express our appreciation to David Lynn. Our sincere thanks also go to the reviewers for the Laboratory Manual:

John Stratton, Lester B. Pearson Collegiate Institute
Doug Wrigglesworth, Sir Sandford Fleming Secondary School

We also thank David Knight, David Humphreys, Jane Forbes, Robert Richardson, and Alan Blizzard for their permission to use selected experiments from the *Structure and Change Laboratory Manual.*

For writing the Putting Chemistry to Work features, we thank Julie E. Czerneda. Further thanks to Barb Baggio, Cheryl Brooks, and Cindy Drouillard for their assistance. And for taking many of the photographs which appear in this book we are grateful to S.T. Fox.

Finally, we especially thank our families. They were always supportive, encouraging, understanding, and patient. Without them, our task would have been impossible.

Herman J. Bruckman and Allan Cruickshanks

To the Student

How This Textbook Can Help You Understand Chemistry

One of the exciting things about chemistry (and science in general) is the fact that, regardless of how much we know about matter and its changes, there is still so much to learn and explain. One very important function of this textbook is to present the basic principles, concepts, and theories of chemistry as part of a framework necessary to explain observations. We consider it important that you understand the usefulness of theories, as well as their limitations – how theories are proposed, amended, and/or rejected in the light of experimental observations. The treatment of atomic theory in Chapter 2 is an excellent example of how a theory becomes amended and expanded to explain what scientists have observed.

Whether you know it or not, chemistry affects your life every day. At the very least, by the time you finish this course, you will have accomplished three things:

1. You will understand basic concepts and processes of chemistry.

2. You will know how to handle chemicals and apparatus safely and intelligently.

3. Both 1 and 2 will help you better appreciate and understand the role chemistry plays in the products and services you use, as well as in potential career choices open to you. In addition, you will benefit from being able to make more informed decisions about sensitive issues involving the impact of science and technology on society.

These basic "goals" can, for example, help you answer questions such as these: Why did the investigations of 19th century chemists enable me to watch my favourite television show last night? What special precautions should I take when I clean a dirty oven? How can understanding the structure and nature of the atomic nucleus help me form intelligent opinions about the use and/or misuse of nuclear energy?

How You Can Help Yourself Understand Chemistry

Understanding anything – whether it is life-saving techniques around a swimming pool, your favourite trumpet solo, or even a chemistry lesson – is an active process. *Understanding Chemistry* is very much a part of this process, providing an approachable and interesting framework for your investigations into chemistry. The end result? Nothing less than developing an understanding of your role in our "chemical" society. To further assist you in this process, we would like to offer the following "study hints" to help you in your studies. (You can also modify these hints to help you in other school courses, as well as in any recreational activity.)

1. Try to *review* material covered in class the same day you learn it. Reinforce your understanding by answering all the questions in the textbook – even if they are not assigned.

2. Of course, *read* material that has been covered in class, but also try to prepare for class by reading material ahead of time. It is also helpful to read or view materials from sources other than this textbook, such as science magazines, newspapers, and appropriate television programs.

3. *Take notes* in class. Devise your own shorthand methods of recording ideas discussed in class. Do not depend only on chalkboard notes written by your teacher.

4. Use class notes and the textbook to *prepare summaries* for study purposes. Studying is most effective with pen or pencil in hand so you can write down the important ideas.

5. *Practice makes perfect* is as true for chemistry as it is for playing piano, shooting baskets, or acquiring any other skill. Skills such as problem solving, naming chemicals, and balancing equations should be practised so that they become automatic.

6. When your teacher assigns experiments from the Laboratory Manual, *prepare* for your laboratory work. You should understand where laboratory experiments "fit" within the context of the course. Laboratory work is an important part of any chemistry program, and is essential for understanding chemical concepts.

7. *Schedule* your study time to avoid that most ineffective of all study methods, "cramming" before tests and examinations.

8. *Understand both your strengths and your weaknesses.* Take advantage of all opportunities to get help with areas in which you may have trouble, and use your strengths to help you (and others) in all areas of the course. Share your talents with your classmates. You may be able to help them in some parts of the course, and, in turn, you may receive help from them.

Herman J. Bruckman and Allan Cruickshanks

To the Teacher

Understanding Chemistry is an introductory course in secondary school chemistry. The textbook emphasizes the empirical, open-ended, and humanistic nature of science, both by following the historical development of scientific models, and by developing and modifying models within the textbook.

Throughout the textbook, the development of scientific literacy is encouraged. This is achieved by discussing in narrative text wherever appropriate the impact and implications of science and technology on society and daily living. In addition, the Special Topics (see Text Features page), as well as thought-provoking questions and assignments, reinforce the science-technology-society inter-relationships.

Students often are concerned about mathematical aspects of chemistry, particularly relationships expressed in a chemical equation. In an attempt to alleviate their anxiety, the stoichiometry of chemical reactions is covered at different points in the textbook. Initially, only mass-mass problems are considered (Chapter 9). Once gases have been discussed, mass-volume and volume-volume problems are introduced (Chapter 12). Finally, after a discussion of solutions, solution stoichiometry is considered (Chapter 14).

Basic Structure of *Understanding Chemistry*

Unit I (Matter) and Unit II (Elements and Chemical Bonding) establish a theoretical base for studying chemistry. The study of matter, its properties, reactions, and classifications lead to the historical development of the atomic model. The study of descriptive chemistry of many elements is used in conjunction with historical information to extend the atomic model and to develop the periodic table and bonding models.

Unit III (Formulas, Moles, and Chemical Equations) and Unit IV (Chemical Reaction Calculations) focus on the chemical vocabulary and calculation skills needed to understand and describe chemical reactions. Stoichiometry is introduced in Unit IV.

Unit V (Gases) and VI (Solutions) review and extend earlier physical and chemical concepts. New laws, techniques, and calculations are developed and integrated with the stoichiometry introduced in Unit IV.

Unit VII (Industry and Society) concentrates on developing an awareness of and appreciation for the relationship that science, and in particular chemistry, has with technology, industry, and society. Geography, politics, and economics are only some of the factors considered in this unit.

Unit VIII (Organic Chemistry) and Unit IX (Nuclear Chemistry) are provided as enrichment to extend ideas, concepts, and issues introduced in earlier chapters (primarily Chapter 7, Formulas and Naming of Compounds, in the case of Unit VIII; and Chapter 2, The Nature of Matter, in the case of Unit IX).

Text Features

- Instructions for the safe use of laboratory equipment and chemicals are provided in the *Safety in the Laboratory* section which precedes Chapter 1. (Similar instructions may also be found in the *Laboratory Manual* accompanying this textbook.) Teachers are encouraged to review these safety rules in detail with students at the beginning of the course.

- Each chapter in this textbook is introduced by means of a photograph or a series of photographs and text which outline in narrative fashion the major topics to be covered in the chapter. As new terms are introduced, they are highlighted in **boldface** type. These terms, along with many others, are included in a comprehensive *Glossary* at the end of the textbook.

- Narrative sections are written in a clear, conversational manner. Reading levels have been carefully monitored to aid in student understanding.

- To develop and reinforce problem solving skills, chapters contain numerous *Sample Problems* to show in detail how problems may be approached and solved. These are usually followed by an *Exercise* with varying numbers of questions which serve to help students gain confidence, to reinforce basic principles, to help students apply principles to their own experience, and to challenge students to test their understanding.

- Over sixty *Special Topics* are included in the textbook. Special Topics serve the following purposes:
  1. To introduce important "foundation" skills and processes, such as selected laboratory skills and the factor-label method of problem solving, when and where they are first required.
  2. To help students appreciate the role of chemistry in other disciplines.
  3. To illustrate further the way that chemistry affects people's lives.
  4. To provide enrichment material concerning particular chemicals, processes, applications, and theories.

- Also interspersed throughout the textbook are features entitled *Putting Chemistry to Work*, which focus on the ambitions, background, and work of people who make use of a knowledge of chemistry in their careers.

- Short paragraphs in the margin beside appropriate text alert both teacher and student to the fact that the accompanying *Laboratory Manual* contains an experiment which develops and/or reinforces the text material under discussion.

- At the end of each chapter, a *Vocabulary Checklist* summarizes all of the terms that appear in boldface type in the chapter. A list of *Chapter Objectives* is also included here, to help review the chapter.

- End-of-chapter questions are divided into three categories: *Review*, *Applications and Problems*, and *Challenge*. Many of these questions deal with situations that may be familiar to students, and also deal with reactions of a practical nature. In addition, many end-of-chapter questions encourage students to apply what they have learned to issues of societal and global importance.

- To help students develop and improve research and communication skills, Chapter 16, Understanding Chemical Industries, contains guidelines and suggestions for preparing and organizing a research report.

Safety in the Laboratory

Chemistry is the study of real materials, so your laboratory experience is at least as important as your study of chemical theory and its applications. A chemistry laboratory is as safe as the people who use it. As a student, you have a responsibility to yourself and your fellow students to maintain high standards of safe work. Use your own common sense and chemical knowledge as you conduct the experiments in your *Laboratory Manual*. For example, before you begin each experiment in the Manual, you should read the sections on Safety in the Laboratory and the General Guidelines for Laboratory Work.

As well, follow the guidelines and general safety rules listed below. This list is not intended to be exhaustive. Your teacher will give you specific information on additional safety procedures recommended in your province or Board of Education, and on routines that apply to your school. You will also be informed about the locations of all safety equipment.

1. Work quietly and carefully. Accidents, and certainly poor results, can be caused by rushing or carelessness.

2. Never attempt any unauthorized experiments.

3. Never work alone in the laboratory. If you have an accident, there will be no one close to help you.

4. Wear face/eye protection at all times in the laboratory. Use aprons for experiments involving chemicals and heat. If any chemical comes in contact with the eye, wash it with flowing water for at least 15 min. Have someone contact medical help immediately. For those who wear contact lenses, you have a special reason for wearing suitable eye protection. Chemicals entering the eyes may become trapped behind the lenses. Advise your teacher if you are wearing contacts, and always wear safety goggles during experiments. If any chemical comes in contact with the eyes, remove contact lenses immediately and follow the same procedures described above.

5. If you spill a chemical on your skin, wash it off immediately with plenty of cool water. This is especially important if chemicals get into a person's eyes.

6. Learn where the nearest fire extinguisher is and how to use it. Learn how to use the other safety equipment and safety procedures for your laboratory, and use them when needed.

7. Report even minor injuries to your teacher, who will decide whether first aid is necessary.

8. When heating materials, make sure that the test tubes you use are Pyrex and are clean and not cracked. Always keep the open end of the test tube pointed away from other people and from yourself. If you are using a burner flame, move the test tube gently through the flame so heat is distributed evenly.

9. Handle hot objects carefully. If you suffer a burn, immediately apply cold water or ice.

10. If it is necessary to smell a substance, use your hand to gently waft the vapours toward you. Do *not* put your nose directly over the container.

11. Do *not* taste anything in the laboratory.

12. Wash your hands after leaving the laboratory.

13. Beware of what may appear to be drops of water on the bench. They may be drops of corrosive liquids.

14. Dilute acid or base spills as follows.
    (a) acid on clothes: use plenty of water
    (b) base on clothes: use plenty of water
    (c) acid on desk: use solid sodium hydrogen carbonate (bicarbonate), then water
    (d) base on desk: use dilute acetic acid or vinegar, then water
    When in doubt, rinse acid or base spills with plenty of cold water.

15. Never use your mouth to draw liquid into a pipet.

16. Reactions involving hazardous gases should be carried out in a fume hood.

17. Do not use a flammable liquid in the presence of an open flame.

18. All potentially dangerous reactions must be carried out behind a shield.

19. Never wear open-toed shoes or sandals. Tie back long hair and loose clothing. Do not bring butane lighters into the classroom.

20. Dispose of waste according to your teacher's instructions.

I

MATTER

1 CHEMISTRY: MATTER AND CHANGES

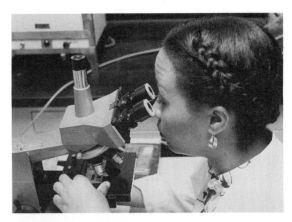

What is chemistry? What do chemists do? As a starting point, a reasonable definition of chemistry will be: Chemistry is the study of matter, its composition, its properties, and the changes that it undergoes. In reality, however, chemistry covers a much wider scope.

The study of **chemistry** provides a means of solving problems and explaining our environment. Chemists try to answer questions such as: Why does iron oxidize (rust) but gold does not? What causes leaves to change colour in the fall? What is the action of soaps and detergents? Why does the colour of some paints fade in sunlight?

The scope of chemistry is so wide and the amount of chemistry knowledge so vast that it is no longer possible for one person to be an expert in all of its areas. There are, however, many basic principles, concepts, and activities of chemistry that you will learn in your chemistry course. These will enable you to understand the more complex ideas that you encounter later.

People working in many different areas make extensive use of the principles of chemistry. Pharmacists must know about the interactions of drugs. Food scientists must know the effects of additives and preservatives on food. Horticulturists need to know about various pesticides and fertilizers. Biologists, dentists, engineers, fuel technologists, agriculturalists, physicians, veterinarians, nutritionists, nurses, and geologists require a knowledge of chemistry to help understand and explain many phenomena in their area of interest.

Throughout this textbook you will come across numerous applications of chemistry to your own daily life. Although your present plans for the future may not include a chemistry-related area of study, a knowledge of chemistry will provide you with a better understanding of the world around you.

1.1 The St. Clair River Blobs – A Case Study of the Scientific Method

In August 1984, a group of divers was conducting a survey of the St. Clair River near Sarnia, Ontario, for the Great Lakes Institute of the University of Windsor. (See Figure 1.1.) During this survey, the divers discovered on the river bed several deposits or "puddles" of a black tarry substance.

After samples of the deposits, or "blobs", as they were referred to by the information media, were taken, a story of scientific detective work began. As you read the following account of the developments of the year and a half after the initial discovery, you should come to understand some of the activities in which scientists are involved when solving problems. In the following paragraphs, the terms shown in **boldface** are terms with which you may be familiar

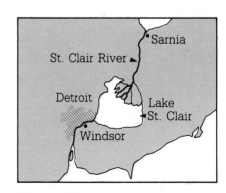

Figure 1.1
Location of St. Clair River

from an earlier science course. Some of these terms are also defined in sections 1.2 and 1.3 of this chapter.

Some important questions that had to be answered about the blobs were:

1. What was the composition of the deposits?
2. Where were they produced?
3. How did they get into the river?
4. How could they be removed from the river?
5. Why did the material not dissolve in the water?
6. Were the blobs likely to contaminate the drinking supplies of downriver communities?

The original samples of material from August 1984 were not completely analyzed until August 1985. Chemists found as many as 37 **compounds** in the samples, some of them known to cause cancer.

The situation became more complicated in mid-August 1985 when Dow Chemical, one of the largest chemical plants in Sarnia, accidentally spilled about 40 000 L of perchloroethylene into the river. Most of the perchloroethylene was recovered, but about 11 000 L entered the St. Clair River.

A great deal is known about perchloroethylene. It is a **pure substance**, a compound of the **elements** carbon, hydrogen, and chlorine. It is insoluble in water, but is able to dissolve **solutes** such as oil and grease. This **property** allows it to be used as a **solvent** in the dry-cleaning industry. Since perchloroethylene is denser than water, it sank to the bottom of the river, where it collected in pools. In late August 1985, Dow began to clean up the perchloroethylene spill by a suctioning technique. When the material they recovered was analyzed, it was found to contain 97% perchloroethylene.

Further samples of the black tarry material were taken from the river bed near Dow Chemical in September and October 1985. Analysis of these samples showed alarmingly high concentrations of a number of toxic chemicals associated with the petrochemical industry. Scientists investigating the problem thought that the contamination was caused either by the dumping of toxic wastes or by the seepage of materials from old underground toxic waste dump sites. In the Sarnia area, about 8×10^9 L of industrial waste had been pumped into underground wells from 1958 to 1976. In addition, Dow disposed of about 4×10^7 L of wastes into underground salt caverns between 1968 and 1984. These wastes contained compounds found in the tarry contamination.

In an attempt to remove the black puddles, Dow resumed vacuuming of the river bed in November 1985. (See Figure 1.2.) One month later, further samples of the contamination were taken. Analysis showed that these samples differed in composition from those samples taken in September and October. (See Figure 1.3.) Scientists from the federal Department of the Environment thought

Figure 1.2
Divers in clean-up operations in December 1985

Figure 1.3
Pie graphs comparing the composition of the samples of contamination collected in September and in December 1985. Because the samples differed in composition, it is almost certain that they came from different sources.

3% other materials

97% perchloroethylene

September samples

35% carbon tetrachloride

65% perchloroethylene

December samples

at this time that the puddles were caused by a small leak from the Dow plant. They discounted the **hypothesis** that underground storage sites were leaking waste materials through the river bed.

In early January 1986, after the clean-up was complete, divers discovered more tarry puddles in the same area. The quantities they found were small, and analysis of samples showed that their composition was different from the samples taken the previous August. This suggested that the contamination was new, rather than something left over from the perchloroethylene spill. Divers as well as video camera images confirmed the existence of more puddles later in January. Scientists were now convinced that the black puddles observed in December and January were not related to the perchloroethylene spill in August 1985.

In late January, Dow discovered a small seepage of a black oil-like material under a sewer pipe close to the site of the original black puddles. Further investigation revealed the presence of more puddles in a drain which ran alongside the sewer. Tests showed that this material was almost identical to the material in the original black deposits. These observations suggested that the contamination resulted from the leakage of industrial waste materials stored on Dow's property or from a leaking pipe at its plant.

In February 1986, further evidence of continued leakage was found. The final report by the federal Department of the Environment and the Ontario Ministry of Environment strongly suggested that the leakage was from a sewer on Dow's property. However, the possibility of the chemicals having seeped up through the river bed cannot be ruled out.

The final report answered some of the questions posed at the beginning of this section. Other questions, however, arose as a result of the report. For example:

1. Were the puddles formed by seepage through the river bed?
2. If so, from where were the materials coming?
3. How can further contamination be prevented?

These questions, and others, can only be answered by further scientific research.

In producing information for the final report, scientists were involved in a number of fundamental scientific activities.

1. Observations, both **qualitative** and **quantitative**, were made; for example, both the identity and concentration of each compound in the puddles were determined.
2. The observations were analyzed to find trends or patterns; for example, the substances found in the puddles were compared with materials used or produced in the chemical plants.
3. A suggested explanation or hypothesis was proposed to explain the pattern; for example, it was suggested that the contamination came from industrial wastes stored in underground salt caverns.
4. The findings, consisting of one or more **theories** to explain the observations, were communicated to others. This communication came in the form of the final report on the incident which was made public.

The series of steps listed above is sometimes referred to as the **scientific method** or the scientific approach to problem solving. While investigating a problem, scientists probably carry out all of these steps, but not necessarily in the order shown. Usually, after making a hypothesis, a scientist will carry out further experiments. The scientist will make observations and look for evidence that either supports or refutes the hypothesis. This may lead to a modification of the hypothesis, then to further experimentation and so on. Figure 1.4 summarizes the scientific method.

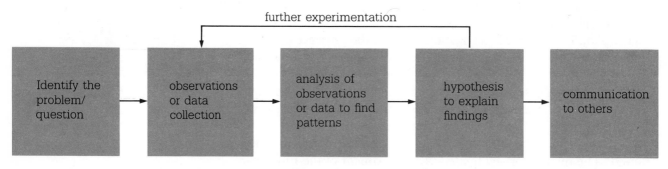

Figure 1.4
Steps involved in the scientific method

SPECIAL TOPIC 1.1

SCIENTIFIC NOTATION

Sciences often require the use of some very large and very small numbers. In chemistry, for example, there is an important number you will be learning about later in this textbook:

602 000 000 000 000 000 000 000

Obviously, this number is quite awkward to use. There is a much more convenient way of expressing the above number:

6.02×10^{23}

This latter number is expressed in what is called *scientific notation*. Using scientific notation, the first digit other than zero is placed before the decimal point and the other digits are placed after it. Then the number is multiplied by the appropriate power of 10.

You can appreciate how useful scientific notation is by studying the examples shown in the table below.

Examples of Scientific Notation

Quantity Measured	Measurement (most values approx.)	Scientific Notation
Distance to Andromeda Galaxy	19 000 000 000 000 000 000 000 m	1.9×10^{22} m
Distance to nearest star (Proxima Centauri)	40 000 000 000 000 000 m	4×10^{16} m
Distance from earth to sun	150 000 000 000 m	1.5×10^{11} m
Diameter of earth	13 000 000 m	1.3×10^{7} m
Length of a football field	100 m	1×10^{2} m
Width of a fingernail	0.01 m	1×10^{-2} m
Thickness of a credit card	0.0005 m	5×10^{-4} m
Thickness of a human hair	0.000 05 m	5×10^{-5} m
Thickness of a spider web strand	0.000 005 m	5×10^{-6} m
Diameter of a tungsten atom	0.000 000 000 1 m	1×10^{-10} m

PUTTING CHEMISTRY TO WORK

HAROLD (HAL) STONE / FORENSIC SCIENTIST

Hal Stone is a Staff Sergeant and forensic scientist in the chemistry section of the Royal Canadian Mounted Police. The role of forensic scientists in the solving of crimes is much greater than most people realize. Television or movie drama may have led us to believe that the police rely more on instinct or tipsters than is actually the case. In reality, detailed and sophisticated work in a laboratory very often produces crucially important information.

"My work," Hal explains, "consists of two general types of examinations: identifying some unknown material from a crime scene or suspect, and comparing a sample of material from a known source to other samples." The materials which come to Hal's laboratory include everything from paints and cosmetics to explosives and soil samples.

Many aspects of chemistry come into use in these examinations. "We receive crime scene items to compare with items from a suspect," Hal says. For example, paint found on a person suspected of break and enter can be compared to the paint from a burglarized home. Arson, the deliberate setting of fires, is also investigated by chemical analysis. "After a fire, we can usually detect if there are any accelerants (chemicals which are used to start fires) in the charred debris."

Hal became interested in chemistry after joining the Royal Canadian Mounted Police. The force sponsored his university training and provided him with an understudy program in the forensic laboratory in Sackville, New Brunswick, where he is now stationed. "The understudy program is essential," Hal cautions. "A key part of being a forensic scientist is presenting your expert opinion and evidence in any courtroom, from the magisterial level up to the Supreme Court.

When you take the stand and are questioned by both the Crown counsel and the defence attorney, you must be able to present accurate information in a clear, impartial, and objective manner. There is no room for error in your work."

Both officers and civilians are employed in the RCMP's forensic laboratories. The basic requirements are a university degree in chemistry or biochemistry, along with the understudy program. "What I like about my work," Hal notes, "is the part I play in helping to determine the guilt or innocence of an individual. We are dealing with human beings and it is immensely satisfying when our evidence assists the victim of a crime obtain justice."

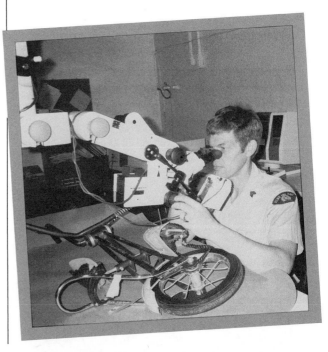

Exercise

1. Give two hypotheses that were suggested to explain the source of the black puddles in the St. Clair River.

2. Why were the investigating scientists sure that the tarry material found in September and October 1985 was not related to the perchloroethylene spill in August 1985?

3. How did the analysis of waste materials stored in an underground salt cavern on Dow property lend support to one of the hypotheses in your answer to question 1?

4. What later observations prompted scientists to discount the hypothesis referred to in question 3?

1.2 Review of Important Terms

In section 1.1, several terms were presented in **boldface**. These are terms that you probably recognize from an earlier science course. These terms, and many others, form part of the language of chemistry, and the study of chemistry requires an understanding of their meanings. A review of some of these important terms follows.

States of Matter

Matter can exist in three **states** – solid, liquid, and gas. The characteristics of the three states are compared in Figure 1.5.

In some solids, the particles making up the solid are arranged in a regular repeating pattern. Such solids are said to be **crystalline**. Salt, sugar, and diamond are examples of crystalline solids. Solids which lack this regular structure are said to be **amorphous**. Putty and rubber are examples of amorphous solids.

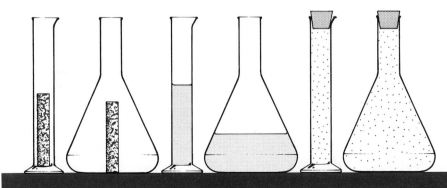

Figure 1.5
Characteristics of the three states of matter. Solids have a definite shape and volume. Liquids have a definite volume, but they take the shape of the container. Gases take on the same shape and volume as the container.

Solid   liquid   gas

The state of a sample of matter can be changed, usually by changing the temperature, but sometimes a change in pressure is also required. Figure 1.6 lists the names of the six changes of state.

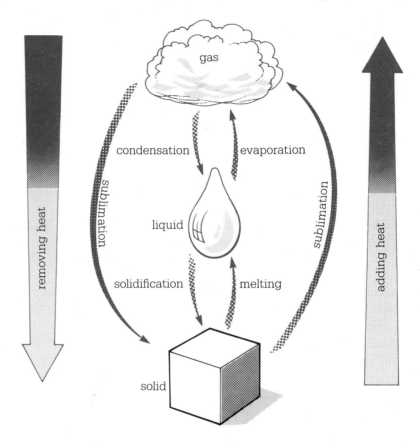

Figure 1.6
Changes of state in matter

Physical and Chemical Changes

Laboratory Manual
In Experiment 1-1, you will observe a variety of reactions, and summarize the evidence for chemical change.

A **physical change** is a change in which the chemical identity or composition of a sample of matter remains the same. There is a change in one or more of the properties of the sample, but no new substance is formed. For example, when candle wax melts, the state of the wax changes, but it is still the same substance. This is a physical change.

A **chemical change** is a change in which at least one new substance is produced. The composition and properties of the substance(s) formed by the chemical change (the **products**) are different from those of the starting materials (the **reactants**). For example, when a candle burns, carbon dioxide and water vapour (and perhaps other substances) are produced. The properties of carbon dioxide and water vapour are obviously quite different from the properties of the original candle wax. Experiments would show that their composition is also different from that of candle wax.

You can usually conclude that a chemical change has occurred if one or more of the following changes are observed:

1. A colour change occurs, which is not simply a blending of the initial colours.
2. A gas is produced, which does not result from the boiling of a liquid.
3. A solid (a precipitate) is formed when two liquids are mixed together.
4. Energy (usually in the form of heat or light) is released or absorbed.

These observations are evidence that a chemical change has taken place, but they are not conclusive proof. The proof that a chemical change has occurred is that at least one new substance, which has a composition different from that of any of the starting materials, has been formed.

Chemical changes, though often conducted and studied in the chemistry laboratory, also occur in and around the home, in your body, in the automobile engine, and in the environment. Some of the changes that occur in cooking and baking are chemical changes. When milk goes sour, that is a chemical change. The bleaching of clothes, the burning of gasoline, the digestion of food, and the formation of the acids that cause acid rain are all examples of chemical changes. In each case, at least one new substance is produced.

Exercise

5. Describe the following as physical or chemical changes.
   (a) the disappearance of a moth ball
   (b) the formation of clouds
   (c) the baking of a cake
   (d) the formation of ice cubes
   (e) the burning of leaves
   (f) the explosion of dynamite

Physical and Chemical Properties

Characteristics or features of matter are called **properties**.

Physical properties are properties that can be determined without altering the chemical composition of the material. Colour, odour, electrical conductivity, and melting point are physical properties.

If the property can be measured, it is called a **quantitative physical property**. For example, a quantitative physical property of water is that it has a density of 1.0 g/mL.

If the property is simply a description and does not have a measurement, it is called a **qualitative physical property**. For example, a qualitative property of gold is that it is a solid at room temperature.

Chemical properties of matter describe how matter behaves in the presence of other substances or when subjected to heat, light,

Laboratory Manual
You will examine the physical and chemical properties of selected elements in Experiment 1-2.

or electricity. For example, a chemical property of baking soda is that when heated, it produces carbon dioxide. A chemical property of hydrochloric acid is that it reacts with magnesium to produce hydrogen. The chemical properties of a sample of matter can be determined only by changing the composition of the matter in some way. In other words, a chemical change must take place.

The properties of matter may be used as a means of identifying substances or telling substances apart. This is possible because scientists have never found two different substances with identical properties. You may have made use of this principle in the kitchen. Sugar and salt are similar in appearance. If they are kept in unmarked containers, a careless cook may produce some interesting recipes! In this case, a difference in taste may be used to identify the correct substance. Note that in the laboratory, you should never use the property of taste to identify substances.

SPECIAL TOPIC 1.2

FOOL'S GOLD

Sir Martin Frobisher was an English explorer who, in 1576, while searching for a passage to Asia, discovered what is now Frobisher Bay. While exploring in the Frobisher Bay area in the Canadian Arctic, he discovered an ore that he thought contained gold. He made two subsequent voyages to the same area, the last one with a fleet of 15 vessels, and returned to England with tons of ore.

Subsequent testing showed that the ore was in fact worthless; the people who had financed Frobisher's voyages were ruined. It is now fairly certain that the ore Frobisher believed to have contained gold was actually fool's gold, an ore of iron and sulphur.

Exercise

6. Refer to section 1.1. List two physical properties of perchloroethylene.

7. Identify each of the following properties of sodium chloride (common salt) as either physical or chemical. Indicate whether the physical properties are qualitative or quantitative.
   (a) It is soluble in water.
   (b) It melts at 801°C.
   (c) It reacts with sulphuric acid to form hydrogen chloride gas.
   (d) It forms cubic crystals.
   (e) It possesses a distinctive (salty) taste.
   (f) It has a density of 2.2 g/cm³.

8. Gold and fool's gold (iron pyrites or iron disulphide) are sometimes difficult to tell apart. The similarity in their appearance has been the cause of great anguish to many gold prospectors. Find one physical property and one chemical property that would allow you to distinguish the two materials. (To answer this question, you could look in a chemistry handbook, such as the *Merck Index* or the *CRC Handbook*, a chemical encyclopedia, or a reference book from your teacher.)

1.3 The Classification of Matter

All samples of matter can be classified into one of two groups – pure substances or mixtures. A **pure substance** is a sample of matter that has a definite, fixed composition and is uniform throughout. If the properties of a sample of matter are the same throughout the sample, the matter is said to be **homogeneous**. Water, iron, oxygen, and sodium chloride are examples of pure substances.

Elements and Compounds

On the basis of chemical properties, chemists have subdivided pure substances into elements and compounds. An **element** is a pure substance that cannot be broken down into simpler substances by any chemical means. Of the examples of pure substances mentioned above, iron and oxygen are elements. Other elements that you may be familiar with include hydrogen, helium, carbon, aluminum, chlorine, copper, silver, and gold.

Chemists have been able to isolate and identify 108 elements. Of these, 91 are found in nature. They can be found in the earth, the sea, and the atmosphere. The other 17 elements are synthetic; that is, they are made by humans. All known compounds, about 6×10^6 of them, are made from these elements. There is presently no evidence that any other elements occur anywhere else in the universe.

The distribution of elements in nature is very uneven. Figure 1.7 shows the percentages of the most common elements in the Earth's crust (the outer layer to a depth of about 15 km).

A **compound** is a pure substance that is formed when two or more elements chemically combine. Sodium chloride and water are examples of compounds. Other examples with which you may be familiar are sugar, baking soda, ammonia, and carbon dioxide.

The physical and chemical properties of a compound are distinctly different from the properties of the elements that make up the compound. Consider the properties of sodium and chlorine, for example. Sodium is a silvery metal which reacts violently with water. It also reacts rapidly with air; to prevent this reaction, sodium

Figure 1.7
Distribution of elements in the earth's crust. Notice that ten elements make up about 99% of the earth's crust. These elements are found for the most part combined with other elements in compounds. We use vast quantities of elements such as iron, aluminum, nickel, zinc, and chlorine. The extraction of elements from their compounds is an extremely important chemical process in industry.

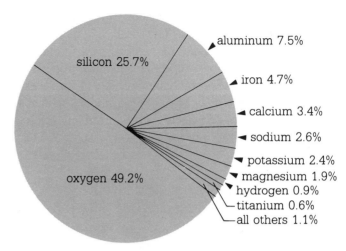

silicon 25.7%
aluminum 7.5%
iron 4.7%
calcium 3.4%
sodium 2.6%
potassium 2.4%
magnesium 1.9%
hydrogen 0.9%
titanium 0.6%
all others 1.1%
oxygen 49.2%

is often stored in oil. Although sodium is an essential element in our diet, we certainly would not sprinkle sodium on our food! Chlorine, on the other hand, is a greenish-yellow gas. It is an extremely poisonous element and must be handled with great care. Now, if a piece of sodium is gently warmed, then placed in a container of chlorine, an obvious chemical reaction occurs. (See Figure 1.8.) After the reaction has subsided, a white crystalline solid remains. Analysis of this solid shows that it is sodium chloride, more familiar to you as table salt. The highly reactive elements, sodium and chlorine, have combined to form the relatively unreactive compound sodium chloride. The properties of sodium chloride are obviously completely different from the properties of both sodium and chlorine.

(a) Sodium

(b) Chlorine

(c) Reaction between sodium and chlorine

Figure 1.8
The chemical reaction between sodium and chlorine produces sodium chloride. The properties of sodium chloride cannot be predicted from the individual properties of sodium and chlorine.

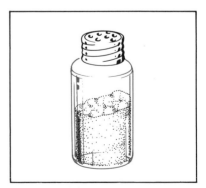

(d) Sodium chloride

Exercise

9. Examine the labels of a number of household products. List the names of the compounds you find on the labels.

Mixtures

A **mixture** is formed when two or more pure substances are mixed without chemically reacting in any way. The individual components of the mixture retain their own physical and chemical properties. A mixture is **homogeneous** if the components are uniformly distributed throughout, and **heterogeneous** if they are not.

A homogeneous mixture is called a **solution**. A stirred mixture of sugar in water is a solution – the sugar is uniformly distributed throughout the water. The terms **solute** and **solvent** are used to describe the components of a solution. The **solute** is the component which dissolves; in the sugar solution, sugar is the solute. The **solvent** is the component that does the dissolving; in the sugar solution, water is the solvent. In a solution, the components cannot be visibly distinguished from each other. A solution is said to have only one **phase** or one visible region of matter. Although the composition of a solution is the same throughout, it is possible to prepare solutions in which the solute and solvent are present in different proportions. Some common examples of solutions that you may find in the home are vinegar, food colouring, window-cleaning fluid, perfume, and mouthwash. (See Figure 1.9.)

Figure 1.9
Variety of solutions found in the home

Solutions may be solid, liquid, or gaseous. Brass (an alloy of copper and zinc) is an example of a solid solution. Sterling silver (copper and silver), solder (tin and lead), and steel (iron and carbon) also fall into this category.

Liquid solutions, such as those shown in Figure 1.9, are by far the most common. Water is the most common solvent in liquid solutions. The solute may be solid, liquid, or gaseous. In salt water, the solute is a solid (sodium chloride); in vinegar, the solute is a liquid (acetic acid); and in carbonated drinks, the solute is a gas (carbon dioxide).

An example of a gaseous solution is dry, pollution-free air. This is a solution of mainly oxygen in nitrogen, with very small amounts of carbon dioxide and other gases. Breathing tanks of deep-sea divers contain a solution of oxygen and helium. (Helium is less soluble than nitrogen in the bloodstream. There is, therefore, less chance of a diver suffering from the bends, a painful affliction caused by nitrogen bubbles in the blood.)

A heterogeneous mixture, often known as a **mechanical mixture**, is a mixture in which the components are not uniformly distributed. In some mechanical mixtures the individual components may be visibly distinguished. In other mixtures, where the particles are very small, it may be necessary to use a microscope to distinguish the components. (See Figure 1.10.) A heterogeneous mixture has two or more phases; these phases may be the same or different.

Figure 1.10
Homogenized milk (a) may appear homogeneous, but a microscopic view (b) reveals that it is a mixture of several substances, including water and globules of fat.

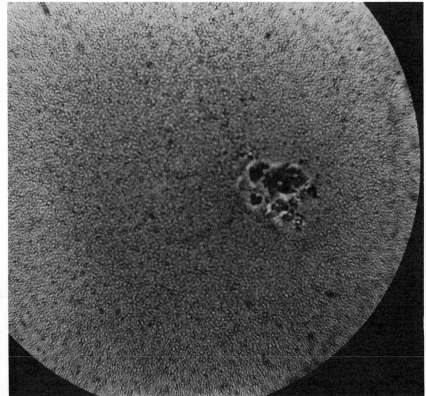

(a)   (b)

Muddy water, sand, salad dressing, and chocolate chip cookies are examples of heterogeneous mixtures. Figure 1.11 summarizes the classification of matter.

Figure 1.11
Classification scheme for matter

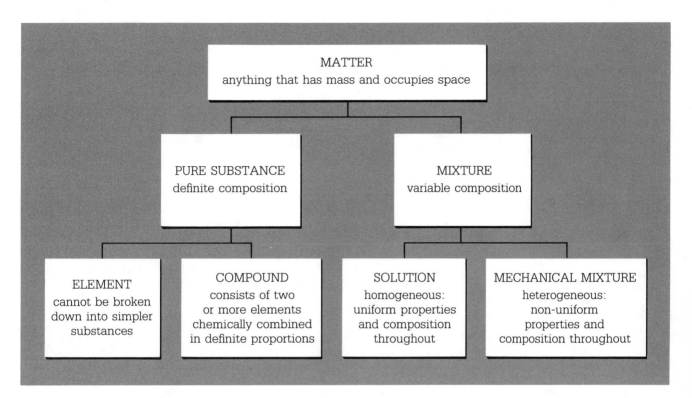

Exercise

10. Some consumer products are claimed by the manufacturers to be 100% pure; for example, 100% pure orange juice or 100% pure raspberry jam. Are these products pure in the same sense as scientists mean by the word? Explain what you think the manufacturers mean by the term 100% pure.

11. Classify each of the following as an element, compound, solution, or mechanical mixture.
    (a) sulphur
    (b) clear tea
    (c) polluted air
    (d) tomato juice
    (e) sulphur dioxide
    (f) nitrogen
    (g) pizza
    (h) windshield washer fluid
    (i) gasoline
    (j) lawn fertilizer

17

MIXTURES IN THE HOME AND THE ENVIRONMENT

Variety of mechanical mixtures and solutions

USEFUL MIXTURES

Many mixtures, both solutions and mechanical mixtures, are extremely useful in and around the home. Most foodstuffs are mixtures, as are gasoline, paints, personal care products, cleaning fluids, and fertilizers.

FLUORIDE IN DRINKING WATER

In many cities and towns in Canada, small amounts of sodium fluoride are added to drinking water in water treatment plants. Evidence has shown that this very dilute sodium fluoride solution can reduce the incidence of tooth decay, particularly in young people. Interestingly, increasing the amount of sodium fluoride in the water beyond a certain level does not seem to lead to any further decrease in tooth decay. Instead, teeth take on a mottled appearance. Experiments have shown that the optimum concentration of sodium fluoride in water is about 1 g of the compound in 10^3 kg of water.

Despite evidence of the beneficial aspects of fluoride treatment in water, some communities refuse to have sodium fluoride added to their water supplies because the compound is poisonous. It should be noted that the substance *is* poisonous if taken in large quantities, but its concentration in treated water is so low that drinking normal quantities of the water should not be a concern. To ingest a lethal dose of sodium fluoride, a person would have to drink about 4000 L of water in one day.

PROBLEM MIXTURES

Large quantities of fuels are used everyday in industry, transportation, and homes. To meet this demand, millions of tons of crude oil are transported across the oceans in tankers every year. Although there have been few serious accidents involving these tankers, oil spills do occur from time to time and damage to the environment can be severe. Crude oil is a thick, black mixture of many substances and it is particularly unpleasant when it finds its way onto beaches. Birds suffer greatly when they come in contact with the oil; the oil causes their feathers to mat together and the birds can no longer fly or even float. In addition, the feathers lose their ability to act as an insulator and as a result, many birds die of exposure, if they have not already drowned.

Birds coated with oil usually die from exposure, unless they are found and cleaned.

Many methods have been attempted to clean up oil spills, but none has been found to be entirely satisfactory. For small spills, mechanical skimming of the surface of the water works well. This method has the advantage of recovering the oil. The oil can sometimes be removed by burning, but this produces large amounts of air pollution.

When materials such as sand and cement are spread on an oil spill, they cause the oil to sink. This removes the problem from the surface but it has been found that marine life at the bottom of the ocean is adversely affected. Another method which is fairly successful for smaller spills is to spread an absorbent material such as sawdust or straw over the oil. The material soaks up the oil and is then removed for disposal. The absorbent method is often used for cleaning beaches fouled by oil.

Contaminants do not have to be present in large amounts to cause serious problems. For example, compounds called dioxins are produced in small quantities during the manufacture of some herbicides and pesticides. When these chemicals are used, minute amounts of the dioxins are introduced into the environment. Some chemists have also suggested that dioxins are produced by the combustion of many common materials, such as wood, coal, oil, municipal wastes, and cigarettes. Evidence suggests that even trace amounts of dioxins can cause serious health problems ranging from skin disorders to liver problems and cancer. Dioxins have also been implicated in birth defects.

Improved detection techniques now allow chemists to detect levels of dioxins in water that previously would have been impossible to detect. Levels of dioxins as low as 1 g in 10^{11} kg of water can now be detected. It is very difficult to imagine a ratio this small. (Compare this ratio with 1 s in 3.3×10^6 a.) Recent testing has shown the presence of dioxins at extremely low concentrations in Ontario farm produce and in the water supplies of a number of municipalities in south western Ontario.

The term dioxin actually refers to a family of about 75 compounds with similar properties. The most toxic of these compounds is 2,3,7,8-tetrachlorodibenzo-p-dioxin (more commonly known as TCDD). It is probably the most toxic substance ever manufactured. Only the toxins from bacteria that are responsible for the diseases botulism, tetanus, and diphtheria are more potent.

Laboratory animals exposed to dioxins have shown increased levels of cancer, miscarriages, and birth defects. The evidence linking dioxins to similar problems in humans is still inconclusive.

The most obvious problem caused by dioxins is chloracne, a particularly unpleasant skin disease. Some evidence suggests that people exposed to low levels of dioxin for two to three years have a lower resistance to disease.

Scientists have found that the amount of dioxin that can kill a single hamster will kill 5000 guinea pigs. Based on these observations, it is impossible to predict a "safe" level for humans, or if in fact such a level exists.

Dioxins are chemically unreactive substances. This chemical stability causes a problem because it means dioxins persist in the environment for a long time. During the Vietnam War, United States forces sprayed herbicides contaminated with dioxins. Between 1962 and 1971, about 4×10^7 L of Agent Orange was sprayed on jungles to remove the foliage. Studies have shown high levels of TCDD in fat tissue of southern Vietnamese people who were exposed to the herbicide. In addition, the same studies showed high levels of dioxins in the soil.

There is no known safe, reliable method for the removal of dioxins from contaminated samples. High temperature incineration at about 1000°C is thought to be effective. Recent research has centred around attempts to create bacteria that would feed on dioxins and break them down into harmless substances. Certainly, one solution to the dioxin problem is to prevent their manufacture and their release into the environment.

1.4 The Separation of Mixtures

Most naturally occurring samples of matter are solutions or mechanical mixtures. To isolate pure substances from these mixtures, the components of the solution or mechanical mixture must be separated.

To separate the components of a mixture, first we have to find a physical property that is different for each of the components. Rock salt, for example, is a mechanical mixture of sodium chloride and soil or sand. The separation of the salt from the soil makes use of the fact that salt is soluble in water, while soil and sand are insoluble. Water is added to the rock salt to dissolve the salt. The mixture is filtered to remove the soil or sand, and the water evaporated from the salt water solution to give solid salt. Large deposits of rock salt are found under the Detroit River near Windsor, Ontario. The techniques described in this paragraph are used by industry to obtain pure sodium chloride from rock salt.

Exercise

12. For each of the mixtures below, state at least one physical property that is different for the components of the mixture which would allow you to separate the components.
    (a) iron powder and sand
    (b) talcum powder and powdered glass
    (c) sugar and sand
    (d) alcohol and water
    (e) pebbles and soil

Industries and the Separation of Mixtures

The separation of mixtures is an extremely important process in many chemical industries. Some of the most common separation methods are described briefly.

Flotation

Most metal ores are found in the ground mixed with soil and rock particles. Before the metal can be extracted from the ore, the ore must be concentrated, or separated from the impurities. A method of separation known as **flotation** is used to concentrate ores of zinc, copper, nickel, and lead.

One of the main nickel-producing refineries in Canada is situated at Thomson, Manitoba. To separate the nickel ore from impurities, water, oil, and a detergent are added to the crushed impure ore. The nickel-containing particles become coated with oil, and the soil and rock particles do not. When the mixture is stirred and air is blown through it, the oil-coated ore particles float to the top, where they are held in a froth. The soil and rock particles sink to the bottom of the container, and are discarded. (See Figure 1.12.) The concentrated ore is skimmed off the surface, then a series of physical and chemical changes is applied to produce nickel.

Fractional Distillation

The process of **fractional distillation** is used extensively in the petrochemical industry. You will learn more about this industry in Chapter 16, but a brief discussion is included here.

Crude oil, or petroleum, contains a large number of substances, each of which has a different boiling point. To separate the components, the crude oil is vaporized and the vapour is allowed to pass up a tall tower. The temperature in the tower decreases with elevation. As the hot vapour rises up the tower, it progressively cools. As a result, substances with different boiling points condense at different levels. Substances with low boiling points condense near the top of the tower, if in fact they condense at all, while substances with high boiling points condense near the bottom of the tower. The material obtained at each level is not pure, but is a mixture of substances with similar boiling points. Some of the products obtained from the fractional distillation of petroleum are gasoline, kerosene, jet fuel, diesel fuel, and lubricating oils. (See Figure 1.13.)

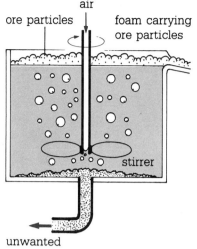

Figure 1.12
Flotation is a method of concentrating the ores of some metals.

(a) Fractional distillation towers

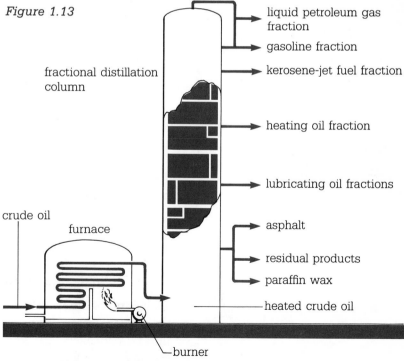

Figure 1.13

(b) Separation of petroleum by fractional distillation

Water Purification

Water treatment plants make use of a number of separation techniques. (See Figure 1.14.) Large objects such as tree branches, fish, and garbage are removed by passing the water from the river or lake through a screen. The water is then treated with alum and lime, which react to form a sticky, jelly-like precipitate of aluminum hydroxide called a floc. As the precipitate forms it sticks to small suspended particles in the water and settles with them to the bottom of a large tank. This settling process is called **sedimentation**. The water is filtered once again by passing it through beds of sand and gravel. This removes most of the remaining small particles. To kill bacteria, which are too small to be removed by sand bed filtration, the water is usually treated with chlorine. As discussed earlier in Special Topic 1.3, the water may be treated with sodium fluoride before it is supplied to homes.

Figure 1.14
A water treatment plant. Various methods of separation are used to remove impurities from the water.

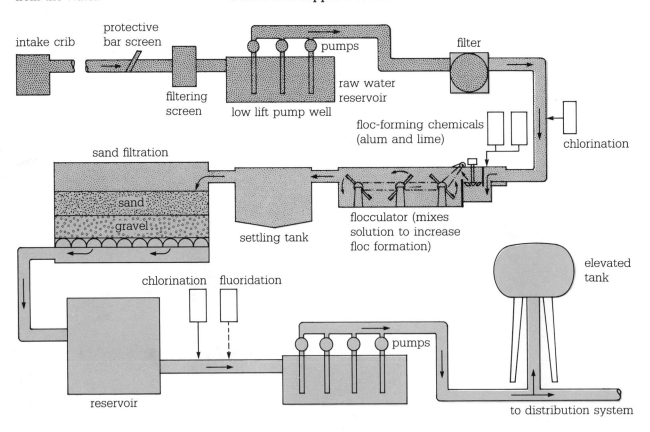

Other Methods of Separation

Conventional filtering methods may be used to remove solid particles from gases. Vacuum cleaner filters, air filters in cars, and air filters in furnaces are common examples. A different method of separation is required to remove solid particles from industrial smokestack gases. A device called an electrostatic precipitator induces an electrical charge on the solid particles, which are then attracted to oppositely charged plates. The particles are then collected for future disposal. (See Figure 1.15.) A similar process occurs in electronic air cleaners, which are sometimes added to forced air furnaces in homes.

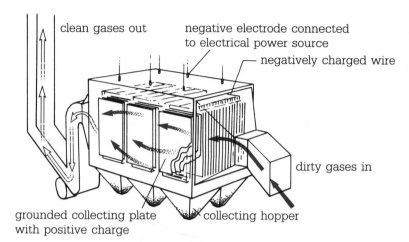

clean gases out

negative electrode connected to electrical power source

negatively charged wire

dirty gases in

grounded collecting plate with positive charge

collecting hopper

Figure 1.15
An electrostatic precipitator removes solid particles from gases by inducing an electrical charge on the particles. The charged particles are then attracted to oppositely charged metal rods.

If one of the components of a mixture is a magnetic substance, then it can be easily separated from the mixture by using a magnet. This separation technique is used in some solid waste treatment plants to separate objects containing iron (such as tin cans) from other objects made of glass, aluminum, and plastic.

The removal of gaseous pollutants from air, such as sulphur dioxide, carbon monoxide, and nitrogen oxides, is much more difficult and usually requires chemical methods of separation. As an example, sulphur dioxide can be separated from a mixture of gases by bringing the mixture in contact with magnesium oxide. A chemical reaction occurs between sulphur dioxide and magnesium oxide, forming magnesium sulphate which is removed. If the magnesium sulphate is allowed to react with ammonia and oxygen, the magnesium oxide is regenerated and ammonium sulphate, which is used to produce fertilizers, is also formed. This series of chemical reactions, in addition to removing a pollutant, therefore also produces an extremely valuable and useful compound.

1.5 Chemical Symbols, Formulas, and Equations

When you learn to speak a new language, you must first learn some of the basic vocabulary of that language. In some ways, learning chemistry is like learning a new language. In this section, you will learn the chemical symbols (a type of chemical shorthand) for several elements. In addition, you will learn the meaning of chemical formulas of compounds and chemical equations for reactions.

BALANCING THE COSTS: ACID RAIN

The release of huge quantities of sulphur dioxide and nitrogen oxides into the atmosphere is responsible for one of the major environmental problems of the 20th century — acid rain. Where does this sulphur dioxide come from? A great deal of it comes from the combustion of sulphur-containing fossil fuels. A significant quantity is also produced during the extraction of some metals from their ores. Canada is one of the world's top metal-producing countries, ranking first in zinc production, second in nickel production, and third in silver and gold production. Canada's reputation as a metal-producer has not been obtained, however, without cost to the environment.

Many metal ores contain sulphur. One of the first steps in the extraction of the metal is to heat the ore in air. This is called roasting the ore. This produces sulphur dioxide, much of which escapes into the air. Further reaction of sulphur dioxide with oxygen produces sulphur trioxide, which combines with moisture in the air to form sulphuric acid, a major component of acid rain. It has been estimated that in Canada, smelters produce about 45% of the total mass of sulphur dioxide produced each year. This represents more than 2×10^6 t of sulphur dioxide.

Lakes severely affected by acid rain can no longer support aquatic life and these lakes eventually "die". A large part of eastern Canada and the United States is affected by acid rain. Although some of this acid rain comes from sulphur dioxide originating in Canada, industries located in the United States also contribute to the problem in Canada. Negotiations between the two countries continue in an attempt to reduce acid rain. In Canada, the federal government and the eastern provinces signed an agreement in 1984 to reduce total sulphur dioxide emissions by 50% by 1994. In Ontario, the provincial government passed legislation in late 1985 that will force the metal smelters and Ontario Hydro to reduce their sulphur dioxide emissions by 67% from the 1.9×10^6 t produced in 1980. But is this reduction sufficient? Should we insist that sulphur dioxide emissions be reduced further? What is an acceptable level of sulphur dioxide emissions? Who should make these decisions? The questions are easy to ask, but the answers are difficult to find. The problem of acid rain has not been solved. Scientists and politicians must continue to look for solutions.

Names and Symbols for Elements

The names of the elements have been derived from many different sources. For example, you can probably tell that germanium, francium, and californium are named after places. Other elements are named after famous scientists: fermium (after Enrico Fermi), einsteinium (after Albert Einstein), and curium (after Marie Curie) are examples. Some elements are named after figures in mythology: promethium after the Greek god Prometheus who stole fire from heaven, and titanium after the Greek Titans, giants who sought to rule heaven and were overthrown by Zeus. And some elements are named after a property they possess. For example, iodine is derived from the Greek word meaning violet and argon comes from the Greek word meaning inactive. (For a complete list of the origin of the names of the elements, see Appendix B.)

Each element has been assigned a unique **chemical symbol** consisting of one or two letters. The first letter is always capitalized.

For two-letter symbols, the second letter is lowercased. Examples of single-letter symbols include O (oxygen), N (nitrogen), P (phosphorus), and C (carbon). Examples of two-letter symbols include Ca (calcium), Mg (magnesium), He (helium), and Cl (chlorine).

While many symbols of the elements are based on the English names, some symbols are based on the Latin names for the elements. Examples include Na (sodium, from *natrium*), Cu (copper, from *cuprum*), Ag (silver, from *argentum*), Au (gold, from *aurum*), and Fe (iron, from *ferrum*).

There are two reasons why so many symbols are based on Latin names. First, some elements were known to the Latin-speaking Romans as early as 2000 years ago. These elements are either relatively easy to isolate from their compounds (for example, copper), or appear in the uncombined state in nature (for example, gold). Secondly, since the Romans dominated the world for such a long time, Latin was used throughout much of the world, and for a long period in history was considered the language of learning. Long after the collapse of the Roman Empire, scholars still used Latin in their writing. As alchemists (as the early "scientists" are called), and later scientists discovered new elements, it was natural for them to give these elements Latin names. To keep the system of naming consistent, the elements discovered even recently are usually given Latinized names; No (nobelium) and Lr (lawrencium) are two examples.

Chemical Formulas

Just as chemical symbols represent elements, **chemical formulas** represent compounds. Since compounds are formed from elements, it is not surprising that the chemical formulas of compounds consist of the symbols of the elements they contain. For example, the formula of carbon monoxide is CO, methane CH_4, and sucrose (table sugar) $C_{12}H_{22}O_{11}$. The formula of a compound indicates not only the elements that are in the compound, but also gives the relative number of atoms of each element in the compound. The formula CH_4, for example, indicates that the ratio of hydrogen atoms to carbon atoms in methane is four to one. Note that if no subscript appears with an element, a subscript of 1 is assumed. Later in this textbook you will see more complex formulas, such as that of ammonium phosphate (a fertilizer). The formula of this compound is $(NH_4)_3PO_4$. The brackets around the NH_4, and the subscript 3 outside the brackets, indicate that three groups of NH_4 (consisting of three nitrogen atoms and twelve hydrogen atoms) are present along with one phosphorus atom and four oxygen atoms.

Chemical Equations

A **chemical equation** summarizes in shorthand the changes that take place during a chemical reaction. The simplest type of equation is a **word equation**, which consists of the names of the starting materials, or reactants, and the resulting materials, or products. The word equation for the burning of magnesium is

$$\text{magnesium} + \text{oxygen} \rightarrow \text{magnesium oxide}$$

The "+" between the reactants means "reacts with" and the arrow is interpreted as "to produce". The equation is therefore read as follows: "Magnesium reacts with oxygen to produce magnesium oxide".

If the names are substituted by symbols and formulas, the equation then becomes

$$\text{Mg} + \text{O}_2 \rightarrow \text{MgO}$$

(You may wonder why oxygen is written as O_2 and not simply O. The answer to this question will have to wait until Chapter 5.)

It is also possible to include information about the physical state of each reactant and product. The symbols (s), (l), (g), and (aq) are used to indicate whether they are solid, liquid, gaseous, or in water solution. (The symbol aq is an abbreviation of aqueous, derived from the Latin word *aqua* for water. An aqueous solution is a solution of something in water.) For example, carbon dioxide is often prepared in the laboratory by reacting calcium carbonate (marble chips) with hydrochloric acid. The equation for the reaction is

$$\text{CaCO}_3(s) + \text{HCl}(aq) \rightarrow \text{CaCl}_2(aq) + \text{CO}_2(g) + \text{H}_2\text{O}(l)$$

If you count the number of atoms of each element on each side of the equation, you will notice that there are more hydrogen atoms and chlorine atoms on the right side of the equation than there are on the left side. The equation is said to be **unbalanced**. The equation can be made into a **balanced equation** by placing a coefficient of 2 in front of the HCl on the left side of the equation. You will learn in Chapter 8 how to balance equations.

Exercise

13. Test your knowledge of element names and symbols.
    (a) Write symbols for the following elements.
        (i)   lithium *Li*
        (ii) neon *Ne*
        (iii) sulphur *S*
        (iv) sodium *Na*
        (v)   iron *Fe*
        (vi) chlorine *Cl*
        (vii) magnesium *Mg*
        (viii) aluminum *Al*
        (ix) phosphorus *P*
        (x)   lead *Pb*
    (b) Name the elements represented by the following symbols.
        (i)   Ag *Silver*
        (ii) Cu *Copper*
        (iii) N *Nitrogen*
        (iv) He *Helium*
        (v)   B *Boron*
        (vi) F *fluorine*
        (vii) Si *Silicon*
        (viii) Ar *Argon*
        (ix) Ca *Calcium*
        (x)   K *Potessium*

PUTTING CHEMISTRY TO WORK

GARITH GUMBS / TECHNICIAN

A knowledge of the characteristics of elements and compounds is the basis of analytical chemistry — the determination of the amounts and kinds of substances in a given sample. Garith Gumbs is a technician in a laboratory which receives a wide variety of samples for analysis. He needs to know a great deal about the physical properties of certain substances.

"In spite of the fact that analytical chemistry uses elaborate and advanced equipment," Garith explains, "it is the principles of chemistry which I first learned in high school that matter most." Among these principles are solubility and boiling point. "For example, I am often given samples of water to analyze for the presence of oil or grease. Before I begin I need to know that oily substances do not easily dissolve in water." Garith adds an organic solvent, such as trichloromethane, to each sample. The oils and greases dissolve readily in the organic solvent. Because the organic solvent does not mix with water, it separates into a distinct layer that can be decanted, or poured, into another container.

The next step is to isolate the chemicals dissolved in the organic solvent. "Knowing that the boiling point of the solvent is much lower than that of the chemicals I'm after, I can isolate the chemicals by evaporating away the solvent." Garith can be confident that any dry residue left behind in the container is oil and grease which come from the original water sample. The quantity can be measured and the residue can be used for further analysis if necessary.

Another means of separating substances in solution that Garith often uses is fractional distillation. "Again, this is basic chemistry," comments Garith, "although there is a lot of gadgetry available to do the labour. Different compounds have different boiling points. What I need to do is to boil a solution at the temperature that is the boiling point of the chemical I want to remove and collect." As the chemical boils, it becomes a gas and is carried away in a system of tubes; cooling the gas returns it to a liquid. "Other compounds can be separated in a similar way by varying the temperature."

Garith enjoyed science in high school and went on to obtain a university degree in chemistry. "It is a rewarding feeling to be able to apply directly my education and interests," he notes. "And in a very real sense, the tools of my work are simply the fundamental characteristics of matter."

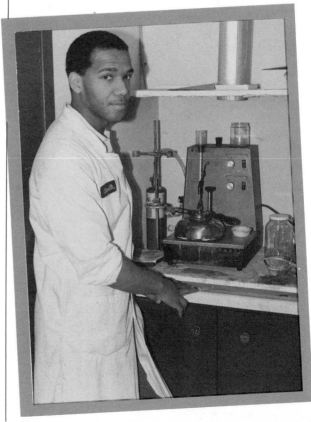

14. Write the number of atoms of each element represented by each of the following formulas.

(a) NH_3

(b) P_2O_5

(c) K_2CO_3

(d) C_2H_6O

(e) H_2SO_4

(f) H_3PO_4

(g) $(NH_4)_2SO_4$

(h) $Fe_3(PO_4)_2$

Looking Back – Looking Ahead

This chapter has presented the groundwork for the rest of the textbook. You have learned about the scientific approach to problem solving, and reviewed some of the basic concepts and vocabulary of chemistry. Answers were provided to some questions, but many new questions probably occurred to you as you read the chapter and carried out experiments in the laboratory.

In Chapter 2 you will learn about one of the greatest achievements of science, the development of the atomic theory. As scientists in the 17th and 18th centuries sought to explain the behaviour of matter, the need for a model or theory of matter became obvious. The account of the search for such a theory is a fascinating part of the history of science.

Vocabulary Checklist

You should now be able to define or explain the meaning of the following terms.

amorphous	element	phase	reactant
balanced equation	flotation	physical change	scientific method
chemical change	fractional distillation	physical property	sedimentation
chemical equation	heterogeneous	product	solute
chemical formula	homogeneous	property	solution
chemical property	hypothesis	pure substance	solvent
chemical symbol	matter	qualitative physical	state
chemistry	mechanical mixture	property	theory
compound	mixture	quantitative physical	unbalanced equation
crystalline		property	word equation

Chapter Objectives

You should now be able to do the following.

1. Describe the methods that scientists use in solving problems.

2. Identify the three states of matter and list the physical characteristics of each state.

▶

3. Name the changes of state and give an example of each.

4. Differentiate between physical and chemical properties of matter and between qualitative and quantitative physical properties.

5. Classify samples of matter as homogeneous or heterogeneous, and as an element, compound, solution, or mechanical mixture.

6. Explain common methods of separating the components of mixtures including filtration, sedimentation, fractional distillation, magnetic separation, electrostatic precipitation, and flotation.

7. Describe practical applications of the methods of separation listed in objective 6.

8. Identify a given change as a physical change or chemical change.

9. Give several examples of chemical reactions in and around the home.

10. State the names and symbols for approximately 30 elements.

11. State the number of atoms of each element in a compound represented by a chemical formula.

12. Recognize that many useful materials are obtained from chemical reactions, that these reactions are often accompanied by negative effects on the environment, and that chemistry has a vital role to play in solving the problems that it itself, sometimes, creates.

Understanding Chapter 1

REVIEW

1. There is no single scientific method, although scientists generally use similar methods in solving problems. What activities are common to all scientific approaches to problem solving?

2. Name the following.
   (a) the type of solid whose particles are in a regular pattern
   (b) the part of a solution that does the dissolving
   (c) the starting materials in a chemical reaction
   (d) the change of state that occurs when a solid air freshener "disappears"
   (e) the kind of physical property involving the use of a measurement
   (f) the substance added to water supplies to reduce tooth decay
   (g) a region of matter that appears to be uniform
   (h) the state of matter with a definite volume but no definite shape

3. State one important difference between each pair of items.
   (a) a mechanical mixture and a solution
   (b) a mechanical mixture and a compound
   (c) an element and a compound

4. The following elements are found in living matter: oxygen, carbon, hydrogen, nitrogen, calcium, phosphorus, chlorine, sulphur, potassium, sodium, magnesium, iodine, and iron. Write the symbols for these elements.

5. Name these elements: F, Li, Ba, Ne, Al, Cu, Ag, Au, Ni, Br.

6. Explain the difference between a balanced chemical equation and an unbalanced chemical equation.

7. State the number of atoms of each element in the following formulas.
   (a) HNO_3 5
   (b) Na_2SO_4 7
   (c) $Ca(ClO_3)_2$ 9
   (d) $Al_2(SO_3)_3$ 2+8+9 = 14
   (e) $Cu_3(PO_4)_2$ 3 2 8 13

8. (a) What two industrial processes result in the release of large amounts of sulphur dioxide gas into the atmosphere?
   (b) How does the sulphur dioxide result in acid rain?
   (c) Suggest two ways to reduce sulphur dioxide emissions.

APPLICATIONS AND PROBLEMS

9. Identify each of the following changes as physical or chemical.
   (a) the burning of natural gas chem. ✓
   (b) the grinding of coffee beans phy. ✓
   (c) the addition of flour to water phy

(d) the tarnishing of silver
(e) the formation of dew
(f) the rotting of food
(g) the explosion of a firecracker
(h) the melting of solder

10. Which of the following properties are physical properties and which are chemical properties? Explain your answer in each case.
    (a) Vinegar tastes sour.
    (b) Aluminum reacts with hydrochloric acid to produce hydrogen.
    (c) A small piece of gallium metal melts if you hold it in your hand.
    (d) Copper is a conductor of electricity.
    (e) The density of ice is 0.92 g/cm^3.
    (f) Sugar turns black when heated for a long time.
    (g) Chlorine combines with sodium to form sodium chloride.
    (h) Calcium carbonate, the major constituent of many antacid tablets, neutralizes stomach acid.

11. List three qualitative physical properties and three quantitative physical properties of water.

12. Identify the following pure substances as either elements or compounds.
    (a) chlorine   (e) baking soda
    (b) carbon monoxide   (f) sodium fluoride
    (c) water   (g) acetic acid
    (d) helium   (h) sugar

13. What method(s) of separation would be suitable to separate the components of each of the following mixtures?
    (a) muddy water
    (b) salt water
    (c) iron filings and aluminum chunks
    (d) flour and powdered glass
    (e) gold nuggets and gravel

14. Identify the following samples of matter as homogeneous or heterogeneous.
    (a) nail polish remover
    (b) vinegar
    (c) soil
    (d) clean air
    (e) fog
    (f) an orange
    (g) salad dressing
    (h) cologne

15. Read each of the following descriptions, then determine if the substance described in each case is an element, a compound, a solution, or a mechanical mixture. Explain your answer in each case.
    (a) A cloudy liquid is left to stand for several hours. At the end of this time, it is more cloudy toward the bottom of the liquid.
    (b) A clear, colourless liquid is heated. It first boils at a constant temperature, but after a period of time, the temperature increases. The liquid continues to boil at a higher fixed temperature.
    (c) A colourless gas is reacted with burning phosphorus. Part of the gas is unaffected, regardless of the amount of phosphorus used.
    (d) A solid has a sharp melting point. When heated strongly, it decomposes into another solid and a gas.

16. Three samples of matter, labelled A, B, and C, were filtered separately. The filtrate (that is, the liquid passing through the filter paper) in each case was then heated to evaporate the liquid. From the results shown below, identify A, B, and C as a pure substance, solution, or mechanical mixture. Explain your choice in each case.

SAMPLE	EFFECT OF FILTRATION	EFFECT OF HEAT ON FILTRATE
A	white residue	no residue
B	no residue	white residue
C	no residue	no residue

CHALLENGE

17. One proposed "solution" to the problem of sulphur dioxide emissions is to build taller smokestacks. INCO, the world's largest nickel producer, now has a smokestack at Sudbury, Ontario that is almost 400 m high. Comment on the advantages and disadvantages of the use of a higher smokestack as a means of reducing air pollution.

18. To separate the components of some dyes or inks, the process of chromatography is used. Find out how this process works. What is the physical property on which this separation depends?

2

THE NATURE OF MATTER

MATTER
matter

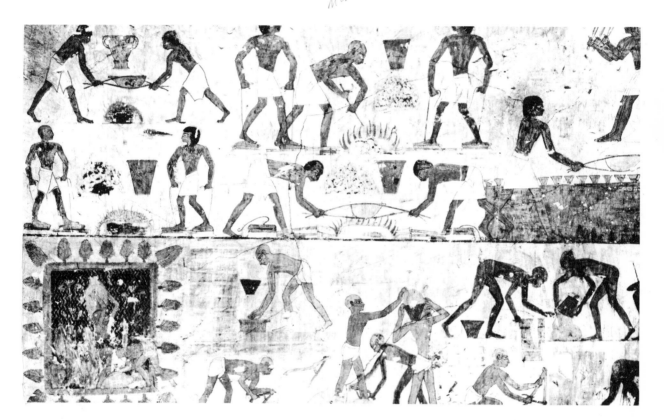

Chemistry as we know it is a relatively new science, dating back only to the 18th century. There is evidence, however, that ancient civilizations knew of certain chemical processes. For example, archaeologists have found that the early Egyptians, around 2000 B.C., were able to extract metals such as copper from their ores. There is also evidence that people during this same period in history extracted perfumes from plants, made glass, and manufactured dyes. In all cases, however, there seemed to be little or no attempt made to explain what was happening or how the chemical processes worked. What was important was to recognize that when certain processes were carried out, useful and perhaps valuable materials were obtained.

Understanding the nature of matter is essential for an understanding of chemistry and other sciences. This chapter traces the historical development of theories and models to explain the nature of matter.

2.1 Early Ideas about Matter

The ancient Greek philosophers, around 600 B.C., were the first people in recorded history to offer theories about the nature of matter. They did not, however, formulate these theories on the basis of experiments. The theories, therefore, could neither be supported nor rejected by experimental evidence. Contrast this approach with the scientific method discussed in Chapter 1.

Empedocles, about 450 B.C., suggested that all matter was composed of four "elements" – earth, air, fire, and water. The four elements, according to this theory, were produced by the four "properties" – hotness, dryness, coldness, and wetness. (See Figure 2.1.) Water was produced by the combination of wetness and coldness. Air was formed by the combination of wetness and hotness.

Around 400 B.C., another Greek philosopher named Democritus proposed what may be considered as the first atomic theory of matter. Democritus reasoned that if a sample of matter was divided into smaller and smaller pieces, a point would eventually be reached where no further subdivision could occur. According to Democritus, this point was reached not because the particles were too small to see, but because they were actually indivisible. He named these extremely small particles **atoms**, from the Greek word *atomos*, meaning indivisible. Democritus argued that different materials had different properties because the atoms making up the materials had different shapes. Materials with spherical atoms, for example, had properties that differed from those of materials made up of cubic atoms.

Aristotle, a well respected and influential Greek philosopher, opposed Democritus' ideas but supported the four-element theory proposed by Empedocles. The influence of Aristotle was so great that his theory dominated thinking among western philosophers, the clergy, and experimenters for almost 2000 years following his death. It was not until the beginning of the 19th century that the four-element theory of matter was replaced by the now universally accepted atomic theory.

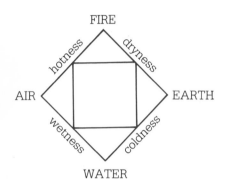

Figure 2.1
Elements and properties proposed by Empedocles. There were two properties associated with each of four elements. Earth, for example, was cold and dry. If the properties of an element were changed, another element was produced. For example, if the coldness of earth was replaced by hotness, fire was formed.

2.2 Alchemy – Science or Magic?

During the period 500–1600 A.D., the practice of **alchemy** flourished. One of the main goals of alchemists was to discover a substance that would turn base metals such as iron and copper into gold. This substance is commonly referred to as the "philosophers' stone". To modern chemists, this may seem like a hopeless task, but it was the focus of perhaps thousands of alchemists in many countries for hundreds of years. Alchemists spent countless hours heating, stirring, and sometimes even tasting mixtures of chemicals in an attempt to find the philosophers' stone.

Toward the end of the era of the alchemists, about 1300–1600 A.D., they directed their efforts to the search for an "elixir of life". This was a potion they hoped to discover which would cure all diseases, and give eternal youth to anyone who drank it. This search, like that for the philosophers' stone, was fruitless.

Modern scientists do not consider alchemy as a science. Alchemists made little or no attempt to explain any changes that occurred during their experiments. There *were* some aspects of chemistry in alchemy, along with some suggestions of magic, with perhaps a hint of religion. The alchemists experimented on what was essentially a trial and error basis. Although they were unsuccessful in their quest, the alchemists did invent some techniques and laboratory equipment still used by modern chemists. (See Figure 2.2.)

Figure 2.2
A woodcut showing an Arabian alchemist with some of his equipment

2.3 The Birth of Chemistry

The alchemists' major goal was to find uses for matter, rather than to explain its behaviour. This was no different from the goal of the early Egyptians. We have to look to the work of Robert Boyle, Joseph Priestley, and Antoine Lavoisier during the 17th and 18th centuries, for the first attempts to explain the properties and behaviour of matter. (See Figure 2.3(a) and (b) below and (c) on page 34.)

Figure 2.3

(a) Robert Boyle (1627–1691) challenged the beliefs of alchemists for years before he was able to prove that the four-element theory of matter was wrong. Boyle pioneered the scientific method.

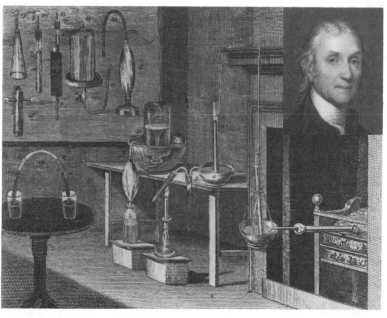

(b) Joseph Priestley (1733–1804) and his laboratory. Priestley was a non-conformist minister whose political views eventually forced him to leave England. He was too poor to afford even a balance, and depended on friends to provide his equipment.

(c) Antoine Lavoisier (1743–1794). Lavoisier's wife, Marie, was an active partner in his research, but her contributions are seldom mentioned. Lavoisier was also a member of a tax-collecting firm, and because of this, he was arrested by anti-monarchists (as shown here), found guilty of treason, and executed during the French Revolution.

Boyle was the first scientist to establish that the four-element theory of matter was wrong. In 1661, Boyle published a book entitled *The Skeptical Chymist*, in which he referred to elements as "simple, unmingled bodies". Boyle's major contribution to science was probably his insistence on carefully controlled experiments. His meticulous observations led to the discovery of Boyle's Law, which deals mathematically with the relationship between the volume of a gas and the pressure applied to the gas. (You will learn more about this law in Chapter 11.)

Priestley is best known for his discovery of oxygen. Lavoisier is recognized for, among other things, his use of the balance to make accurate measurements of the masses of substances involved in a chemical reaction. Lavoisier's careful measurements led to the **law of conservation of mass**. This important law states that in a chemical reaction, there is no difference in mass between the reactants and products. Because of his extensive use of the balance in experiments, Lavoisier is often referred to as the "father of modern chemistry".

2.4 Laws, Hypotheses, Theories, and Models

A **law** in science is a statement of a regularity that has been observed, usually in a large number of experiments. For example, the law of conservation of mass, mentioned in section 2.3, is a statement of what has been observed. A law offers no explanation for the observed regularity.

A **hypothesis** is a tentative explanation for an observation or pattern. For example, when Democritus suggested that the properties of a material depended on the shape of the atoms making up the material, he was offering a hypothesis.

A **theory** is a hypothesis which has been tested by experiments and which has been shown to be consistent with many observations. For example, the kinetic molecular theory, which you have probably studied in earlier courses and which will be reviewed in Chapter 10, is extremely useful in explaining and predicting the behaviour of solids, liquids, and gases.

A **model** is a physical representation or an idea of a system that cannot be observed directly. For example, atoms cannot be seen directly. To enable us to discuss the behaviour of atoms, scientists have developed various models of the atom. The simplest of these is that the atom can be thought of as a tiny, indestructible sphere. In this chapter, you will learn about other more complex atomic models.

2.5 The Law of Constant Composition

The quantitative experiments carried out by Lavoisier encouraged other scientists to investigate mass relationships in chemical reactions. In 1799, another French chemist, Joseph Proust, found that the chemical composition of a compound – that is, the proportion by mass of each element in the compound – was always the same, regardless of how the compound was formed. For example, water is always composed of hydrogen and oxygen in the ratio 1 g of hydrogen to 8 g of oxygen. It does not matter how much hydrogen and oxygen are available to react with each other; they always react in the same ratio.

Proust's findings are summarized in the **law of constant composition**, which states that in a particular compound the elements are always present in the same proportion by mass. The composition of a compound is often given in terms of percentages by mass. The method of calculating the percentage composition of a compound is shown in Sample Problem 2.1 on page 36. Study the Sample Problem, then do the Exercise below.

Laboratory Manual
You will have an opportunity to demonstrate the law of constant composition for yourself in Experiment 2-1.

Exercise

1. Calculate the percentage composition of water by mass, given that 1.00 g of hydrogen combines with 8.00 g of oxygen to form 9.00 g of water.

2. Analysis of sucrose (table sugar) showed that a 9.79 g sample contained 4.12 g of carbon, 0.63 g of hydrogen, and 5.04 g of oxygen. Determine the percentage composition of sucrose by mass.

Sample Problem 2.1

Sodium fluoride is added to toothpaste and to water supplies to help reduce tooth decay. When a 3.65 g sample of the compound was analyzed, it was found to consist of 2.00 g of sodium and 1.65 g of fluorine. Calculate the percentage by mass of each element in the compound.

Solution

$$\text{Percentage of sodium by mass} = \frac{\text{mass of sodium}}{\text{mass of sodium fluoride}} \times 100\%$$

$$= \frac{2.00 \text{ g}}{3.65 \text{ g}} \times 100\%$$

$$= 54.8\%$$

$$\text{Percentage of fluorine by mass} = \frac{\text{mass of fluorine}}{\text{mass of sodium fluoride}} \times 100\%$$

$$= \frac{1.65 \text{ g}}{3.65 \text{ g}} \times 100\%$$

$$= 45.2\%$$

Alternatively, the percentage by mass of fluorine can be found by subtracting the percentage by mass of sodium from 100% ($100\% - 54.8\% = 45.2\%$).

2.6 Dalton's Atomic Theory

By the end of the 18th century, scientists had available to them large amounts of information concerning mass relationships in chemical reactions. Both the law of conservation of mass and the law of constant composition had been discovered. These laws summarized the known facts about how matter behaved in chemical reactions. What was needed was a theory to explain this behaviour. In 1808, John Dalton, an English school teacher (see Figure 2.4), revived and greatly expanded the atomic theory proposed by Democritus almost 2000 years earlier. Dalton's theory was able to explain the laws which summarized the behaviour of matter. It can be stated as follows.

1. Matter is made of tiny indivisible particles called atoms.
2. All atoms of a particular element are identical in mass, size, and other properties.
3. Atoms of different elements have different characteristics, such as mass and size.

4. Atoms of different elements combine in small whole number ratios to form compounds; for example, 1:1, 1:2, 3:2, and so on. The combination of atoms forms a **molecule**. (Dalton used the term "compound atom".)
5. In chemical reactions, atoms are not destroyed; they simply join together or separate from each other.

Some of Dalton's ideas were later modified in light of subsequent experimental observations. But the essential feature of Dalton's theory – that matter is made up of tiny particles – is still valid today. Dalton's theory was undoubtedly a major milestone in the history of chemistry and it influenced the direction of research for many years. It is recognized as one of the most important theories in all of science.

Dalton's Theory and the Law of Conservation of Mass

According to Dalton's theory, atoms are neither created nor destroyed in a chemical reaction. All of the atoms that were present at the beginning of the reaction must still be there at the end. There is, therefore, no change in mass during the reaction.

Dalton's Theory and the Law of Constant Composition

According to Dalton's theory, atoms combine to form molecules in a fixed ratio in a given chemical reaction. Since all the atoms of an element are identical, the proportion by mass of each element in

Figure 2.4
John Dalton (1766–1844) became a schoolteacher by the age of 12. He was particularly interested in meteorology, and this led to a study of the properties of air and other gases. Dalton is shown here collecting methane gas (known as swamp or marsh gas). His examination of gases developed into a more general interest in chemistry. And this, in turn, eventually led to Dalton's statement of the atomic theory in 1808.

the molecule must also be fixed. For example, suppose that sodium atoms and chlorine atoms combine in a 1:1 ratio to form sodium chloride. Every time one atom of sodium combines with one atom of chlorine, sodium contributes a certain fixed proportion to the total mass, as does chlorine. The formation of a sample of sodium chloride involves an extremely large number of such combinations of atoms. Since each combination of atoms results in the same relative proportions of each element, the compound has a constant composition.

2.7 Sub-Atomic Particles

Dalton stated that atoms were indestructible. This was a perfectly reasonable assumption on the basis of the evidence available to him. However, within a century after Dalton proposed his theory, other scientists had shown his assumption to be incorrect. In this section you will learn about some of the experiments which disproved Dalton's assumption.

Sir Humphrey Davy and Michael Faraday, two English scientists (see Figure 2.5), carried out a number of experiments during the

Figure 2.5

(a) Michael Faraday (1791–1867) was at one time an assistant to Sir Humphrey Davy. Our present electrical industry grew from his experiments with electricity and magnetism. Faraday was the first scientist to isolate benzene, an important organic compound. He also developed a method for liquefying gases. His main contribution to chemistry is his pioneering work in electrochemistry.

(b) Sir Humphrey Davy (1778–1829) was a great English inventor. By experimenting with electric current, he was able to develop a method for extracting metals from compounds. Davy discovered sodium and potassium. He also isolated barium, calcium, strontium, boron, and magnesium. One of his best known inventions is the miners' safety lamp.

first half of the 19th century using electrical batteries. They showed that electricity passed through many aqueous solutions and liquid or molten compounds. The passage of electricity also resulted in the formation of new substances. Davy used this process and discovered the metals sodium, potassium, calcium, and magnesium. Faraday suggested that solutions and molten compounds were able to conduct electricity because of the presence of electrically charged atoms, some positive and some negative.

Scientists were unable to obtain conclusive evidence of the electrical nature of matter until they had the means to generate very high electrical voltages and to produce extremely low pressures or vacuums. These became available during the second half of the 19th century, and provided the tools for scientists to reveal the inner structure of atoms.

Discovering the Electron

During the 1860s, many scientists experimented with discharge tubes. A **discharge tube** is a glass tube which is sealed at both ends. A metal plate is sealed into the tube at each end. These plates, called **electrodes**, are connected to a source of high voltage, usually between 10 kV and 20 kV. The tube is also connected to a vacuum pump to remove the gas inside it. (See Figure 2.6.) Discharge tubes are often known as Crookes tubes, after Sir William Crookes, the English scientist who did much of the early work with them. (See Figure 2.7.)

Laboratory Manual
You will use a discharge tube to predict the effects of a magnetic field on cathode rays in Experiment 2-2.

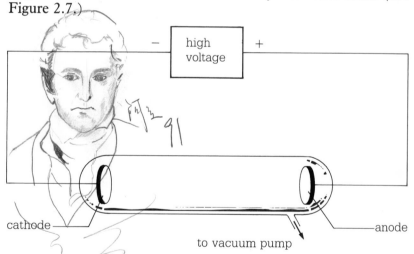

Figure 2.6
A simple discharge tube. A high voltage of several thousand volts is applied across the electrodes. The vacuum pump can remove most of the gas inside the tube.

Experiments showed that when some of the air in the discharge tube was removed, and a very high voltage applied to the electrodes, electricity flowed through the tube and the air remaining in the tube glowed. Experiments also showed that different gases in the tube produced different colours. Air gave a pink glow, neon red, and mercury green. If almost all of the gas in the tube was removed, the electricity continued to flow, but the coloured glow disappeared. In its place, a faint, green glow appeared on the walls of the tube.

Figure 2.7
William Crookes (1832–1919) was a pioneer in the study of discharge tubes.

The glow could be made more visible if the inside of the tube was coated with the compound zinc sulphide. When a screen with a slit in it, and coated with zinc sulphide, was placed in the discharge tube as shown in Figure 2.8, it was seen that the glow originated from the negative electrode, called the **cathode**, and not from the positive electrode, called the **anode**.

Figure 2.8
The screen in this discharge tube is coated with zinc sulphide to make the glow more visible. The fact that the green glow passes through the slit shows that the glow originates at the cathode.

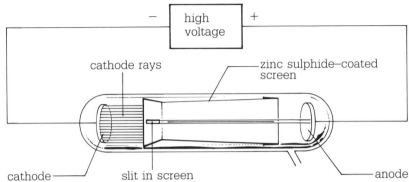

Scientists concluded that the glow was caused by invisible rays coming from the cathode. The rays were named **cathode rays**. Further experiments showed that cathode rays were attracted toward a positively charged plate held outside the tube. (See Figure 2.9.) This showed that they carried a negative charge. Scientists further discovered that cathode rays consisted of tiny, negatively charged particles. These particles were named **electrons**.

Figure 2.9
Cathode rays are attracted toward a positively charged plate, indicating that the rays are negatively charged.

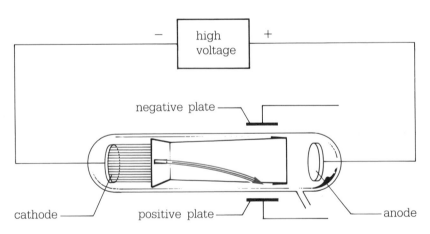

Figure 2.10
J.J. Thomson (1856–1940) discovered the electron, using a specially built discharge tube. He was awarded the Nobel prize in physics in 1906.

In 1897, J.J. Thomson (see Figure 2.10), an English physicist, carried out quantitative experiments on the nature of cathode rays. Thomson constructed a discharge tube similar to the one shown in Figure 2.11. Although he was unable to measure the charge or the mass of an electron, he was able to determine the charge-to-mass ratio of the particles. Thomson found that in all cathode ray experiments the same charge-to-mass ratio was obtained, regardless of the gas in the tube or the material which made up the cathode. He concluded that electrons were part of all matter.

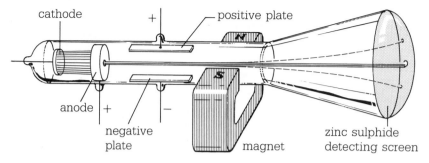

cathode — positive plate

anode + — negative plate — magnet — zinc sulphide detecting screen

Figure 2.11
Thomson carried out quantitative experiments with cathode rays. He investigated the deflection of the rays by electrical and magnetic fields. The electrons that made up the rays were deflected in one direction by the electrical field, and in the opposite direction by the magnetic field. The amount of deflection varied directly as the charge on the electron (the greater the charge, the greater the deflection). The amount of deflection also varied inversely as the mass of the electron (the greater the mass, the smaller the deflection). Thomson determined the size of the electrical field that was needed to exactly balance the effect of a known magnetic field. From this he was able to calculate the charge to mass ratio for the electron. Neither the charge nor the mass of the electron could be determined separately by this method.

Discovering the Proton

In 1886, Eugen Goldstein, a German physicist, carried out experiments with a modified discharge tube. In Goldstein's experiments, the cathode had a hole in the centre. (See Figure 2.12.) When the tube was in operation, a faint glow appeared in the part of the tube behind the cathode. Goldstein concluded that the glow in the tube was caused by charged particles moving through the hole in the cathode. Since these particles moved in a direction opposite to that of the cathode rays, it seemed reasonable to conclude that these particles were positively charged. Goldstein was able to confirm this when he found that the particles were attracted toward a negatively charged electrical plate held outside the tube.

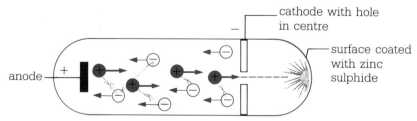

cathode with hole in centre

surface coated with zinc sulphide

anode

Figure 2.12
Goldstein's modified discharge tube. The zinc sulphide-coated inner surface of the tube was positioned where the electron beam could not reach it, yet it still glowed. Goldstein hypothesized that a stream of particles was moving through the hole in the cathode, in a direction opposite to that of the cathode rays. This, and other experiments, showed that there were positively charged particles in the discharge tube.

APPLICATIONS OF DISCHARGE TUBES

Discharge tubes have several practical applications today, as you can see in the photographs below.

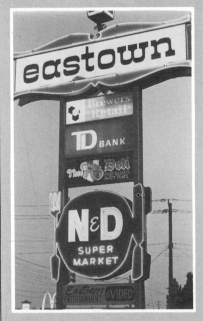

"Neon" advertising signs are simply discharge tubes bent into various shapes. The familiar different colours associated with these signs are produced by using the gases helium, neon, argon, and krypton in the discharge tubes.

Street lights are discharge tubes containing mercury or sodium vapour. Sodium vapour lamps in particular are more economical than normal incandescent lamps. In sodium vapour lamps, most of the electrical energy supplied to the lamp appears as visible yellow light. In an incandescent lamp – a light bulb, for example – much of the electrical energy is wasted as heat.

Fluorescent lighting tubes are discharge tubes containing mercury vapour. When electricity is passed through the tube, the mercury gives off light. The inside of the tube is coated with a substance that emits a white glow when struck by this light.

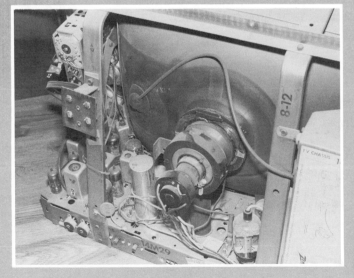

In principle, a television tube is similar in many ways to the cathode ray tube used by Thomson. The deflection of electrons by electrical and/ or magnetic fields produces the picture.

Scientists proposed that the positive particles were formed when electrons in the cathode rays collided with neutral atoms of gas in the tube. The force of the collision was able to knock one or more electrons off the atom, leaving behind a positively charged particle, known as a positive ion. These positive ions travelled toward the cathode and some passed through the hole in the cathode, causing the glow in the tube.

About 20 years later, scientists were able to show that the masses of these positive particles, unlike electrons, varied with the type of gas in the tube. They discovered that the lightest mass was obtained when the tube contained hydrogen. They also found that multiples of this mass were obtained when gases other than hydrogen were used. This led to the suggestion that a hydrogen atom without its electron was a fundamental particle in all matter. This positively charged particle, called a **proton**, was found to be 1837 times heavier than an electron.

Discovering the Neutron

Around 1920, scientists began to suggest the existence of a neutral particle in atoms, with a mass approximately equal to that of a proton. In 1932, Sir James Chadwick, another British scientist (see Figure 2.13), confirmed the existence of the **neutron**.

Chadwick was investigating the bombardment of beryllium by alpha particles. Alpha particles are helium atoms with their electrons removed. Chadwick found that a beam of rays was given off from the beryllium when it was bombarded. The rays were not deflected by a magnetic or an electrical field. This suggested that the rays carried no charge. Chadwick showed that the beam of rays consisted of neutral particles with a mass approximately equal to the mass of a proton.

The relative masses and charges of electrons, protons, and neutrons are shown in Table 2.1. Since atoms are electrically neutral, the number of protons in an atom must be equal to the number of electrons.

Research into atomic structure did not stop with the discovery of the electron, proton, and neutron. Scientists have predicted and

Figure 2.13
Sir James Chadwick (1891–1974) discovered the neutron. He was awarded the Nobel prize in physics in 1935.

Table 2.1   Relative Masses and Charges of Electrons, Protons, and Neutrons

Particle	Relative Mass	Relative Charge
Electron	$\dfrac{1}{1837}$	-1
Proton	1	$+1$
Neutron	1	0

confirmed the existence of many other sub-atomic particles. You may have heard of some of these: neutrinos, positrons, quarks, and mesons. Even now, the detailed structure of the atom is the subject of a great deal of research. Within the scope of this textbook, however, we will consider the atom as consisting of three main particles: electrons, protons, and neutrons.

2.8 The Nuclear Atom

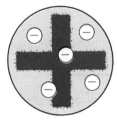

Figure 2.14
Thomson model of the atom. Thomson thought that the positive charge was spread uniformly over the atom, and that the electrons were embedded in this positively charged sphere.

In 1904, after the discovery of the electron, but before the proton and the neutron were discovered, J.J. Thomson proposed a model for the structure of the atom. This model is now known by the somewhat strange, but highly descriptive name of the "plum pudding", or "raisin bun", or "chocolate chip muffin" model. Thomson imagined the atom as a sphere with the positive charge more or less uniformly spread over the whole atom. (See Figure 2.14.) He pictured the electrons as being embedded in the atom much like plums in a pudding, raisins in a bun, or chocolate chips in a muffin.

Ernest Rutherford (Figure 2.15) was a New Zealand physicist who joined J.J. Thomson in 1895 as a research assistant at Cambridge University in England. A few years later, Rutherford came to McGill University in Montreal, where he stayed for nine years, before returning to England.

In 1909, Rutherford designed an experiment to test Thomson's model of the atom. He bombarded a very thin sheet of gold foil with positively charged alpha particles. He expected to find that the alpha particles, which were very small and moving at high speed, would pass through the foil more or less unaffected. While he observed that most of the alpha particles passed straight through the gold foil, he found that a few alpha particles were slightly deflected from a straight path. An even smaller number were deflected sharply or even bounced back from the foil. (See Figure 2.16.)

Figure 2.15
Ernest Rutherford (1871–1937) proposed the nuclear model for the atom.

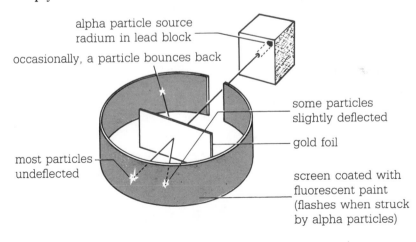

Figure 2.16
Rutherford's alpha particle scattering experiment

Rutherford was astounded by this observation. He is reported to have said, "It was almost as incredible as if you had fired a 15 inch shell at a piece of tissue paper and it came back and hit you."

To explain these observations, Rutherford proposed a model of the atom as consisting of an extremely small, very dense, positively charged region at the centre of the atom. According to Rutherford, virtually all of the mass of the atom was concentrated in this region, which he called the **nucleus** of the atom. The electrons circled the nucleus, much like the planets orbit the sun. Rutherford reasoned that most of the alpha particles passed through the gold foil unaffected, because they did not approach close enough to the nucleus to be influenced by its positive charge. Those alpha particles passing close by the nucleus were repelled by it and were therefore slightly deflected. Alpha particles approaching directly at the nucleus were repelled by the much heavier, positive nucleus and reflected from the gold foil. (See Figure 2.17.)

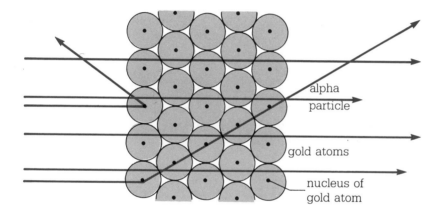

alpha particle

gold atoms

nucleus of gold atom

Figure 2.17
Rutherford's explanation of the results of his alpha particle scattering experiment. According to Rutherford, the alpha particles bounced back from the gold foil because they came very close to the tiny, positively charged centre of a gold atom, and were repelled by it.

According to Rutherford's model of the atom, most of the atom consists of empty space, with electrons orbiting in it. Rutherford calculated that the diameter of an atom is about 10^4 times the diameter of the nucleus. A useful analogy might be to think of the nucleus as about the size of a small ball bearing. On this scale, the atom would be about the size of a football stadium.

Exercise

3. The mass of the nucleus of a hydrogen atom has been determined to be 1.67×10^{-24} g, and the volume of the nucleus to be 1.7×10^{-39} cm³.
   (a) Calculate the density of the hydrogen atom nucleus in g/cm³.
   (b) How many times denser than water is the hydrogen atom nucleus?

4. A neutron has approximately the same density as a proton. A neutron star, which is a star that has neared the end of its "life", consists only of neutrons. Calculate the mass of 100 cm³ of the material in a neutron star.

2.9 Atomic Number and Mass Number

With the discovery of the proton and neutron, the structure of the nucleus became much clearer in 1932 than it had been at the time of Rutherford's alpha particle scattering experiment. We now know that the nucleus of an atom consists of protons which are positively charged and neutrons which are neutral. For each element, the number of protons in the nucleus is characteristic of that element. Helium is the only element which has 2 protons in each atomic nucleus; uranium atoms, and no other atoms, have 92 protons per nucleus. The number of protons in the nucleus of an atom of an element is called the **atomic number** of the element. Note that because an atom is neutral, the atomic number is also equal to the number of electrons in the atom. The **mass number** of an element is defined as the total number of protons and neutrons in the nucleus of the atom.

Chemists use a shorthand notation to indicate the symbol for an element, its atomic number and its mass number. This is shown below.

$$\begin{matrix} \text{mass number} \longrightarrow A \\ \text{atomic number} \longrightarrow Z \end{matrix} X \longrightarrow \text{symbol for element}$$

If both the atomic number (Z) and the mass number (A) of an element are known, then the numbers of the three fundamental particles can easily be calculated. The number of neutrons is determined by subtracting the atomic number from the mass number (A − Z).

Sample Problem 2.2

Determine the number of electrons, protons, and neutrons in a $^{19}_{9}F$ atom.

Solution

The number of protons   = 9 (from the atomic number)
The number of electrons = 9 (same as the number of protons)
The number of neutrons = 10 (19 − 9)

Exercise

5. Copy the following table into your notebook, then fill in the blanks in the table. *Do not write in the textbook.*

Element	Number of Protons	Number of Electrons	Number of Neutrons
$^{12}_{6}C$	6	6	6
$^{56}_{26}Fe$	26	26	30
$^{40}_{18}Ar$	18	18	22
$^{235}_{92}U$	92	92	143

2.10 Isotopes

One of the points of Dalton's atomic theory is that all atoms of a particular element are identical in mass. Using a **mass spectrometer**, scientists have shown that atoms of the same element can in fact differ in mass. A mass spectrometer is a device for determining the relative masses of individual atoms. (See Special Topic 2.4 on page 56.)

All neutral atoms of an element must have the same number of protons and electrons. The most reasonable explanation for atoms of the same element to have different masses is that the number of neutrons in the nucleus of the atom may vary. For example, in a sample of naturally occurring magnesium (Mg), *three* different kinds of magnesium atoms have been identified. Some magnesium atoms have 12 neutrons, some have 13 neutrons, and some have 14 neutrons. These atoms have different mass numbers (24, 25, and 26, respectively); they also differ in mass. Atoms of the same element with the same atomic number but with a different mass number are called **isotopes**.

Most elements have two or more naturally occurring isotopes. Tin has 11 such isotopes while some elements such as aluminum and fluorine each have only a single isotope. The proportion, or percent abundance, of each isotope present in a sample of an element is essentially constant. A sample of magnesium from anywhere in the world consists of the same isotopes in essentially the same proportions. Table 2.2 on page 48 lists the natural isotopes for several elements, along with their percent abundance.

Because the atomic number of an element is unique, you will often find the atomic number omitted from the symbol of an isotope. For example, the isotopes of magnesium can be represented by ^{24}Mg, ^{25}Mg, and ^{26}Mg. They may also be referred to as magnesium-24 (Mg-24), magnesium-25 (Mg-25), and magnesium-26 (Mg-26).

Table 2.2 Naturally Occurring Isotopes for Selected Elements

Element	Isotope	% Abundance	Atomic Mass*
Hydrogen	^1_1H	99.985	1.0078
	^2_1H	0.015	2.0141
Helium	^3_2He	0.00014	3.0161
	^4_2He	99.99986	4.0026
Carbon	$^{12}_6C$	98.89	12.0000
	$^{13}_6C$	1.11	13.0034
Oxygen	$^{16}_8O$	99.758	15.9949
	$^{17}_8O$	0.038	16.9991
	$^{18}_8O$	0.204	17.9992
Fluorine	$^{19}_9F$	100	18.9984
Neon	$^{20}_{10}Ne$	90.92	19.9924
	$^{21}_{10}Ne$	0.257	20.9930
	$^{22}_{10}Ne$	8.82	21.9914
Magnesium	$^{24}_{12}Mg$	78.99	23.9850
	$^{25}_{12}Mg$	10.00	24.9858
	$^{26}_{12}Mg$	11.01	25.9826
Silicon	$^{28}_{14}Si$	92.23	27.9769
	$^{29}_{14}Si$	4.67	28.9765
	$^{30}_{14}Si$	3.10	29.9738
Chlorine	$^{35}_{17}Cl$	75.77	34.9689
	$^{37}_{17}Cl$	24.23	36.9659
Silver	$^{107}_{47}Ag$	51.83	106.9051
	$^{109}_{47}Ag$	48.17	108.9047

*Atomic mass is measured in atomic mass units (symbol u). $1 u = ^1/_{12}$ the mass of a carbon-12 atom. (See page 55 for more information about atomic mass units.)

There are two naturally occurring isotopes of hydrogen, ^1H and ^2H. The latter isotope is often known as deuterium or heavy hydrogen. Water containing deuterium atoms instead of the ordinary hydrogen atoms is known as heavy water. It is produced in large quantities for use in some types of nuclear reactor, including the Canadian developed CANDU reactor. A heavy water production plant is located near Kincardine, Ontario. (See Figure 2.18.)

Figure 2.18
Bruce heavy water production plant near Kincardine, Ontario

Radioisotopes

All elements have one or more isotopes that are unstable. Unstable isotopes undergo radioactive decay. This means that they give off radiation, and in the process, the composition of their nuclei changes. This spontaneous emission of radiation from the nucleus of an atom is known as **radioactivity**. Isotopes which are radioactive are known as **radioisotopes**.

Radioactivity was discovered in 1896 by Henri Becquerel, a French scientist. (See Figure 2.19.) Becquerel's experiments were carried out using an ore of the element uranium. Marie and Pierre Curie (see Figure 2.20) investigated this ore and isolated from it two new, extremely radioactive elements, polonium and radium. As a result of their studies, Becquerel and both Marie and Pierre Curie were jointly awarded the Nobel prize for physics in 1903.

Experiments by a number of scientists, including Ernest Rutherford, the discoverer of the nucleus, established that there are three types of radiation given off by radioisotopes. These are alpha particles, beta particles, and gamma rays. (*Alpha*, *beta*, and *gamma* are the first three letters of the Greek alphabet.)

Figure 2.19
Henri Becquerel discovered radioactivity in 1896, during experiments to find out if the exposure of certain minerals to sunlight could make them glow and give off X rays. One cloudy day, Becquerel stored his mineral samples (uranium compounds among them) on top of some covered, unexposed photographic film kept in a drawer. Four days later, he developed the film to find that it had been completely exposed. He concluded that uranium produced a form of energy, which we now know as radioactivity.

Figure 2.20
Marie and Pierre Curie discovered two new radioactive elements in their experiments with uranium. They named the first polonium, which is hundreds of times more radioactive than uranium. The second, called radium, is even more radioactive than polonium. Three years after they shared the Nobel prize with Becquerel, Pierre was killed in a traffic accident, and Marie carried on their research alone. In 1911, she became the first of only two scientists in history to win a second Nobel prize. (The other is Linus Pauling.)

Alpha particles are the nuclei of helium atoms and they travel at high speed. They each consist of two protons and two neutrons and have a 2 + charge. Although alpha particles are able to penetrate thin sheets of metal or paper, they will not pass through human skin. However, they can cause skin to burn, and if taken internally can cause serious problems such as cancer and birth defects.

Beta particles are electrons travelling at a very high speed. They have about 100 times the penetrating ability of alpha particles. A thin sheet of lead will usually stop their passage.

Gamma rays do not consist of particles. They are a form of high-energy radiation, somewhat similar to X rays. The penetrating ability of gamma rays is about 1000 times that of alpha particles. A sheet of lead several centimetres thick is required to stop their passage. (See Figure 2.21.)

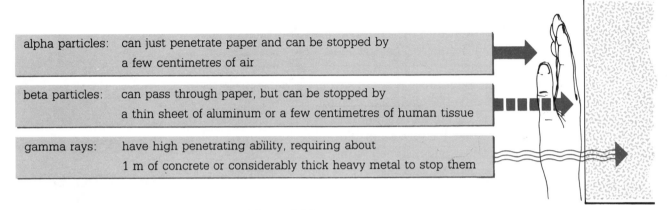

alpha particles:	can just penetrate paper and can be stopped by a few centimetres of air
beta particles:	can pass through paper, but can be stopped by a thin sheet of aluminum or a few centimetres of human tissue
gamma rays:	have high penetrating ability, requiring about 1 m of concrete or considerably thick heavy metal to stop them

Figure 2.21
Penetrating abilities of alpha particles, beta particles, and gamma rays

The radiation given off by a radioisotope can be detected using a device known as a Geiger counter. This allows radioisotopes to be used as tracers in many applications. For example, you can check a buried pipeline for leaks by adding a radioactive isotope to the liquid or gas in it and using a Geiger counter to detect if radiation is spreading from any point along the pipeline. (See Figure 2.22.)

Figure 2.22
Radioisotopes can be used to detect leaks in underground pipes. (Diagram not to scale)

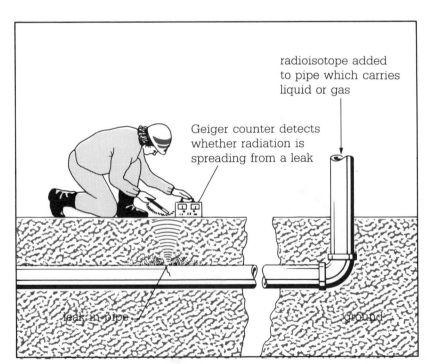

radioisotope added to pipe which carries liquid or gas

Geiger counter detects whether radiation is spreading from a leak

leak in pipe

ground

RADON

Radon is the densest gaseous element. Its density of 9.73 g/L is almost seven times that of oxygen. Radon is a colourless, odourless gas that occurs in extremely small concentrations in the atmosphere. There are 20 known isotopes of radon, all of them radioactive. The most stable isotope is radon-222.

Radon is formed when the radioactive element radium decays. It has been estimated that in every square kilometre of soil measured to a depth of about 40 cm, there is approximately 1 g of radium. As this radium decays, it releases radon into the atmosphere. The amount of radon formed is small. Since it is usually dispersed in the atmosphere, there is little danger to people. In places where the concentration of radium is higher than normal, however, such as uranium mining areas, radon, and hence the level of radioactivity, can build up to a level dangerous to humans.

During the 1970s, scientists found alarmingly high levels of radon in homes in Elliot Lake, Ontario, a uranium-mining town, and in Port Hope, Ontario, the site of a former uranium refinery. It was revealed that the waste material from uranium mines, often known as uranium tailings, was commonly used as land fill material. Radon was given off by this material and collected in the homes built on the filled land.

When radon gas decays, it releases alpha particles and forms other atoms which are known as radon daughters. These can attach to airborne dust particles. Further radioactive decay of the radon daughters also releases alpha particles. When radon or the radon daughters are inhaled, the alpha particles released in the body cause damage to the lung tissue and may even cause lung cancer.

Radon is used in the treatment of some cancerous tissues. Small amounts of radon are sealed in hollow, gold needles, which are then implanted in the tumour. The radon releases alpha particles which destroy the cancerous tissue.

Half-life of a Radioactive Element

Scientists are able to measure the rate at which a radioisotope decays. Every radioisotope has a characteristic property called its **half-life**. This is the time taken for one half of the original number of radioactive atoms to decay. The half-life of a radioisotope is independent of the amount of radioisotope initially present. It is also unaffected by temperature and pressure.

Half-lives of radioisotopes vary from fractions of a second to millions of years. For example, the half-life of polonium-214 is 1.6×10^{-4} s and that of uranium-238 is 4.5×10^9 a. After one half-life, the number of radioactive atoms reduces to one half of the original number; after two half-lives, only one quarter of the original number of radioactive atoms remains; after three half-lives, only one eighth of the initial number remains, and so on. The decaying atoms become atoms of other elements. (See Figure 2.23.)

The half-life of carbon-14 is approximately 5700 a. This radioisotope occurs naturally in the atmosphere. Plants take in carbon-14 along with non-radioactive carbon-12 and carbon-13 during photosynthesis. The carbon eventually finds its way into animals including human beings. As long as the plant or animal is alive, there is

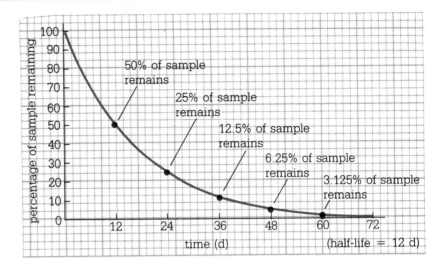

Figure 2.23
Typical half-life graph of a radioactive element. During a period of time equal to the half-life of the radioisotope, one half of the initial sample decays.

a continual interchange between the carbon in it and the carbon in the atmosphere. The ratio of carbon-14 atoms to carbon-12 atoms in the plant or animal remains essentially constant. When the plant or animal dies, this exchange of carbon ceases. The stable carbon-12 atoms remain unchanged but the carbon-14 atoms begin to decay. After 5700 a, only one half of the original number of carbon-14 atoms remain. By finding the ratio of carbon-14 to carbon-12 in a sample of the remains of the plant or animal, scientists can estimate the number of carbon-14 half-lives that has elapsed, and therefore the length of time since the plant or animal died. This process is known as **carbon dating**. It has been used to determine the approximate age of such objects as Viking ships, fossils, California redwood trees, and ancient artifacts such as the famous Dead Sea scrolls. (See Figure 2.24.)

Figure 2.24
This remarkable fossil, from the Burgess Shales, near Field, British Columbia, closely resembles animals that might have been ancestors to insects. Radioisotope dating places the age of the fossil at about 5.3×10^8 years ago.

PUTTING CHEMISTRY TO WORK

JEANNE JOHNSON / ISOTOPE LAB TECHNICIAN

Just south of Saudi Arabia, along the Arabian Sea, is a small country called Oman. When the government of Oman decided to plan large-scale irrigation of its arid land, what it needed was a detailed understanding of its water resources — past as well as present. To find out the age of the underground water upon which irrigation would depend, thousands of samples of water from across the entire country were sent to the isotope laboratory at the University of Waterloo, Ontario. Jeanne Johnson is the technician responsible for dating such samples using isotope techniques.

"We determined the age of the Oman water samples by detecting the presence of isotopes such as oxygen-18, carbon-14, tritium, and deuterium in the water," Jean explains. Rain water picked up the isotopes in the atmosphere on its way to the ground. This naturally-occurring radioactivity gradually decays over time. "By measuring the radioactivity left in a sample, an expert can tell how much time has passed since the water was last in the atmosphere." Some of the water from Oman's deeper wells turned out to be 20 000 years old. Using this water for irrigation would therefore be a last resort, since it is as much a non-renewable resource as any of Oman's oil fields.

Dating water samples is just one example of the work done in the isotope lab. "Carbon isotope dating is useful for providing dates for organic material up to 50 000 years old," Jeanne says. "It is especially useful in archaeology and geology." One of the most interesting samples Jeanne worked on was a mammoth bone. "On my first try, the bone turned out to be only 4000 years old. This was crazy, since mammoths haven't been around for 10 000 years." The answer? "The original bone sample was porous enough to have allowed recent atmospheric carbon to penetrate. I was then given a piece of the animal's tusk. This sample dated correctly but the tusk, which was to be used in a display, lost half a kilogram in the process!"

Jeanne enjoys working in the university environment. "The people are friendly and very interesting," she says with a smile. "Although the work is routine, the problems we solve are varied and often quite exciting. And I find I can easily schedule my work hours to match the demands of my growing young family. It's an ideal arrangement."

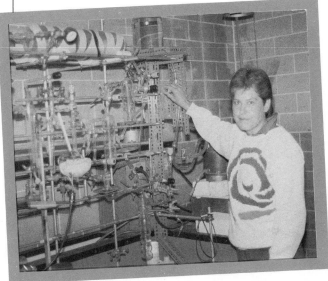

Other Applications of Radiosotopes

Radioisotopes have many other applications. For example, they may be used to determine wear in engines. A small amount of a radioisotope is incorporated into the metal in a moving part of the engine, such as a piston ring. The amount of radioactivity that appears in the oil is a measure of how much the piston ring has worn.

Another interesting application is in the determination of the thickness of a piece of metal – sheet steel or aluminum foil for example – during its manufacture. The amount of radioactivity passing through the metal depends on the thickness of the metal. By measuring the radioactivity transmitted by the metal, the thickness of the metal can be constantly monitored. If there is any deviation from the required thickness, the manufacturing process can be altered to correct the problem. (See Figure 2.25.) The same principle is used in determining whether sealed containers are filled to the correct level or not. This is much faster than measuring the mass of each container or opening each one up to determine the amount of the contents.

Figure 2.25
Radioisotopes can be used to monitor the thickness of sheet material such as metal, rubber, plastic, and paper. The amount of radiation reaching the detector is affected by the thickness of the sheet material. An electrical circuit from the detector (via a control device) adjusts the rollers to maintain correct thickness.

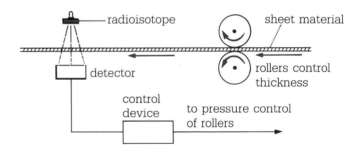

SPECIAL TOPIC 2.3

RADIATION AND STERILIZATION

During the 1960s, in the United States, there was a very interesting and unusual application of radioisotopes. Screwworm flies were causing a major problem with cattle. The flies fed on the open wounds of the cattle, making them unfit for human consumption, or in extreme cases, killing the cattle. Scientists were able to sterilize the male flies by treating them with radioactive cobalt. Since the female flies mated only once in their lifetime, the population of screwworm flies decreased dramatically.

Radioisotopes are also used to preserve certain foodstuffs. For example, meat is exposed to certain types of radiation to kill any bacteria in it. The meat stays fresh much longer than if untreated. Wheat crops in storage are often badly damaged by insects. These insects can be killed by treating the wheat with radiation.

Radioisotopes are used widely in medicine for both diagnosis and treatment of diseases. Table 2.3 summarizes a number of these applications.

Table 2.3 Uses of Radioisotopes in Medicine

Isotope	Application
Arsenic-74	Determining the location of brain tumours
Chromium-51	Red blood cell studies
Cobalt-60	Radiation treatment of cancer
Gallium-67	Diagnosis of some types of cancer
Indium-113	Liver, lung, and brain scans
Iodine-131	Determination of thyroid problems
Iron-59	Measuring blood flow and blood volume
Phosphorus-32	Treatment of leukemia
Sodium-24	Detecting blockages in the circulatory system

2.11 Determining Atomic Mass

Atoms are so tiny that it is impossible to measure their mass by conventional methods. Scientists are able to measure atomic masses using a mass spectrometer. (See Special Topic 2.4.)

Before it was possible to measure the masses of individual atoms, scientists established a relative atomic mass scale. This scale simply compared atomic masses to a standard, without giving an actual mass for any one atom. Relative atomic masses were calculated from experiments in which the masses of combining elements were determined. For example, suppose that an experiment showed that in a compound of hydrogen and chlorine, 35.5 g of chlorine combined with 1 g of hydrogen. If it is known or assumed that 1 atom of chlorine combines with 1 atom of hydrogen, then it follows that an atom of chlorine is 35.5 times heavier than an atom of hydrogen. Relative atomic masses for other elements could be obtained by carrying out similar experiments.

Although relative atomic masses have no units, it is common practice to use a unit called an **atomic mass unit** (symbol u). The size of the atomic mass unit is chosen arbitrarily. In fact, past definitions of the atomic mass unit are different from its present definition. Since 1961, the basis of the atomic mass scale has been the carbon-12 atom, the most common isotope of carbon. This isotope is assigned a mass of exactly 12 u. In other words, 1 u is

THE MASS SPECTROMETER

In a mass spectrometer, molecules of a gas are bombarded by electrons. This bombardment knocks one or more electrons off the neutral molecules, forming positive ions. These ions are accelerated toward a negatively charged plate with a hole or slit in it. When they pass through this hole, the ions form a sharply defined beam.

The beam of ions then passes through a magnetic field. As the ions pass through, they are deflected by the magnetic field. The amount of deflection experienced by the ions depends on their charge and mass as well as the strength of the magnetic field. If all the ions have the same charge, heavier ions are deflected less than lighter ions. If the beam of ions consists of ions of different masses, then the magnetic field separates them into two or more beams, each one consisting of ions of the same mass. The intensity of each ion beam is a measure of the number of ions and can be measured experimentally. From this information and the strength of the magnetic field, isotopic masses and abundances can be calculated.

The mass spectrometer is also an extremely powerful analytical tool for the chemist. When complex molecules are bombarded with electrons, they fragment (break up) and form positive ions. Usually several different fragment ions are produced from a particular molecule. The masses of the fragment ions and their relative abundances are characteristic of the original molecule. This is known as the *mass spectrum* of the molecule. It is possible to identify a molecule by comparing its mass spectrum with the mass spectra of known molecules. In addition, the structure of new molecules may be determined by carefully piecing together the fragment ions to produce the original molecule. (This is rather like creating a picture from the pieces of a jigsaw puzzle.)

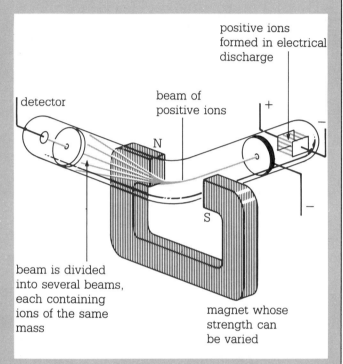

In a mass spectrometer, positive ions are deflected by a magnetic field. Since the amount of deflection depends on the mass of the ion, a beam of ions consisting of ions of different masses can be separated.

equal to $\frac{1}{12}$ of the mass of a carbon-12 atom. This is obviously a very small unit; the relationship between atomic mass units and kilograms is:

$$1 \text{ u} = 1.661 \times 10^{-27} \text{ kg}$$

The masses of a proton, electron, and neutron based on this atomic mass scale are shown in Table 2.4. Table 2.5 lists the atomic masses of several atoms.

Table 2.4 Masses of Protons, Electrons, and Neutrons

Particle	Mass in u	Mass in kg
Proton	1.0073	1.672×10^{-27}
Electron	0.0005486	9.107×10^{-31}
Neutron	1.0087	1.675×10^{-27}

Table 2.5 Atomic Masses of Selected Atoms

Atom	Mass in u
^{1}H	1.00783
^{14}N	14.00307
^{16}O	15.9949
^{24}Mg	23.9850
^{32}S	31.9721

Notice that the mass of an atom in atomic mass units is approximately the same as the mass number of the atom. This is not surprising since a proton and a neutron each have a mass of approximately 1 u, and the mass of an electron, compared to the mass of the proton or the neutron, can be ignored. Unless high accuracy is required, the numerical value of the atomic mass may be considered to be equal to the mass number.

Average Atomic Mass

The values of atomic masses given in atomic mass tables and in the periodic table of the elements are the **average atomic masses** of the elements. Most elements consist of two or more naturally occurring isotopes. This means that the element is made up of atoms of at least two different masses. The average atomic mass of the element depends on the relative proportions of the isotopes in a sample of the element. Suppose, for example, that an element consists of equal proportions of two isotopes, one of atomic mass 150.0 u, and the other of atomic mass 160.0 u. Then the average atomic mass of this element is $(150.0 + 160.0) \div 2$ or 155.0 u. This situation, with two isotopes present in equal amounts, is unusual. When the isotopes are present in other than equal proportions, the average atomic mass must be calculated. Study Sample Problem 2.3, then do the Exercise below.

Exercise

6. Lithium consists of two isotopes in these proportions:
       92.50% of ^{7}Li, atomic mass 7.016 u
        7.50% of ^{6}Li, atomic mass 6.015 u
   Calculate the average atomic mass of lithium.

7. Use the information in Table 2.2 to calculate the average atomic mass of magnesium.

Sample Problem 2.3

A naturally occurring sample of chlorine consists of two isotopes in these proportions:

75.77% of ^{35}Cl, atomic mass 34.9689 u
24.23% of ^{37}Cl, atomic mass 36.9659 u
Calculate the average atomic mass of chlorine.

Solution

$$\text{Average atomic mass} = \left(\begin{array}{c}\text{atomic mass} \\ \text{of } ^{35}\text{Cl}\end{array} \times \begin{array}{c}\text{fraction of} \\ ^{35}\text{Cl in sample}\end{array}\right) + \left(\begin{array}{c}\text{atomic mass} \\ \text{of } ^{37}\text{Cl}\end{array} \times \begin{array}{c}\text{fraction of} \\ ^{37}\text{Cl in sample}\end{array}\right)$$

$$= \left(34.9689 \text{ u} \times \frac{75.77}{100}\right) + \left(36.9659 \text{ u} \times \frac{24.23}{100}\right)$$

$$= 26.4959 \text{ u} + 8.9568 \text{ u}$$

$$= 35.4527 \text{ u}$$

$$= 35.45 \text{ u (to 4 significant digits)}$$

Alternative Solution

Suppose that we have a sample of chlorine containing 10 000 atoms.

7577 atoms in the sample (75.77%) will each have a mass of 34.9689 u.

The total mass of these atoms $= 7577 \times 34.9689$ u
$= 264\ 959$ u

2423 atoms in the sample (24.23%) will each have a mass of 36.9659 u.

The total mass of these atoms $= 2423 \times 36.9659$ u
$= 89\ 568$ u

The total mass of the 10 000 atoms $= 264\ 959$ u $+ 89\ 568$ u
$= 354\ 527$ u

The average mass of a chlorine atom $= \dfrac{354\ 527 \text{ u}}{10\ 000}$

$= 35.45$ u

PUTTING SCIENCE TO WORK

TOM MOY / TECHNICAL SALES REPRESENTATIVE

New scientific instrumentation is constantly being developed and improved. Tom Moy is a technical sales representative for a company which makes instruments such as spectrophotometers used to analyze and identify chemicals in the environment. His job is to link his firm's latest technology with the needs of potential customers.

"Selling scientific equipment is very different from other types of sales," Tom explains. "I am more a consultant than a salesperson. Scientists and researchers want the best instruments available so their own work can be as effective as possible. So my job is to let my customers know about the newest advances in our equipment so that they can choose the apparatus which best suits their needs. At the same time, I find out what the scientific community will want from our company's future products."

Tom decided to pursue a scientific sales career because he wanted to keep up with the latest in science. "It is definitely not a routine job. It is exciting to be involved both with the best in technology and with the people who are using that technology to break new ground in science.

"I was interested in chemistry, biology, drafting, and mathematics while in secondary school," Tom reflects. Tom followed these interests by obtaining a degree in chemistry at university. But it is not just his science background that Tom finds helpful in his everyday work. "My high school history, geography, and arts courses have given me the ability to talk about a wide range of topics. This is important because people want to talk about themselves and what interests them as well as about the requirements of their laboratories. It seems strange to me now that I once thought these courses would be of no use to me."

Tom often becomes involved in assisting scientists to solve their research problems. "Although scientists know a great deal more about their fields than I do, they may not have the necessary knowledge in analytical chemistry to use the instrumentation in the best way possible." Consider a person who must drive a car to reach a certain destination. This person knows where to go, but does not necessarily know how the engine works.

"In some way," Tom says with satisfaction, "I have contributed to a very wide variety of research; something I could never have done as a research scientist. This is the best part of my job."

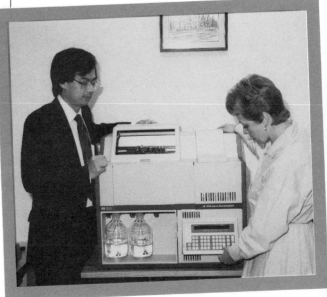

2.12 A Revised Atomic Model: The Bohr Atom

Figure 2.26
Niels Bohr (1885–1962) proposed that electrons in atoms were restricted to certain energy levels. When electrons moved from one energy level to another, they either absorbed or gave off a specific amount of energy. Bohr was awarded the Nobel prize in physics in 1922 for his work in the determination of atomic structure.

Laboratory Manual
In Experiment 2-3, you will have a chance to observe the line spectra of several elements.

Although the work of Rutherford gave scientists a useful model of the atomic nucleus, it did not answer the question of how the electrons in the atom were arranged. Rutherford suggested that the electrons orbit the nucleus in much the same way that the planets orbit the sun. According to the laws of physics, a moving charged particle, in this case the electron, should give off energy. As the electron gives off energy, it should slow down and eventually fall into the nucleus. Since this does not happen, the "planetary" model of the atom had to be modified.

In 1913, Niels Bohr, a Danish physicist who had worked with Rutherford in England, proposed a model for the hydrogen atom. (See Figure 2.26.) The Bohr model was based on a series of experiments involving the spectra of elements.

You have probably seen a **continuous spectrum** before – a rainbow is a good example. In a continuous spectrum, the colours merge into each other with no obvious dividing line between them. If the light from a regular light bulb is passed through a prism or diffraction grating, a continuous spectrum is observed. White light is a combination of all of the colours in the continuous spectrum. (See Colour Plate 1.)

When electricity is passed through gaseous elements at low pressure, light of a characteristic colour is produced. For example, neon produces a red colour, and hydrogen produces a violet colour. If this coloured light is passed through a prism or diffraction grating, you will obtain a series of coloured lines separated by black spaces. This pattern of coloured lines is called a **line spectrum**. Each element has its own unique line spectrum, which may be used as a means of identification. (See Colour Plate 1.)

The analysis of spectra from stars has enabled scientists to determine the elements present in the stars. One of the most interesting examples of this resulted in the discovery of helium. During a solar eclipse in 1868, scientists analyzed the light from the outer layers of the sun. They observed spectral lines that did not correspond to any known element. The lines were attributed to the presence of a new element, which was given the name helium, derived from the greek word *helios*, meaning sun. Helium was not identified on Earth until 1895. Line spectra are also useful in analyzing alloys (mixtures of metals). A small sample of the alloy is vapourized and the light emitted is viewed through a spectroscope. (A spectroscope is basically a diffraction grating.) The lines present in the line spectrum identify the metals in the alloy.

In the case of hydrogen, the visible part of its line spectrum consists of four lines. (See Figure 2.27 and Colour Plate 1.) Each line corresponds to a specific amount of energy.

violet   green   red

Figure 2.27
Line spectrum of atomic
hydrogen. Each line in the
spectrum corresponds to a
specific amount of energy.

To explain line spectra, Bohr proposed a revolutionary hypothesis. He suggested that the electrons in atoms can move in a certain number of allowed **energy levels**. An electron can change its energy by going from one energy level to another, but cannot be between energy levels. According to Bohr, an electron may move from a lower energy level to a higher level by absorbing energy, for example, from a flame or from an electric current. When the electron falls back to a lower level, energy is released. The energy given out is the difference in energy between the two energy levels. Since there are only certain allowed energy levels, the energy released therefore has specific allowed values. Each of these allowed energy values corresponds to a line in the line spectrum; some of these lines are in the visible part of the spectrum. (See Figure 2.28.)

When the electrons in an atom are in the lowest possible energy levels, the atom is said to be in its **ground state**. If an electron moves to a higher energy level by absorbing energy, the atom is now said to be in an **excited state**. An atom in its excited state will eventually return to its ground state, and in doing so release energy.

To help you understand Bohr's proposal, you might find it useful to think of a marble on a staircase. The marble can come to rest only on a step, but not between steps. If the marble is moved to a higher step, its potential energy increases. If the marble falls to a lower step, its potential energy decreases. The marble can possess only certain amounts of energy, much like an electron in an atom.

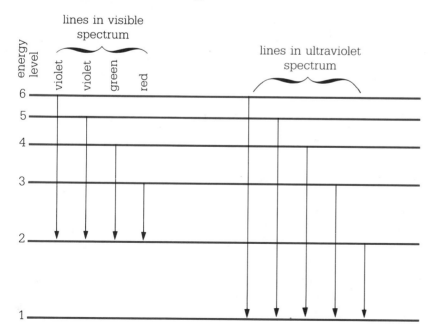

Figure 2.28
Energy levels and the line spectrum of hydrogen. The lines in a line spectrum are caused by electrons falling from a higher energy level to a lower energy level. In the case of a hydrogen atom, the lines in the visible spectrum are caused by electrons falling from a higher level to energy level 2. (Diagram not to scale)

FLAME TESTS AND FIREWORKS

Some elements show one or more particularly prominent lines in their line spectrum. An application of this is in the use of flame tests to identify the presence of these elements. For example, the line spectra of all sodium compounds have two very noticeable yellow lines.

When sodium compounds are placed in a Bunsen burner flame, the flame takes on a distinct yellow colour. A similar test with barium compounds results in a green flame, while calcium compounds produce a red flame. This same phenomenon is put to use in fireworks. Small amounts of compounds containing ele-ments with characteristic flame colours are added to the fireworks. When the fireworks are ignited, beautiful colours are produced. You can make colourful firelogs by tightly rolling some newspapers and tying them into bundles with string, and then soaking the bundles in a pail of water containing small amounts of sodium chloride, copper sulphate, and strontium chloride. After soaking for about two weeks, remove the newspaper bundles and let them dry for two or three months. When the dried bundles burn, the flame will appear yellow, green, and red.

Further studies of line spectra enabled scientists to predict that there was a maximum number of electrons that could be present in a given energy level. The maximum number of electrons is equal to $2n^2$, where n is the number of the energy level. (See Table 2.6.)

Table 2.6 Maximum Number of Electrons in Each Energy Level

Energy Level	Maximum Number of Electrons
1	$2 \times 1^2 = 2$
2	$2 \times 2^2 = 8$
3	$2 \times 3^2 = 18$
4	$2 \times 4^2 = 32$
5	$2 \times 5^2 = 50$
n	$2 \times n^2 = 2n^2$

Limitations of Bohr's Theory

Bohr was able to explain all of the lines in the line spectrum of atomic hydrogen, and even to predict the existence of other lines which had not yet been observed. However, his theory cannot explain the observed line spectra of other elements.

In a later chemistry course, you will learn about a theory known as quantum mechanics or wave mechanics. This theory is able to explain the line spectra of elements other than hydrogen. Although this theory is much more complex than Bohr's theory, it still retains the basic concept of energy levels in atoms. Modern atomic theory visualizes the atom as consisting of a nucleus surrounded by regions of space where electrons are likely to be found. These regions are not sharply defined like the orbits of Bohr's theory, but are somewhat "fuzzy". In these fuzzy regions, there is a very high probability of finding electrons. Modern atomic theory refers to these regions as **orbitals**.

2.13 Electron Arrangement in Atoms

Bohr's theory of atomic structure was eventually discarded as other more reasonable ideas were developed. The concept of electrons being in distinct energy levels, however, is critical to an understanding of chemistry. The distribution of electrons among the different energy levels in the atoms of an element is largely responsible for the properties of the element. This relationship between electron arrangement and chemical properties will be investigated further in Chapters 3 and 4.

You will sometimes see diagrams showing the arrangement of electrons in energy levels. These diagrams are often referred to as Bohr diagrams or Bohr-Rutherford diagrams. A few examples are shown in Figure 2.29. You should not interpret these diagrams too literally. The electrons do not move around the nucleus in fixed paths, as the diagrams seem to suggest.

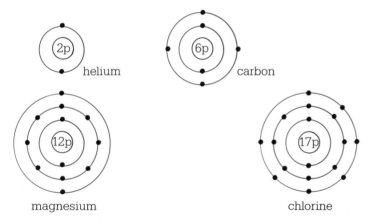

Figure 2.29
Bohr diagrams for several atoms

In Chapter 4 you will learn that chemists are particularly interested in the electrons in the highest energy level of an atom. These are the electrons that primarily determine the chemical properties of elements.

Looking Back – Looking Ahead

In this chapter, we have traced the development of one of the most powerful theories in science. You have seen how the atomic theory has been repeatedly modified as new experimental evidence became available. This is true of all scientific theories – a theory is accepted as long as it is able to explain experimental observations. When there are observations which cannot be explained by the theory, then the theory needs to be modified to accommodate the new information. If the theory cannot be modified, it must be rejected and a new theory proposed. You will see later that although Bohr's theory of atomic structure is limited, a great deal of chemistry can be explained using a slightly modified version of Bohr's theory.

Vocabulary Checklist

You should now be able to define or explain the meaning of the following terms

alchemy	cathode ray	hypothesis	model
alpha particle	continuous spectrum	isotope	molecule
anode	discharge tube	law	nucleus
atom	electrode	law of conservation	neutron
atomic mass unit	electron	of mass	orbital
atomic number	energy level	law of constant	proton
average atomic mass	excited state	composition	radioactivity
beta particle	gamma rays	line spectrum	radioisotope
carbon dating	ground state	mass number	theory
cathode	half-life	mass spectrometer	

Chapter Objectives

You should now be able to do the following.

1. State the key ideas of Dalton's atomic theory, and indicate why they had to be modified later.

2. Describe how Dalton's theory explains the laws of conservation of mass and constant composition.

3. Determine the percentage composition of a compound given the mass of each element in a sample of the compound.

4. Give at least three practical applications of radioisotopes.

5. Compare protons, neutrons, and electrons with respect to mass, charge, and their location in the atom.

6. Describe the atomic model proposed by J.J. Thomson.

7. Outline the alpha particle scattering experiment carried out by Ernest Rutherford.

8. Describe the nuclear model of the atom proposed by Rutherford.

9. Appreciate that a scientific theory may be tentative, and explain why the atomic theory had to be modified a number of times.

10. Calculate the average atomic mass of an element given the percentage and mass of each isotope in a sample of the element.

11. Describe the Bohr theory of atomic structure and explain the experimental evidence on which Bohr developed his theory.

12. Explain the line spectra of elements using the concept of energy levels in atoms.

Understanding Chapter 2

REVIEW

1. Alchemists carried out thousands of experiments, yet alchemy is not considered to be a science. What is the major difference between alchemy and science?

2. State the major contribution of each of the following scientists to our understanding of the nature of matter.
   (a) John Dalton
   (b) J.J. Thomson
   (c) Ernest Rutherford
   (d) James Chadwick
   (e) Niels Bohr

3. Indicate whether each of the following statements is true or false. If the statement is false, rewrite it to make it correct.
   (a) Protons are negatively charged particles.
   (b) The mass of a proton is approximately equal to the mass of an electron.
   (c) The neutron is an uncharged particle found in the nucleus of atoms.
   (d) John Dalton discovered the electron.
   (e) In all atoms, the number of neutrons is equal to the number of protons.
   (f) Most of the mass of an atom is contained in the nucleus.
   (g) Isotopes of the same element have different atomic numbers.

4. In your notebook, match the definitions in column A with the correct scientific term in column B.

COLUMN A	COLUMN B
A mental picture or description	Law
A first or tentative explanation	Theory
A statement of a regularity based on experimental results	Hypothesis
A tested explanation as to why nature behaves the way it does	Model

5. Explain what is meant by the term "heavy water".

6. Explain the difference between atomic mass and mass number for an individual atom.

7. With the discovery of isotopes, which part(s) of Dalton's theory required modification? Explain why.

8. (a) What is required for an electron in an atom to move from a lower energy level to a higher energy level?
   (b) What is observed when the electron returns to a lower level?

9. How did Bohr's theory explain the line spectrum of hydrogen?

10. An atom has only one ground state, but has several excited states. Explain this statement.

11. Explain why it was necessary to modify the theories of atomic structure proposed by Dalton, Thomson, Rutherford, and Bohr.

APPLICATIONS AND PROBLEMS

12. Solid calcium chloride has the ability to absorb water vapour from the air. For this reason, it is often spread on dirt roads to reduce dust during the summer. The calcium chloride absorbs water from the air, and eventually forms a solution which mixes with the dust. This helps to keep the road damp and less dusty. Analysis of a 3.18 g sample of calcium chloride showed that it consisted of 1.15 g calcium and the rest chlorine. Determine the percentage composition of calcium chloride by mass.

13. Two students each prepared a compound containing only the elements copper and oxygen. A sample of each compound was then analyzed; the results are shown in the table below.

STUDENT	MASS OF SAMPLE	MASS OF COPPER IN SAMPLE	MASS OF OXYGEN IN SAMPLE
A	1.56 g	1.25 g	0.31 g
B	3.24 g	2.88 g	0.36 g

   (a) Calculate the percentage composition by mass of each sample.
   (b) Do you think that these were samples of the same compound? Explain your answer.

14. Copy and complete the following table in your notebook. *Do not write in the textbook.*

ATOM	ATOMIC NUMBER	MASS NUMBER	NUMBER OF PROTONS	NUMBER OF NEUTRONS	NUMBER OF ELECTRONS
$^{40}_{20}Ca$ *20*	*40*	20	20	*20*
$^{58}_{28}Ni$ *28*	58	*28*	*30*	28
$^{84}_{36}Kr$ *36*	*84*	36	48	*36*

15. Naturally occurring silicon consists of three isotopes as follows:
    92.23% silicon-28 (atomic mass 27.977 u)
    4.67% silicon-29 (atomic mass 28.976 u)
    3.10% silicon-30 (atomic mass 29.974 u)
    Calculate the average atomic mass of silicon.

16. Naturally occurring chromium consists of the following:
    4.31% ^{50}Cr (atomic mass 49.946 u)
    83.76% ^{52}Cr (atomic mass 51.941 u)
    9.55% ^{53}Cr (atomic mass 52.941 u)
    2.38% ^{54}Cr (atomic mass 53.939 u)
    Calculate the average atomic mass of chromium.

17. Oxygen consists of three isotopes: oxygen-16, oxygen-17, and oxygen-18. Show how many different masses an oxygen molecule, O_2, may have.

18. Suppose that a carbon-12 atom is assigned a mass of exactly 18 u instead of 12 u. Based on this new standard, what is the atomic mass of each of the following elements?
    (a) H   (b) O   (c) Mg   (d) Pb

19. A certain radioisotope has a half-life of 24 d. If the initial mass of a sample of this isotope is 4.8 g, calculate the mass of the isotope remaining after (a) 48 d, (b) 96 d, and (c) 168 d.

20. Explain how Bohr's theory of atomic structure can be considered both a success and a failure.

21. When electricity is passed through a sealed glass tube containing neon at low pressure, a reddish coloured glow appears in the tube. Explain this observation in terms of Bohr's theory of atomic structure.

22. Sodium vapour street lamps emit a characteristic yellow light. Suggest a reason for this.

23. Isotopes of a particular element have identical chemical properties, even though they have different masses. Suggest a reason for this observation.

CHALLENGE

24. Would you expect isotopes of the same element to have the same line spectra? Explain your answer.

25. The mass of a proton and of a neutron is approximately 1 u; the mass of an electron is negligible. All atoms are composed of whole numbers of protons and neutrons. It might be expected, then, that the atomic masses of the elements would be very close to whole numbers. This is true for many elements, but the atomic masses of some elements (magnesium and chlorine, for example) are far from being whole numbers. Suggest an explanation for this observation.

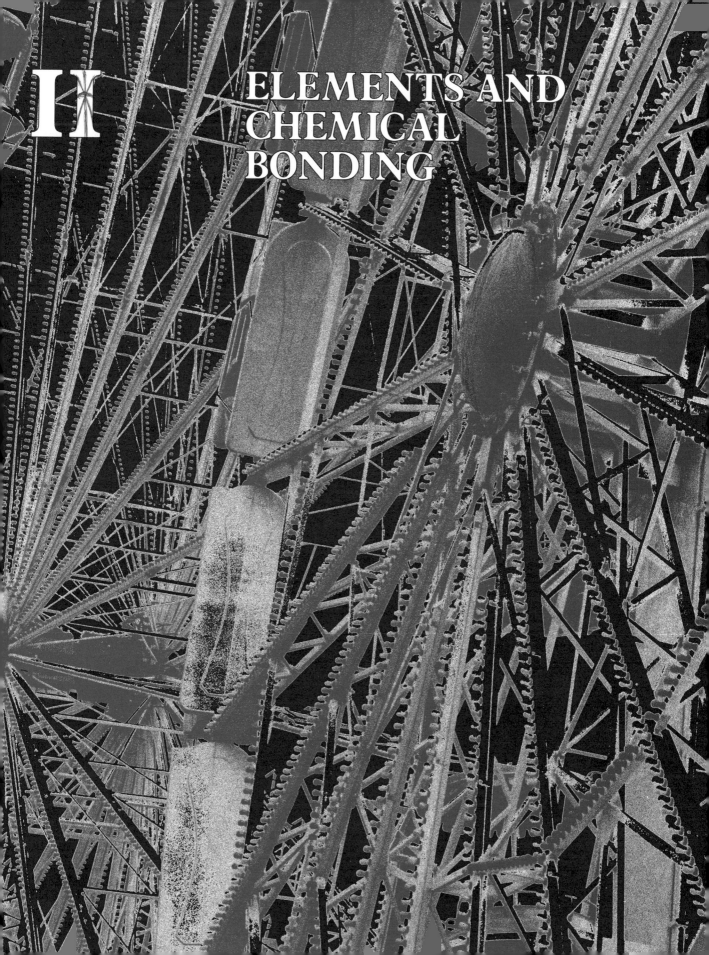

II
ELEMENTS AND CHEMICAL BONDING

3 PATTERNS

Examine the illustration that introduces this chapter. You will no doubt easily identify the fish and the geese. As you look at the picture more closely and accumulate more information, you instinctively begin to organize the information. Images of fish and geese permeate the picture and seem to meld together. You start to appreciate the entire picture when, perhaps unconsciously, you begin to organize the visual images into a pattern. As your eyes move diagonally down the picture, the image of the fish emerges from the background air and becomes dominant. At the same time, as you move your eyes diagonally up the picture, the background water gradually evolves into images of geese.

The steps you followed in making sense of this picture are similar to those a scientist uses to study and understand nature. You will use these same steps – observation, organization, and pattern seeking – in your study of the chemical elements, which are the building blocks of our physical environment.

In Chapter 2 you followed the development of an atomic model. You used the model to explain some of the differences among the elements. In this chapter you will examine the behaviour and properties of some elements and will try to relate these properties to atomic structure. As with your experience of the introductory picture, you will try to find a pattern or principle that can be used to classify and organize the elements. Then, in Chapter 4, you will continue those efforts by trying to explain the patterns in terms of the atomic model.

3.1 Elements – Similarities and Differences

The elements hydrogen and helium are colourless gases with densities very much lower than that of air. Both gases have been used to support "lighter than air" ships. (See Figure 3.1.) Hydrogen, however, burns vigorously in air, while helium is completely unreactive. This difference has a dramatic impact on the uses we have for the two gases. (See Figure 3.2.) Similarities and differences like those between hydrogen and helium occur among other elements. It is useful to have a method of organizing the elements so that their properties and behaviour can be more easily understood and predicted. To develop such a method, an essential first step is to determine the physical and chemical properties of the elements.

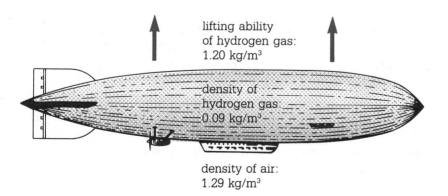

lifting ability
of hydrogen gas:
1.20 kg/m³

density of
hydrogen gas:
0.09 kg/m³

density of air:
1.29 kg/m³

Figure 3.1
Lifting ability of hydrogen. Hydrogen's density (at 0°C and normal atmospheric pressure) is 0.09 kg/m³. Under the same conditions, air's density is 1.29 kg/m³. Thus, each cubic metre of hydrogen in the airship can lift 1.2 kg.

Figure 3.2
Hydrogen versus helium. In May 1937, while landing at Lakehurst New Jersey, the hydrogen-filled Hindenberg caught fire (probably due to a discharge of lightning near a hydrogen leak). Thirty-six passengers and crew died in that inferno. Modern blimps, which in contrast are filled with helium gas, will not catch fire in this way.

Laboratory Manual
You will observe the physical properties of elements and look for patterns to help you classify them in Experiment 3-1.

Then you need to find answers to questions such as these: What similarities do the elements have? What are their differences? (See Colour Plate 2.) Are there any trends or gradual changes in specific properties among them? Are there any relationships among the elements that you could use to group them? To answer these questions, you need to obtain information by doing experiments.

Obtaining Information: Knowledge and Science

Direct observations result in the accumulation of what is called **empirical knowledge**. The word empirical means pertaining to experiments, observations, or experiences. The inspiration that results in pattern recognition or hypotheses is often the result of instinct or intuition. Intuition is the ability to perceive "truth" without apparent reasoning or analysis – a person "just knows" that something is correct. Many great advances in science, as well as in other areas, have been the result of intuition. Einstein's theory of relativity is one such example. However, a scientific theory is never accepted unless there is experimental (empirical) evidence to support it.

Another aspect of science is its "open-endedness". In attempting to solve one scientific problem many more questions arise. For example, in your attempts to classify elements you may have asked yourself additional questions such as these: Why is sodium stored under oil while phosphorus is stored under water? Why are so many elements shiny? Why are there so few liquid elements? This open-endedness is part of the interest and excitement of scientific investigation.

HYDROGEN

Hydrogen is the simplest and the most abundant element in the universe. It is the major constituent of the sun, the stars, and the larger planets of our solar system. On earth, it is not found as the element, but in the form of compounds in materials like water, some crustal rock, and substances that make up, or come from, living organisms. Hydrogen is a flammable gas first distinguished from other gases by Henry Cavendish in 1766. It was not until 1781, however, that the name hydrogen (meaning "maker of water") was coined by Antoine Lavoisier.

Large quantities of hydrogen are used in petroleum refining and in the synthesis of ammonia and methanol. (You will learn more about this in Chapter 16.) As well, hydrogen is chemically added to some substances by means of a process called hydrogenation. For example, liquid vegetable and animal oils are hydrogenated to produce solid margarine and vegetable shortening.

Some of the physical, chemical, and nuclear properties of hydrogen have made it a key player in the advance, but perhaps also in the eventual destruction, of our civilization.

USE OF PHYSICAL PROPERTIES
People made use of hydrogen's low boiling point and low density more than 70 years ago. They used hydrogen as a lifting gas for balloons and airships.

USE OF CHEMICAL PROPERTIES
Today, hydrogen's "lifting" capability is based on its chemical reactivity. Hydrogen's high reactivity with oxygen makes it a prime fuel for high performance rocket ships such as the space shuttle. The flaming Hindenberg disaster (see Figure 3.2) has its parallel in the equally shocking disintegration of the shuttle, Challenger, 49 years later in 1986. Despite such disasters, hydrogen is still regarded as a fuel with much potential. Hydrogen's very rapid reaction with air makes it the only practical fuel for engines designed to power the hypersonic aircraft now being developed. (These will travel at 3 to 25 times the speed of sound.)

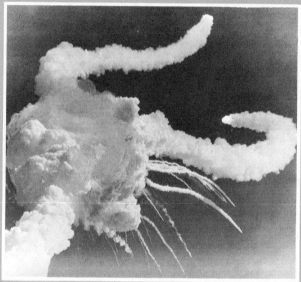

The United States space shuttle, Challenger, was destroyed during a launch on January 28, 1986, when a seal in its right booster rocket burned through and caused the rupture of the main liquid hydrogen fuel tank. Challenger's seven crew members were killed in the explosive fireball formed by this hydrogen-oxygen reaction.

MISUSE AND USE OF NUCLEAR PROPERTIES
Hydrogen's tritium isotope, ^3_1H, presently has the most serious implications for our society. This isotope is able to undergo a nuclear reaction by fusing (merging) with a deuterium nucleus, ^2_1H, to form a helium nucleus and a neutron. This reaction produces about 7×10^6 times the energy of burning an equivalent mass of hydrogen in oxygen. Since the early 1950s, this scientific knowledge has been applied to build the tremendously destructive hydrogen bombs that threaten our very existence.

→

On the constructive side, this same scientific knowledge is being applied to the research and development of safe nuclear fusion power reactors to generate electricity. Commercially successful nuclear fusion power reactors would tap the enormous energy reserves of deuterium in our oceans.

The hydrogen fireball formed after the first hydrogen bomb was loosed from an American airplane at dawn over the small Pacific island of Bikini Atoll.

Exercise

1. Identify two pairs of elements, other than hydrogen and helium, that have some properties which are similar and some which are different. List these similarities and differences.

2. Identify a property or behaviour of matter about which you are curious. Propose a possible hypothesis that you think would explain this property or behaviour. How would you attempt to prove or disprove your theory?

3.2 Data and Organization

Elements can be organized according to either their physical or chemical properties or a combination of both. Consider the elements that you have examined and for which you have recorded physical properties. You can organize them in different ways based on those properties.

Organization by State

Elements may be organized according to their physical state. Most elements are solids, eleven are gases, and only two are liquids at room temperature and pressure. This organization is actually based on two physical properties of the elements: their boiling points and melting points.

Organization by Colour and Appearance

Examine Colour Plate 2. The majority of elements fall into one category – silvery and shiny solids. A few are both shiny and coloured. Some are not shiny but are coloured. Except for chlorine and fluorine, all of the gaseous elements are colourless. Liquid bromine and solid iodine have intensely coloured vapours.

Organization into Metals and Non-metals

The two types of organization discussed above are only two of many ways of subdividing the elements. Instead of using only one property as a basis for organization, you could, for example, select a "package" of related properties and perhaps obtain a more useful subdivision. By doing experiments on a group of elements which have some similar properties, you can determine further characteristics of the group. These characteristics help to identify further the elements as a group.

When you examine the results of experiments such as those shown in Figure 3.3, you will find that there is a pattern. The elements that are shiny are also good conductors of heat and electricity. As well, they tend to bend and "flow" under mechanical stress. By contrast, the non-shiny elements are mostly poor conductors of heat and electricity. These elements tend to break or shatter under mechanical stress. These observations are in fact a "package" of properties that sorts the elements into metals and non-metals.

Laboratory Manual

In Experiment 3-2, you will compare the heat and electrical conductivity of selected elements, and compare the response of selected elements to mechanical stress.

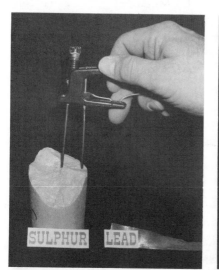

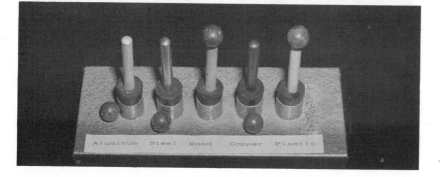

Figure 3.3
The electrical and thermal conductivity of substances can be determined in the laboratory. Metals conduct both heat and electricity relatively well; non-metals do not. Wax is used to attach a plastic ball to one end of each rod. The opposite ends of the rods are placed in hot water at the same time. The wax on the more heat conductive materials will melt sooner and the plastic balls will fall off first.

73

Figure 3.4
Reaction of different elements to
mechanical stress

(a) Solid metals tend to bend
and "flow" under stress. Metals
are said to be malleable.

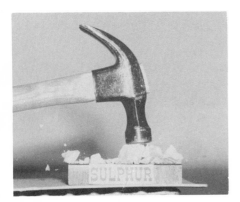

(b) Solid non-metals tend to
fracture or break under stress.
Non-metals are said to be
brittle.

Metals are substances that normally have these physical properties: they are shiny, they are good conductors of electricity and heat, and they tend to bend or "flow" rather than break under mechanical stress. The property of "flowing" rather than breaking when a substance is squeezed or hammered is called malleability. The ability of metals to be drawn out into threads or wire without breaking is called ductility. Metals are **ductile and malleable. Non-metals**, on the other hand, are not normally shiny, they are poor conductors of heat and electricity, and in the solid state, they usually break or shatter under mechanical stress. Materials that break or shatter under mechanical stress are described as **brittle**. Non-metals are brittle. (See Figure 3.4.)

If the elements are organized using the criteria of shininess, good thermal and electrical conductivity, and malleability, you will find about four times as many metals as non-metals. You will also discover some "doubtful" elements. These elements have some metallic properties as well as some non-metallic properties. For example, graphite, a form of pure carbon, is not shiny, and it breaks under mechanical stress. It is, however, an excellent conductor of electricity. Thus, you will recognize the need for a new classification for elements such as these. Those elements which possess some, but not all, of the properties associated with metals are called **semi-metals** or **metalloids**. (See Table 3.1.) Metalloids such as silicon and germanium are used to make semi-conductors—substances that conduct electricity, but not nearly as well as metals. Semi-conductors are extensively used in electronic components such as computer chips and diodes. (See Figure 3.5.)

Now what differences are there among the metals? In what ways do the non-metals differ from each other? How else are metals different from non-metals? In the next section, you will seek answers to these questions by investigating the chemical properties of the elements.

Table 3.1 Physical Properties of Selected Elements

	Sodium	Sulphur	Neon	Mercury	Germanium	Silicon
State (at room temperature and pressure)	Solid	Solid	Gas	Liquid	Solid	Solid
Appearance	Shiny	Yellow	Colourless	Shiny	Shiny	Grey-black lustrous crystals
Electrical Conductivity	Good	Poor	Poor	Good	Poor	Poor
Heat Conductivity	Good	Poor	Poor	Good	Moderate	Good
Reaction to Stress	Bends	Breaks	–	–	Breaks	Breaks
Classification	Metal	Non-metal	Non-metal	Metal	Metalloid	Metalloid

Figure 3.5
Semi-conductors

Exercise

3. Identify an element other than aluminum that is used extensively
   in our society. On what properties does the use of this element
   depend?

SPECIAL TOPIC 3.2

ALUMINUM – PROPERTIES AND USES

The physical properties of an element determine some of the uses for it. Aluminum metal, for example, with its high electrical conductivity and moderate cost, is used as an electrical conductor on long distance power lines. Aluminum's good heat conductivity and low density make it an ideal material for pots and pans. Because it is malleable and ductile, inexpensive, and has a low density, aluminum alloys are used extensively as structural materials; for example, for making aircraft bodies, window frames, and house siding.

Pure aluminum metal is produced by passing electricity through a molten solution of aluminum ore (bauxite) dissolved in a solvent (cry-olite). Although no aluminum ore is mined in Canada, the availability of large quantities of inexpensive hydroelectric power, chiefly in British Columbia and Quebec, has made Canada one of the world's leading producers of this versatile element. Canada obtains most of its aluminum ore from Jamaica where many rich bauxite deposits are found.

Thus, the many uses for aluminum depend on its properties. Aluminum's low density, durability, malleability, and reasonable cost make it an ideal structural material for products as diverse as aircraft and trains, recyclable food containers, and works of art. See page 76.

→

Alcan's aluminum smelter in Grande Baie, Quebec. The two major inputs needed for the production of aluminum are bauxite (aluminum ore) and electricity. The bauxite is brought in by ship to the harbour (visible in the background), and the electricity is brought in via electrical towers (middle left side of the photograph).

ologant

3.3 Chemical Properties of Elements

In Chapter 1, you learned that the chemical properties of a substance describe how it behaves in the presence of other substances. For example, some of the chemical properties of the element iron can be determined by observing how it reacts with substances such as water, hydrochloric acid, or oxygen. (See Figure 3.6.) Chemical properties can be used to classify the elements.

To classify the elements on the basis of their chemical properties, you will have to build up a bank of chemical data. You do this by conducting experiments or by using information that is available in the **scientific literature**. This literature is made up of science journals and reference books. These publications describe and summarize the experiments, the experimental results, and the theories of scientists over the years. (See Figure 3.7.) This ever-changing and expanding body of knowledge is one of the major tools scientists use in their search for new knowledge. You can use the literature to find the properties of some of the elements which are dangerous or unavailable.

Figure 3.6
Some chemical properties of iron. In (a), iron reacts with hydrochloric acid to produce a gas; there is no immediate visible reaction when iron is immersed in water. When heated, the iron wool in (b) burns very quickly in pure oxygen.

(a)

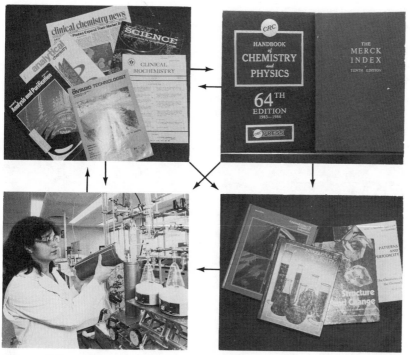

Figure 3.7
The scientific literature. The scientific literature is made up of science journals and reference books published by various scientific organizations and publishers. The results of research work and scientific thinking (hypotheses, models, and theories) are published in journals and summarized in reference books. Scientists use the information in the literature to design further experiments and develop theories. Authors use the literature for reference when they write textbooks for training future generations of scientists.

anonymous

(b)

CHLORINE

Chlorine was discovered in 1774 by the Swedish chemist Karl Wolfgang Scheele. It was not until 1810 that Sir Humphrey Davy named this element chlorine from the Greek word *chloros* meaning light green. Chlorine, a yellowish green gas, is a very reactive element and is a constituent of many compounds. It is used so extensively in the modern world that it is sometimes considered a barometer of a nation's economic health. The production of paper, dyestuffs, textiles, petroleum products, plastics, rubber, medicines, antiseptics, insecticides, foodstuffs, solvents, paints, and many other products depends on chlorine.

Chlorine is an extremely hazardous element. Liquid chlorine burns skin and chlorine gas causes choking. Breathing more than a few breaths of air containing 1% or more of chlorine gas is likely to be fatal. This toxic property led to the use of chlorine as a deadly weapon in World War I.

There is, however, a valuable positive side to chlorine's toxic nature. Chlorine is used to treat water for drinking and for swimming pools because of its ability to kill harmful bacteria. Many municipal water treatment plants add small quantities of chlorine directly to the water. Swimming pools are usually treated with compounds that release chlorine on contact with water. Chlorine-releasing chemicals can be quite hazardous and should be handled with great care.

Some of the many products containing the element chlorine

Chlorine was first used for chemical warfare in the second battle of Ypres, France in 1915. It caused many deaths and even more life-long injuries.

SPECIAL TOPIC 3.4

METAL REACTIVITY AND COOKING

Aluminum metal reacts with both acidic and basic substances to form soluble aluminum compounds. Soluble aluminum compounds may be detrimental to health. Aluminum foil wrap and cooking pots react with acidic foods such as sauerkraut, tomato sauce, and dishes containing vinegar or wine. For this reason good quality aluminum pots and pans have their cooking surfaces coated with an inert layer of ceramic or teflon in order to keep the food separate from the aluminum metal surface.

Copper metal, by contrast, is not very reactive; but copper compounds are very toxic — even a very slow reaction between copper and food can build up dangerous amounts of poisonous compounds. Thus, copper cooking utensils are coated with the relatively slow-reacting, non-toxic metal, tin.

Unprotected metals can react with some foods producing undesirable or toxic substances. Metal cooking vessels are often coated with non-reactive and non-toxic materials.

Organization by Reactivity

The data that you have accumulated, both by experiment and from reference sources, should be carefully examined for trends or patterns. For example, which of the metals is the most reactive? Which is the least reactive?

Are there any similarities or differences when metal and non-metal elements react with oxygen? Are there any trends apparent in the reactivity of these two classes of elements? This pattern-seeking activity can be the most difficult but the most creative part of scientific investigation. It can be most exhilarating and exciting when new patterns or explanations are discovered!

Laboratory Manual
In Experiment 3-3, you will further your attempts to classify elements by observing the reactivity of several elements with air, water, and hydrochloric acid.

Exercise

4. Use experimental results and data from reference sources to answer the following questions.
   (a) In terms of their reaction with water and dilute hydrochloric acid, what is the relative order of reactivity (from most reactive to least reactive) for the following metals: Fe, K, Al, Cu, Na, Ca, Sr, Li, Mg, and Zn? (Remember that water, compared to the acid, is a relatively mild reagent.)
   ▶

(b) What is the relative order of reactivity with iron for the following non-metals: Br, S, Cl, O, I, and P?

(c) How do the metals and non-metals differ in terms of the following?
  (i) the substances with which they react
  (ii) the type of compounds their oxides form in water

The Value of Negative Results

You will have done many experiments with different elements and found that most of them reacted with one or more of the testing reagents used. One particular group of elements, however, is so unreactive that nothing that can be done in a high school laboratory can cause those elements to react! They have the chemical property of inertness or non-reactivity with almost all other substances. These chemically unreactive elements are helium, neon, argon, krypton, xenon, and radon. They are called the **noble gases**.

Often, negative results such as those obtained with the noble gases are as important as the positive ones. As you will see in Chapter 5, the discovery of this set of unreactive elements was the key in developing a theory to explain the reactivity of the elements.

Chemical inertness is very often a valuable property. For example, argon gas (air contains about 1% of this gas) is used at low pressure to fill light bulbs because argon will not react with the bulb's white hot tungsten filament. The oxygen in air *will* react

Figure 3.8
Jewellery made from unreactive gold and platinum stays bright and shiny without special attention.

with tungsten and burn out the filament. Gold and platinum are metals used to make jewellery partly because they do not react with most other substances and therefore they will not "rust" or tarnish in air. Both metals will stay bright and shiny year after year without any special attention. (See Figure 3.8.)

Organization by Atomic Mass

One other key piece of information that is used to further organize the elements is their relative atomic masses. Atomic masses of the elements have been determined very precisely using a mass spectrometer. Historically, however, the relative atomic masses of the elements were determined by measuring the mass of one element that combined with a known mass of a second element (refer to Chapter 2). From empirical results such as this, the relative atomic masses of the elements were deduced and became essential scientific knowledge now listed in chemistry texts, reference books, and data banks. (See Table 3.2.) The arrangement of the elements in order of their relative masses is one of the key pieces of information used to produce a useful organization of the elements. The data you have now accumulated, and their preliminary organization will be used in the following section to show how the greatest organizing pattern in chemistry – the periodic table – was developed.

Table 3.2   The Relative Atomic Mass of Selected Elements

Element	Atomic Mass	Element	Atomic Mass	Element	Atomic Mass
Hydrogen	1.0	Scandium	45.0	Niobium	93.0
Helium	4.0	Titanium	47.9	Molybdenum	95.9
Lithium	6.9	Vanadium	50.9	Technetium	97.0
Beryllium	9.0	Chromium	52.0	Ruthenium	101.1
Boron	10.8	Manganese	54.9	Rhodium	102.9
Carbon	12.0	Iron	55.9	Palladium	106.4
Nitrogen	14.0	Nickel	58.7	Silver	107.9
Oxygen	16.0	Cobalt	58.9	Cadmium	112.4
Fluorine	19.0	Copper	63.6	Indium	114.8
Neon	20.2	Zinc	65.4	Tin	118.7
Sodium	23.0	Gallium	69.7	Antimony	121.8
Magnesium	24.3	Germanium	72.6	Iodine	126.9
Aluminum	27.0	Arsenic	74.9	Tellurium	127.6
Silicon	28.1	Selenium	79.0	Xenon	131.3
Phosphorus	31.0	Bromine	79.9	Cesium	132.9
Sulphur	32.1	Krypton	83.8	Barium	137.3
Chlorine	35.5	Rubidium	85.5	Lanthanum	138.9
Potassium	39.1	Strontium	87.6	Cerium	140.1
Argon	39.9	Yttrium	88.9	Praseodymium	140.9
Calcium	40.1	Zirconium	91.2	Neodymium	144.2

3.4 The Periodic Table of the Elements

If you consider together the chemical and physical properties as well as the atomic masses of the elements, you will find even more regularities and patterns. A regularity in the atomic masses of some chemically and physically similar elements was first noticed by a German chemist, Johann Döbereiner, in 1829. Consider the three elements lithium, sodium and potassium. You may recall that they are all soft, very reactive metals that must be stored under oil. The difference between the relative atomic masses of lithium and sodium is 16.1. The same difference exists between the relative atomic masses of sodium and potassium. Similar results are found in other groups of three similar elements, for example, the coloured and reactive non-metals chlorine, bromine, and iodine. Döbereiner's discovery is sometimes known as the law of triads. (See Table 3.3 below.)

The peak period of new element discovery occurred between 1800 and 1850. (See Appendix B.) As information concerning these new elements accumulated, a problem arose. Döbereiner's regularity did not generally apply to all of the elements in a group. Nor did it apply to all groups of similar elements. This law of triads is an example of an organizing principle that was discarded as the search for a more general law continued.

Table 3.3   Döbereiner's Triads

Triad	Properties	Relative Atomic Mass	Atomic Mass Difference
Lithium Sodium Potassium	Soft, very reactive metals	6.9 23.0 39.1	16.1 16.1
Chlorine Bromine Iodine	Coloured, highly reactive non-metals	35.5 79.9 126.9	44.4 47.0

It was a British chemist, John Newlands, 35 years later in 1864, who arranged the known elements in the order of increasing relative atomic mass. He noticed that every eighth element seemed to have similar chemical and physical properties. Newlands called this regular repetition the law of octaves. (See Table 3.4.)

However, when more elements were organized using the law of octaves, many inconsistencies arose. This organization only worked well for the first 20 elements. In fact, it was found that to include more elements, several sets of three elements had to be listed in one spot. (See column VIII in Table 3.5.) In addition, the properties of the elements in groups were not consistent. For example, group one combined the very reactive metallic elements Li, Na, K, Rb, and Cs with the much less reactive metals, Cu, Ag, and Au. New-lands' law of octaves had provided a means of organizing the elements, but the many inconsistencies meant that his organization needed modification.

Table 3.4 Periodic Repetition of Properties in Selected Elements Known to Newlands

Element	Atomic Mass	Element	Atomic Mass	Element	Atomic Mass	Group Properties
H	1.0					
Li	6.9	Na	23.0	K	39.1	Very reactive metals
Be	9.0	Mg	24.3	Ca	40.1	Reactive metals
B	10.8	Al	27.0			
C	12.0	Si	28.1			
N	14.0	P	30.0			
O	16.0	S	32.1			Reactive non-metals
F	19.0	Cl	35.5			Very reactive non-metals

Table 3.5 Elements Organized Using The Law of Octaves

Element Groups								
	I	**II**	**III**	**IV**	**V**	**VI**	**VII**	**VIII**
1							H	He★
2	Li	Be	B	C	N	O	F	Ne★
3	Na	Mg	Al	Si	P	S	Cl	Ar★
4	K	Ca	Sc★	Ti	V	Cr	Mn	Fe Co Ni
5	Cu	Zn	Ga★	Ge★	As	Se	Br	Kr★
6	Rb	Sr	Y	Zr	Nb	Mo	Tc★	Ru Rh Pd
7	Ag	Cd	In	Sn	Sb	Te	I	Xe★
8	Cs	Ba	La	Hf★	Ta	W	Re★	Os Ir Pt
9	Au	Hg	Tl	Pb	Bi	Po★	At★	Rn★

(★ elements not yet discovered in Newlands' time)

Replacing the Law of Octaves

Although several scientists worked on modifying Newlands' law it was Dmitri Mendeleev, a Russian scientist, who from 1869 onwards was the driving force behind many of the refinements in organizing the elements. (See Figure 3.9.) Mendeleev proposed and used this hypothesis as his organizing principle: *The properties of the elements are periodic functions of their relative atomic masses*. We refer to this

Figure 3.9
Dmitri Mendeleev (1834–1907). Born in Siberia, Mendeleev was the youngest of 17 children. After studying in Europe, he set up a graduate school of chemistry in Russia. His organization of the elements into a periodic table was a stroke of genius that made the study of chemistry manageable.

principle as Mendeleev's periodic law. Mendeleev did not restrict himself to a repetition pattern of eight. He was able to organize some of the elements into subgroups. (See Figure 3.10(a).) Direct adaptations of this **periodic table** were used up to about 1950. The "long form" of this table has also been widely used. (See Figure 3.10(b).) The most recent internationally accepted revisions have been included in the periodic table shown in Figure 3.11 and on the inside back cover of this textbook.

Groups:		I		II		III		IV		V		VI		VII		VIII
Subgroup or family:		A	B	A	B	A	B	A	B	A	B	A	B	A	B	
	1	H														
	2	Li		Be		B		C		N		O				
	3	Na		Mg		Al		Si		P		S		Cl		
Period	4	K		Ca			—		Ti		V		Cr		Mn	Fe, Co, Ni
			Cu		Zn	—		—		As		Se		Br		
	5	Rb		Sr			Y		Zr		Nb		Mo		—	Ru, Rh, Pd
			Ag		Cd	In		Sn		Sb		Te		I		
	6	Cs		Ba			La		—		Ta		W		—	Os, Ir, Pt
			Au		Hg	Tl		Pb		Bi		—		—		

Figure 3.10

(a) Mendeleev's periodic table of the elements, 1871. By splitting the element groups into subgroups, and by allowing for two "subperiods" for periods past period 3, Mendeleev greatly improved on Newlands' original "octave" model. Note that the dashes represent elements not known to Mendeleev at the time. Also, the entire family of noble gases was not known to Mendeleev, and have thus been left out.

Mendeleev's original work was done using the 63 elements known at the time. One of his greatest inspirations was to leave a blank space in his table when it appeared that the known element having the next higher relative atomic mass did not have the properties to fit the regular periodic pattern. By arranging and rearranging the order of the known elements, Mendeleev was able to identify these blank spots where, he predicted, elements that had yet to be discovered should be. (See Figure 3.12.)

The value of any law or hypothesis is very often proven by its usefulness in making predictions. Mendeleev was spectacularly successful in applying his periodic law to predict not only the existence of undiscovered elements, but their physical and chemical properties as well! For example, in 1871, the known element that had the next higher relative atomic mass after zinc was arsenic. But arsenic's properties were not at all similar to those of aluminum, beneath which it should have been placed according to the periodic law. In fact, arsenic did not have the properties to fit under silicon either.

IA																VIIA	VIIIA
H 1	IIA											IIIA	IVA	VA	VIA	H 1	He 2
Li 3	Be 4											B 5	C 6	N 7	O 8	F 9	Ne 10
Na 11	Mg 12	IIIB	IVB	VB	VIB	VIIB	VIII			IB	IIB	Al 13	Si 14	P 15	S 16	Cl 17	Ar 18
K 19	Ca 20	Sc 21	Ti 22	V 23	Cr 24	Mn 25	Fe 26	Co 27	Ni 28	Cu 29	Zn 30	Ga 31	Ge 32	As 33	Se 34	Br 35	Kr 36
Rb 37	Sr 38	Y 39	Zr 40	Nb 41	Mo 42	Tc 43	Ru 44	Rh 45	Pd 46	Ag 47	Cd 48	In 49	Sn 50	Sb 51	Te 52	I 53	Xe 54
Cs 55	Ba 56	La 57	Hf 72	Ta 73	W 74	Re 75	Os 76	Ir 77	Pt 78	Au 79	Hg 80	Tl 81	Pb 82	Bi 83	Po 84	At 85	Rn 86
Fr 87	Ra 88	Ac 89															

Lanthanide Series	Ce 58	Pr 59	Nd 60	Pm 61	Sm 62	Eu 63	Gd 64	Tb 65	Dy 66	Ho 67	Er 68	Tm 69	Yb 70	Lu 71
Actinide Series	Th 90	Pa 91	U 92	Np 93	Pu 94	Am 95	Cm 96	Bk 97	Cf 98	Es 99	Fm 100	Md 101	No 102	Lw 103

(b) The periodic table – long form. This expansion of Mendeleev's table separated the ''A'' and ''B'' subgroups and provided for a continuous ordering of the elements up to element 57, La. Groups or columns run up and down vertically and are identified by Roman numerals. Periods or rows run across the periodic table.

Period

Figure 3.11
The modern periodic table

1* IA	2 IIA							

Period 1

1.008   1
1.31
H   2.1
1
Hydrogen

KEY

** atomic mass	•12.001 4•	common valences
	1.09•	ionization energy (MJ/mol)
symbol	**C** 2.5•	electronegativity
atomic number	•6	
name	•Carbon	

Period 2

6.939 1	9.012 2
0.52	0.90
Li 1.0	**Be** 1.5
3	4
Lithium	Beryllium

Period 3

22.99 1	24.31 2	3 IIIB	4 IVB	5 VB	6 VIB	7 VIIB	8	9 VIIIB
0.42	0.74							
Na 0.9	**Mg** 1.3							
11	12							
Sodium	Magnesium							

Period 4

39.10 1	40.08 2	44.96 3	47.90 3,4	50.94 2,3	52.00 2,3	54.94 2,3	55.85 2,3	58.93 2,3
0.42	0.59	0.63	0.66	0.65	0.65	0.72	0.76	0.76
K 0.9	**Ca** 1.1	**Sc** 1.2	**Ti** 1.3	**V** 1.5	**Cr** 1.6	**Mn** 1.6	**Fe** 1.7	**Co** 1.7
19	20	21	22	23	24	25	26	27
Potassium	Calcium	Scandium	Titanium	Vanadium	Chromium	Manganese	Iron	Cobalt

Period 5

85.47 1	87.62 2	88.91 3	91.22 4	92.91 3,5	95.94 2,3	98.91 7	101.1 3,4	102.9 3,4
0.40	0.55	0.62	0.66	0.66	0.69	0.70	0.71	0.72
Rb 0.9	**Sr** 1.0	**Y** 1.1	**Zr** 1.2	**Nb** 1.3	**Mo** 1.3	**Tc** 1.4	**Ru** 1.4	**Rh** 1.5
37	38	39	40	41	42	43	44	45
Rubidium	Strontium	Yttrium	Zirconium	Niobium	Molybdenum	Technetium	Ruthenium	Rhodium

Period 6

132.9 1	137.3 2	175.0 3	178.5 4	180.9 5	183.9 6	186.2 4,7	190.2 3,4	192.2 3,4
0.38	0.50	0.52	0.65	0.76	0.77	0.76	0.84	0.88
Cs 0.9	**Ba** 0.9	**Lu** 1.2	**Hf** 1.2	**Ta** 1.4	**W** 1.4	**Re** 1.5	**Os** 1.5	**Ir** 1.6
55	56	71	72	73	74	75	76	77
Cesium	Barium	Lutetium	Hafnium	Tantalum	Tungsten	Rhenium	Osmium	Iridium

Period 7

(223) 1	226.0 2	(260) —	(261)	(262)	(263)	(262)	(265)	(266)
—	0.51	—						
Fr 0.9	**Ra** 0.9	**Lr** —	**Unq**	**Unp**	**Unh**	**Uns**	**Uno**	**Une**
87	88	103	104	105	106	107	108	109
Francium	Radium	Lawrencium						

Alkali Metals   Alkaline Earth Metals

Lanthanide Series

138.9 3	140.1 3,4	140.9 3,4	144.2 3	(147) 3	150.4 2,3	152.0 2,3
0.54	0.53	0.52	0.53	0.54	0.54	0.55
La 1.1	**Ce** 1.1	**Pr** 1.1	**Nd** 1.2	**Pm** —	**Sm** 1.2	**Eu** —
57	58	59	60	61	62	63
Lanthanum	Cerium	Praseodymium	Neodymium	Promethium	Samarium	Europium

Actinide Series

(227) 3	232.0 4	(231) 4,5	238.03 4,5,6	(237) 3,4,5,6	(242) 3,4,5,6	(243) 3,4,5,6
0.47	0.59	0.57	0.59	0.60	0.59	0.58
Ac 1.1	**Th** 1.2	**Pa** 1.5	**U** 1.7	**Np** 1.3	**Pu** 1.3	**Am** 1.3
89	90	91	92	93	94	95
Actinium	Thorium	Protactinium	Uranium	Neptunium	Plutonium	Americium

*There are, at this time, two systems for labelling groups in the periodic table. The more traditional system uses the Roman numerals I to VIII, along with the letters A and B to designate subgroups. More currently, another system which uses Arabic numerals, 1 to 18, has come into use. To avoid possible confusion, this textbook has tended to refer to groups by name, rather than number; for example, "noble gas group" for the last group in the table. In those rare cases when referring

					18 VIIIA
					4.003 — / 2.37 / **He** / 2 / Helium / —
13 IIIA	14 IVA	15 VA	16 VIA	17 VIIA	
10.81 3 / 0.80 / **B** 2.0 / 5 / Boron	12.011 4 / 1.09 / **C** 2.5 / 6 / Carbon	14.01 3,5 / 1.40 / **N** 3.1 / 7 / Nitrogen	16.00 2 / 1.31 / **O** 3.5 / 8 / Oxygen	19.00 1 / 1.68 / **F** 4.1 / 9 / Fluorine	20.18 — / 2.08 / **Ne** / 10 / Neon
26.98 3 / 0.58 / **Al** 1.5 / 13 / Aluminum	28.09 4 / 0.79 / **Si** 1.8 / 14 / Silicon	30.97 3,5 / 1.01 / **P** 2.1 / 15 / Phosphorus	32.06 2,4,6 / 1.00 / **S** 2.4 / 16 / Sulphur	35.45 1 / 1.25 / **Cl** 2.9 / 17 / Chlorine	39.95 — / 1.52 / **Ar** / 18 / Argon

Left transition-metal groups for the rows below:

10	11 IB	12 IIB

10	11 (IB)	12 (IIB)	13 (IIIA)	14 (IVA)	15 (VA)	16 (VIA)	17 (VIIA)	18 (VIIIA)
58.71 2,3 / 0.74 / **Ni** 1.8 / 28 / Nickel	63.55 1,2 / 0.74 / **Cu** 1.8 / 29 / Copper	65.38 2 / 0.91 / **Zn** 1.7 / 30 / Zinc	69.72 3 / 0.58 / **Ga** 1.8 / 31 / Gallium	72.59 4 / 0.76 / **Ge** 2.0 / 32 / Germanium	74.92 3,5 / 0.94 / **As** 2.2 / 33 / Arsenic	78.96 2,4,6 / 0.94 / **Se** 2.5 / 34 / Selenium	79.91 1 / 1.14 / **Br** 2.8 / 35 / Bromine	83.80 — / 1.35 / **Kr** / 36 / Krypton
106.4 2,4 / 0.81 / **Pd** 1.4 / 46 / Palladium	107.9 1 / 0.73 / **Ag** 1.4 / 47 / Silver	112.4 2 / 0.87 / **Cd** 1.5 / 48 / Cadmium	114.8 3 / 0.56 / **In** 1.5 / 49 / Indium	118.7 2,4 / 0.71 / **Sn** 1.7 / 50 / Tin	121.8 3,5 / 0.83 / **Sb** 1.8 / 51 / Antimony	127.6 2,4,6 / 0.87 / **Te** 2.0 / 52 / Tellurium	126.9 1 / 1.01 / **I** 2.2 / 53 / Iodine	131.3 — / 1.11 / **Xe** / 54 / Xenon
195.1 2,4 / 0.87 / **Pt** 1.5 / 78 / Platinum	197.0 1,3 / 0.89 / **Au** 1.4 / 79 / Gold	200.6 1,2 / 1.01 / **Hg** 1.5 / 80 / Mercury	204.4 1,3 / 0.59 / **Tl** 1.5 / 81 / Thallium	207.2 2,4 / 0.72 / **Pb** 1.6 / 82 / Lead	209.0 3,5 / 0.70 / **Bi** 1.7 / 83 / Bismuth	(209) 2,4 / 0.81 / **Po** 1.8 / 84 / Polonium	(210) 1 / — / **At** 2.0 / 85 / Astatine	(222) — / 1.04 / **Rn** / 86 / Radon

Halogens   Noble Gases

157.3 3 / 0.59 / **Gd** 1.1 / 64 / Gadolinium	158.9 3,4 / 0.56 / **Tb** 1.2 / 65 / Terbium	162.5 3 / 0.57 / **Dy** — / 66 / Dysprosium	164.9 3 / 0.58 / **Ho** 1.2 / 67 / Holmium	167.3 3 / 0.59 / **Er** 1.2 / 68 / Erbium	168.9 2,3 / 0.60 / **Tm** 1.2 / 69 / Thulium	173.0 2,3 / 0.60 / **Yb** 1.1 / 70 / Ytterbium
(247) 3 / 0.58 / **Cm** — / 96 / Curium	(247) 3,4 / 0.60 / **Bk** — / 97 / Berkelium	(251) — / 0.61 / **Cf** — / 98 / Californium	(252) — / 0.62 / **Es** — / 99 / Einsteinium	(257) — / 0.63 / **Fm** — / 100 / Fermium	(258) — / 0.64 / **Md** — / 101 / Mendelevium	(259) — / 0.64 / **No** — / 102 / Nobelium

to the number of a group, we have used the Roman system, followed by the Arabic system in parenthesis; for example, Group VIIIA (18).

**Based on $^{12}C = 12.00000$. Values in parentheses are for the most stable or best known isotopes.

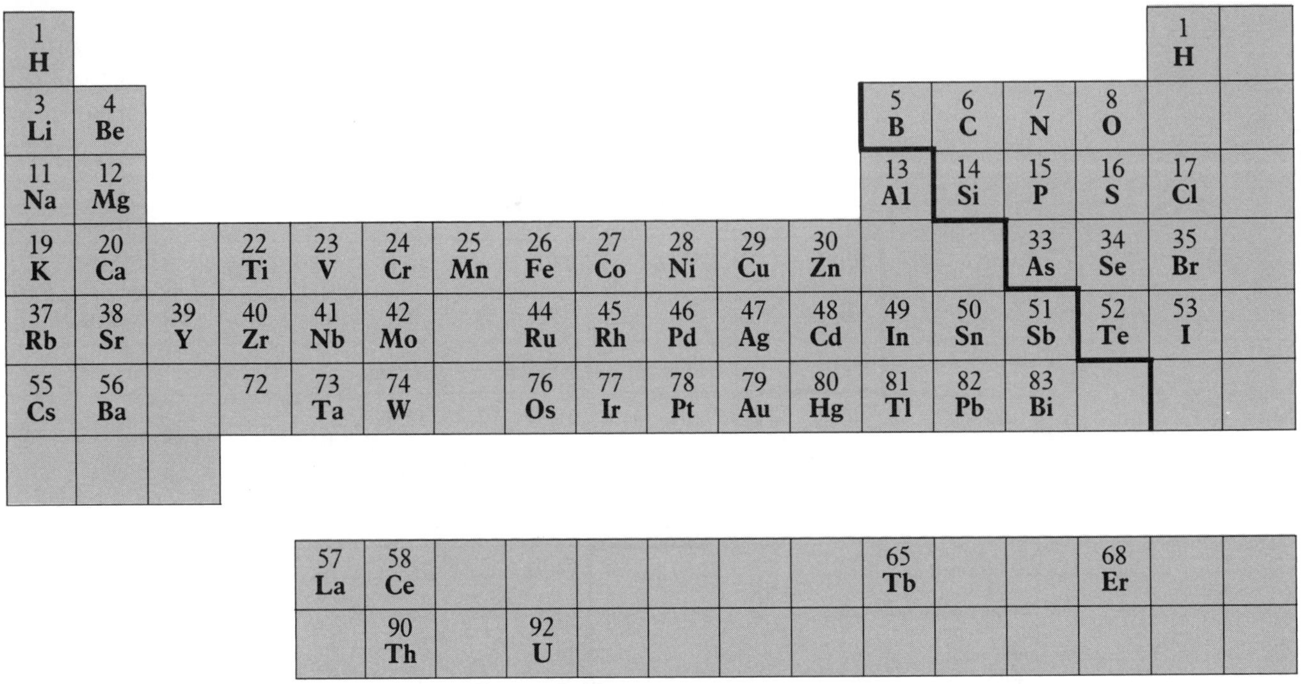

1 H																1 H	
3 Li	4 Be										5 B	6 C	7 N	8 O			
11 Na	12 Mg										13 Al	14 Si	15 P	16 S	17 Cl		
19 K	20 Ca		22 Ti	23 V	24 Cr	25 Mn	26 Fe	27 Co	28 Ni	29 Cu	30 Zn			33 As	34 Se	35 Br	
37 Rb	38 Sr	39 Y	40 Zr	41 Nb	42 Mo		44 Ru	45 Rh	46 Pd	47 Ag	48 Cd	49 In	50 Sn	51 Sb	52 Te	53 I	
55 Cs	56 Ba		72	73 Ta	74 W		76 Os	77 Ir	78 Pt	79 Au	80 Hg	81 Tl	82 Pb	83 Bi			

57 La	58 Ce						65 Tb			68 Er	
	90 Th	92 U									

Figure 3.12
The periodic table showing only elements known to Mendeleev. Mendeleev's feat of organizing the elements is all the more remarkable considering he was working with incomplete information.

Arsenic actually had properties more like phosphorus. Mendeleev placed it under phosphorus and predicted that there were two "missing" elements that should go beneath aluminum and silicon. (See Figure 3.12.) He predicted their properties by examining the elements surrounding them in his periodic table. (See Table 3.6.) The first of these missing elements, gallium, was discovered four years later in 1875; the second, germanium, fifteen years later in 1886.

Table 3.6 Two of Mendeleev's Predicted Elements

	Gallium		Germanium	
Property	Predicted 1871	Discovered 1875	Predicted 1871	Discovered 1886
Relative Atomic Mass	68	69.9	72	72.3
Density (g/cm³)	5.9	5.94	5.5	5.47
Melting Point	Low	30°C	High	2830°C
Solubility in:				
Acids	Will slowly dissolve	Slowly dissolves	Will be slightly soluble	Insoluble
Bases	Will slowly dissolve	Slowly dissolves	Will resist dissolving	Slowly dissolves

GALLIUM

The existence of an element with the properties of the element now known as gallium was predicted by Mendeleev in 1871. He called this as yet undiscovered element eka-aluminum. (The Greek word *eka* means first after.) Gallium was discovered spectroscopically in a compound by the French chemist Lecoq de Boisbaudran in 1875. He named the new element gallium in honour of France (from the old Latin name for France, Gallia). That same year, de Boisbaudran isolated a sample of the metal and determined its properties.

When Mendeleev learned of this new element, he claimed that it was his predicted eka-aluminum. Mendeleev, however, refused to believe that the density of 4.7 g/cm³ that de Boisbaudran had determined was correct. He maintained that it had to be closer to his own predicted value of 5.9 g/cm³. As it turned out, Mendeleev was right; de Boisbaudran's sample of gallium had been contaminated with potassium which had reduced the density. Mendeleev's confidence in his predictions was well founded since a number of his other predicted properties for gallium were shown to be more accurate than those first described by de Boisbaudran.

Gallium has a beautiful, silvery appearance and in the liquid state resembles mercury. Unlike mercury, gallium is non-toxic. It has the third lowest melting point (30°C) among the metals (after mercury, with a melting point of −39°C, and cesium, with a melting point of 28°C). Like water, and unlike most metals, liquid gallium expands when it solidifies.

Gallium's major use is in the preparation of the semiconductor gallium arsenide. Gallium arsenide in turn is used to manufacture computer chips, transistors, solar cells, and LEDs (light emitting diodes – a semiconducting laser).

Gallium's melting point of 30°C is so low that it can be melted simply by holding it in your hand.

Despite Mendeleev's remarkable ingenuity, several inconsistencies in the organization of his periodic table were still apparent. One such inconsistency was the apparent misordering of the elements iodine and tellurium. The atomic mass of the element tellurium (127.6 u) is greater than that of iodine (126.9 u). The physical and chemical properties of both elements, however, suggested that their placement in the periodic table should be reversed. Iodine is very much like bromine and not at all like selenium. The reverse is true for tellurium. The explanation for this and other similar inconsistencies had to wait for the developing atomic model.

In 1912, the British scientist Henry Moseley discovered a relationship between X-ray energies emitted by the highly excited atoms of an element and the element's nuclear charge. His discovery was used to determine the positive charge on each element's nucleus (that is, the atomic number). This additional data led to a modification of Mendeleev's original periodic law. Whereas Mendeleev had the elements arranged in order of increasing atomic mass, the modern periodic law has the elements arranged in order of increasing *atomic number*. The **periodic law** now states that *when the elements are arranged in order of increasing atomic number, there is a periodic repetition of elements with similar properties.*

Exercise

5. Using a periodic table, identify three sets of elements where arranging them in order of their atomic masses would have placed them in the wrong group.

6. "The development of the periodic table is an excellent example of the evolution and use of scientific laws." Discuss the validity of this statement.

7. Often in the development of scientific laws, models, and theories, new data and an accumulation of ideas provide the right climate for a breakthrough. This is what happened in the development of the periodic table. Along with Mendeleev, other chemists were working on the elements and their classification. At about the same time, several of them suggested periodic laws very similar to that proposed by Mendeleev. His, however, was the first to appear in the scientific literature. Thus Mendeleev received most of the credit for producing the periodic table. Use reference books to find another chemist who had independently developed a periodic table similar to Mendeleev's. Should he have been given the same credit and fame for his efforts? Discuss your reasons.

3.5 Trends Within Groups of Elements

Consider the elements in the lithium group: lithium, sodium, potassium, rubidium, and cesium. This group is called the **alkali metals**. All alkali metals have very high chemical reactivity. They all react vigorously with water. However, when potassium reacts with water, the hydrogen gas produced always ignites in air. When sodium reacts with water, the hydrogen gas produced sometimes ignites in air, and on occasion initiates a violent explosion if the hydrogen-air mixture has the right composition. The temperature reached during the potassium and water reaction is always high enough to ignite the hydrogen in air; with sodium the temperature is lower

and hydrogen ignition is inconsistent. Lithium reacts less violently with water than sodium and the hydrogen gas produced in the reaction never ignites in air. (See Figure 3.13.)

These results indicate a trend in chemical reactivity. As shown in Figure 3.14, the reactivity of metals increases as you go down a column of the periodic table. Some physical properties such as hardness and melting point also follow a trend. Lithium is harder than sodium and has a higher melting point. Lithium is somewhat difficult to cut with a knife while sodium can be easily sliced. Potassium is softer than sodium and has a lower melting point.

Figure 3.13
Reactivity of alkali metals with water.

(a) The temperature reached in the potassium-water reaction is always high enough to cause the hydrogen produced to ignite in air.

(b) The temperature reached in the sodium-water reaction is high enough to melt the sodium and sometimes cause the hydrogen produced to ignite in air.

(c) The temperature reached in the lithium-water reaction is never high enough to cause the hydrogen produced to ignite in air.

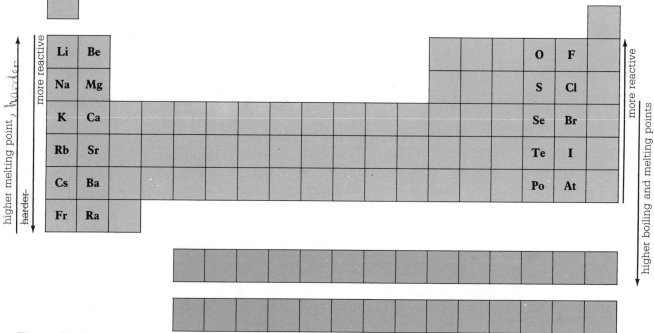

Figure 3.14
Elements within groups exhibit trends in their chemical and physical properties. Note that there are exceptions to these trends.

Similar property trends can be observed with the beryllium group of elements, called the **alkaline earth metals**. This group consists of beryllium, magnesium, calcium, strontium, and barium. Figure 3.15 shows the relative rates of reaction of three of these alkaline earth metals with water.

Figure 3.15
Reactivity of alkaline earth metals with water.

(a) Strontium reacts very quickly in water, producing large streams of hydrogen gas.

(b) Calcium reacts quickly enough to produce visible streams of hydrogen gas bubbles.

(c) Magnesium produces hydrogen gas so slowly that a wait of an hour or more is needed to detect its formation.

You can also see trends in the properties of the non-metallic fluorine group of elements called the **halogens** (fluorine, chlorine, bromine, and iodine). The reactivity of these non-metallic elements decreases as you go down the periodic table. (See Colour Plate 3.) This is the opposite trend to that shown by the metallic groups.

The physical properties of non-metals also show trends opposite to those of the metals. Chlorine, a gas at room temperature and pressure, has a lower boiling point than bromine, which is a liquid under the same conditions. Iodine is a solid at room temperature.

Similarly, in the oxygen group, oxygen is more reactive with iron than is sulphur. Selenium does not readily react with iron. In terms of physical properties, oxygen is a gas at room temperature. Sulphur is a solid with a melting point of about 119°C. Selenium, also a solid, has a higher melting point of 217°C. It was these trends within element groups that enabled Mendeleev to predict the properties of the elements which had yet to be discovered.

Sample Problem 3.1

1. What would you predict the physical state of the element astatine to be at room temperature?

2. Which element would you expect to be harder, calcium or magnesium?

Solutions

1. Astatine is in the non-metallic halogen group. The melting points of the elements in the group increase down the periodic table. Since astatine is below iodine, which is a solid at room temperature, then astatine will also be a solid. (The actual melting point of iodine is 113.5°C and that of astatine is 302°C.)

2. The elements in the metal groups tend to be harder going down the period table. Since magnesium is above calcium, magnesium is expected to be the harder of the two metals. (The hardness of calcium is 1.5 and magnesium is 2.0 on the Mohs hardness scale of which gives talc a value of 1 and diamond a value of 10.)

Exercise

Explain how you arrive at your prediction for each of the following questions.

8. What is the physical state of the element fluorine at room temperature?

9. Which element has the higher melting point, selenium or tellurium?

10. Which element would react more vigorously in water, cesium or rubidium?

11. Which element would react more vigorously with dilute hydrochloric acid, beryllium or barium?

3.6 Trends Across a Row of Elements

The elements also show trends in physical and chemical properties in any one row across the periodic table. Each row is called a **period**. Period 1 begins with hydrogen and period 2 with lithium. In period 3, the metal sodium is much more reactive with dilute hydrochloric acid than is the metal magnesium. Magnesium, in turn, is more reactive with acid than is aluminum. The element silicon is a metalloid and does not react with hydrochloric acid. If the noble gas argon is ignored, an opposite trend is observed among the non-metals of period 3. Chlorine reacts more vigorously with iron than does sulphur. Sulphur in turn is more reactive with iron than is phosphorus. In general the reactivity is lowest in the middle of a period and increases toward either end, disregarding the noble gases. (See Figure 3.16.)

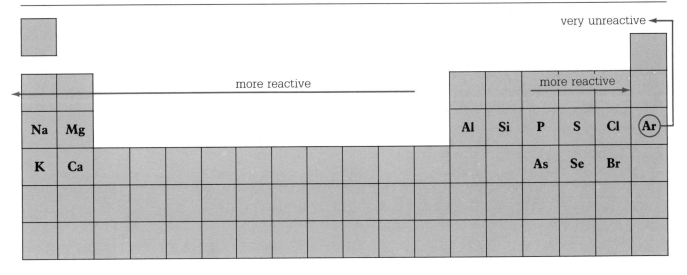

Figure 3.16
Trends across a period. With the exception of the noble gases, the reactivity of elements in a period in the perodic table tends to increase as one moves out toward the edges.

Remember, trends are just that – trends; they are not 100% predictors. For example, the element sulphur is fairly reactive with iron, but is stable in air at room temperature. On the other hand, phosphorus reacts slowly with iron, but is extremely reactive in air. Phosphorus can ignite spontaneously in air at room temperature and for this reason white phosphorus must be stored under water at all times. The trends within the periods of the periodic table allow us to predict with reasonable accuracy some of the chemical and physical properties of unfamiliar elements.

Sample Problem 3.2

Predict which of the following elements would react the most vigorously with iron: Se, As, Br, or Kr.

Solution

Krypton is a noble gas and would not react at all with iron. Arsenic, selenium, and bromine are non-metals in the same period of the periodic table. Chemical reactivity tends to increase toward the end of the table. Since bromine is closest to the end, it would be the element that reacts most vigorously with iron.

Exercise

Answer the following questions and explain how you arrive at your answers.

12. Which of the following elements would be most reactive with iron: I, Sb, or Te?

13. Which of the following metals would be most reactive with oxygen at room temperature: Al, Na, or Mg?

14. Which of the following metals would produce the most vigorous reaction when put in water: Rb, Sr, or K?

Looking Back – Looking Ahead

The periodic law provided the framework for the organization of the elements. The periodic table which resulted from this organization summarizes a great deal of experimental data and is an excellent method of organizing the elements. As yet, however, we have not developed a model or theory to explain why the differences in elements exist, why the trends occur at all, or why the trends go in the directions that they do. We will consider these questions in the next chapter.

Vocabulary Checklist

You should now be able to define or explain the meaning of the following terms.

alkali metals	empirical knowledge	metalloid	periodic law
alkaline earth metals	halogens	noble gases	periodic table
brittle	malleable	non-metal	scientific literature
ductile	metal	period	semi-metal

Chapter Objectives

You should now be able to do the following.

1. Appreciate the periodic table as a tool for grouping elements and summarizing their properties.

2. Identify elements as metals, non-metals, or metalloids on the basis of their chemical and physical properties, and list some examples of each.

3. Perform experiments to determine the physical and chemical properties of a number of elements.

4. Use a set of physical and chemical properties of some elements to establish several similarities, trends, and differences among them.

5. Use the periodic table and the trends in groups or periods to predict the properties of some of the elements not tested.

6. Identify and state the trends in the reactivity of some metals with water, dilute acids, and oxygen.

7. Use experimental data to deduce and state the order of reactivity for some metals.

8. Identify and state the trends in the reactivity of some non-metals with oxygen and copper.

9. Trace the historical attempts of the chemists Döbereiner, Newlands, and Mendeleev to organize the elements in some logical fashion.

10. Relate the uses of some elements to their physical and chemical properties.

11. Appreciate some of the methods, tools, and problems associated with scientific investigation.

Understanding Chapter 3

REVIEW

1. Differentiate between metals, non-metals, and metalloids. List at least three examples of each.

2. What do the terms malleable, ductile, and brittle mean?

3. List two physical and two chemical properties for (a) sulphur, (b) iron, and (c) sodium.

4. List the elements that belong to the following groups in the periodic table.
   (a) the halogens
   (b) the alkali metals
   (c) the noble gases
   (d) the alkaline earth metals

5. Differentiate between empirical and intuitive knowledge.

6. What is the scientific literature?

7. What contribution did each of the following early scientists make to the development of the periodic table?
   (a) Döbereiner
   (b) Newlands
   (c) Mendeleev

8. State the periodic law that is used to organize the elements today.

9. (a) List the elements that are gases at room temperature.
   (b) List the elements that are liquids at room temperature.

10. In general, which element in each of the following element sets is the most reactive?

Explain your answers.
   (a) Li, Cs, Na, Rb
   (b) O, N, Ne, F
   (c) O, Se, S

11. In general, which element in each of the following element sets is the least reactive? Explain your answers.
   (a) Ba, Mg, Sr, Be
   (b) Mg, Al, Na
   (c) I, Cl, F, Br
   (d) Ar, Cl, S

APPLICATIONS AND PROBLEMS

12. Arrange the following metals in order of decreasing chemical reactivity (that is, from most reactive to least reactive): Zn, Cs, Au, Fe, Li, Ba, Cu, Mg, Rb, Sr.

13. (a) Explain how chemical unreactivity can be a desirable property.
    (b) Identify three chemically unreactive substances and list one use for each that depends on chemical unreactivity.

14. Why do most metals occur as compounds or ores?

15. (a) Identify the periods and columns in the periodic table.
    (b) Which of these refers to groups of elements with similar properties?

16. Use Colour Plate 3 and any experimental observations you might have to list two similarities and two differences among the halogen group of elements.

17. Refer to Colour Plate 2 and any experimental observations you might have to identify two similarities among the elements in each of the following element groups:
    (a) alkali metals
    (b) noble gases
    (c) coinage metals (Cu, Ag, Au)

18. Arrange the following elements in order of decreasing reactivity with copper: Br, S, F, I, Cl, P.

19. Name a metal and identify three uses for it. State which property or properties are responsible for each use.

20. Name a non-metal and identify two uses for it. State which property or properties are responsible for each use.

21. Name a metalloid and identify a use for it. State which property or properties are responsible for its use.

22. Use the periodic table to predict which element in each of the following sets would be the most reactive with the reagent given.
    (a) lithium, beryllium, or boron; with hydrochloric acid
    (b) fluorine, oxygen, or nitrogen; with copper
    (c) potassium, rubidium, or cesium; with water
    (d) silicon, germanium, or tin; with hydrochloric acid.

CHALLENGE

23. Describe how a scientist might determine the answer to a question concerning matter and its behaviour.

24. Suggest two possible reasons why Newlands and Mendeleev were more successful than Döbereiner in finding a pattern with which to organize the elements.

25. Bromine compounds are used in "hot tubs" to disinfect the water while chlorine compounds are usually used for swimming pools. Explain why the much more expensive bromine compounds are used in "hot tub" water.

26. The radioactive gas radon is very unreactive, yet it can be very dangerous. If buildings are constructed on or with materials that produce radon, there must be adequate ventilation to prevent a build-up of radon gas inside. Explain why the chemically unreactive gas radon is hazardous to health.

27. Suppose you were searching for two undiscovered elements with atomic numbers 118 and 119 respectively. What relative chemical and physical properties would you predict for each of these elements? Use the elements that would be near elements 118 and 119 on the periodic table as a basis for comparison.

28. Investigate and write a report on the progress science and technology have made toward the use of tritium, ^3_1H, as a source of electrical power.

29. The hydrogen isotope, tritium, is radioactive and half of it decays every 12.3 a. (See Chapter 2.)
    (a) Why do fusion or hydrogen bombs eventually become ineffective and "die of old age"?
    (b) What possibilities, if any, does this open up for a practical scheme for nuclear disarmament?

30. A fairly recent scientific model predicts that after a nuclear war, the world would be gripped by extremely cold weather for an extended period of time. This is called the "nuclear fall" model. Find out about this theory and any experiments that have been done to test it. Is it likely that this theory will ever be tested fully? Do you think that all experiments are worth performing? Explain.

4

MERGING MODELS: THE ATOMIC MODEL AND THE PERIODIC TABLE

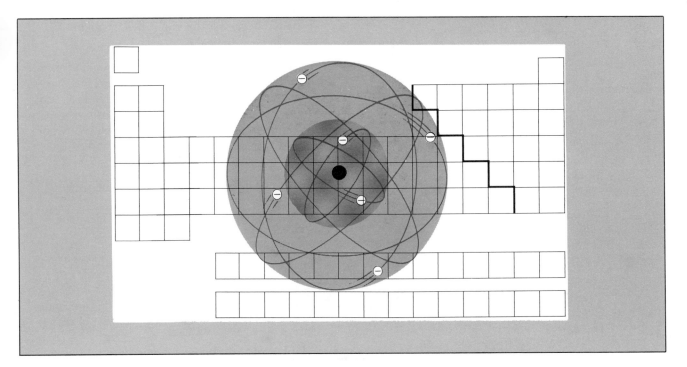

Science can be compared to a fabric with a weave of many theories and models. Different disciplines are interwoven, and theories developed in one area often contribute to development in other areas. The atomic model is one thread that winds its way throughout the structure of science. For example, scientists use current models of the extremely small atomic nucleus to probe and explain the behaviour of objects of enormous size such as supernovas (stars which explode violently as they near "death") and quasars (intensely energetic, galaxy-like objects). Scientists even discuss the behaviour of the universe immediately after its formation in terms of these nuclear models.

Similarly, we can examine the periodic table of the elements in light of the atomic model studied in Chapter 2. Is the underlying atomic structure of the elements related to their chemical behaviour? If so, what is it about their structures that makes periodic organization possible? In this chapter, you will examine the periodic table through the "eyes" of the atomic model and search for relationships between these two fundamental concepts of chemistry.

4.1 Elements – Atomic Characteristics

As you learned in Chapter 2, atoms of each element consist of a different combination of protons, neutrons and electrons. Each element also has its own "light fingerprint" or characteristic line spectrum. Can these characteristics be used to explain the differences in chemical reactivity among the elements? Are there any other atomic characteristics that might add to such an explanation? How, for example, does the size of atoms affect their reactivity? We will examine this intriguing question later in this chapter. Since chemical reactions can involve considerable energy (recall the reaction of sodium in water!), it is useful to investigate the way in which energy interacts with atoms.

Ionization Energy

An **ion** is an electrically charged particle formed when an atom or molecule loses or gains electrons. Ions have a positive charge if the atoms or molecules lose electrons, and a negative charge if the atoms or molecules gain electrons. (See Figure 4.1.) A lithium atom that has lost an electron has three protons in its nucleus and two electrons in its outer energy levels, leaving the lithium ion with a net positive charge of one unit. Conversely, an atom that has gained an electron becomes an ion with a negative charge of one unit. The ionic charge is the net electrical charge on the particle—that is, the difference between the numbers of protons and electrons. An atom that has gained two extra electrons, for example, becomes an ion with an ionic charge of $2-$.

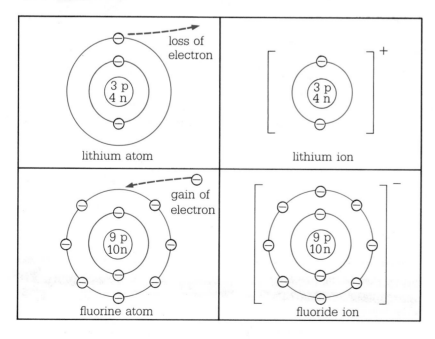

Figure 4.1
Formation of ions. The formation of ions involves either the loss or gain of electrons. The ionic charge is the net electrical charge (difference between the number of protons and electrons) in the ion.

The minimum energy needed to remove the most loosely bound electron from an atom in the gas phase is called the first **ionization energy**. (See Figure 4.2.) This energy can be supplied in a variety of ways: by light, by fast moving particles (such as electrons or alpha particles), or by energetic atomic collisions (for example, "hot" atoms colliding in a flame).

Figure 4.2
First ionization energy. Energy is required to remove an electron from an atom. The energy required to remove the first electron is an atom's first ionization energy.

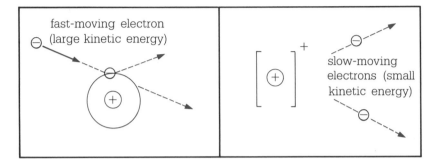

SPECIAL TOPIC 4.1

IONS AND PEOPLE

Do ions affect human health and behaviour? This question has been discussed both in the scientific literature and in the popular press for three decades. Research results seem to indicate that negative ions in the air are beneficial to health and happiness, while positive ions are responsible for many ills and feelings of depression.

Clinical studies, which deal with the study of patients as opposed to laboratory experiments, have demonstrated that positive ions in the air can aggravate sinus infections, asthma, and other respiratory afflictions, as well as induce headaches, fatigue, and dizziness. Negative ions, on the other hand, have been shown to relieve these conditions, as well as to induce a sense of well being.

Are some of the circumstances that tend to make us feel better, such as open fires and the outside air immediately after a thunderstorm, due to an increase in negative ions in the air? Is the gloomy feeling we have just before a thunderstorm or on a drizzly day due to positive ions? (High humidity tends to result in an excess of positive ions over negative ions in the air.)

Interestingly, the relationship between negative ions and good mood has been noted commercially. Engineers have designed and built negative ion generators. Some electronic air cleaners have negative ion generators built into them. People buy these generators because they believe that the negative ions they produce will tend to promote a better mood in their homes.

If you were provided with both negative and positive ion generators, how would you go about proving or disproving these claims?

Measuring Ionization Energy

There are two methods used to measure ionization energy. **Photoionization** uses light energy to remove electrons from atoms. The other method, **electron bombardment,** uses fast moving free electrons to form ions.

Photoionization

As you have learned earlier, electrons in atoms can absorb light. In order to be absorbed, however, this light must have a specific energy (colour). (See Figure 4.3.)

Figure 4.3
Absorption of energy by electrons

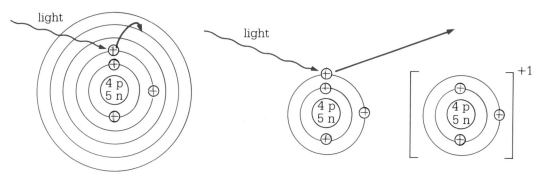

(a) Light energy lower than the ionization energy can shift an electron to a higher Bohr energy level.

(b) Light energy at or above the ionization energy removes an electron completely.

Figure 4.4
Measuring ionization energies. If light or bombarding electrons have the minimum energy needed, they will knock out the most loosely held electrons of the atoms filling the gas chamber. These atoms then break up into free electrons and positive ions. The electons are attracted to the positive plate in the detection circuit and the positive ions are attracted to the negative plate. The flow of charged particles to the plates sets up a flow of electricity that can be measured by an ammeter.

(a) Photoionization

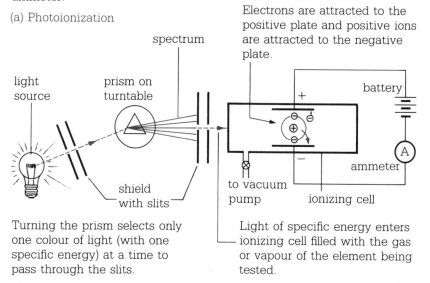

Turning the prism selects only one colour of light (with one specific energy) at a time to pass through the slits.

Light of specific energy enters ionizing cell filled with the gas or vapour of the element being tested.

Figure 4.4(a) shows an instrument that can be used to measure ionization energy using photoionization. White light passing through a prism forms a spectrum which is slowly moved across a slit. This slit allows light of only one colour (that is, with a specific energy) to enter a cell containing a vapour of the element under study. The energy is slowly increased by scanning from red light (lower energy) toward blue light (higher energy). The apparatus contains a circuit that detects ions. The energy of the light that first produces ions is the first ionization energy of the element under study.

Electron Bombardment

A second method of ionizing atoms is to bombard them with electrons. Only if these bombarding electrons have a certain minimum energy will they "knock" an electron completely away from the atom and produce an ion. Figure 4.4(b) shows an apparatus that will generate bombarding electrons of known energy. Like the photoionization apparatus, it also contains a circuit that detects ions. The accelerating voltage of the bombarding electrons is slowly increased until ions are just detected. This voltage can be used to calculate the first ionization energy. (Ionization energy in MJ/mol = accelerating voltage × 0.0963.) Electron bombardment is capable of much higher energies than photoionization.

(b) Electron bombardment

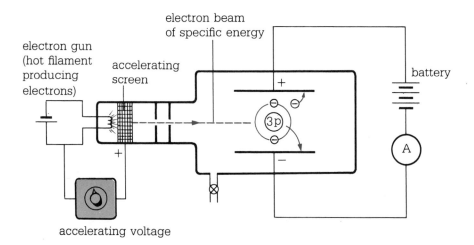

Using both photoionization and electron bombardment, chemists have empirically determined the first ionization energies of all the elements. The first ionization energies of the first 20 elements are graphed in Figure 4.5.

Exercise

1. How do the ionization energies of the first 20 elements (see Figure 4.5) relate to the organization of the elements in the periodic table? Can you use this extra data to explain why the elements exhibit a periodic repetition of reactivity?

2. Photoionization alone cannot be used to determine the first ionization energies of all of the elements. Why?

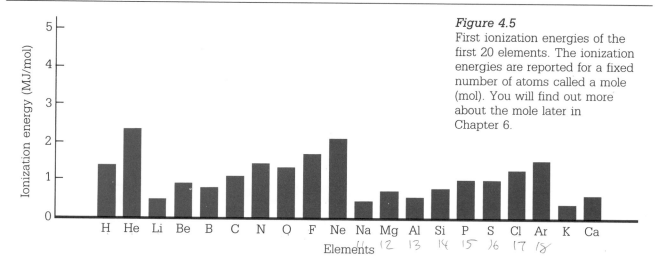

Figure 4.5
First ionization energies of the first 20 elements. The ionization energies are reported for a fixed number of atoms called a mole (mol). You will find out more about the mole later in Chapter 6.

SPECIAL TOPIC 4.2

THE GEIGER COUNTER

Radioactive elements eject very energetic particles from their nuclei when they undergo decay. These particles (alpha, beta, and gamma radiation) readily ionize atoms in their path. One of the most common instruments used to detect radioactive substances, the Geiger counter, utilizes this ionization principle. The radiation detector in the Geiger counter is much like the devices used to measure ionization energy. · A sealed tube containing a gas, such as argon, has a wire running down its centre. A metal cylinder lines the inside of the tube and surrounds the wire.

In operation, the wire in the Geiger tube has a large positive charge and the metal cylinder has a large negative charge. Any fast moving particle that enters the tube ionizes the gas molecules along its path. The positive ions and free electrons produced are drawn to the metal cylinder and wire respectively. This produces a burst of electrical current. This brief pulse of electricity is converted into a meter reading or into sound. The sound is the characteristic "click" associated with the operation of a Geiger counter.

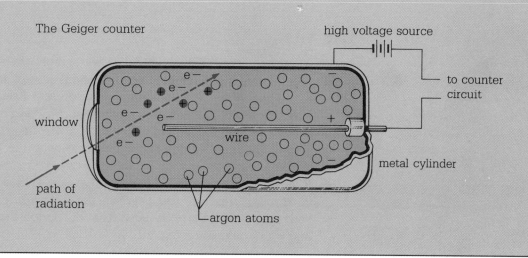

4.2 Analyzing First Ionization Energies

What would be the best way to analyze or organize first ionization energy data? The periodic pattern developed earlier seems to have been an excellent organizing tool, so perhaps listing the first ionization energies using the periodic form will be valuable. This is shown in Figure 4.6.

Figure 4.6
First ionization energies of the elements (MJ/mol)

1 H 1.31																	2 He 2.37
3 Li 0.52	4 Be 0.90											5 B 0.80	6 C 1.09	7 N 1.40	8 O 1.31	9 F 1.68	10 Ne 2.08
11 Na 0.50	12 Mg 0.74											13 Al 0.58	14 Si 0.79	15 P 1.01	16 S 1.00	17 Cl 1.25	18 Ar 1.52
19 K 0.42	20 Ca 0.59	21 Sc 0.63	22 Ti 0.66	23 V 0.65	24 Cr 0.65	25 Mn 0.72	26 Fe 0.76	27 Co 0.76	28 Ni 0.74	29 Cu 0.74	30 Zn 0.91	31 Ga 0.58	32 Ge 0.76	33 As 0.94	34 Se 0.94	35 Br 1.14	36 Kr 1.35
37 Rb 0.40	38 Sr 0.55	39 Y 0.62	40 Zr 0.66	41 Nb 0.66	42 Mo 0.69	43 Tc 0.70	44 Ru 0.71	45 Rh 0.72	46 Pd 0.81	47 Ag 0.73	48 Cd 0.87	49 In 0.56	50 Sn 0.71	51 Sb 0.83	52 Te 0.87	53 I 1.01	54 Xe 1.11
55 Cs 0.38	56 Ba 0.50	71 Lu 0.52	72 Hf 0.65	73 Ta 0.76	74 W 0.77	75 Re 0.76	76 Os 0.84	77 Ir 0.88	78 Pt 0.87	79 Au 0.89	80 Hg 1.01	81 Tl 0.59	82 Pb 0.72	83 Bi 0.70	84 Po 0.81	85 At	86 Rn 1.04
87 Fr	88 Ra 0.51	103 Lr															

57 La 0.54	58 Ce 0.53	59 Pr 0.52	60 Nd 0.53	61 Pm 0.54	62 Sm 0.54	63 Eu 0.55	64 Gd 0.59	65 Tb 0.56	66 Dy 0.57	67 Ho 0.58	68 Er 0.59	69 Tm 0.60	70 Yb 0.60
89 Ac 0.47	90 Th 0.59	91 Pa 0.57	92 U 0.59	93 Np 0.60	94 Pu 0.59	95 Am 0.58	96 Cm 0.58	97 Bk 0.60	98 Cf 0.61	99 Es 0.62	100 Fm 0.63	101 Md 0.64	102 No 0.64

Exercise

3. (a) Another method of analyzing data is to graph it. Use graph paper (or a computer graphing program) to plot the ionization energy versus atomic number for the first 40 elements. Label the peaks and troughs with the elements they represent.

   (b) Search for and list any regularities you can detect on the graph. Do these trends relate in any way to the element property trends that were established earlier?

   (c) Refer to your graph and the regularities you detected as you go through the next few pages in the textbook. Compare your results with those deduced from the periodic table listing of first ionization energies.

First Ionization Energy: Metals and Non-metals

In Chapter 3, you learned that metals and non-metals have very different properties and behaviour. If you examine Figure 4.6 carefully, you will find several regularities in the ionization energies of metals and non-metals. The first ionization energies of the metals are all much lower than those of the non-metals. In fact, there is a trend toward increasing first ionization energies across any one period (row) of the table. There are some instances in each period where this is not the case, but in general the trend is increasing first ionization energies within a period. For example, the first ionization energy increases from a value of 0.52 MJ/mol for lithium to a value of 2.08 MJ/mol for neon. (You can see exceptions to this in going from beryllium to boron and from nitrogen to oxygen.) Ionization energy is at a minimum for the alkali metals and reaches a maximum for the noble gases.

Within a group in the periodic table, the lower down an element is, the lower its first ionization energy tends to be. For example, for the alkaline earth metals, first ionization energies drop consistently from beryllium's 0.90 MJ/mol to barium's 0.50 MJ/mol. You can relate this trend with the reactivity trends within a group for metals and non-metals studied in Chapter 3. (See Figure 3.14.) For metals, reactivity tends to increase down a group. This means that reactivity increases as the first ionization energy decreases. For non-metals, with the exception of the noble gases, reactivity tends to increase up a group. This means that reactivity increases as the first ionization energy increases. (See Figure 4.7.)

Figure 4.7

Ionization energy and trends. Note that the reactivity of metals tends to increase as their first ionization energy decreases; the reactivity of non-metals (except for the noble gases) tends to increase as their first ionization energy increases; and the reactivity of the elements across a period tends to follow these trends.

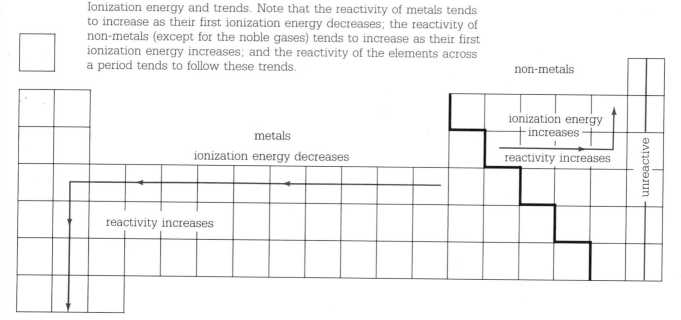

A Tentative Model for Chemical Reactivity of Elements

The identification of regularities and trends is often the inspiration that leads to the formation of hypotheses. In the case of metals, you have found that a lower first ionization energy generally means a more reactive element. A reasonable hypothesis might be the following: "Metals react by losing electrons. The easier it is for them to lose their electrons, the more vigorously they will react."

How can you explain the reactivity of the non-metals? Ignoring the noble gases for the moment, you will find that the non-metals are generally more reactive as their first ionization energy increases. Since metals and non-metals seem to behave so differently in many other respects, perhaps they differ in how they react chemically as well. If metals react by losing electrons, it might not be surprising that non-metals react by gaining electrons. This would make sense if you consider the first ionization energy data. If an element holds on to its electrons very strongly (a high first ionization energy), then it might just have a very strong attraction for extra electrons. The energy released (or absorbed) when a neutral atom attracts an extra electron is called its **electron affinity.** If the above hypothesis is correct, then the higher the first ionization energy is, the greater the electron affinity should be (more energy released), and the more reactive a non-metal should be.

If you combine this idea of non-metals reacting by accepting electrons with the earlier hypothesis that metals react by losing electrons, you will have a model to explain the chemical reactivities of the elements. Figure 4.8 illustrates how this model can explain the reaction between sodium and fluorine.

This model works superbly for the alkali metals, alkaline earth metals, halogens, and the oxygen group elements. It is, however, a disaster for the noble gases. The noble gases all have very high first ionization energies, but they are all very unreactive. Helium in particular has the highest first ionization energy of all the elements but it is also the most unreactive.

Figure 4.8
A model explaining chemical reactivity. Metals react by losing electrons and non-metals react by gaining electrons.

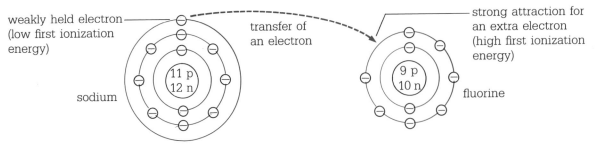

weakly held electron
(low first ionization energy)

transfer of an electron

strong attraction for an extra electron
(high first ionization energy)

sodium

fluorine

I.E. = 0.50 MJ/mol
(therefore, relatively low electron affinity)

I.E. = 1.68 MJ/mol
(therefore, relatively high electron affinity)

THE ION ROCKET ENGINE

The ion rocket engine is fundamentally different from the chemical engines used to lift rockets out of the Earth's atmosphere. The production of large numbers of ions requires a vacuum. Therefore, an ion rocket engine is only practical in outer space. It is not a very powerful engine, and produces a much smaller push than even a very small chemical rocket. An ion engine, however, is very efficient and can operate for a long period of time and ultimately result in a much greater rocket speed. Such an engine was successfully tested in space in the mid-1960s. Ion engines would likely be selected to power any very long-distance space craft such as a probe to the solar system's nearest star system, Alpha Centauri.

An ion engine generates its thrust by first forming, and then electrostatically accelerating, ions out of its exhaust. Cesium, with its low boiling point and its low first ionization energy, is an ideal "fuel" for such an engine. Cesium, in an ion engine, is capable of providing an effective rocket speed 140 times greater than that produced by a chemical rocket engine using the same mass of any known combination of chemicals.

In operation, the ion engine heats cesium to form a vapour which is ionized on contact with a heated grid. The ions generated are accelerated by an electric (or magnetic) field. They are then neutralized by returning the previously removed electrons to them, and discharged as the engine's exhaust at a speed close to 3×10^8 m/s, almost the speed of light. A rocket's thrust depends on the speed of its exhaust. A chemical rocket's fiery exhaust has a speed of less than 8×10^3 m/s. The near light speed of the cesium ion exhaust is therefore about 40 000 times greater than a chemical rocket's exhaust. This difference results in the ion engine's considerably greater thrust per kilogram of fuel used.

Operation of chemical and ion rocket engines

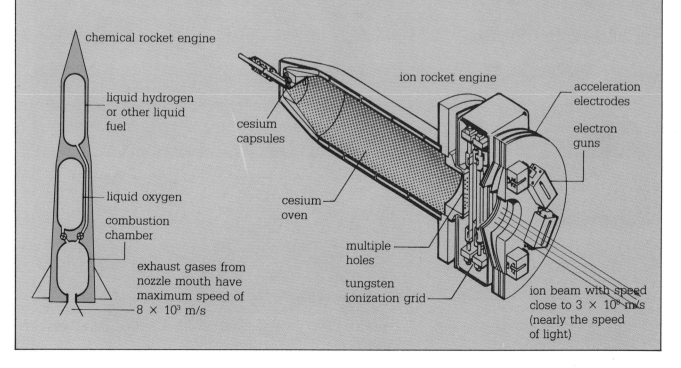

chemical rocket engine

liquid hydrogen or other liquid fuel

liquid oxygen

combustion chamber

exhaust gases from nozzle mouth have maximum speed of 8×10^3 m/s

cesium capsules

cesium oven

multiple holes

tungsten ionization grid

ion rocket engine

acceleration electrodes

electron guns

ion beam with speed close to 3×10^8 m/s (nearly the speed of light)

THE NOBLE GASES AND NEIL BARTLETT

The helium group of elements – helium, neon, argon, xenon, and radon – were all discovered between 1894 and 1900. Their main characteristic is their lack of chemical reactivity. Until the early 1960s, they were thought to be truly inert since no one had ever been able to get them to react.

In 1962, the British chemist, Neil Bartlett, working at the University of British Columbia, jolted the world of chemistry. He reacted oxygen with platinum hexafluoride, PtF_6, to form the compound O_2PtF_6. This compound behaved as if it were composed of ions, with the oxygen part being positive. Oxygen has such a high ionization energy that the discovery of an O_2^+ ion was startling. Bartlett noticed that the noble gas xenon had a lower ionization energy than oxygen. Would PtF_6 also take electrons away from xenon? Only an experiment would provide the answer. Bartlett combined Xe and PtF_6 gases in a reaction vessel. They reacted instantly, forming a crystalline yellow solid, $XePtF_6$.

Other chemists followed Bartlett's lead, and many more xenon compunds have since been synthesized. A krypton compound, KrF_2, has also been synthesized. However, the other noble gases, helium, neon, and argon have proved to be much more inert than xenon. To date no compounds of these elements have been formed.

The work done by Bartlett is just one example of an old and accepted theory of science being challenged by new experimental evidence. The knowledge of science is constantly changing. It seems that the only certainty in science is constant change. In fact, one of the major theories underlying particle physics is Heisenberg's uncertainty principle, which states that there is an ultimate "unknowability" concerning the location and energy of subatomic particles such as electrons or protons.

Neil Bartlett

The noble gas compound KrF_2

Modifying The Model

Can this model – metals lose electrons, non-metals gain electrons – be modified to fit the observed behaviour of the noble gases? If the atoms that are gaining and losing electrons to form ions are examined more closely, perhaps another regularity can be found. For example, the ion resulting when sodium loses an electron has the same electron structure as an atom of the noble gas neon. (See Figure 4.9.) A fluorine atom, on the other hand, also achieves an electron structure the same as that of neon when it gains an electron to become a negative fluoride ion.

Both positive sodium ions and negative fluoride ions form so readily that sodium and fluorine do not exist in their elemental form in nature. They are only found in compounds. What about neon? If the neon atom gained an electron, it would have an electron structure like the very reactive element sodium. Similarly, if neon lost an electron its electron structure would be the same as the very reactive fluorine atom. Stable neon ions, positive or negative, are never found in nature. The electrons in the Bohr energy levels in neon represent an exceptionally stable arrangement. It is very hard to remove an electron from neon, yet neon has very little attraction for an extra electron. All of the noble gases share these characteristics.

How can these stable or noble gas electron arrangements be worked into your chemical reactivity model? A new concept is needed. Examine Figure 4.10. The number of electrons that each energy level can accommodate is limited; otherwise, the attraction of these electrons to the nucleus would eventually be overwhelmed by the repulsion among the increasing number of electrons. Does the lack of chemical reactivity and the high first ionization energy of the noble gases indicate that they have a stable electron arrangement?

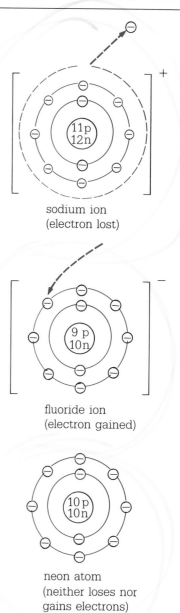

sodium ion
(electron lost)

fluoride ion
(electron gained)

neon atom
(neither loses nor
gains electrons)

Figure 4.9

Noble gas electron structures are stable. Sodium atoms and fluorine atoms react by losing and gaining an electron respectively. Both produce ions which have the same electron structure as the neutral neon atom. A neon atom's electron structure is very stable.

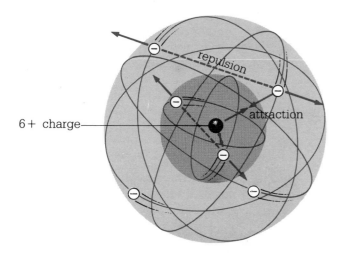

Figure 4.10
The electrostatic forces of attraction (between electrons and the nucleus) and repulsion (between electrons) determine the maximum number of electrons in the electron energy levels.

Sample Problem 4.1

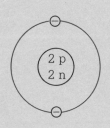

Figure 4.11
Bohr diagram of helium atom. Helium is very unreactive. Its first electron energy level is full with only two electrons.

If the noble gas helium has a stable electron arrangement how many electrons are there in its energy levels? Draw a Bohr model representation of the ${}_{2}^{4}He$ atom.

Solution

Helium's atomic number (the number of protons in its nucleus) is 2. Thus a neutral helium atom must have two electrons. Since helium is the first noble gas, it probably has only one electron energy level which contains both of these electrons. Therefore, the energy level is filled when it contains two electrons. A Bohr diagram of the ${}_{2}^{4}He$ atom is shown in Figure 4.11.

Exercise

4. Assuming that noble gases have stable electron arrangements, determine for the noble gas argon: (a) the number of energy levels, and (b) the number of electrons in each of these energy levels. (c) Then draw the Bohr diagram for argon.

Testing the Model: Predictions

So far, our model for chemical reactivity includes the following important ideas:

1. Metals react by losing electrons. The lower the first ionization energy, the more reactive the metal is.

2. Non-metals (except for the noble gases) react by gaining electrons. The higher the first ionization energy, the more these non-metals attract extra electrons (the higher the electron affinity is) and the more reactive they are.

3. Noble gases are very unreactive because they have stable electron arrangements and do not easily lose or gain electrons. Helium has two electrons in its full outer energy level and others have eight electrons.

How can you test the validity of this model? Like Mendeleev using his new periodic law in 1871, you can use the chemical reactivity model to make predictions. Remember, one of the best tests of a model or hypothesis is its use in making accurate predictions.

Exercise

5. Use the chemical reactivity model discussed above to make any prediction you can about the chemical properties and behaviour of the elements (a) beryllium and (b) tellurium.

If the noble gas atoms do have a "full outer energy level," other properties and data concerning the atoms should reflect this. Using your chemical reactivity model, try to make some general predictions concerning the relative atomic sizes of the elements. What should happen each time a new energy level is started? How should the size of a noble gas atom such as neon compare to the size of the atom with the next higher atomic number, sodium?

Chemists have estimated the atomic sizes of the elements by using the measured distances between the centres of atoms in compounds. One estimate is given in Figure 4.12. (The atomic sizes of the noble gases which have not yet been made to react – helium, neon, and argon – have been approximated from the sizes of nearby elements in the periodic table.)

Figure 4.12
Relative atomic sizes and atomic radii (in pm) for the elements

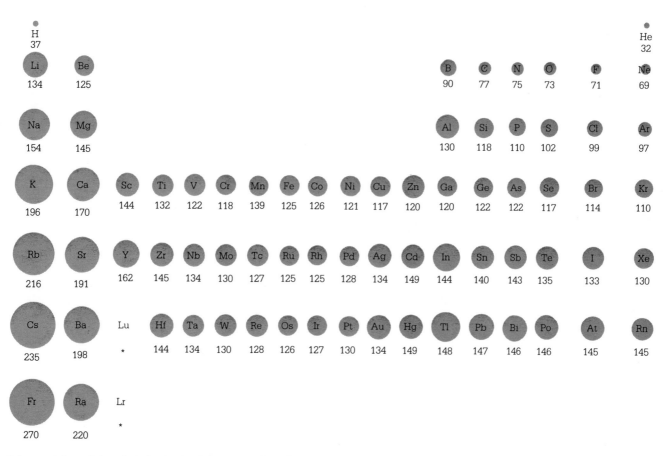

* Lu and Lr exist so briefly in the laboratory that determinations of the atomic sizes and radii are not useful for our purposes here.

Exercise

6. Use graph paper to plot a graph of atomic radius versus atomic number for the first 40 elements. If possible, do this on your ionization energy versus atomic number graph from Exercise 3. (Alternatively, you could use a computer to do both on the same graph.) Label any prominent points on your graph with the symbols of the elements they represent.

7. Identify any regularities or trends you can find on your graph. What comparisons can you make between the ionization energy and atomic radius graphs? Try to explain these trends and comparisons.

If you examine the size of the successive atoms, fluorine, neon, and sodium (see Figure 4.13), you will see that fluorine and neon atoms are approximately the same size, but a sodium atom has about twice the diameter of a neon atom. The sodium atom has only one electron more than the neon atom. An explanation for this, as we hypothesized earlier, is that in the sodium atom a whole new electron energy level has been started. Thus, atomic size confirms the earlier idea that neon has a full outer electron energy level.

Figure 4.13
Atomic sizes of fluorine, neon, and sodium

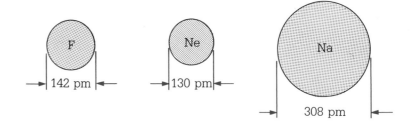

Electron Structure

By using this idea that the unreactive noble gases have full outer electron shells, the electron arrangement for the first 20 elements can be determined. (See Table 4.1.) The electrons first occupy the lowest energy level available until it is full and then start to fill up the next lowest level.

Exercise

8. Draw Bohr diagrams for atoms of the following elements.
   (a) lithium   (d) carbon
   (b) chlorine   (e) aluminum
   (c) calcium   (f) nitrogen

Table 4.1 Electron Arrangement for First 20 Elements

Element Symbol	Atomic Number	Electron Distribution			
		Energy Level 1	Energy Level 2	Energy Level 3	Energy Level 4
H	1	1			
He	2	2			
Li	3	2	1		
Be	4	2	2		
B	5	2	3		
C	6	2	4		
N	7	2	5		
O	8	2	6		
F	9	2	7		
Ne	10	2	8		
Na	11	2	8	1	
Mg	12	2	8	2	
Al	13	2	8	3	
Si	14	2	8	4	
P	15	2	8	5	
S	16	2	8	6	
Cl	17	2	8	7	
Ar	18	2	8	8	
K	19	2	8	8	1
Ca	20	2	8	8	2

In place of a Bohr diagram, the electron arrangement in an atom can be shown simply by writing down the number of electrons in each energy level. For example, the electron arrangement of an aluminum atom is written as Al-2,8,3. The number of electrons in the lowest energy level is always written first.

Exercise

9. Write down the electron arrangement for atoms of the following elements.
   (a) oxygen
   (b) neon
   (c) sodium
   (d) phosphorus

Using the Model

In merging the atomic model with the periodic table's organization of the elements, you have developed a chemical reactivity model that can be used to predict the chemical properties of unfamiliar elements. (See Figure 4.14.) Remember that like all models, this one also has its limitations. Some properties may be affected by factors that the model has not taken into account.

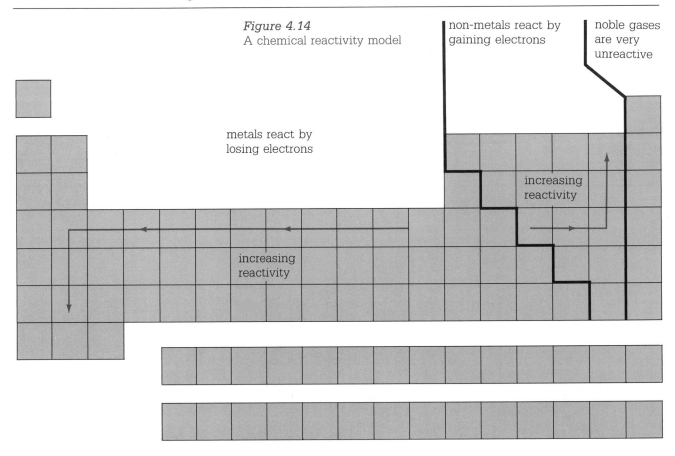

Figure 4.14
A chemical reactivity model

metals react by
losing electrons

non-metals react by
gaining electrons

noble gases
are very
unreactive

increasing
reactivity

increasing
reactivity

Sample Problem 4.2

The element scandium is difficult to isolate and therefore too expensive to experiment with in a high school laboratory. Use the data in Figure 4.6 to help you predict how scandium would react with iron and with oxygen.

Solution

Scandium has a first ionization energy of 0.63 MJ/mol. This relatively low first ionization energy and scandium's position in the periodic table indicate that it is a metal. Therefore, scandium should react by losing electrons. It should not react chemically with the metal iron which also reacts by losing electrons. Scandium's chemical reactivity can be estimated by comparing it to elements with which you may have experimented that are close to it in the periodic table, for example magnesium, calcium and aluminum. Its first ionization energy is greater than that of calcium (0.59 MJ/mol) and aluminum (0.58 MJ/mol), but it is smaller than magnesium's (0.74 MJ/mol). Thus, scandium's reactivity with oxygen should be similar to magnesium, calcium, or aluminum.

Exercise

10. The element fluorine is a very dangerous element to work with. Use the first ionization energies shown in Figure 4.6 and the reactivities of those elements near fluorine in the periodic table to predict the reactivity of fluorine with iron. How would this predicted reactivity compare to the reactivities of oxygen and chlorine with iron?

11. Recall the chemical reactivities of the elements copper and zinc with hydrochloric acid. Compare their first ionization energies shown in Figure 4.6. Is first ionization energy always a good predictor of chemical reactivity? Explain.

4.3 Examining the Chemical Reactivity Model: Graphical Analysis

Patterns in science are often discovered by categorizing and grouping items. Numerical analysis, which is seeking patterns in quantitative data, is another way of analyzing experimental results. In Chapter 3, for example, you saw how Döbereiner used numerical analysis in discovering his element "triads". Another analysis technique you used earlier in this chapter was graphing.

You can examine the chemical reactivity model developed in the last section (Figure 4.14) in light of atomic size and ionization energy information that you have organized in graphical form. Examine the graph you made for Exercise 6 and refer to Figure 4.15.

Figure 4.15
Bar graph showing atomic radius (pm) versus atomic number for the first 22 elements. The atomic radius of the elements shows a regular repeating pattern.

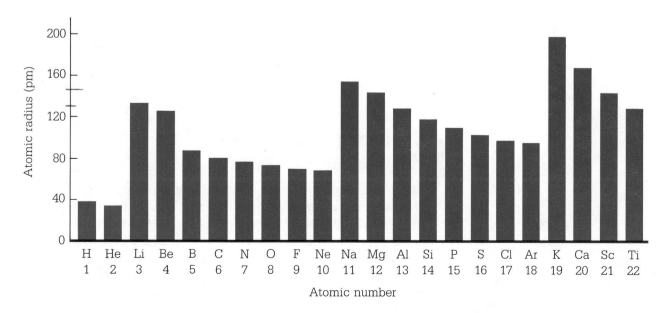

In Figure 4.15, the bar graph of atomic radius of the elements versus their atomic number shows that the atomic radius generally decreases as atomic number increases until a noble gas atom is reached. The first atom after a noble gas atom has a very much larger radius. This is consistent with the starting of another electron energy level. Can these regular patterns in atomic radius be used to further develop our model to explain chemical reactivity? What determines an atom's size in the first place?

Atoms and the Electrostatic Force

What force causes the particles in atoms to hold together or to form different arrangements? Calculations have shown that gravitational force is much too weak to be a factor. What about electrostatic force? Since atoms are mostly made up of electrically charged particles (protons and electrons), the electrostatic force is most likely. In actual fact, it is the electrostatic attraction between oppositely charged particles and the repulsion between similarly charged particles that determine an element's chemical behaviour and such physical properties as hardness and melting point.

Electrostatic Force Laws

Laboratory Manual
Experiment 4-1 will allow you to determine some basic electrostatic force laws.

Several of the forces found in nature, such as gravitational, electrostatic, and magnetic forces, operate according to an *inverse square law*. Scientists have experimentally determined that the strength of each of these forces varies inversely as the square of the distance between the objects exerting the force. (That is, as the distance between the charged objects in doubled, the force drops by four; when it is tripled, the force drops to $\frac{1}{9}$ its former value.) The electrostatic force between two charged objects also varies directly as the charge on the objects. (See Figure 4.16.)

Electrostatic forces operating in atoms vary directly as the quantity of positive charge on the nucleus, and inversely as the square of the distance between the nucleus and electrons. Both the atomic number (number of protons in the nucleus) and the atomic radius (distance between the nucleus and outer electrons) are important in determining an element's physical and chemical properties.

Explaining Atomic Radius and First Ionization Energy

The variation of atomic size as the number of protons in the nucleus increases can be explained using the principles of electrostatic force. As seen in Figure 4.17, the nucleus of the boron atom contains five protons, one more than in the nucleus of the beryllium atom. Since five positive charges will attract an electron more strongly than four positive charges will, both the outer and inner electrons of boron will be pulled in closer than those of beryllium.

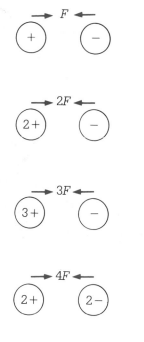

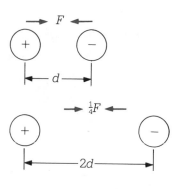

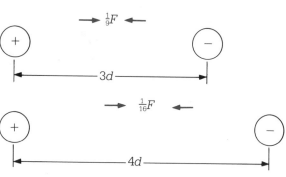

Figure 4.16
Characteristics of electrostatic force. The electrostatic force between two charged objects depends on (a) the amount of charge on each of the objects and (b) the distance between the objects.

(a) As the charge on one of two attracting charged objects is doubled, the force is doubled; as the charge is tripled the force is tripled, etc. In other words, the force varies directly as the amount of charge involved.

(b) As the distance between the charged objects is doubled, the force drops to $\frac{1}{4}$ of its former value; when it is tripled, the force drops to $\frac{1}{9}$ of its former value, etc. In other words, the force varies inversely as the square of the distance between the objects.

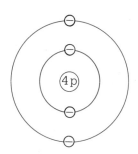

beryllium: radius 125 pm

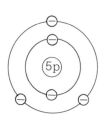

boron: radius 90 pm

Figure 4.17
Nuclear charge and electron energy level size. The greater the positive nuclear charge, the closer the electron energy levels are pulled in toward the nucleus and the smaller the atom will be.

The Shielding Effect

What effect do the electrons in an inner energy level of an atom have on the attraction of the nucleus for the outer electrons? In the boron atom shown in Figure 4.18, the two negative inner electrons tend to shield the outer electrons from the full positive charge on the nucleus. This partial cancelling of the nuclear positive charge by inner electrons is called the **shielding effect**. The five positive charges in the boron nucleus are partially cancelled by the two ⟶ *(continued on page 120)*

FORCES IN NATURE

Scientists recognize that there are four types of forces acting in the universe – gravitational, electromagnetic, and the weak and strong nuclear forces.

THE GRAVITATIONAL FORCE

Gravitational force is an attractive force. It is the attraction between two objects that is due to the amount of matter they possess – the greater their masses, the greater the gravitational force between them. Gravitational force also depends on the distance between the attracting objects – the further apart they are, the weaker the force of attraction. Gravitational force is a relatively weak force. For example, the electrostatic attraction between an electron and a proton separated by a distance equal to the radius of a hydrogen atom is about 10^{39} times stronger than the gravitational force of attraction between them. Yet gravitational force is a force that operates throughout the universe, holding very massive objects such as the stars to their galaxies and the planets in their orbits.

THE ELECTROMAGNETIC FORCE

Electromagnetic force has two forms – electrostatic and magnetic (discussed later on page 119). Electrostatic force is a very strong force, arising between objects or particles that possess electrical charges. The force depends on how much charge each object has and how far the charged objects are apart. It can be a force of attraction or repulsion. Since all atoms contain positive protons and negative electrons, it is electrostatic forces that hold the atoms and molecules of our material world together.

THE WEAK NUCLEAR FORCE

The weak nuclear force operates inside the nucleus of atoms and is responsible for the natural radioactivity (alpha, beta and gamma radiation) of unstable isotopes. In the nucleus, the electrostatic force is about 100 times stronger than the weak nuclear force.

Gravitational forces hold galaxies together. This is the Andromeda galaxy.

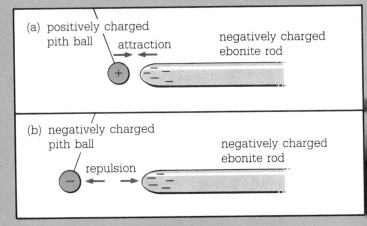

(a) positively charged pith ball — attraction — negatively charged ebonite rod

(b) negatively charged pith ball — repulsion — negatively charged ebonite rod

Electrostatic forces. Unlike electrical charges attract each other and like electrical charges repel.

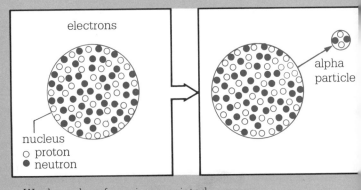

electrons

nucleus
○ proton
● neutron

alpha particle

Weak nuclear force is associated with radioactive decay of nuclei.

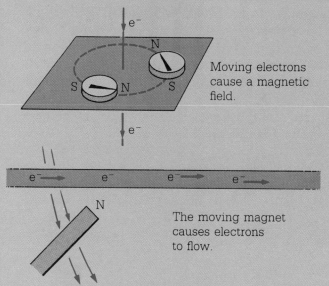

Nuclear power plant at
Pickering, Ontario

Moving electrons
cause a magnetic
field.

The moving magnet
causes electrons
to flow.

Electromagnetic force—a
combined force. Two apparently
separate forces, electrostatic
and magnetic, are intimately
connected.

THE STRONG NUCLEAR FORCE

The strong nuclear force holds the nucleus, with its many positive protons, together. It is a force that operates among quarks, the subatomic particles that make up protons and neutrons. (See Chapter 18.) This force is extremely strong, but only becomes effective at very close distances of about the diameter of the protons or neutrons themselves. Since it is about 100 times stronger than the electrostatic forces in the nucleus, the strong nuclear force can easily overcome the proton – proton electrostatic repulsions, at least up to about 100 protons in the nucleus. This is the force that is operating in nuclear reactions that release energies millions of times greater than chemical (electrostatic) energies.

COMBINING FORCES

Scientists have shown that two apparently different forces, magnetic and electrostatic forces, are in reality different forms of the same force — electromagnetic force. A moving electrical charge generates a magnetic field (this is how electric motors work) and a moving magnet generates an electrical field (this is how electrical generators work).

Albert Einstein devoted the last years of his life to finding a theory which would unite the electromagnetic force with the other three universal forces. As a result of an ongoing effort since then, scientists have had success at devising a unification theory. At much higher energies than are common on the earth, experiment and theory have shown the weak nuclear force to have features in common with the electromagnetic force. (Scientists have dubbed this combined force the ''electroweak force.'') Theory also predicts that, at the enormous energies generated in the first few seconds after the universe came into existence, even the strong nuclear force had features in common with the electroweak force. These theories go under the name GUTS, which stands for ''grand unified theories.'' The theorists, however, have not had equal success in explaining how the gravitational force could be associated with the other types of forces.

negative charges of the inner electrons. As a result, the outer electrons experience a reduced positive charge in the core of the atom. The term *atomic core* refers to the nucleus and all inner electrons.

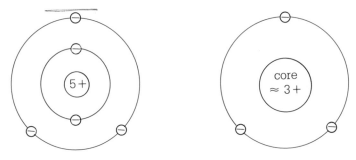

Figure 4.18
Shielding effect in the boron atom. The inner electrons tend to partially cancel the effect of the positive charge on the nucleus. The outer electrons of the boron atom experience an approximate 3+ charge from the atom's core (nucleus and all inner energy levels.)

Electrons in an atom repel each other. This repulsive force will partially cancel out the attractive pull of the atom's core. The more electrons there are in the outer energy level and the closer they are to each other, the greater this repulsion will be. An atom's radius will be partially determined by these two opposing electrostatic forces.

The Atomic Radius Graph

You can now explain most of the features of the atomic radius versus atomic number graph (from Exercise 6 or Figure 4.15) by using the electrostatic force principles. As the nuclear positive charge increases, the electrons in any one energy level are attracted more and more to the nucleus. As a consequence, the electrons are drawn in closer and closer. Repulsion among the electrons tends to limit this effect at smaller atomic radii. Only when a new outer electron energy level is started will the size of the atom suddenly become very much larger again. (See Figure 4.19.)

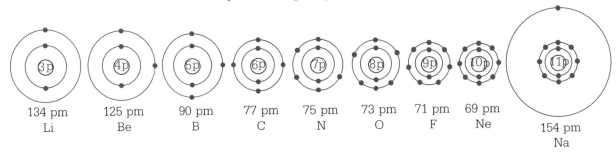

| 134 pm | 125 pm | 90 pm | 77 pm | 75 pm | 73 pm | 71 pm | 69 pm | 154 pm |
| Li | Be | B | C | N | O | F | Ne | Na |

Figure 4.19
The radius of electron energy levels tends to decrease as the positive charge on the nucleus (number of protons) increases. The atomic radius increases suddenly when a new energy level (shown here with sodium) is started.

First Ionization Energy Graph

Why do metals have such low first ionization energies and non-metals have such high first ionization energies? An atom's first ionization energy is partially determined by its nuclear charge, its radius, and the number of electrons in the atom. As you move to the right across a period (row) in the periodic table, both the nuclear charge and the number of outer electrons increase. The increasing nuclear charge tends to pull the electrons in more closely, decreasing the atomic radius and further increasing the electron – nucleus attraction. (Remember, a small change in distance causes a much larger change in electrostatic force.) At the same time the increasing number of outer electrons and their decreasing electron – electron distance cause a counterbalancing repulsion among the electrons. The net effect is a tendency to increasing first ionization energy as you move to the right across a period.

Metals, with relatively large radii and small atomic core charges, have low first ionization energies. In contrast, non-metals with relatively small radii and large atomic core charges, have large first ionization energies. (See Figure 4.20.) Since non-metals have small radii, extra electrons can approach very closely to a highly charged atomic core. This also explains the high electron affinity of non-metals.

Figure 4.20
Atomic radii and first ionization energies. Metals, with relatively low atomic core charges and large radii, have low first ionization energies. Non-metals, with relatively large core charges and small radii, have large first ionization energies.

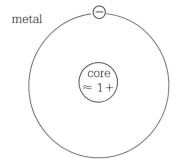
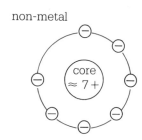

sodium atom
– small core charge
– large distance from the core
  to the outer electron

fluorine atom
– large core charge
– small distance from the core
  to the outer electrons

Figure 4.21 shows a combined first ionization energy and atomic radius versus atomic number graph. As expected the first ionization energy and atomic radius trends of the elements are opposite – as the radius decreases, the first ionization energy increases. (The small zig-zags within periods can only be adequately explained using the more advanced atomic model discussed in Appendix C.)

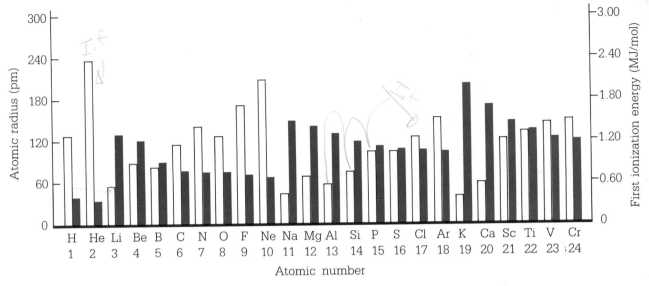

Figure 4.21
Ionization energy (open bar) and atomic radius (solid bar) versus atomic number for the first 24 elements. The atomic radius and first ionization energy trends are opposite to each other.

You have seen how the electrostatic force principles can be used to explain the chemical reactivity model developed earlier. The electrostatic force model can also be used to predict more element characteristics.

Exercise

Where necessary, use the graph in Figure 4.21 to answer the following questions.

12. Explain why the first ionization energy usually increases across a row of the periodic table (from left to right).

13. Explain why the first ionization energy of an atom generally increases as its atomic size (radius) decreases.

14. Identify some specific unexplained or contrary trend features on the first ionization energy versus atomic number graph.

15. Why is potassium more reactive than lithium?

16. Which element should be more reactive, argon or chlorine? Explain your answer.

17. Which element should be more reactive, sulphur or bromine? Explain why.

Successive Ionization Energies

What should the successive values of ionization energy be if all of the electrons are removed, one by one, from an atom? The electrostatic force model is used to answer this question as shown in Table 4.2. The energy required to remove the first electron from an atom is called the first ionization energy, the energy required to remove the second is called the second ionization energy, and so on.

Table 4.2 Predicting Successive Ionization
Energies for Beryllium

Be neutral atom

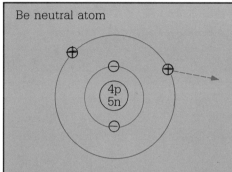

First Ionization Energy, *I.E.*$_1$
Removing the first electron from a neutral Be atom takes 0.90 MJ/mol. The repulsion exerted by the other outer electron helps to keep this energy relatively low.

Be$^+$ ion

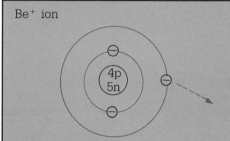

Second Ionization Energy, *I.E.*$_2$
Removing the second electron from a Be$^+$ ion should be more difficult because the second outer electron is missing and cannot assist by repelling the electron to be removed.

Be^{2+} ion

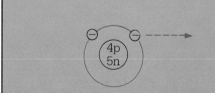

Third Ionization Energy, *I.E.*$_3$
Removing the third electron from a Be^{2+} ion involves a much greater energy. The third electron is very close to the attracting nucleus. (Remember how the electrostatic force changes with distance – halve the distance and the force increases fourfold!) In addition, since there are no shielding inner electrons, the full effect of the positive charge on the nucleus is felt by these remaining electrons. These factors would probably add up to produce a very large ionization energy.

Be^{3+} ion

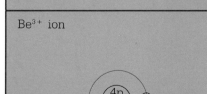

Fourth Ionization Energy, *I.E.*$_4$
The energy needed to remove the last electron should be higher yet. There is no other electron to help the ionization process by repulsion, and the lone remaining electron will have been drawn in even closer to the nucleus.

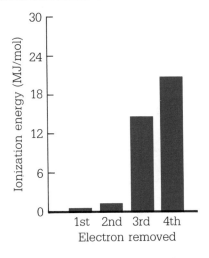

Figure 4.22
Successive ionization energies
for beryllium

The following is a summary of the prediction made in Table 4.2 for the successive ionization energies for the beryllium atom.

$$I.E._1 = 0.90 \text{ MJ/mol (actual)}$$
$$I.E._2 = \text{higher than } I.E._1$$
$$I.E._3 = \text{very much higher than } I.E._2$$
$$I.E._4 = \text{higher than } I.E._3$$

The biggest jump in predicted ionization energy comes when an electron from a full inner energy level is removed. A graph showing the experimentally determined successive ionization energies for beryllium is given in Figure 4.22.

The experimentally determined successive ionization energies for the beryllium atom certainly agree with our predictions using the principles of electrostatics. This, combined with the experimental evidence of trends in first ionization energies and atomic radii discussed earlier, gives us confidence in using this model to explain and predict other characteristics of atoms.

Sample Problem 4.3

An element has the following successive ionization energies:

$$I.E._1 = 1.1 \text{ MJ/mol} \qquad I.E._4 = 6.2 \text{ MJ/mol}$$
$$I.E._2 = 2.4 \text{ MJ/mol} \qquad I.E._5 = 37.8 \text{ MJ/mol}$$
$$I.E._3 = 4.6 \text{ MJ/mol} \qquad I.E._6 = 47.3 \text{ MJ/mol}$$

How many outer electrons do the atoms of this element have?

Solution

Calculate the factor differences between pairs of successive ionization energies. In other words, find how many times greater the next ionization energy is compared to the one before.

Ionization Energies	Factor Difference
$I.E._1 = 1.1$ MJ/mol	
	2.2
$I.E._2 = 2.4$ MJ/mol	
	1.9
$I.E._3 = 4.6$ MJ/mol	
	1.4
$I.E._4 = 6.2$ MJ/mol	
	6.1 (largest)
$I.E._5 = 37.8$ MJ/mol	
	1.3
$I.E._6 = 47.3$ MJ/mol	

The largest factor difference occurs between electrons 4 and 5. This indicates that the fifth electron is considerably more difficult to remove than the first four electrons were. It means that the fifth electron comes from an inner energy level. Therefore, this atom is likely to have four electrons in its outer energy level.

Exercise

18. An element has the following successive ionization energies:

$$I.E._1 = 0.5 \text{ MJ/mol}$$
$$I.E._2 = 4.6 \text{ MJ/mol}$$
$$I.E._3 = 6.9 \text{ MJ/mol}$$
$$I.E._4 = 9.5 \text{ MJ/mol}$$
$$I.E._5 = 13.4 \text{ MJ/mol}$$

How many outer electrons does an atom of this element likely have?

19. To which group of the periodic table does the element with the following successive ionization energies likely belong?

$$I.E._1 = 0.58 \text{ MJ/mol}$$
$$I.E._2 = 1.82 \text{ MJ/mol}$$
$$I.E._3 = 2.74 \text{ MJ/mol}$$
$$I.E._4 = 11.58 \text{ MJ/mol}$$
$$I.E._5 = 14.83 \text{ MJ/mol}$$
$$I.E._6 = 18.38 \text{ MJ/mol}$$

4.4 Representing Atoms: Lewis Symbols

The chemical reaction – electrostatic force model developed in this chapter has been used to predict and explain many of the observed properties and behaviour of the elements. The most complex and detailed model, however, is often not the best one to use in all applications. A simpler model is sometimes sufficient when many repetitive predictions, illustrations, or explanations are required. In Chapter 5 you will need to explain and predict the chemical behaviour of many sets of elements. A simple, yet still reasonably accurate model enabling you to make these predictions and explanations is valuable.

Gilbert Lewis, an American chemist, was one of the major contributors to the theory of chemical reactivity. In 1916, he devised a very simple method of illustrating atoms and their reactions with one another, now known as **Lewis symbols.**(See Figure 4.23.) Note that with Lewis symbols, the atom's nucleus and all inner electrons (the atom's core) are replaced by the element symbol. The outer electrons are represented by dots around the symbol in a square arrangement. These dots are arranged in twos if more than four outer electrons are present.

For elements of the same group in the periodic table, their Lewis symbols all show the same number of dots. For example, the Lewis symbol for bromine has seven dots around its element symbol, the same number that fluorine or chlorine has. (See Table 4.3.)

Bohr model: Na

Bohr model: Cl

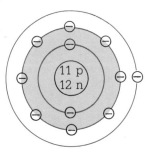

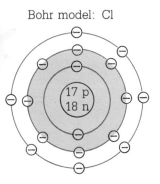

Figure 4.23
A Lewis symbol replaces the core of the Bohr atomic model with the element's symbol. Outer energy level electrons are represented by dots around the symbol in a square arrangement.

Lewis Symbol

Na·

Lewis Symbol

.C̤l:

Table 4.3   Lewis Symbols for Selected Elements

L̇i	Be·	Ḃ·	·Ċ·	·N̈·	:Ö:	:F̈:	:N̈e:
Ṅa	Mg·					:C̈l:	:Är:
K̇						:B̈r:	

The Lewis symbol is a simple yet very effective model for explaining and illustrating much of the chemistry of the elements. You will be using it extensively in Chapter 5.

Exercise

20. Draw Lewis symbols for each of the following:
    Al, Ba, He, Xe, S, and H.

Looking Back – Looking Ahead

Historically, as you have seen in Chapter 3, the periodic table was developed to organize the elements according to their properties. In this chapter, you have come to appreciate that the periodic table also reflects the underlying atomic structure of the elements. The periodic table is truly one of the greatest tools we have to explain chemical behaviour. You will use the periodic table in the next chapter to identify the type of bonding expected between different elements.

Vocabulary Checklist

You should now be able to define or explain the meaning of the following terms.

ion
ionization energy

photoionization
electron bombardment

electron affinity
shielding effect

Lewis symbols

Chapter Objectives

You should now be able to do the following.

1. Appreciate the activities and processes that scientists use to establish and develop models and theories.

2. Organize experimental data concerning the elements and their reactivities by preparing tables and graphs.

3. Use the tables and graphs (objective 2) to
   (a) seek out trends and regularities,
   (b) relate the trends and regularities to the properties of the elements, and
   (c) propose hypotheses to explain these trends and regularities.

4. Trace the development of a chemical reactivity model based on the regularities and trends established in objective 3.

5. Relate the metals and non-metals to their relative atomic sizes and ionization energies.

6. Identify the electrostatic force as the major force operating within atoms and use it to explain the relative atomic sizes and the ionization energies of the elements.

7. Show the arrangement of electrons in atoms of the first 20 elements in two ways.
   (a) listing the number of electrons in each electron energy level
   (b) drawing Bohr diagrams

8. Use the successive ionization energies of an unknown element to predict the number of outer energy level electrons in its atoms.

9. Use a simplified version of the atomic model to draw Lewis symbols.

10. Use the periodic table, along with the atomic model and principles of electrostatics, to predict the relative reactivities between given sets of elements or between a list of elements and a given reagent.

Understanding Chapter 4

REVIEW

1. (a) What is an ion?
   (b) Describe two ways in which an ion can form from an atom.
   (c) What is meant by ionization energy?

2. Describe two ways that electrons in atoms can absorb energy.

3. (a) List two methods of measuring first ionization energy.
   (b) Briefly describe how one of these methods works.

4. Use either Figure 4.21 or your own first ionization energy versus atomic number graph done in Exercise 3 to answer the following questions. How do the first ionization energies of the metals compare to those of the non-metals? What is the trend in first ionization energies from left to right across a period in the periodic table? Examine period 3 and identify two specific exceptions to the trend.

5. What is meant by the term electron affinity? If the noble gases are ignored, how is electron affinity related to ionization energy?

6. Identify how each of the following characteristics tends to change in the indicated direction in the periodic table.

CHARACTERISTIC	DIRECTION OF MOVEMENT
(a) first ionization energy	from left to right *Incr.*
(b) first ionization energy	up *family of elem.* *In*
(c) atomic size	from left to right *dec.*
(d) atomic size *more energy levels*	down *Inc.*
(e) reactivity of metals	down *Inc.*
(f) reactivity of metals *more diff. to rem. e−*	from left to right *dec*
(g) reactivity of non-metals	from left to right *In*
(h) reactivity of non-metals	down *dec*

7. Briefly describe a first model explaining chemical reactivity. Why do the chemically unreactive noble gases not fit into this model?

8. List three methods of analyzing scientific data in order to find regularities and patterns. Give one example of each method.

9. (a) What is the major type of force operating within atoms?
   (b) Under what conditions does this force exert an attraction?
   (c) Under what conditions does this force exert a repulsion?
   (d) What two factors affect the strength of this force?

10. Use diagrams and electrostatic force theory to explain the following.
    (a) Why do metals have relatively large atoms and relatively low ionization energies?
    (b) Why do non-metals have relatively small atoms and relatively large ionization energies?

APPLICATIONS AND PROBLEMS

11. Use Figure 4.21 (or your own graphs for Exercises 3 and 6), and experimental results concerning chemical reactivity (from Chapter 3) to answer the following questions.
    (a) What general correlation between ionization energy and chemical reactivity can you make for the following?
      (i) metallic elements
      (ii) non-metallic elements
    (b) What general correlation between atomic radius and chemical reactivity can you make for the following?
      (i) metallic elements
      (ii) non-metallic elements
    (c) What correlation is there between the atomic radii and the first ionization energies of the elements?

12. Two objects, one having a negative charge of one unit, and the other a positive charge of one unit, are separated by 12 cm. The force of attraction between them (F) is 1 N (one newton). Use this information to calculate the force of attraction between the objects in each of the situations shown in the table.

Electrostatic Force Variation

Description	Diagram
Two objects, one having a negative charge of one unit, and the other a positive charge of one unit, are separated by 12 cm.	$\rightarrow F = 1$ N$\leftarrow$ (1−) (1+) $\leftarrow$12 cm$\rightarrow$
(a) Object A with a 2− charge and object B with a 1+ charge are separated by 12 cm.	$\rightarrow F = ?\leftarrow$ (2−) (1+) $\leftarrow$12 cm$\rightarrow$
(b) Object A with a 1− charge and object B with a 4+ charge are separated by 12 cm.	$\rightarrow F = ?\leftarrow$ (1−) (4+) $\leftarrow$12 cm$\rightarrow$
(c) Object A with a 1− charge and object B with a 1+ charge are separated by 6 cm.	$\rightarrow F = ?\leftarrow$ (1−) (1+) $\leftarrow$6 cm$\rightarrow$
(d) Object A with a 1− charge and object B with a 1+ charge are separated by 24 cm.	$\rightarrow F = ?\leftarrow$ (1−) (1+) $\leftarrow$24 cm$\rightarrow$
(e) Object A with a 2− charge and object B with a 2+ charge are separated by 4 cm.	$\rightarrow F = ?\leftarrow$ (2−) (2+) $\leftarrow$4 cm$\rightarrow$

13. In an effort to establish accepted theories to explain scientific phenomena, many physical and mental activities are carried out. Arrange the following list of activities into a reasonable

order in which they might be carried out: hypothesizing, pattern recognition, stating laws, using theories, experimenting, proposing models, organizing data, predicting, observing, data analysis, problem recognition, publishing. (Note: Some activities may occur more than once.)

14. Draw Bohr diagrams for the following atoms: K, P, B, Cl, and Be.

15. Write down the electron arrangement of each of the following atoms: Mg, He, Si, Cl, and Ne.

16. List the elements that would likely have the following successive ionization energies.

    $I.E._1 = 1.01$ MJ/mol $I.E._6 = 21.17$ MJ/mol
    $I.E._2 = 1.90$ MJ/mol $I.E._7 = 25.40$ MJ/mol
    $I.E._3 = 2.91$ MJ/mol $I.E._8 = 29.85$ MJ/mol
    $I.E._4 = 4.96$ MJ/mol $I.E._9 = 35.87$ MJ/mol
    $I.E._5 = 6.27$ MJ/mol

17. How many outer energy level electrons does the atom with the following successive ionization energies have?

    $I.E._1 = 0.55$ MJ/mol $I.E._4 = 5.50$ MJ/mol
    $I.E._2 = 1.06$ MJ/mol $I.E._5 = 6.91$ MJ/mol
    $I.E._3 = 4.21$ MJ/mol $I.E._6 = 8.76$ MJ/mol

18. Sketch Lewis symbols for the following elements: phosphorus, iodine, selenium, aluminum, cesium, and strontium.

19. Predict which of the following elements would have the most vigorous reaction with chlorine: Kr, Be, Ga, Rb, or Co. Explain your choice.

20. Predict which of the following elements would have the most vigorous reaction with zinc: Ba, Br, C, Xe, or I. Explain your choice.

CHALLENGE

21. The following Lewis symbols represent atoms of two elements that lie on the same row of the periodic table: X·   :Z:

    Compare X and Z with respect to each of the following properties.

    (a) chemical reactivity with iron
    (b) first ionization energy
    (c) chemical reactivity with oxygen
    (d) atomic radius
    (e) state at 200°C

22. Sketch the probable Lewis symbol for an element that has the following properties.
    (a) relatively low first ionization energy
    (b) solid at room temperature
    (c) conducts electricity both as a solid and a liquid
    (d) reacts fairly rapidly with sulphur
    (e) relatively large atomic radius
    (f) relatively large third ionization energy

23. Use the periodic table to predict which atom or ion in each of the following has the greater radius.
    (a) sodium ion, aluminum ion
    (b) sodium ion, cesium ion
    (c) fluoride ion, iodide ion
    (d) sodium atom, chlorine atom
    (e) sodium ion, chlorine ion

24. The atomic number of the hypothetical element, andromadine, is 117.
    (a) Which element group does andromadine belong to?
    (b) Compare the atomic radius and first ionization energy of andromadine with the atomic radius and first ionization energy of radium and astatine.

25. Ionic compounds conduct electricity when they are molten or when they are dissolved in water because their ions are free to move. Propose a hypothesis to explain how solid metals can conduct electricity.

26. Predict which of the following pairs of elements would react most readily and which set would react the least readily. Explain your choices.
    (a) calcium and oxygen
    (b) cesium and fluorine
    (c) lithium and phosphorus
    (d) sodium and bromine
    (e) beryllium and selenium
    (f) potassium and chlorine

5 CHEMICAL BONDING

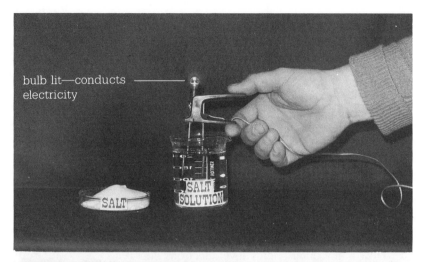

bulb lit—conducts electricity

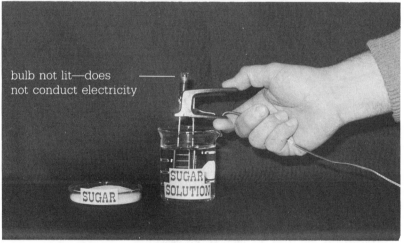

bulb not lit—does not conduct electricity

The entire universe is made up of about 90 elements. However, most of the substances that you deal with every day are not elements. The elements in these substances are "hidden". They have combined to form the compounds, solutions, and mixtures with which you are more familiar.

Compounds, as you know, are substances composed of a fixed ratio of particles bound together. The forces holding particles together in a compound are called **chemical bonds**. These forces form the basis of chemical properties and reactions. Why, for example, do the similar looking compounds, salt and sugar, behave so differently when dissolved in water and tested for electrical conductivity? Is it due to the different way their elements are held together? In this chapter, you will develop a bonding model by using the same scientific techniques you previously used in developing the chemical reactivity model in Chapters 3 and 4. In doing so, you will discover the nature of chemical bonds, and how and why they form.

5.1 Compounds –
Data and Regularities

Information is the fuel needed to drive the search for explanations. In Chapter 3, your starting point was information about the elements. To develop a bonding model, you need information about the properties and behaviour of compounds. Since there are so many more compounds than elements, this task can be a daunting one. Chemical reference books, such as the *Handbook of Chemistry and Physics* and the *Merck Index,* list the properties of thousands of the more common or useful compounds. Compounds may be solids, liquids, and gases. They come in different colours but the majority are white in appearance. (See Colour Plate 4.) Many, such as salt, are hard and brittle; some, such as stearic acid, a wax-like compound, are soft and malleable. They may or may not be good conductors of electricity or heat.

How are you to organize such a wealth of information? Perhaps reflecting on how the elements were classified would provide an inspiration. The sorting of elements into metals and non-metals was most useful in starting the development of the periodic table. Can compounds be classified in terms of the metallic or non-metallic elements they contain?

Electrical Conductivity

One characteristic property of substances is the ability or inability to conduct electricity. Very few compounds can conduct electricity at room temperature. Many compounds, however, do conduct electricity when they are molten (heated to become a liquid) or when they are dissolved in water. (See Table 5.1.)

Laboratory Manual
You will test the electrical conductivity of the water solutions of several compounds in Experiment 5-1.

Table 5.1 Conductivity of Selected Compounds in Water Solution

Solute		Conduct Electricity
Name	Formula	
Pure water	H_2O	No
Salt	NaCl	Yes
Sugar	$C_6H_{12}O_6$	No
Methanol	CH_3OH	No
Calcium chloride	$CaCl_2$	Yes
Zinc bromide	$ZnBr_2$	Yes

DEVELOPING SCIENTIFIC MODELS

Often the inspiration for a hypothesis to explain new phenomena comes from an already well established scientific model.

An explanation for the rather complicated motions of the planets in the night sky was proposed by the Polish astronomer Copernicus in 1543. His simple model placed the sun at the centre and the planets, including the earth, revolving around it. By the late 19th century, when atoms were found to consist of charged particles, the Copernican model of the solar system was well established. The results of Ernest Rutherford's alpha particle scattering experiment were interpreted in terms of a "solar system" model for the atom. The heavy nucleus of the atom, like the sun, was at the centre and attracted the lighter orbiting electron "planets" by electrostatic, rather than gravitational, forces.

Rutherford's atomic model was widely accepted because of its simplicity. Linking the very small (atoms) to the very large (solar system) still appeals to scientists today. For example, sub-nuclear particles (such as quarks, mesons, and neutrinos) are linked to the behaviour of the entire universe through cosmological models.

In the 1860s Mendeleev organized the approximately 60 elements known at the time into the periodic table. This activity had its parallel in the 1960s when the American physicists, Yuval Ne'eman and Murray Gell-Mann, invented a classification system for the dozens of sub-nuclear particles that the "atom-smashers" of the time had discovered. Their classification brought to the confused world of particle physics the same kind of order that Mendeleev's periodic table brought to the confusion of chemistry in the 1860s. Just as Mendeleev ordered known elements and predicted the existence of previously unknown ones, Ne'eman and Gell-Mann ordered the known sub-nuclear particles and predicted the existence of others. The discovery of these predicted particles confirmed their model and led, within ten years, to the current theory of quarks as the basic building blocks of matter. (This theory is discussed in more detail in Chapter 18.)

The Copernican solar system and the nuclear atom models. Similar models are often used in widely different fields of science. (Diagrams not to scale)

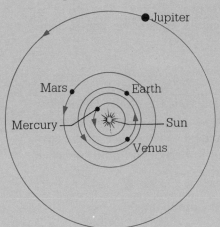

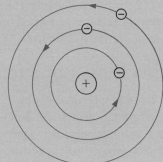

Gravitational force attracts the relatively light planets to the heavy central sun.

Electrostatic force attracts the relatively light electrons to the heavy central nucleus.

Exercise

1. With your lab partner or assigned discussion group, examine the data in Table 5.1.
   (a) Identify any regularities your group can find.
   (b) In consultation with your group members, propose hypotheses to explain these regularities.

(Note: The exchange of ideas and information in exercises such as these is a very important feature of science. International and regional scientific seminars, meetings, and conferences are frequently held to provide forums for scientists to test their models and exchange ideas with their colleagues.)

Explaining Conductivity of Compounds in Solution

Water solutions made with compounds containing a metallic element conduct electricity. Many of the compounds composed only of non-metallic elements form water solutions that do not conduct electricity. What hypothesis can you make to explain this regularity?

Think about the compounds that form conductive solutions. How can this conductivity be explained? Recall the methods used to measure ionization energy discussed in Chapter 4. An electrical current started to flow between the two detection plates only after ions had been produced and were moving between them. Perhaps the conduction of electricity in solutions has the same explanation. A reasonable hypothesis would be: Compounds containing metallic elements consist of ions. When such compounds are dissolved in water these ions become free to move and allow an electrical current to flow through the solution. Now what predictions can you make and use to test this hypothesis?

Conductivity of Compounds in the Liquid State

If metal-containing compounds consist of ions, what other conditions would free those ions to conduct electricity? Recall that in the liquid state, the particles of a substance are mobile. (The kinetic molecular theory is reviewed in Chapter 10.) If compounds containing metallic elements are indeed made up of ions, then these compounds should conduct electricity in the liquid state when their ions are free to move about. (See Table 5.2 and Figure 5.1.)

Laboratory Manual
In Experiment 5-2, you will test the electrical conductivity of several liquid compounds.

Table 5.2   Conductivity of Compounds in the Liquid State

Compound		Conduct Electricity
Name	Formula	
Potassium iodide	KI	Yes
Stearic acid	$C_{18}H_{36}O_2$	No
Strontium chloride	$SrCl_2$	Yes
Lead(II) bromide	$PbBr_2$	Yes
Phosphorus(V) sulphide	P_4S_{10}	No
Glycerol	$C_3H_8O_3$	No

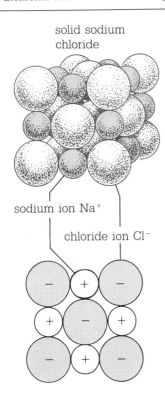

solid sodium chloride

sodium ion Na⁺

chloride ion Cl⁻

Figure 5.1
A conductivity hypothesis. Compounds containing a metallic element consist of ions. When melted, the ions are free to move. These moving ions can carry an electrical current.

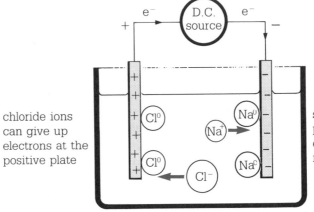

melted sodium chloride

chloride ions can give up electrons at the positive plate

sodium ions can pick up extra electrons at the negative plate

In Table 5.2 the liquid compounds which contain metallic elements do conduct electricity. This regularity supports our earlier hypothesis that those compounds consist of ions. Another regularity noted is that liquid compounds consisting only of non-metallic elements do not conduct electricity.

SPECIAL TOPIC 5.2

CONDUCTIVITY AND METAL PRODUCTION

Most metals are quite reactive and are never found in their elemental form in nature. They are only found as compounds in deposits called ores. One way of extracting the metal from these compounds is to pass electricity through the molten or dissolved compound in a process called electrolysis. For example, the production of aluminum metal depends on dissolving its purified ore, bauxite (Al_2O_3), in a solvent called cryolite (Na_3AlF_6), at about 1000°C. When electricity is passed through this conducting solution, aluminum metal is produced. In another example, the highly reactive metal sodium is produced by passing electricity through molten sodium chloride.

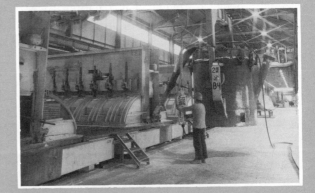

Siphoning liquid aluminum in Alcan's Grande Baie, Quebec smelter

A Bonding Model

We can now modify the chemical reactivity model developed in Chapter 4 to account for chemical bonding. This revised model includes the following ideas:

1. Metals react chemically by losing electrons, forming positive ions.

2. Non-metals react chemically by gaining electrons, forming negative ions.

3. Chemical bonds in compounds containing a metallic element and a non-metallic element consist of the electrostatic attraction between these oppositely charged ions.

4. Metallic atoms lose all the electrons in their outer energy level when they react chemically; the resulting ion has an electron structure of a noble gas. Non-metallic atoms gain enough electrons to form ions also having an electron structure of a noble gas.

Since all of the noble gases except helium have eight electrons in their outer electron energy levels, this last concept (4) of atoms reacting chemically to obtain a noble gas electron structure is referred to as the **octet rule.** (See Figure 5.2.)

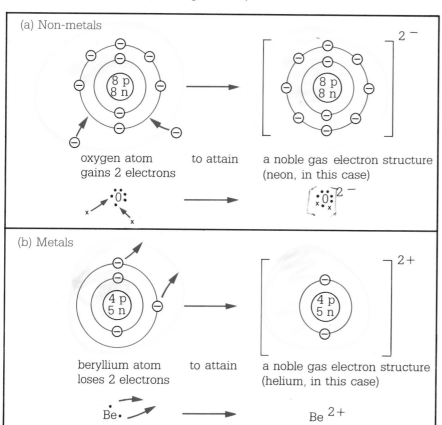

(a) Non-metals

oxygen atom   to attain   a noble gas electron structure
gains 2 electrons   (neon, in this case)

(b) Metals

beryllium atom   to attain   a noble gas electron structure
loses 2 electrons   (helium, in this case)

Figure 5.2
The octet rule. In a chemical reaction, atoms lose or gain enough electrons to obtain a noble gas electron structure.

The octet rule does not explain why bonding occurs; it only provides a model that attempts to explain what happens to electrons when bonding does take place. It predicts whether or not bonding will occur; and as you will see later, it *predicts* how many atoms of different elements are likely to react with each other.

Why Bonding Occurs

Why do atoms tend toward a noble gas electron structure? When atoms react chemically, they follow one of the great tendencies in nature – "things tend to roll downhill." Put more scientifically, systems (or collections of objects) tend to move toward the state of lowest possible potential energy (Figure 5.3). If a force of attraction exists between two objects, and the objects are far apart, the force of attraction will tend to pull the objects together. The energy possessed by the two objects due to their separation and the attraction between them is potential energy.

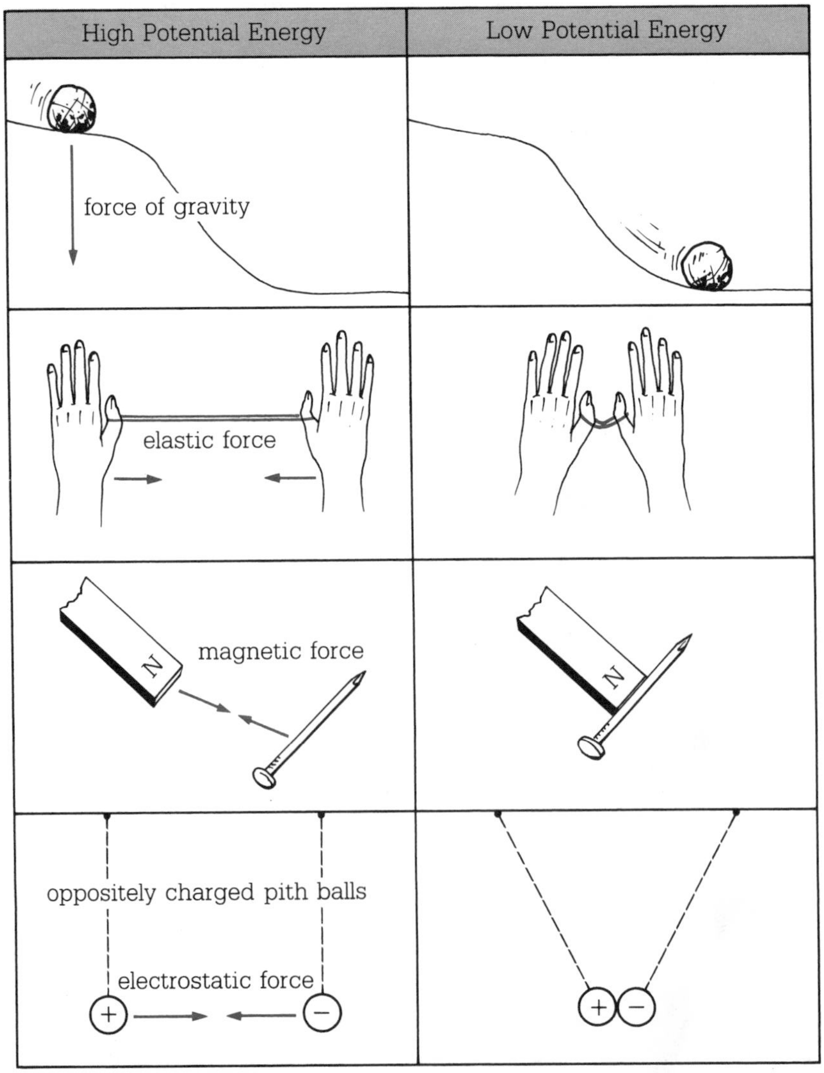

Figure 5.3
Examples of potential energy being lowered. The energy possessed by two objects due to their relative positions and the attraction (or repulsion) between them is their potential energy. Potential energy is lowered when attracting objects approach each other or when repelling objects move apart.

High Potential Energy	Low Potential Energy
force of gravity	
elastic force	
magnetic force	
oppositely charged pith balls — electrostatic force	

CHEMICAL REACTIONS AND ENERGY

When potential energy is lowered, the excess potential energy is often converted into kinetic energy, the energy of motion. For example, a rock falling toward the ground loses potential energy which is converted into kinetic energy and the rock increases in speed. When atoms combine in a chemical reaction, their potential energy is lowered and the excess potential energy is converted into kinetic energy of the product molecules. This molecular kinetic energy often appears as heat.

One of the most common and widely used chemical reactions is the combustion of fuel — the reaction of fuel with oxygen. A major use of this reaction is to produce heat. The starting materials, oxygen and the fuel such as coal (mainly carbon) or natural gas (mainly methane, CH_4), have a very high chemical potential energy. When the fuel combines with oxygen,

the potential energy of the system is lowered. The excess chemical energy is converted into kinetic energy in the product molecules and appears as heat.

Chemical potential energy being transformed into molecular kinetic energy (heat)

water vapour and carbon dioxide (low chemical potential energy)

candle and oxygen (high chemical potential energy)

5.2 Bonds Involving Metallic Elements: Ionic Bonding

How can we use the bonding model to explain the formation of compounds that conduct electricity when they are molten? The atoms of metallic elements have relatively low first ionization energies, and those of non-metallic elements have relatively large electron affinities. Therefore, we can predict that chemical reactions between metallic and non-metallic elements would involve the transfer of electron(s) from metallic to non-metallic atoms, as shown in Figure 5.4 on page 138. This electron transfer results in positively charged metal ions and negatively charged non-metal ions. The electrostatic attraction between these oppositely charged ions is called an **ionic bond**. Using the octet rule we can also predict that enough electrons are lost or gained to produce ions having noble gas electron structures.

As discussed in Chapter 4, the ionization energy of inner electrons is very high. Therefore, only the outer electrons from a metal atom are transferred in forming an ionic bond. Similarly, since the electron affinity of a noble gas electron structure is very low, a non-metal atom will only accept electrons in its outer energy level when bonding takes place. The outer electrons of an atom are called

Figure 5.4
Formation of an ionic bond

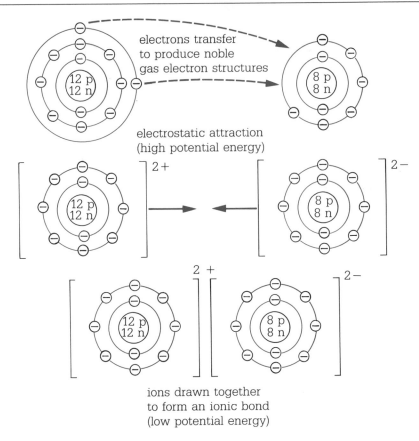

electrons transfer
to produce noble
gas electron structures

electrostatic attraction
(high potential energy)

ions drawn together
to form an ionic bond
(low potential energy)

valence electrons. These are the dots shown in the Lewis symbols for elements. The major bonding characteristics of an element are determined by the number of valence electrons in their atoms.

Electron Transfer Limits

Energy is required to remove electrons from an atom (the ionization energy). As determined earlier, the ionization energy increases as each successive electron is removed. The potential energy available during bond formation is limited. Thus, when chemical bonds form, there is a limit to the number of electrons that can be removed from the outer energy level of an atom. Similarly, there is a limit on the number of electrons an atom can accept, since the placement of extra electrons into the outer energy level of an atom becomes progressively more difficult as the negative charge on the atom increases. Evidence has shown that, in general, atoms can lose or gain no more than three electrons. Using the energy considerations discussed above, we can simplify the bonding model:

1. Those atoms that have one, two, or three valence electrons (less than half of an outer noble gas electron structure (stable octet)) will react chemically by losing them. This results in a noble gas electron structure. The atoms that lose these electrons become positively charged ions.

Sample Problem 5.1

Use Lewis structures and the simplified bonding model to illustrate how the following sets of atoms bond, and write the formula for each of the compounds that are formed.

(a) sodium and chlorine
(b) magnesium and oxygen
(c) lithium and nitrogen

Solution

(a)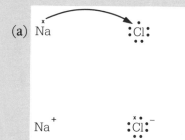

The sodium atom has one valence electron. It loses this electron to become a positive ion and attain the electron structure of the noble gas neon. Chlorine, with seven valence electrons, gains one more electron. It becomes a negative ion and attains the electron structure of the noble gas argon. Note: Electrons are all identical. The use here of different electron symbols is just a method of keeping track of them for the atoms of different elements.

Therefore, when sodium and chlorine react, one sodium atom bonds with one chlorine atom. The formula for the resulting compound is Na_1Cl_1 or $NaCl$.

(b)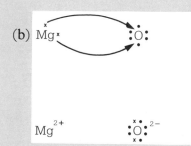

The magnesium atom has two valence electrons and loses them to attain a noble gas electron structure. It becomes an ion with a 2+ charge. The oxygen atom, with six valence electrons, gains two electrons to attain a noble gas electron structure. It becomes an ion with a 2− charge.

Therefore, one magnesium atom bonds with one oxygen atom. The formula for the resulting compound is MgO.

(c)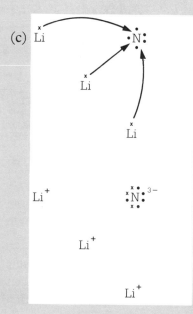

The lithium atom has one valence electron and loses it to the nitrogen atom to attain a noble gas electron structure, becoming a 1+ ion. The nitrogen atom, with five valence electrons, needs three more electrons to attain a noble gas electron structure. It gains one electron from each of three lithium atoms, becoming a 3− ion.

Therefore, three lithium atoms bond with one nitrogen atom. The formula of the resulting compound is Li_3N.

RADIOACTIVE FALLOUT

In the spring of 1986 experiments that were being conducted in a nuclear power reactor in Chernobyl, USSR, went awry. A chemical explosion and subsequent fire exposed the reactor's hot core and released a considerable amount of radioactive material to the atmosphere. It was reported that the accident had caused the death of over 30 people at and near the site of the reactor. Millions of others were exposed to varying levels of radiation and it is estimated that more than 6000 of these will develop radiation-induced cancer in the years to come. What are the dangerous elements in radioactive fallout? How are they ingested, how do they cause harm, and what protective measures exist?

The radioactive materials released in a nuclear accident vary depending on the temperature in the reactor core and on the boiling points of the substances involved. Some fission products are noble gas isotopes, the most important being xenon-133. (Xenon-133 was the dominant source of radioactivity released in the nuclear reactor accident at Three Mile Island in the United States in 1979.) The noble gas isotopes disperse very quickly. But since they are chemically quite inert, they do not deposit on the ground or enter food chains. The danger they present diminishes rapidly as they are diluted in the atmosphere.

The most dangerous radioactive isotopes are iodine-131, cesium-137, and to a lesser extent, strontium-90. Iodine and cesium have relatively low boiling points and readily vaporize out of the reactor's core into the atmosphere. Iodine-131, with a half-life of 8 d, represents the more immediate danger. If a person inhales iodine-131 or ingests contaminated food or water, this isotope can become concentrated in the thyroid gland. The radioactivity can destroy the

Destroyed nuclear power reactor at Chernobyl, USSR, April 1986

2. Elements whose atoms have five, six, or seven valence electrons (more than half of a full outer noble gas electron structure (stable octet)) have the ability to gain enough electrons to form a noble gas electron structure and become negatively charged ions.

Metals characteristically have three or fewer valence electrons and form ionic bonds with non-metals that have five, six, or seven valence electrons. As you will see later, the hydrogen atom, with a single valence electron, is a notable exception to the above "rules".

thyroid gland function or induce cancer. One protective measure is to take tablets of non-radioactive iodine-127 (in the form of potassium iodide). This swamps the thyroid gland with non-radioactive iodine, diluting the amount of radioactive iodine the gland can absorb. The danger from iodine-131, with half of it decaying every 8 d, fades within a month or so. During this time, foodstuffs that absorb iodine, such as milk and fresh vegetables, are monitored and discarded if their radiation level is too high.

Cesium-137 and strontium-90 have half-lives of about 30 a and can contaminate soils and foodstuffs for a long time. Cesium has the same outer electron structure as sodium and potassium. Cesium is taken up by the body as if it were sodium or potassium. Cesium-137 causes radiation damage throughout the body, increasing the chance of cancer and genetic defects. Strontium has the same outer electron structure as calcium and is incorporated into the body's bone structure along with calcium. The presence of strontium-90 in the bone can affect the production of red blood cells by the bone marrow and cause blood cancer (leukemia).

Radioactive fallout—protective measures after Chernobyl

◄(a) Disposing of vegetables for fear of radioactive contamination in Italy

(b) Flying uncontaminated milk► into Poland

(c) In the USSR, a young girl►► stands beside a sign listing precautions against fallout from the Chernobyl power station, including a warning to limit children's playtime outside and to beware of dust on tree leaves.

Exercise

2. Use Lewis symbols and the octet rule to illustrate how each of the following sets of atoms bond. Write the formulas for the compounds formed.
   (a) potassium and fluorine
   (b) lithium and sulphur
   (c) calcium and sulphur
   (d) aluminum and bromine
   (e) sodium and hydrogen
   (f) potassium and phosphorus

5.3 Bonds Involving Non-metallic Elements: Covalent Bonding

Most of the compounds consisting of only non-metallic elements do not conduct electricity even when melted. We might conclude therefore that the particles making up these compounds are not ions but are neutral combinations of atoms. Electrically neutral combinations of atoms in fixed ratios are called **molecules**.

Pure water does not conduct electricity very well. Analysis shows that water consists of two hydrogen atoms for each oxygen atom. Thus, water consists of electrically neutral molecules, H_2O. When hydrogen and oxygen react chemically, large amounts of heat are released. This indicates that the potential energy of the hydrogen and oxygen atoms is lowered and a bond is formed. We can use our Lewis symbol–octet rule model to explain this type of bonding, but a new hypothesis is needed.

Electron Sharing

Perhaps atoms can attain the electron arrangement of a noble gas (stable octet) in a more co-operative manner by sharing electrons. Consider the case of two fluorine atoms combining:

$$:\!\overset{\displaystyle ..}{\underset{\displaystyle .}{F}}\!\cdot \qquad \overset{\displaystyle x\,x}{\underset{\displaystyle x\,x}{{}^{x}F^{x}}}$$

Since both fluorine atoms have the same ionization energies and electron affinities, neither is capable of removing an electron from the other. A compromise occurs when the two fluorine atoms share a pair of electrons, each atom providing one electron to make up the pair.

$$:\!\overset{\displaystyle ..\,x\,x}{\underset{\displaystyle ..\,x\,x}{F\,{}^{x}_{x}F}}\!{}^{x}$$

Let us examine this electron sharing hypothesis in the light of the electrostatic laws discussed earlier in Chapter 4.

What forces arise as two atoms approach each other? Figure 5.5 illustrates the forces on two hydrogen atoms as they approach each other to form a bond. Forces of attraction exist between the nucleus of each hydrogen atom and the electrons in both atoms. Forces of repulsion exist between the two nuclei and between the two electrons as well. When there is a relatively large separation between the atoms, the average distance between the nuclei and the electrons is shorter than the distance between the two repelling nuclei and the distance between the two repelling electrons. Therefore, the forces of attraction are greater than the forces of repulsion. (Recall that electrostatic forces are greater the shorter the distance between the charged objects – the inverse square law.) As the two atoms approach each other, a net electrostatic attractive force pulls them

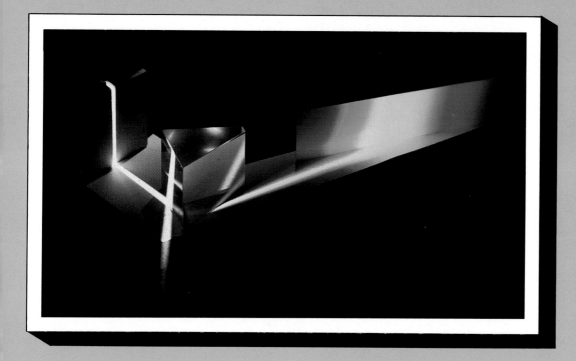

(Top) A narrow beam of white light is split by a prism into a continuous rainbow of colours corresponding to the frequencies of light in the visible part of the electromagnetic spectrum. (Bottom) The continuous spectrum produced by an incandescent lamp, as well as the line spectra produced by sodium and hydrogen.

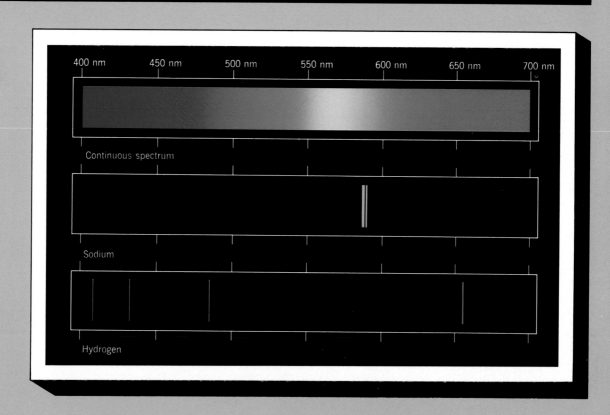

◁ What are the similarities among some of these elements? What differences can you see?

A representative selection of elements sorted by state. Eleven of the elements are gases, and two are liquids at room temperature and pressure. The rest of the elements are solids.

COLOUR PLATE 2

Trends within a non-metal group, the halogens

(a) Hot copper glows bright red and melts as it reacts with chlorine.

(b) In bromine, hot copper ▷ produces thick smoke.

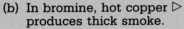

(c) Hot copper is heavily tarnished in iodine.

(d) At room temperature and pressure, chlorine is a gas, bromine is a liquid, and iodine is a solid.

COLOUR PLATE 3

Compounds come in all forms and have a wide variety of properties.
COLOUR PLATE 4

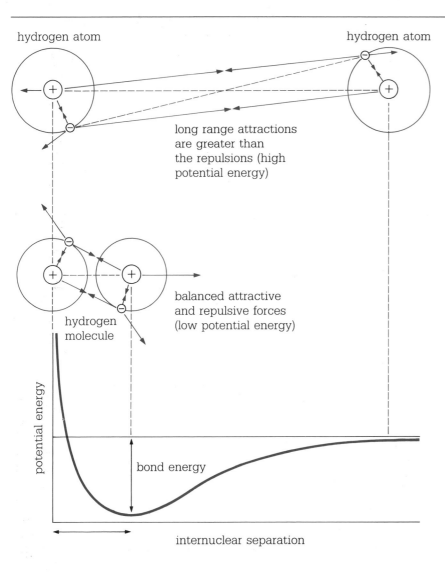

hydrogen atom   hydrogen atom

long range attractions
are greater than
the repulsions (high
potential energy)

balanced attractive
and repulsive forces
(low potential energy)

hydrogen
molecule

potential energy

bond energy

internuclear separation

Figure 5.5
Formation of H-H covalent
bond. As two hydrogen atoms
approach each other, electrostatic forces of attraction and
repulsion arise between their
electrons and nuclei. A net
attraction results in potential
energy being lowered and in the
formation of the bond.

closer together, lowering their potential energy. The forces of repulsion between the nuclei and between the electrons increase as the
atoms come closer together. At some point they become large enough
to balance the attractive forces. The inter-atomic distance at which
the forces of attraction are balanced by the forces of repulsion is
called the **bond length**.

Covalent Bonds

When atoms combine by sharing electrons, the bond formed is
called a **covalent bond**. A covalent bond results from the simul-
taneous attraction of two positive nuclei to the same shared pair of
electrons. Just as in the formation of an ionic bond, the potential
energy of the bonded atoms is usually less than the potential energy
of the individual atoms. Experimentally, when hydrogen atoms are
allowed to combine with each other to form hydrogen molecules,
heat is released. The decrease in potential energy appears as heat.

Sample Problem 5.2

Use Lewis symbols and the octet rule to illustrate how the following sets of atoms would bond.
(a) hydrogen and fluorine
(b) oxygen and chlorine
(c) nitrogen and iodine

Solution

(a) Both hydrogen and fluorine have half or more than half of a stable octet of electrons. They will share electrons and form covalent bonds. In order to attain a noble gas electron structure, both hydrogen and fluorine need an extra electron. This is achieved by a hydrogen atom sharing electrons with a fluorine atom, forming a HF molecule as shown.

(b) Both chlorine and oxygen have more than half of a stable octet of electrons. They will form covalent bonds. A chlorine atom needs an extra electron to attain a noble gas electron structure. The oxygen atom, on the other hand, needs two electrons to complete its stable octet. This is achieved by an oxygen atom sharing electrons with two chlorine atoms, forming a Cl_2O molecule as shown.

(c) Both the iodine and nitrogen atoms have more than half of a stable octet of electrons. They will form covalent bonds. An iodine atom needs an extra electron to complete its stable octet, while a nitrogen atom needs three electrons. This is achieved by a nitrogen atom sharing electrons with three iodine atoms, forming a NI_3 molecule as shown.

Note that in each molecule shown above, all of the atoms have achieved a noble gas electron structure. The shared electrons are counted as though they belonged to each of the bonding atoms.

A covalent bond is usually formed if both bonding atoms have half or more than half of a full noble gas outer electron structure (stable octet). Neither atom is likely to lose its large number of valence electrons; thus, ions tend not to form. Non-metals, except for hydrogen, have four or more valence electrons – they tend to form covalent bonds when they react with each other.

Exercise

3. Use Lewis symbols and the octet rule to illustrate how each pair of elements would bond.
   (a) hydrogen and iodine   (d) chlorine with itself
   (b) carbon and fluorine   (e) oxygen with itself
   (c) phosphorus and bromine (f) hydrogen and rubidium

Multiple Bonds

Sometimes the only way that bonding atoms can attain a stable octet of electrons is by sharing two or more pairs of electrons. A **double bond** occurs when two atoms share two pairs of electrons between them. For example, two oxygen atoms can each attain a stable octet of electrons if they form a double bond by sharing two pairs of electrons between them.

two pairs of shared electrons
(a double bond)

Similarly, carbon and sulphur can both attain a stable octet if each of two sulphur atoms forms a double bond with a carbon atom.

one   ×C×   and two   :S·   form   :S::C::S:

The maximum number of covalent bonds that is commonly found between two bonding atoms is three. This is referred to as a **triple bond**. The nitrogen molecule, for example, contains a triple bond.

:N·   and   ×N×   form   :N⫶N×

Note that both of the bonding nitrogen atoms end up with a stable octet of electrons.

Atoms of the element carbon have the added ability of forming covalent bonds with other carbon atoms to form molecules that contain long chains of carbon atoms. This ability, called **catenation**, makes carbon unique among the elements in the vast number and variety of compounds it can form. (The study of carbon compounds constitutes an entire branch of chemistry called organic chemistry.)

Exercise

4. Use Lewis symbols to illustrate the bonding for each of the following molecules.
   (a) CO_2 (b) HCN   (c) C_2H_2 (d) CH_2O

Valence

An element's bonding capacity – the number of electrons its atoms are capable of losing, gaining or sharing – is called its **valence**. It is the number of bonds (ionic or covalent) which an atom is able to form. If the atoms of an element react with others by either losing or gaining electrons (forming ions and ionic bonds), its bonding capacity is referred to as its **electrovalence**; if the atoms share electrons, the bonding capacity is described as its **covalence**. For example, the valence of magnesium is two, but since it always loses two electrons when it reacts, it is said to have an electrovalence of $2+$. Oxygen also has a valence of two, but can either gain two electrons, exhibiting an electrovalence of $2-$, or share two electrons and have a covalence of 2. The usual valences of the elements near the ends of the periodic table can be determined using Lewis symbols and the octet rule. (See Figure 5.6.) The valences of some of the elements near the middle of the table, however, are not easily predicted using this model. A more complex model, which is beyond

Figure 5.6
Most common bonding capacities (valences) of elements

H 1																	He
Li 1	Be 2											B 3	C 2,4	N 3,5	O 2	F 1	Ne
Na 1	Mg 2											Al 3	Si 4	P 3,5	S 2,4,6	Cl 1	Ar
K 1	Ca 2	Se 3	Ti 3,4	V 4,5	Cr 2,3	Mn 2,3	Fe 2,3	Co 2,3	Ni 2,3	Cu 1,2	Zn 2	Ga 3	Ge 4	As 3,5	Se 2,4,6	Br 1	Kr
Rb 1	Sr 2	Y 3	Zr 4	Nb 3,5	Mo 2,3	Tc 7	Ru 3,4	Rh 3,4	Pd 2,4	Ag 1	Cd 2	In 3	Sn 2,4	Sb 3,5	Te 2,4,6	I 1	Xe
Cs 1	Ba 2	Lu	Hf 4	Ta 5	W 6	Re 4,7	Os 3,4	Ir 3,4	Pt 2,4	Au 1,3	Hg 1,2	Tl 1,3	Pb 2,4	Bi 3,5	Po 2,4	At 1	Rn
Fr 1	Ra 2	Lr															

the scope of this course, is required to explain the two or three different valences that some elements exhibit.

The valences or bonding capacities of the elements are used in writing formulas and in naming compounds. You will find out about this in Chapter 6.

5.4 Predicting Bond Type

You have seen that the differences in electrical conductivity of compounds may be understood by using two different mechanisms to explain bonding: electron transfer and electron sharing. We can further simplify this model by establishing a principle that would accurately predict which bond type would form between two bonding atoms. The key is the number of valence electrons each of the bonding atoms possesses.

With very few exceptions, if one of the bonding atoms has less than half the maximum number of valence electrons (as in the case of metals), it will tend to lose them to an accepting atom, forming ions and an ionic bond. An atom with less than half of its maximum number of valence electrons does not usually attain a stable octet of electrons by sharing. Consider the bonding between calcium and oxygen as shown in Table 5.3. If the atoms bond by sharing electrons, the calcium atom does not attain noble gas electron structure. If, however, calcium loses its two valence electrons to oxygen, both atoms end up with a noble gas electron structure.

Laboratory Manual
Experiment 5-3 will give you an opportunity to examine and predict the shapes of molecules.

Table 5.3   Bonding between Calcium and Oxygen

Covalent Attempt	Ionic Attempt
Ca is still unstable. It has access to only four outer electrons. This type of bonding for CaO is inconsistent with our model.	Both atoms have stable octets. This must be how CaO forms.

Bonding atoms that have half or more than half of their maximum number of valence electrons (as in the case of non-metals) can attain stable octets by sharing electrons. These ideas are summarized in Table 5.4 on page 148.

Table 5.4 Predicting Bond Type

Number of Valence Electrons	Bond Type
One of the bonding atoms has less than half the maximum number of valence electrons (1, 2, or 3) except for hydrogen; i.e., metals are involved.	IONIC
Both of the bonding atoms have half or more of their maximum number of valence electrons; i.e., both are non-metals.	COVALENT

The periodic table (see Figure 5.7) can be used to determine the number of valence electrons for many elements. Using this information and applying the rules in Table 5.4, you can predict the type of bonding between two atoms with very good, but not total accuracy. For example, the model predicts that boron should form ionic bonds when it reacts, but experimental evidence indicates that boron forms covalent bonds! The stable octet model we are using cannot explain this anomaly. In fact, bonding predictions concerning the elements near the metal–non-metal line shown on the periodic table in Figure 5.7 are often inaccurate. Other more advanced models do adequately explain the bonding of those elements. You will learn about these in future chemistry courses.

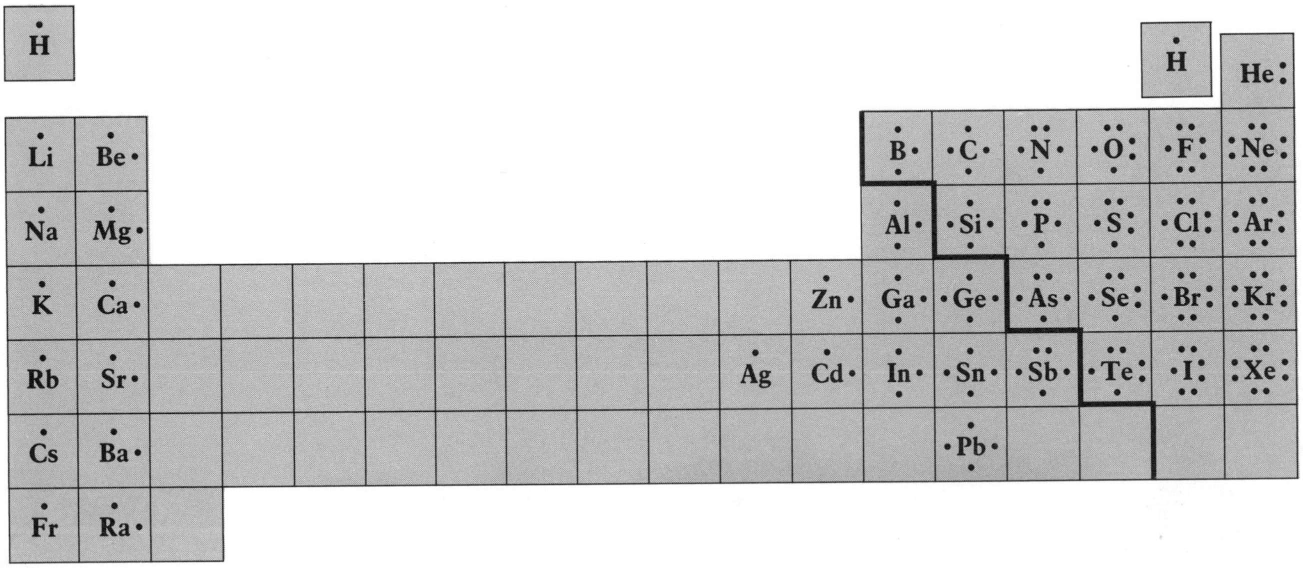

Figure 5.7

Periodic table and bonding. Metals have 3 or fewer electrons in their outer energy levels. They tend to lose electrons and form ionic compounds. Non-metals have 4 or more electrons in their outer energy levels and tend to share electrons when they bond with each other, forming covalent compounds.

Sample Problem 5.3

Predict the type of bonding that would occur between the elements in each of the following sets.
(a) barium and sulphur
(b) phosphorus and bromine

Solution

(a) Atoms of barium have two valence electrons, $\overset{x}{Ba}x$; those of sulphur, a non-metal, have six, $:\overset{\cdot}{\underset{\cdot}{S}}:$. Since atoms of one of the elements have less than half the maximum number of valence electrons, they will lose electrons in bonding. The bonding between barium and sulphur will be ionic.

(b) Atoms of phosphorus have five valence electrons, $\overset{x\ x}{x\underset{x}{P}x}$, and those of bromine have seven, $:\overset{\cdot}{\underset{\cdot\cdot}{Br}}:$. Since the atoms of both elements have more than half the maximum number of valence electrons, they will form bonds by sharing electrons. The bonding between phosphorus and bromine will be covalent.

Exercise

5. Use the information given in Figure 5.7 to predict the type of bond that would form between the elements in each of the following sets.
   (a) rubidium and oxygen
   (b) silver and sulphur
   (c) nitrogen and oxygen
   (d) carbon and chlorine
   (e) zinc and fluorine
   (f) calcium and phosphorus

5.5 In-between Bonds

Are there any other properties of compounds that might shed more light on the nature of the particles (that is, ions and molecules) that make them up? Consider solubility. Oil does not dissolve in water very well, yet it easily dissolves in other solvents like varsol. (Varsol is a mixture of carbon-hydrogen compounds.) What is it about the molecules of water and varsol that makes such a difference?

Unequal Sharing of Electrons

Some chemical bonds are not distinctly ionic or covalent. Instead, they have both ionic and covalent characteristics. How can we have a transfer and a sharing of electrons at the same time? As discussed

Figure 5.8
Unequal sharing of electrons

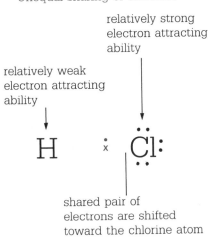

relatively strong
electron attracting
ability

relatively weak
electron attracting
ability

shared pair of
electrons are shifted
toward the chlorine atom

earlier, atoms have differing abilities to attract electrons. Therefore, within a covalent bond between two different atoms, one of the atoms can have a larger attraction for the shared pair of electrons than the other atom does. In such a bond the shared electron pair will be drawn closer to one of the bonding atoms than to the other. In effect, the electrons will be unequally shared. (See Figure 5.8.)

Electronegativity

The relative power of an atom to attract shared electrons in a covalent bond is called the atom's **electronegativity**. In the 1930s, the American chemist, Linus Pauling, developed a table of electronegativities for the elements. These electronegativities, listed in Figure 5.9, represent only a comparative scale and no units are associated with them.

Except for the noble gases, which have not been assigned any electronegativity value, note that the electronegativities tend to follow the same pattern as ionization energies. They tend to increase as you go to the right and up on the periodic table.

Figure 5.9
A table of electronegativities

H 2.1																	He
Li 1.0	Be 1.5											B 2.0	C 2.5	N 3.1	O 3.5	F 4.1	Ne
Na 1.0	Mg 1.3											Al 1.5	Si 1.8	P 2.1	S 2.4	Cl 2.9	Ar
K 0.9	Ca 1.1	Sc 1.2	Ti 1.3	V 1.5	Cr 1.6	Mn 1.6	Fe 1.7	Co 1.7	Ni 1.8	Cu 1.8	Zn 1.7	Ga 1.8	Ge 2.0	As 2.2	Se 2.5	Br 2.8	Kr
Rb 0.9	Sr 1.0	Y 1.1	Zr 1.2	Nb 1.3	Mo 1.3	Tc 1.4	Ru 1.4	Rh 1.5	Pd 1.4	Ag 1.4	Cd 1.5	In 1.5	Sn 1.7	Sb 1.8	Te 2.0	I 2.2	Xe
Cs 0.9	Ba 0.9	Lu 1.2	Hf 1.2	Ta 1.4	W 1.4	Re 1.5	Os 1.5	Ir 1.6	Pt 1.5	Au 1.4	Hg 1.5	Tl 1.5	Pb 1.6	Bi 1.7	Po 1.8	At 2.0	Rn
Fr 0.9	Ra 0.9	Lr															

Lanthanides: 1.0 – 1.2
Actinides: 1.0 – 1.2

Polar Covalent Bonds

How does the differing ability to attract electrons within a molecule affect the bond? In the hydrogen fluoride molecule (see Figure 5.10), the shared electrons are attracted much more by the fluorine atom (electronegativity of 4.1) than by the hydrogen atom (electronegativity of 2.1). The electrons are therefore pulled much closer

Figure 5.10
Polar hydrogen fluoride bond. The electrons shift away from the hydrogen atom (low electronegativity) toward the fluorine atom (high electronegativity). This results in hydrogen fluoride having small opposite charges at opposite ends of its molecules. (The symbol δ^+ or δ^- is used to indicate these small charges.)

(slightly positive end) δ^+ (H $\overset{\cdot}{\underset{\cdot}{\times}}$ F̈) δ^- (slightly negative end)

SPECIAL TOPIC 5.5

LINUS PAULING

The two-time Nobel prize winning scientist, Linus Pauling, is responsible for much of our current knowledge about molecular structure. In the 1930s, this American chemist developed the idea of the electronegativity scale. On this scale, fluorine, the most electronegative element, has a value of 4.1. Francium, the least electronegative element, has a value of 0.9.

Having provided valuable information about the structure of inorganic molecules, Pauling began to apply his ideas to organic molecules — specifically in the biochemistry of living matter. He was able to propose models for understanding the structure of protein molecules. One of these models was responsible for determining the cause of the genetic disease sickle cell anemia. For these, and his other contributions to the understanding of chemical bonds and molecular structure, Pauling received the Nobel Prize in chemistry in 1954.

Pauling has always held the view that scientists have an important social responsibility. He was among the first scientists to oppose the testing of nuclear weapons. His considerable efforts in this regard won him the Nobel Peace Prize in 1962.

In 1970, Pauling came into the spotlight once again for his controversial views on the benefits of vitamin C. His claims include the suggestion that massive doses of vitamin C can not only help prevent the common cold, but also help prevent many forms of cancer. Most of Pauling's work since 1970 has focussed on vitamins and their relation to human health, and he has written both scholarly reports and popular books on the subject.

Linus Pauling

SPECIAL TOPIC 5.6

THE BLOB REVISITED

As you have seen, polar and non-polar substances tend not to dissolve in each other. This property contributed to the formation of the St. Clair River "blob" discussed in Chapter 1. A spill of thousands of litres of the dense (1.63 g/mL), non-polar dry cleaning fluid, perchloroethylene, made its way into the river in Canada's "chemical valley" near Sarnia, Ontario. Being insoluble in polar river water, the fluid collected in blob-like pools on the river bed. Other non-polar pollutants such as toxic dioxins dissolved in the perchloroethylene, forming the black blob that lurked on the St. Clair River bottom. Powerful liquid vacuum cleaners were used to remove these pools of contaminated perchloroethylene from the river bottom during the summer and fall of 1986.

to the fluorine atom than to the hydrogen atom. This electron shift results in the fluorine end of the molecule acquiring a small negative charge and the hydrogen end a small positive charge. An object having two oppositely charged electrical "poles" is called a **dipole**; a molecule like HF, having opposite partial charges at either end, is called a **polar molecule.** The bond is called a **polar bond.**

Polar Compounds

Laboratory Manual
In Experiment 5-4, you will relate solubility to bond type.

What experimental evidence is there that polar bonds actually exist? Many of the properties of compounds predicted to be polar can be explained only if these molecules have small positive and small negative charges on their ends. Figure 5.11 illustrates a simple experiment to show the difference between a polar liquid and a non-polar liquid. The molecules in both liquids are free to move and turn. When a positively charged rod is brought close to a stream of polar liquid, the molecules turn so that their negatively charged ends are closer to the positively charged rod. The attraction between the positive rod and the slightly negative ends of the molecules causes the stream to be deflected toward the rod. Similarly, if a negatively charged rod is used, the polar molecules turn so that their positive ends are closer to the rod and the stream is again deflected toward the rod. Non-polar molecules have no similar orientation and a stream of non-polar liquid is not significantly deflected toward a charged rod.

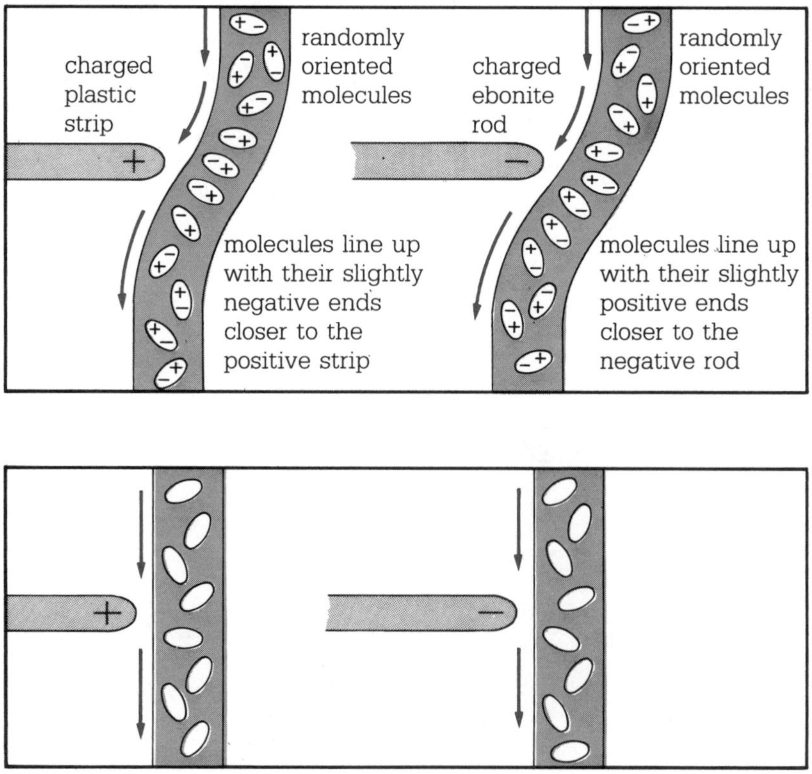

Figure 5.11
Polar liquids are attracted to both a positively and a negatively charged rod. Non-polar liquids are not significantly attracted to either.

(a) Falling stream of a polar liquid

(b) Falling stream of a non-polar liquid

A comparison of the boiling points of substances provides more evidence for the existence of polar compounds. Examine the data in Table 5.5. Can you propose an explanation for the higher (less negative) boiling point of NO compared to that of O_2 and N_2?

Table 5.5 Boiling Points of N_2, NO, and O_2

Compound	Molecular Mass (u)	Boiling Point (°C)
N_2	28	-196
NO	30	-152
O_2	32	-183

Both nitrogen and oxygen are non-polar substances. Since their molecules are made up of identical atoms their electron pairs must be equally shared. In nitrogen(II) oxide, NO, on the other hand, the electron pairs are shared unequally because nitrogen and oxygen have different electronegativities (3.1 and 3.5 respectively). Therefore, the NO molecules are polar. The attractions between oppositely charged ends of the molecules cause NO to have a higher boiling point than N_2 and O_2.

SPECIAL TOPIC 5.7

SOAP

The ability of soaps and detergents to remove dirt is due to the structure of their molecules. Each molecule has both a polar and a non-polar section. The non-polar parts of the soap or detergent molecules dissolve into the non-polar greasy or oily film which holds dirt to skin or fabric. The polar parts of the soap or detergent molecules remain in the water which is polar. With a little agitation and heat, the oily film breaks up into globules, each surrounded by soap or detergent molecules adhering like pins in a pin-cushion.

water soluble end of the soap molecule

grease soluble end of the soap molecule

soap solution

soap solution

A combination polar and non-polar molecule: soap. Soap molecules break up a greasy deposit into dispersed particles. The non-polar part of the soap molecule dissolves in the non-polar grease and the polar part dissolves in water which is polar.

non-polar grease

polar water

153

To summarize, we can now identify three types of bonds – ionic, covalent, and polar covalent bonds. This bond spectrum ranges from strictly ionic bonds, to polar convalent bonds, to strictly covalent bonds. (See Figure 5.12.) When both atoms forming a bond have identical electronegativities, the bond is a true covalent bond. As the electronegativity difference between the bonding atoms increases, the polarity of the bond also increases. Bonds are usually classified as ionic when the electronegativity difference between the bonding atoms exceeds about 1.7. There are a few exceptions to this. For example, hydrogen fluoride has a polar covalent bond but the difference in electronegativity of hydrogen and fluorine is 2.

Figure 5.12
Bond spectrum. The greater the electronegativity difference between atoms, the more polar (the higher % ionic character) the bond between them becomes.

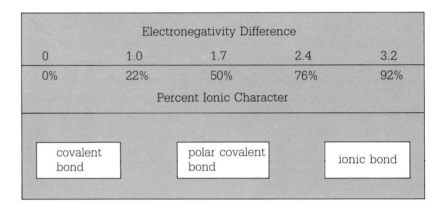

Sample Problem 5.4

Use the electronegativities provided in Figure 5.9 to rank the following bonds in order of increasing polar character: the O–H bond, the N–Br bond, and the Ca–Cl bond.

Solution

Bond	Electronegativity Difference	Bond Type
O–H	3.5 – 2.1 = 1.4	Highly polar
N–Br	3.1 – 2.8 = 0.3	Slightly polar
Ca–Cl	2.9 – 1.1 = 1.8	Very highly polar (ionic)

Therefore, listed in order of increasing polar character, the bonds are N–Br, O–H, and Ca–Cl.

5.6 Molecular Shape and Polar Molecules

Some compounds which have more than one polar bond in their molecules act like non-polar compounds. Carbon tetrachloride, CCl_4, is an example. This can be explained by the hypothesis that, depending on molecular shape, bond dipoles can cancel out each other.

How can we predict molecular shape? The electron pairs around a central bonding atom occupy definite regions in space. (These regions, called orbitals, are discussed in Appendix C.) Since electron pairs are all negatively charged, they repel each other and tend to move as far away from each other as possible. This electron pair repulsion theory can be used to predict molecular shape. For example, the four bonding pairs of electrons around the central carbon atom shown in Figure 5.13 can obtain maximum separation by taking a pyramid-like shape called a tetrahedral shape.

The difference in electronegativity between chlorine and carbon atoms is 0.4 (2.9 − 2.5). The Cl–C bond is a slightly polar covalent bond. The carbon tetrachloride molecule carries a slightly negative charge on each chlorine atom and a slightly positive charge on the central carbon atom. However, the effect of these four dipoles cancels out because of the symmetrical shape of the molecule. In other words, the net dipole is zero and the molecule is non-polar.

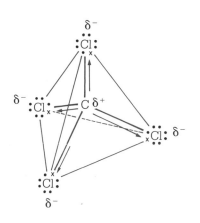

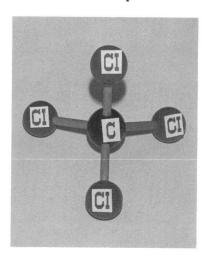

Figure 5.13
Shape of the carbon tetra-chloride molecule. The carbon tetrachloride molecule has a tetrahedral shape due to the repulsion among the four pairs of electrons surrounding the carbon atom.

Are the predictions made in Sample Problem 5.5 on page 156 accurate? Much experimental evidence exists that does indeed support these conclusions. For example, the melting point of water is a relatively high 0°C compared to −79°C, the temperature at which solid carbon dioxide changes from a solid to a gas. The polar water molecules attract each other more strongly than the non-polar carbon dioxide molecules attract each other. Thus, a higher temperature is needed to separate water molecules from each other than is needed to separate carbon dioxide molecules.

Sample Problem 5.5

Predict whether the following molecules are polar or non-polar.
(a) water (H_2O)   (b) carbon dioxide (CO_2)

Solution

(a) The Lewis structure for water is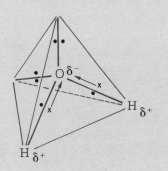

There are four pairs of repelling electrons around the central oxygen atom. They will have maximum separation if they take the tetrahedral form as illustrated in Figure 5.14. Since there are only two bonds, the molecule is actually "V" shaped. The electronegativity difference between oxygen and hydrogen is 1.4 (3.5 − 2.1) and the O−H bond has considerable polar character. The "V" shape of the molecule means that one end of the molecule (the oxygen atom) is slightly negative and the other end (the two hydrogen atoms) is slightly positive. The water molecule is therefore polar.

Figure 5.14
Polar water molecule. The repulsion among the four pairs of electrons around the oxygen atom leads to a "V" shaped molecule. Since the oxygen end of the molecule has a slight negative charge and the hydrogen atoms have a slight positive charge, the water molecule is polar.

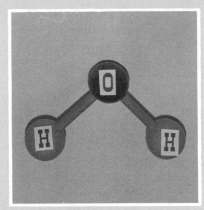

Exercise

6. Determine the shape of the following molecules and the electronegativity differences between their bonding atoms. Use this information to predict whether the molecules will be polar or non-polar.
   (a) ammonia (NH_3)
   (b) chloroform ($CHCl_3$)
   (c) phosphine (PH_3)

(b) The Lewis structure for carbon dioxide is

$$:\!\overset{..}{O}::C::\overset{..}{O}:$$

Carbon forms a double bond with each of the oxygen atoms. There are two regions of negative charge around the central carbon atom. (In a double bond both pairs of electrons remain somewhere between the bonding atoms.) They acquire maximum separation by pointing in opposite directions along a straight line. The carbon dioxide molecule thus assumes a linear shape with all three atoms on a straight line as illustrated in Figure 5.15.

The electronegativity difference between the oxygen and carbon atoms is 1.0 (3.5 − 2.5) and the C=O double bond has considerable polar character. However, since the dipoles point in exactly the opposite direction to each other, they cancel. We would therefore predict that the carbon dioxide molecule is non-polar.

Figure 5.15
Carbon dioxide molecule. The two mutually repelling double bonds around the carbon atom assume a linear arrangement. The oxygen atoms at either end of the molecule are slightly negative and the central carbon atom is slightly positive. Since the molecule does not have opposite charges at opposite ends, it behaves as a non-polar molecule.

Looking Back – Looking Ahead

The Lewis symbol–stable octet model for bonding developed in this chapter can be used to account for the type of bonding between different atoms. It can also be used to predict some of the properties of compounds, such as electrical conductivity and molecular polarity. You will use this model in the next two chapters to write molecular formulas and name compounds.

In a more advanced chemistry course the bonding model developed in this chapter can be modified and extended to explain and predict the physical behaviour of large aggregates of ions, atoms, and molecules.

Vocabulary Checklist

You should now be able to define or explain the meaning of the following terms.

bond length	dipole	molecule	valence
catenation	double bond	polar bond	valence electron
chemical bond	electronegativity	polar molecule	
covalence	electrovalence	octet rule	
covalent bond	ionic bond	triple bond	

Chapter Objectives

You should now be able to do the following.

1. Appreciate the activities and processes that scientists use to establish models and theories.

2. Understand that the formation of bonds between atoms results in net lowering of potential energy of the bonding atoms.

3. Use Lewis symbols and the octet rule to illustrate ionic bonding.

4. Use Lewis symbols and the octet rule to illustrate covalent bonding.

5. Identify the differences between simple covalent and simple ionic compounds.

6. Predict the bond type (ionic or covalent) most likely to form between two given elements.

7. Use the concept of electronegativity to identify polar bonds. Identify strictly covalent and strictly ionic bonds as the extremes of a bond spectrum.

8. Use simple electron pair repulsion theory to predict the shape of molecules.

9. Combine molecular shape with bond polarity to predict the polarity of molecules.

Understanding Chapter 5

REVIEW

1. In terms of potential energy, explain why bonding between atoms takes place.

2. In terms of the octet rule bonding model and valence electrons, explain what happens to atoms when they form bonds.

3. List three ways an atom can achieve a noble gas electron structure.

4. Define or explain each of the following.
   (a) an ionic bond   (c) a double bond
   (b) a covalent bond   (d) a triple bond

5. Explain the meaning of the following terms.
   (a) valence electrons
   (b) the octet rule

6. What is meant by each of the following?
   (a) valence
   (b) electrovalence
   (c) covalence

7. Define or explain the following terms.
   (a) electronegativity   (c) a polar bond
   (b) electric dipole   (d) a polar molecule

APPLICATIONS AND PROBLEMS

8. Identify at least three situations where a system or object tends toward a minimum potential energy.

9. (a) Use a periodic table and sketch Lewis symbols for each of the following atoms: K, S, H, Al, N, Mg, Cl, Si.

(b) Explain how each atom in (a) would probably obtain a noble gas electron structure when it bonds with other atoms. Explain all possibilities if there are more than one.

10. Make a table in your notebook with the following headings. Complete the table with the aid of a periodic table for these elements: Ba, Si, N, F, C, S, Al, Li, Mg, O, H, P, Br, and Ne. Note that not all columns will have an entry.

ELEMENT	NUMBER OF VALENCE ELECTRONS	VALENCE	COVALENCE	ELECTRO-VALENCE

11. Make a table in your notebook with the following headings. Complete the table with the aid of a periodic table for the given pairs of bonding atoms. Consider the number of valence electrons in both bonding atoms in predicting the most probable type of bonding.

BONDING ATOMS	LEWIS SYMBOLS FOR BOTH ATOMS	BOND TYPE

(a) Ca and Cl
(b) S and Br
(c) Si and F
(d) Rb and H
(e) N and I
(f) P and H
(g) O and Ne
(h) Be and S

12. Use Lewis symbols and the octet rule to illustrate how the atoms in the following compounds are bonded together.
(a) LiF
(b) H_2S
(c) $SrBr_2$
(d) NCl_3
(e) AlI_3
(f) CH_4
(g) CS_2
(h) N_2
(i) Ca_3N_2
(j) HOCl
(k) KH

13. Make a table with the following headings. Complete the table with the aid of a table of electronegativities for the given pairs of bonding atoms. Predict whether the bonds would be classed as ionic, strictly covalent, or polar covalent.

BONDING ATOMS	ATOM ELECTRO-NEGATIVITIES	ELECTRO-NEGATIVITY DIFFERENCE	BOND TYPE

(a) H and P
(b) H and S
(c) Na and N
(d) C and S
(e) Cl and F
(f) C and Se

CHALLENGE

14. Explain which of the bonds formed in the following compounds has the greatest covalent character: Mg=O, Li—F, Cl—F, Cl—Na, Rb—S—Rb, S=Sr.

15. Explain which of the following compounds have non-polar covalent bonds: H—Cl, F—K, H—O—H, I—I, Br—F, O=C=O, O=O.

16. Explain which of the following compounds contain both ionic and covalent bonds: F—Sr—Br, Na—O—H, H—O—H, Cl—O—K, Li—S—K, S=C=S.

17. Identify four pairs of element groups on the periodic table that are most likely to form ionic bonds when their elements react.

18. Explain which of the following bonds is the most polar: H—F, O=C=O, F—Br, H—O—H, F—O—F, H—I.

19. Arrange the following compounds in order of decreasing polarity of their bonds: H—Br, H—O—H, H—F, O=C=O, H—I, H—Cl, I—S—I.

20. List the elements that are not assigned an electronegativity value and explain why these elements do not have an electronegativity value.

21. For each of the following, use the periodic table to identify the element:
(a) largest first ionization energy
(b) smallest atomic radius
(c) smallest electronegativity
(d) least chemical reactivity
(e) largest electronegativity
(f) largest atomic radius
(g) largest second ionization energy

22. (a) List three radioactive isotopes likely to be released in a serious nuclear power plant accident.
    (b) Explain why these isotopes are harmful to humans.
    (c) What precautions can be taken to minimize the effect of radioactive fallout from a damaged nuclear power reactor?

23. Explain the differences between polar compounds and non-polar compounds in terms of the following.
    (a) boiling point (if both compounds have a similar molecular mass)
    (b) ability to dissolve:
        (i) polar compounds
        (ii) non-polar compounds
        (iii) ionic compounds

24. Sketch the shape of each of the following molecules. Start with a Lewis structure and use the idea of electron pair repulsion.
    (a) CH_4 (d) H_2S
    (b) PF_3 (e) NBr_3
    (c) HCl   (f) HI

25. (a) Calculate the electronegativity difference for each of the bonds in the molecules listed in question 24.
    (b) Considering the molecular shape of the molecules listed in question 24, predict whether each of the molecules is polar or non-polar.

26. The melting and boiling points of two substances are shown below.

SUBSTANCE	MELTING POINT ($^\circ$C)	BOILING POINT ($^\circ$C)
Cl_2	-101	-35
BrF	-33	-20

The molecules Cl_2 and BrF have approximately the same mass, so the difference in melting and boiling points is not likely to be due to a mass difference. Give an explanation as to why the melting and boiling points of BrF are significantly higher than those of Cl_2.

27. Some ions, called polyatomic ions, consist of groups of atoms covalently bonded together, but missing some of their electrons or containing extra ones. Predict and sketch the shape of the following polyatomic ions.
    (a) NH_4^+
    (b) NO_2^-
    (c) H_3O^+

28. The molecules PCl_5 and SF_6 exist and have been experimentally confirmed. Even though our current bonding model does not adequately explain the bonding in these molecules, predict and sketch the molecular shape of PCl_5 and SF_6.

29. The elements in the hypothetical universe, Omega Orion, have many similarities to the elements found in our own universe. The periodic table of the elements in Omega Orion is given below. Explain which of the following compounds in Omega Orion would not have the following formulas: G_2, CB, IG, DM_2, CF_2, JM_3, GM, I_2L.

A					B
C	D	E	F	G	H
I	J	K	L	M	N

III

FORMULAS, MOLES, AND CHEMICAL EQUATIONS

6 FORMULAS AND MOLES

How do chemists know that the formula of water is H_2O, that the formula of sucrose is $C_{12}H_{22}O_{11}$, or that the formula of sodium chloride is NaCl? Formulas can be determined only by experiment. This involves determining how many atoms (or ions) of each element combine together. But how can this be done, since atoms are so small that we cannot see them or count them?

The answer to this question lies with one of the most important concepts in chemistry – the mole concept. In this chapter, you will learn how to use the mole to count atoms. You will also see how the number of moles of a substance is related to the mass of the substance. In addition, you will find out how chemists experimentally determine chemical formulas.

6.1 The Mole

When chemists work with chemicals in the laboratory, they do not experiment with individual atoms or molecules. Visible and measurable amounts of substances contain enormously large numbers of atoms, ions, or molecules. The relationship between the mass of a sample of an element or compound and the number of atoms or molecules in the sample is expressed in the *mole* (symbol mol). The mole was first defined historically as the mass of an element in grams that is numerically equal to the atomic mass of the element in atomic mass units. For example, the atomic mass of magnesium is 24.3 u; 1 mol of magnesium is 24.3 g. The atomic mass of gold is 197.0 u; 1 mol of gold is 197.0 g.

Later, it was determined experimentally that the number of atoms in 1 mol of an element is 6.02×10^{23}. (This will be discussed further in section 6.2.) It is this relationship that forms the basis for the SI definition of the mole. The **mole** is one of the seven base units of SI. It is presently defined as the amount of substance that contains the same number of elementary particles (that is, atoms, molecules, or ions) as the number of atoms in 1.200×10^{-2} kg (12.00 g) of carbon-12.

Molar Mass – Elements

Even though the definition of the mole has changed over the years, the relationship between atomic mass and the mole still exists. The mass of 6.02×10^{23} atoms of an element in grams is numerically equal to the atomic mass of the element. The same relationship is true for every element, whether we consider a single isotope of an element, or a mixture of isotopes. Four examples, demonstrating the relationship between the atomic mass and the mass of 1 mol of atoms, are shown in Table 6.1.

Table 6.1   Atomic Mass and Mass of 1 mol of Atoms for Some Elements

Atom	Atomic Mass	Mass of 1 mol of Atoms
^{23}Na	23.0 u	23.0 g
^{40}Ar	40.0 u	40.0 g
^{235}U	235.0 u	235.0 g
$\star Cl$	35.5 u	35.5 g

*Note that in the case of chlorine, the average atomic mass is given, not the mass of a particular isotope.

The mass of 1 mol of atoms of an element is known as the **molar mass** of the element.

Exercise

1. Use the periodic table to determine the molar mass of the following.
   - (a) sodium, Na
   - (b) aluminum, Al
   - (c) silicon, Si
   - (d) argon, Ar
   - (e) lead, Pb
   - (f) iodine, I

Molar Mass – Covalent Compounds

For compounds consisting of molecules, there is a relationship between their molar mass and their **molecular mass**. The mass in grams of 1 mol of a compound (6.02×10^{23} molecules) is the molar mass.

The molar mass is numerically equal to the molecular mass in atomic mass units.

For example, the molecular mass of water, H_2O, is simply the sum of the masses of the atoms in the molecule.

$$\text{Molecular mass of water} = (2 \times 1.0 \text{ u}) + 16.0 \text{ u}$$
$$= 18.0 \text{ u}$$

The molar mass of water is therefore equal to 18.0 g. Table 6.2 shows three other examples of this relationship.

Table 6.2 Molecular Mass and Molar Mass for Some Compounds

Compound	Molecular Mass	Molar Mass
CO	28.0 u	28.0 g
NH_3	17.0 u	17.0 g
$C_2H_2F_2$	64.0 u	64.0 g

Exercise

2. Calculate the molar mass of the following.
   (a) carbon dioxide, CO_2
   (b) sulphur trioxide, SO_3
   (c) benzene, C_6H_6
   (d) methanol, CH_3OH
   (e) sulphuric acid, H_2SO_4
   (f) glucose, $C_6H_{12}O_6$
   (g) sucrose, $C_{12}H_{22}O_{11}$
   (h) nitroglycerin, $C_3H_5N_3O_9$

Molar Mass – Ionic Compounds

For ionic compounds, it is incorrect to refer to their atomic mass or molecular mass. The more general expression **formula mass** is used. Formula mass of a substance is the mass of the formula unit of the substance, expressed in atomic mass units. The formula unit may consist of atoms, as in the case of magnesium, Mg. It may consist of molecules, as in the case of methane, CH_4. Or it may consist of ions, as in the case of sodium chloride, NaCl. (See Figure 6.1.)

The molar mass of a substance (in grams) is numerically equal to the formula mass (in atomic mass units). For instance, the molar mass of barium nitrate, $Ba(NO_3)_2$, is found as follows.

$$\text{Molar mass} = 137.3 \text{ g} + 2(14.0 \text{ g} + [3 \times 16.0 \text{ g}])$$
$$= 137.3 \text{ g} + 124.0 \text{ g}$$
$$= 261.3 \text{ g}$$

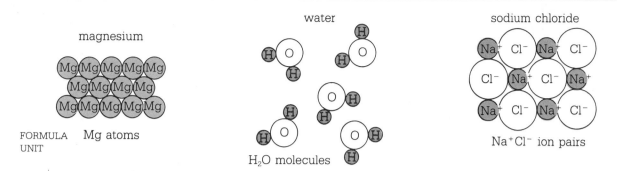

magnesium

water

sodium chloride

FORMULA
UNIT   Mg atoms

H_2O molecules

Na^+Cl^- ion pairs

Figure 6.1
The formula units of magnesium
are atoms, those of water are
molecules, and those of sodium
chloride are ions.

Whether a substance is made of atoms, molecules, or ions, 1 mol of the substance contains 6.02×10^{23} formula units of the substance. (See Figure 6.2.)

Figure 6.2
One mole of any substance
contains the same number of
formula units.

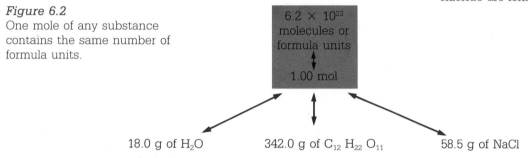

6.2×10^{23}
molecules or
formula units
↕
1.00 mol

18.0 g of H_2O 342.0 g of $C_{12}H_{22}O_{11}$ 58.5 g of NaCl

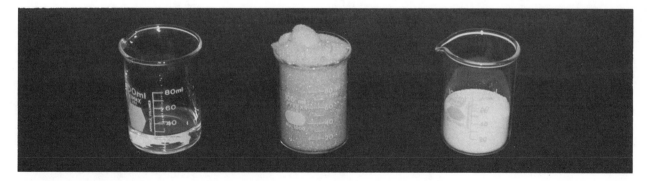

Exercise

3. Calculate the molar mass of the following.
   (a) barium oxide, BaO
   (b) calcium chloride, $CaCl_2$
   (c) potassium sulphide, K_2S
   (d) sodium sulphate, Na_2SO_4
   (e) potassium dichromate, $K_2Cr_2O_7$
   (f) magnesium hydroxide, $Mg(OH)_2$
   (g) calcium phosphate, $Ca_3(PO_4)_2$
   (h) ammonium carbonate, $(NH_4)_2CO_3$

6.2 The Mole and Number of Particles

In section 6.1 you learned how to determine the molar mass of an element or compound. According to the definition of the mole, this mass of substance contains 6.02×10^{23} particles of which the substance is composed. For example, 24.3 g of magnesium contains 6.02×10^{23} magnesium atoms; 18.0 g of water contains 6.02×10^{23} water molecules. The number 6.02×10^{23} is known as **Avogadro's constant**. This value was named in honour of Amedeo Avogadro, a 19th century Italian scientist. You will read about some of Avogadro's contributions to chemistry in Chapter 11.

The relationship between the number of particles and the mass of substance means that when you measure out a mass of substance equal to its molar mass, you are also indirectly counting out 6.02×10^{23} formula units of the substance. If the formula unit contains more than one atom, you can calculate the number of different atoms present in 1 mol of the substance. For example, consider 1 mol of carbon dioxide, CO_2. The molar mass of carbon dioxide is 44.0 g. This mass of carbon dioxide contains 6.02×10^{23} CO_2 molecules. This means that it contains 6.02×10^{23} carbon atoms, and $2(6.02 \times 10^{23})$ or 1.20×10^{24} oxygen atoms. Table 6.3 summarizes the relationship between the mole and number of atoms present for several substances.

Table 6.3   Moles and Number of Particles

Substance	Formula Unit	Number of Particles in 1 mol
helium	He	6.02×10^{23} helium atoms
oxygen gas	O_2	6.02×10^{23} oxygen molecules 1.20×10^{24} oxygen atoms
ammonia	NH_3	6.02×10^{23} ammonia molecules 6.02×10^{23} nitrogen atoms 1.81×10^{24} hydrogen atoms
sodium fluoride	NaF	6.02×10^{23} NaF formula units 6.02×10^{23} sodium ions 6.02×10^{23} fluoride ions
sodium oxide	Na_2O	6.02×10^{23} Na_2O formula units 1.20×10^{24} sodium ions 6.02×10^{23} oxide ions

It is not always necessary to consider 1 mol of a substance. Multiples or fractions of a mole may be used. We can talk about 2 mol of water molecules or 0.25 mol of magnesium atoms. Sample Problem 6.1 shows how to calculate the number of formula units in a sample of matter.

Sample Problem 6.1

A balloon is inflated with 0.50 mol of helium atoms. How many helium atoms are in the balloon?

Solution

1 mol of helium atoms $= 6.02 \times 10^{23}$ atoms

0.50 mol of helium atoms $= 0.50 \text{ mol} \times \dfrac{6.02 \times 10^{23} \text{ atoms}}{1 \text{ mol}}$

$\qquad\qquad\qquad\qquad = 3.0 \times 10^{23}$ atoms

Exercise

4. How many atoms are represented by the following?
   (a) 2.0 mol of neon
   (b) 0.42 mol of magnesium
   (c) 0.0025 mol of copper
   (d) 5.5 mol of gold

5. How many molecules are represented by the following?
   (a) 10.0 mol of water
   (b) 0.002 mol of carbon dioxide
   (c) 1.225 mol of ammonia
   (d) 0.85 mol of sulphur dioxide

6. Calculate the number of atoms ~~of each element~~ present in each of the following.
   (a) 3.24 mol of sulphuric acid, H_2SO_4
   (b) 0.062 mol of acetic acid, CH_3COOH
   (c) 100.0 mol of octane, C_8H_{18}

7. To get some idea of the size of Avogadro's constant, try the following calculations.
   (a) Assume that you can spend money at the rate of $1 million per day. How many years will it take you to spend 6.02×10^{23}? If this seems to be too long a time, do the same calculation but assume that your spending rate is now $1 million per second!
   (b) The population of Canada is approximately 25 million (2.5×10^7). If 6.02×10^{23} pennies (1 mol) were to be shared by all Canadians, how much money would each person receive?
   (c) Imagine a computer that is capable of counting at the rate of 1×10^9 numbers per second. How many years would it take the computer to count to 6.02×10^{23}?
   (d) The average mass of an apple is approximately 100 g. The mass of the earth is approximately 6×10^{24} kg. About how many apples are equivalent in mass to the earth?

THE FACTOR-LABEL METHOD

One of the areas of chemistry with which students often have some difficulty is problem solving. If the problem involves a conversion from one unit to another, the *factor-label method* will help you to carry out the conversion correctly. The key to this method is the selection of conversion factors that will change the units given in the question to the units required in the answer. To make this clear, consider an example unrelated to chemistry.

Imagine that your family will be travelling in the USA this summer. While planning the trip, your parents notice that the road maps they are using have distances marked in miles. They calculate the total distance to be covered as 3450 miles, and ask you to convert that distance to kilometres. This problem, which is simply a conversion from one unit (miles) to another (kilometres), is solved using the factor-label method as follows.

$$\text{kilometres} = \text{miles} \times \frac{\text{kilometres}}{\text{miles}}$$

DESIRED QUANTITY = GIVEN QUANTITY × CONVERSION FACTOR

At this point, no numbers have been included. The conversion factor, kilometres/miles, has been chosen so that the given units (miles) cancel, and the desired units (kilometres) are left. If we had written the conversion factor the other way around (miles/kilometres), we would have obtained the wrong units for the answer. If the units are wrong, then the answer must be wrong.

Now let us put some numbers into the expression. The number of miles is given. The conversion factor expresses a relationship between units. The relationship between miles and kilometres is that 1 mile = 1.61 km or 1 km = 0.621 miles. The conversion factor is

$$\frac{1.61 \text{ km}}{1 \text{ mile}} \quad \text{or} \quad \frac{1 \text{ km}}{0.621 \text{ mile}}$$

The numerator and denominator of each conversion factor represent exactly the same distance expressed in different units. Either conversion factor may be used. What is important is that the conversion factor must be kilometres/miles when you want to convert miles to kilometres.

The distance your family will be travelling this summer is therefore

$$3450 \text{ miles} \times \frac{1 \text{ km}}{0.621 \text{ miles}}$$
$$= 5556 \text{ km}$$

Practise the factor-label method on the following problems by first determining the conversion factor.

1. How many cm are there in 6.12 m?

2. How many s are in 37.5 min?

3. How many sheets of paper are there in 17 packages, if each package contains 500 sheets?

4. You intend to drive from Kingston, Ontario to Windsor, Ontario, a distance of 625 km, at an average speed of 87.5 km/h. How long will the journey take?

5. The density of gold is 18.88 g/cm³. What is the mass of a bar of gold measuring 10 cm × 5 cm × 3 cm?

When we talk about 1 mol of a substance, we are referring to both a mass of substance and to a number of formula units of that substance. For example, 1 mol of glucose, $C_6H_{12}O_6$, has a mass of 180.0 g and contains 6.02×10^{23} glucose molecules. In addition, the composition of 1 mol of glucose can be described in a number of ways. This is shown in Table 6.4 on page 170.

PUTTING CHEMISTRY TO WORK

RAM SWAROOP SADANA / ANALYTICAL CHEMIST

Today's chemical laboratories are very different from those used by scientists twenty or thirty years ago. New instruments as well as new processes have changed every aspect of analyzing – the search for "what" and "how much" has become very sophisticated. The laboratories of the next five or ten years are expected to be even more incredibly advanced, in part due to the work of analytical chemists like Ram Sadana.

"Laboratory work contains many steps, and many repetitions of steps, all of which must be extremely carefully done if the answers are to be trustworthy," notes Ram. "In the past, technicians had to spend time ensuring that even the monotonous portions of their work were done to high standards. This was a difficult task, and reduced the number of analyses that could be done each day."

One answer for Ram's lab has been to use microcomputers to 'smarten up' the instruments themselves. Older instruments would analyze a prepared sample but a technician would have to take the reading from the instrument, prepare graphs and do statistics, then find the answer by comparing the results to known standards. "Today's instruments," Ram explains, "begin by doing the same task of analyzing a prepared sample, but then their computerized portion does the graphs and calculations. The technician obtains a final answer at the push of a button."

"Producing high quality data is one phase of my work, but the other is being able to keep up with the vast numbers of analyses," Ram says. His particular lab specializes in the detection of even very low concentrations of toxic metals (such as mercury, arsenic, selenium, antimony, cadmium, and lead) in fish and other organisms. "To protect people who consume fish, we need to sample populations of fish in critical rivers and lakes frequently. This means large numbers of samples to be analyzed as quickly as possible.

To maintain the pace, Ram will be making increasing use of the latest in chemical technology – automated chemistry or robotics – to design the most efficient analytical procedures. Automated instruments prepare samples, take the necessary readings, give the answers in numerical or graphical form, and then dispose of the samples. Hundreds of accurate analyses can be done in hours rather than weeks. "To the public," Ram says with satisfaction, "this means that they will know exactly what they are eating – before they eat it."

Table 6.4 Composition of 1 mol of Glucose

Carbon	Hydrogen	Oxygen
6 mol of C atoms 3.61×10^{24} C atoms 72.0 g	12 mol of H atoms 7.22×10^{24} H atoms 12.0 g	6 mol of O atoms 3.61×10^{24} O atoms 96.0 g

6.3 Simple Calculations Using the Mole

Changing Grams to Moles

In the laboratory, chemists normally measure amounts of substances in mass units of grams or kilograms. Suppose a chemist wishes to know the number of moles of each reactant used in a chemical reaction. The measured mass of each reactant must be converted into moles. As you will see in this and later chapters, this type of information is important in determining the formula of a compound and the balanced equation for a chemical reaction. As shown in Sample Problem 6.2, we make use of molar mass in calculating the number of moles in a known mass of substance.

Sample Problem 6.2

Sodium hydroxide, NaOH, is used in many commercial drain cleaners. If a can of a certain drain cleaner contains 25.0 g of sodium hydroxide dissolved in water, how many moles of sodium hydroxide are in the solution?

Solution

To solve this problem, you have to convert an amount of substance expressed in grams to an amount expressed in moles. The original mass in grams is multiplied by a conversion factor to change the units to moles. The conversion factor in this case is 1 mol NaOH/ 40.0 g NaOH. The 40.0 g in the denominator of the conversion factor is the molar mass of NaOH; that is, the conversion factor is simply the reciprocal of the molar mass.

$$\text{Number of moles of NaOH} = 25.0 \text{ g NaOH} \times \frac{1 \text{ mol NaOH}}{40.0 \text{ g NaOH}}$$

$$= 0.625 \text{ mol NaOH}$$

Therefore, there are 0.625 mol of NaOH in the solution.

Exercise

8. Calculate the number of moles of the stated substance in each of the following samples.
   (a) 225 g of aspirin, $C_9H_8O_4$
   (b) 35.0 g of baking soda, $NaHCO_3$
   (c) 1.45 kg of gold
   (d) 0.84 g of hydrogen, H_2
   (e) 163 g of sodium fluoride, NaF
   (f) 46.7 g of methanol, CH_3OH

Changing Moles to Grams

Sometimes it is necessary to convert a known number of moles of a substance to the mass in grams. Balances cannot be calibrated in moles since different substances have different molar masses. The molar mass of a substance is used to calculate the mass of a given number of moles of the substance.

Sample Problem 6.3

A dilute solution of hydrogen peroxide, H_2O_2, in water is used as a bleach and as a disinfectant. For a particular experiment, 3.50 moles of hydrogen peroxide are required. What mass of hydrogen peroxide is this?

Solution

The molar mass of H_2O_2 is 34.0 g. Therefore the conversion factor is 34.0 g H_2O_2/1 mol H_2O_2.

$$\text{Mass of hydrogen peroxide} = 3.50 \text{ mol } H_2O_2 \times \frac{34.0 \text{ g } H_2O_2}{1 \text{ mol } H_2O_2}$$

$$= 119 \text{ g } H_2O_2$$

Therefore, the mass of the hydrogen peroxide is 119 g.

Exercise

9. Calculate the mass of each of the following.
   (a) 12.4 mol of helium gas, He(g)
   (b) 0.26 mol of butane, C_4H_{10}
   (c) 255 mol of ammonia, NH_3
   (d) 1.8 mol of magnesium hydroxide, $Mg(OH)_2$
   (e) 0.001 mol of calcium carbonate, $CaCO_3$
   (f) 1.3×10^{-3} mol of calcium phosphate, $Ca_3(PO_4)_2$
   (g) 6.14 mol of silver nitrate, $AgNO_3$
   (h) 2.5×10^{-4} mol of carbon monoxide, CO

SIGNIFICANT DIGITS

Associated with every measurement made in the laboratory is some degree of uncertainty. For instance, you might measure the length of the object shown in the diagram as 23.1 cm. The digits 2 and 1 are certain – there is no doubt that the length is "21 point something" cm. The 3 is uncertain – it might be a little less or a little more. The 2, 1, and 3 are all said to be *significant digits*.

Significant digits in a measurement or calculation consist of all those digits that are certain, plus one uncertain digit. Although your calculator may give you an answer to as many as eight decimal places or more, you cannot include all of these digits in your answer.

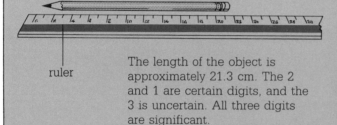

ruler

The length of the object is approximately 21.3 cm. The 2 and 1 are certain digits, and the 3 is uncertain. All three digits are significant.

RULES FOR DETERMINING THE NUMBER OF SIGNIFICANT DIGITS

1. All digits from 1 to 9 are considered to be significant.
2. Zeros between significant digits and following significant digits are significant.
3. Zeros to the left of significant digits, serving only to locate the decimal point, are not significant.
4. Any zero printed to the right of a non-zero digit is to be taken as significant unless there is information given to the contrary. For example, 3600 m has 4 significant digits, unless you are told the measurement was taken to the nearest 100 m.

Study the following examples to see if you understand these rules.

Measurement	Number of significant digits
343 kg	3
0.64 g	2
0.6004 mL	4
22.400 L	5
10.5°C	3

Now try some examples yourself. Determine the number of significant digits in each of the following measurements.

(a) 205.00 m   5   (e) 38.4°C   3
(b) 0.005 mol   1   (f) 137.3 g/mol   4
(c) 10.008 g   5   (g) 402.320 mL   6
(d) 7 602 050 km   7   (h) 20 s   2

Exact numbers, for example, the number of millilitres in a litre or numbers obtained by counting (3 test tubes, 5 beakers), are said to have an infinite number of significant digits.

The use of significant digits in calculations follows the rules described in the following paragraphs.

ADDITION AND SUBTRACTION

When adding or subtracting numbers, the answer should have the same number of decimal places as the quantity in the problem that has the *least* number of decimal places. Here is an example for addition.

Figure 6.3
Types of calculations that can be carried out using the mole. Grams can be converted to moles, and moles to grams. Number of particles (atoms, molecules) can be converted to moles, and moles to number of particles. To convert grams to number of particles, or vice versa, you must first convert to moles.

Mole Calculations – A Summary

Figure 6.3 summarizes the types of calculations that you have carried out so far using the mole.

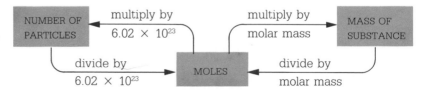

$$17.456$$
$$3.28$$
$$\underline{406.1234}$$
$$426.8594$$

The number 3.28 has the least number of decimal places – two. Therefore, the answer should be rounded to two decimal places—426.86.

MULTIPLICATION AND DIVISION

When multiplying or dividing numbers, the answer should have the same number of significant digits as the quantity in the problem that has the *least* number of significant digits. Here is an example.

Some of the digits on this calculator display may not be significant digits.

three significant digits two significant digits

$$\frac{22.4 \times 1.2}{4.654}$$

four significant digits

$= 5.7769181$ (calculator answer)

$= 5.8$ (to two significant digits)

In a multi-step calculation, you should keep one more significant digit for each step than the rules allow. Then at the final step of the solution, round off to the correct number of significant digits.

Try the following questions to see if you understand the use of significant digits in calculations.

(a) $423 \div 0.058$
(b) $23.456 + 12.11 + 0.33$
(c) 124.5×6.9
(d) 0.0001×2346
(e) $456.7 + 21 - 5.67$

SIGNIFICANT DIGITS AND SCIENTIFIC NOTATION

Recall that scientific notation is a method used to express very large or very small numbers. Here are two examples.

125 000 000 m is written as 1.25×10^8 m

0.00087 cm is written as 8.7×10^{-4} cm

When numbers are expressed in scientific notation, only one digit comes before the decimal. All digits in the first part of the number are considered to be significant. In the first example above, there are three significant digits; the second example has two significant digits.

6.4 Percentage Composition

The composition of a compound is often expressed in terms of its **percentage composition by mass**. This means the percentage of the total mass of the compound contributed by each element. The percentage composition of a compound can be determined using the mole concept. See the Sample Problems on page 174.

Sample Problem 6.4

Potassium chloride, KCl, is often used as a substitute for salt in sodium-restricted diets. What is the percentage by mass of potassium in the compound?

Solution

1 mol of KCl, 74.6 g, contains 1 mol of K, 39.1 g, and 1 mol of Cl, 35.5 g.

Therefore, the percentage by mass of potassium in the compound

$$= \frac{39.1 \text{ g K}}{74.6 \text{ g KCl}} \times 100\%$$

$$= 52.4\%$$

The percentage by mass of chlorine in the compound

$$= \frac{35.5 \text{ g Cl}}{74.6 \text{ g KCl}} \times 100\%$$

$$= 47.6\%$$

Since the compound contains only potassium and chlorine, the percentage by mass of chlorine can also be found by subtraction.

Percentage by mass of chlorine $= (100 - 52.4)\%$
$$= 47.6\%$$

Sample Problem 6.5

Calculate the percentage by mass of each element in sodium sulphate, Na_2SO_4.

Solution

Molar mass of sodium sulphate $= (2 \times 23.0 \text{ g}) + 32.1 \text{ g} + (4 \times 16.0 \text{ g})$
$$= 142.1 \text{ g}$$

1 mol of sodium sulphate, 142.1 g, contains
2 mol of Na atoms, 46.0 g
1 mol of S atoms, 32.1 g
4 mol of O atoms, 64.0 g

Percentage by mass of sodium in sodium sulphate

$$= \frac{46.0 \text{ g Na}}{142.1 \text{ g Na}_2\text{SO}_4} \times 100\%$$

$$= 32.4\%$$

Percentage by mass of sulphur

$$= \frac{32.1 \text{ g S}}{142.1 \text{ g Na}_2\text{SO}_4} \times 100\%$$

$$= 22.6\%$$

Percentage by mass of oxygen

$$= \frac{64.0 \text{ g O}}{142.1 \text{ g Na}_2\text{SO}_4} \times 100\%$$

$$= 45.0\%$$

Exercise

10. Calculate the percentage by mass of each element in the following compounds.
    (a) CH_4 (methane)
    (b) C_3H_8 (propane)
    (c) $NaHCO_3$ (baking soda)

11. Iron and oxygen combine to form two different compounds. The formulas of the compounds are FeO and Fe_2O_3. Calculate the percentage by mass of iron and oxygen in each compound.

12. Ammonium nitrate, NH_4NO_3, and ammonium sulphate, $(NH_4)_2SO_4$, are both used as fertilizers. Show by calculation which compound has the greater percentage by mass of nitrogen.

6.5 Empirical Formula

An **empirical formula** is a formula that gives the relative numbers of atoms of each element present in the formula unit of a compound. The numbers are expressed in their simplest whole-number ratio. For example, FeO, KCl, CO_2, and CH_3 are empirical formulas. C_2H_4, N_2O_4, and $C_6H_{12}O_6$ are not empirical formulas, since the numbers of atoms are not expressed in their simplest whole-number ratios. They are called **molecular formulas** since they represent the actual number of atoms in each molecule. These three formulas can be simplified to CH_2, NO_2, and CH_2O respectively. Empirical formulas are sometimes referred to as **simplest formulas**.

In the introduction to this chapter, you read that chemists use

Laboratory Manual
You will prepare a binary compound, find its percentage composition by mass, and determine its formula in Experiment 6-1.

the mole concept to determine the formula of a compound. How is this done? The first step is to determine the empirical formula of the compound. This is achieved by carrying out an experiment to find the mass of each element present in a sample of the compound. The number of moles of each element is then calculated. The ratio of the numbers of moles gives the empirical formula of the compound. Let us consider an example.

Sample Problem 6.6

Analysis of a compound of sulphur and oxygen, produced during the burning of sulphur-containing fuels, showed that it contained 50.0% sulphur and 50.0% oxygen by mass. What is the empirical formula of the compound?

Solution

If we assume that the mass of a sample of the compound is 100 g, then the mass of each element in the sample simply has the same numerical value as the percentage of the element. What we have to calculate is the number of moles of each element in the 100 g sample, then express them as a whole-number ratio.

$$\text{Number of moles of S atoms} = 50.0 \text{ g S} \times \frac{1 \text{ mol S}}{32.1 \text{ g S}}$$

$$= 1.56 \text{ mol S}$$

$$\text{Number of moles of O atoms} = 50.0 \text{ g O} \times \frac{1 \text{ mol O}}{16.0 \text{ g O}}$$

$$= 3.13 \text{ mol O}$$

Therefore, the ratio of mol S atoms to mol O atoms is 1.56:3.13.

The ratio of moles of S atoms to moles of O atoms is exactly the same as the ratio of S atoms to O atoms. Since only whole numbers of atoms are found in a compound, we must express the ratio 1.56:3.13 as a whole-number ratio. This is done by dividing both numbers of moles by the smaller number, which is 1.56 in this case.

$$\text{Mol of S atoms} = \frac{1.56}{1.56} = 1.00$$

$$\text{Mol of O atoms} = \frac{3.13}{1.56} = 2.01$$

The ratio of the number of S atoms to the number of O atoms is 1:2. The empirical formula of the compound is therefore SO_2.

You may find it more convenient to organize the solution in a table. This is done in the next example.

Sample Problem 6.7

A compound of carbon, chlorine, and fluorine was found in an old air-conditioning unit. The compound was analyzed and found to contain 16.3% carbon, 32.1% chlorine, and 51.6% fluorine by mass. Determine the empirical formula of the compound.

Solution

Assume 100 g of compound so that the percentages give the numerical values of the masses of the elements.

Element	Mass	Number of Moles	Ratio of Moles	Whole-Number Ratio
C	16.3 g	$16.3 \text{ g} \times \dfrac{1 \text{ mol}}{12.0 \text{ g}} = 1.36$	$\dfrac{1.36}{0.904} = 1.50$	$\times 2 = 3$
Cl	32.1 g	$32.1 \text{ g} \times \dfrac{1 \text{ mol}}{35.5 \text{ g}} = 0.904$	$\dfrac{0.904}{0.904} = 1.00$	$\times 2 = 2$
F	51.6 g	$51.6 \text{ g} \times \dfrac{1 \text{ mol}}{19.0 \text{ g}} = 2.72$	$\dfrac{2.72}{0.904} = 3.00$	$\times 2 = 6$

Since an empirical formula shows the simplest whole-number ratio of atoms, we cannot write $C_{1.5}ClF_3$ as the empirical formula. The numbers 1.50, 1.00, and 3.00, obtained in the second last column, are multiplied by 2 to produce whole numbers. The empirical formula is therefore $C_3Cl_2F_6$.

Exercise

13. Determine the empirical formula of each of the following compounds. The percentage compositon by mass is given.
    (a) 85.7% carbon, 14.3% hydrogen
    (b) 52.9% aluminum, 47.1% oxygen
    (c) 62.6% lead, 8.4% nitrogen, 29.0% oxygen

14. A compound of carbon, hydrogen, and chlorine consists of 49.0% carbon, 2.75% hydrogen, and 48.3% chlorine by mass. What is the empirical formula of the compound?

15. Chemical analysis of rubbing alcohol shows that it consists of 59.97% carbon, 13.35% hydrogen, and 26.68% oxygen by mass. What is the empirical formula of rubbing alcohol?

6.6 Molecular Formula

In the previous section, you learned how to determine the empirical formula of a compound. The empirical formula represents the simplest whole-number ratio of the atoms (or ions) of the different elements in the compound. It may or may not be the actual formula of a molecule of the compound. For example, if the empirical formula of a compound is CH_2, this tells us only that the ratio of carbon atoms to hydrogen atoms is 1:2. The actual formula might be C_2H_4, or C_3H_6, or C_4H_8, or generally C_nH_{2n}. (See Figure 6.4.)

To determine the actual or molecular formula of a compound, in addition to the empirical formula we need to know the molecular mass (or formula mass or molar mass) of the compound. Sample Problem 6.8 shows how this is done.

Figure 6.4
The empirical formula CH_2 can represent any of the molecules shown. In each case, the ratio of carbon atoms to hydrogen atoms is 1:2.

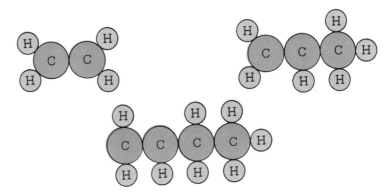

Sample Problem 6.8

The empirical formula of a compound of carbon and hydrogen was found by experiment to be CH_2. Its molecular mass was also found by experiment to be 42 u. What is the molecular formula of the compound?

Solution

Empirical formula mass = 12.0 u + (2 × 1.0 u)
                       = 14.0 u

Since the molecular formula must be a whole-number multiple of the empirical formula, the molecular mass is the same whole-number multiple of the empirical formula mass. We can write

$$\frac{\text{molecular mass}}{\text{empirical formula mass}} = \frac{42 \text{ u}}{14 \text{ u}} = 3$$

This tells us that the molecular formula is three times the empirical formula – three times CH_2, that is, C_3H_6.

Exercise

16. Chemical analysis of one of the constituents of gasoline showed that it consists of 92.29% carbon and 7.71% hydrogen by mass. If the molecular mass of the compound is 78 u, determine its molecular formula.

17. A compound of silicon and fluorine was analyzed and found to consist of 33.0% silicon and 67.0% fluorine by mass. The molecular mass of the compound was determined by experiment to be 170 u. What is the molecular formula of the compound?

18. Glucose is an important source of energy for organisms. Analysis of a sample of glucose shows that it consists of 39.95% carbon, 6.71% hydrogen, and the remainder oxygen. In a separate experiment, the molar mass of glucose is found to be 180 g. Determine the molecular formula of glucose.

6.7 Empirical Formula and Molecular Formula – An Application

Each year thousands of new chemical compounds are synthesized in laboratories. The identification of every new compound requires that, at the very least, its empirical formula, its molecular formula, and its molecular structure be established. The first two requirements involve calculations similar to those you carried out in this chapter. It is important to recognize that empirical formulas and molecular formulas are determined only by experiment.

Empirical Formula Determination

The actual experiment used to determine the percentage composition by mass of a compound varies according to the elements involved. By far the greatest proportion of new compounds contain carbon and hydrogen. Let us consider, therefore, how the empirical formula of a compound containing carbon, hydrogen, and oxygen is experimentally determined.

When a compound containing carbon and hydrogen is burned in a plentiful supply of oxygen, carbon dioxide and water vapour are produced. All of the carbon atoms from the original compound end up in the carbon dioxide molecules which are produced. Similarly, all of the hydrogen atoms in the original compound end up in the water produced. The experiment involves measuring the masses of carbon dioxide and water formed during the combustion of a known mass of the compound. (See Figure 6.5.)

PUTTING CHEMISTRY TO WORK

WAYNE SMITH / LABORATORY TECHNOLOGIST

Large chemical laboratories can be hectic places, especially those which analyze large numbers of samples or which operate according to strict deadlines. Matters are even more complex when different types of analyses are done in the same laboratory. Yet the quality of the scientific work must be maintained along with efficiency. These laboratories depend on the skills of laboratory technologists such as Wayne Smith to oversee and organize their operation.

Wayne is a laboratory technologist for an environmental engineering firm. "My job includes chemical analysis on soil and water samples. Also, I'm in charge of ordering all the supplies for the lab from scientific equipment and chemical suppliers." But this is not all. Wayne's laboratory must use the best technology available, so part of his responsibility is to be aware of the latest advances in equipment. "I ensure that every aspect of the processes done in the lab is accurate and dependable," Wayne concludes. "If environmental decisions are to be based on my results, those results must be trustworthy."

Wayne chose this line of work because recent developments in toxic waste clean-up were leading to the opening of more laboratories — especially ones using highly sophisticated equipment. Wayne's particular job also relates to his concern for the environment.

"By doing my work well, I feel I play an important role in helping to protect the environment. In addition, the opportunity now exists for small laboratory-based businesses to become profitable. With the experience I am gaining from my job, I hope one day to operate such a business."

To become a laboratory technologist, Wayne took advantage of a new type of college program. In this program, courses were combined with practical on-the-job experience. "Although a university degree in chemistry would also have been good preparation for my work, I felt the practical experience of this program would help me find the type of job I wanted."

Wayne decided upon his career by examining his own interests. "I enjoyed chemistry, physics, and biology in high school. It made sense to look for a job which would let me continue to enjoy subjects I knew interested me. Certainly my job does that for me."

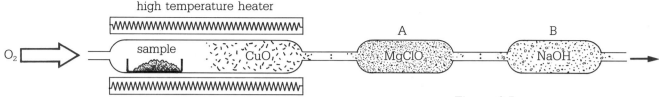

The percentages by mass of carbon in carbon dioxide, and of hydrogen in water are easily calculated. When these percentages are applied to the masses of carbon dioxide and water produced, the masses of carbon and hydrogen present in the original compound can be determined. The mass of oxygen is found by subtracting the total mass of carbon and hydrogen from the mass of the sample. Each mass is then converted to a number of moles to determine the ratio of atoms and hence the empirical formula of the compound.

Figure 6.5
The combustion of a compound containing hydrogen and carbon produces water and carbon dioxide. The water is absorbed by a substance in tube A, and the carbon dioxide by a substance in tube B. From the masses of each compound produced, the masses of hydrogen and carbon in the original compound can be determined.

Molecular Formula Determination

To determine the molecular formula of a compound, its formula mass (or molar mass) must first be found by experiment. Using the empirical formula and formula mass, the molecular formula is easily determined, as described in section 6.5.

Looking Back – Looking Ahead

In this chapter, you have learned how the chemist uses the mole concept to relate the mass of a substance to the number of atoms, molecules, or ions. You have also seen how the mole concept is valuable in determining the formula of a compound. In Chapter 7, you will learn some rules that will enable you to predict the formulas of many compounds. Remember, however, that the only way to verify the formula of a compound is by experiment.

Vocabulary Checklist

You should now be able to define or explain the meaning of the following terms.

Avogadro's constant	molar mass	molecular mass	simplest formula
empirical formula	mole	percentage composition	
formula mass	molecular formula	by mass	

Chapter Objectives

You should now be able to do the following.

1. Calculate the number of atoms, molecules, or ions in a known number of moles of an element or compound.

2. Calculate the number of moles of substance present in a sample of known mass.

3. Calculate the mass of a number of moles of a given substance.

4. Calculate the percentage composition by mass of a compound, given the formula of the compound.

5. Determine the empirical formula of a substance, given information concerning the composition of the substance.

6. Determine the molecular formula of a substance, using the empirical formula and the formula mass or molar mass of the substance.

7. Use the factor-label method in solving problems.

8. Give answers to calculations to the correct number of significant digits.

Understanding Chapter 6

REVIEW

1. Why is the mole concept so important for chemistry?

2. Differentiate among the following: molar mass, molecular mass, and formula mass.

3. Carry out the following calculations and give each answer to the correct number of significant digits.
   (a) 268×0.025
   (b) $4.345 + 567.1 + 0.1234$
   (c) $75.98 - 1.027$
   (d) $5678 \div 14.38$

4. Calculate the number of atoms of each element present in the following samples of matter.
   (a) 12.5 g of carbon dioxide
   (b) 0.84 g of methane, CH_4
   (c) 465 g of methanol, CH_3OH.

5. Calculate the mass of each of the following:
   (a) 150 mol of silicon
   (b) 2.43 mol of acetylene, C_2H_2
   (c) 0.065 mol of aluminum hydroxide, $Al(OH)_3$.

APPLICATIONS AND PROBLEMS

6. Sodium fluoride, NaF, is added to many brands of toothpaste because research has shown that it reduces the incidence of tooth decay, partic-ularly in children. If a tube of toothpaste con-tains 0.012 g of sodium fluoride, how many moles is this?

7. The minimum recommended daily intake of vitamin C, $C_6H_8O_6$, is 60 mg. (This is the amount found to be the minimum required to prevent the occurrence of scurvy.) How many moles of vitamin C is this?

8. If a spoonful of table sugar, $C_{12}H_{22}O_{11}$, has a mass of 10.0 g, calculate the number of moles of sugar in a spoonful.

9. How many moles of sodium chloride, NaCl, are in 0.500 kg package?

10 Cholesterol is a fat-like compound found in many of the foods we eat, such as eggs, butter, and red meat. Cholesterol is one of the major components of fatty deposits found clinging to the walls of blood vessels in some people. There is evidence that these deposits increase the risk of heart disease. The formula of cho-lesterol has been found to be $C_{27}H_{46}O$. Scien-tists estimate that the average human has about 250 g of cholesterol, which is naturally produced in the body by the liver. How many moles of cholesterol is this?

11. (a) A typical glass of water contains approximately 150 g of water. Calculate the number of water molecules in the glass.
    (b) If this glass of water were shared by everyone on the earth, which has approximately 5×10^9 people, how many water molecules would each person receive?

12. Which of the following samples of matter contains the greatest number of molecules? Show clearly how you arrive at your answer.
    (a) 252 g of ethanol, C_2H_5OH
    (b) 300 g of acetic acid, CH_3COOH
    (c) 102 g of ammonia, NH_3

13. Which of the following represents the greatest amount of silver? Show clearly how you arrive at your answer.
    (a) 25 g of silver
    (b) 0.25 mol of silver
    (c) 2×20^{23} silver atoms

14. Sodium hypochlorite, $NaOCl$, is the active ingredient in a number of commercial bleaches. A typical bleach contains approximately 5 g of sodium hypochlorite in every 100 mL of bleach. How many moles of sodium hypochlorite are in a 1.0 L bottle of bleach?

15. Vinegar is a dilute solution of acetic acid, CH_3COOH. A 2.0 L bottle of vinegar contains approximately 2 mol of acetic acid. What mass of acetic acid is present in the bottle?

16. In 1982, Canada produced 62 456 kg of gold and 5196 t of potash, K_2CO_3. Calculate the number of moles of each substance produced.

17. Calculate the mass of
    (a) 9.225 mol of vitamin D_1, $C_{56}H_{88}O_2$
    (b) 1.62 mol of monosodium glutamate, $NaC_5H_8O_4N$.

18. Patients who have to have X rays taken of their intestinal tract are given, prior to the X ray, a drink containing barium sulphate, $BaSO_4$. Since X rays cannot pass through the barium sulphate, an image of the intestinal tract appears on the X ray film. If 0.482 mol of barium sulphate is used, what mass of barium sulphate is added to the drink?

19. Rolaids tablets contain sodium dihydroxyaluminum carbonate, $Na(OH)_2AlCO_3$, which neutralizes excess stomach acid. If each tablet contains 335 mg of this ingredient, calculate the number of moles of the compound in a 20-tablet package of Rolaids.

20. Propane, C_3H_8, used in gas barbecues is usually sold by massing the empty tank, then adding propane until a certain mass is reached. If 2.5 kg of propane is purchased, how many moles of propane does this represent?

21. An artificial flavouring agent that simulates the flavour of peaches has the formula $C_6H_{12}O_2$.
    (a) How many moles of molecules are present in 1.00 g of the compound?
    (b) How many moles of carbon atoms, hydrogen atoms, and oxygen atoms are present in the same sample?

22. The formula of the artificial sweetener, saccarin, is $C_7H_5NO_3S$. Determine the percentage by mass of each element in the compound.

23. Lithium carbonate, Li_2CO_3, and lithium sulphate, Li_2SO_4, are drugs used in the treatment of manic-depression. If the effectiveness of the medication depends on the percentage of lithium in the compound, show by calculation which compound is more effective.

24. Analysis of a compound of potassium, sulphur, and oxygen gave the following results: K 41.02%; S 33.69%; O 25.29%. What is the empirical formula of the compound?

25. A 3.000 g sample of a compound was analyzed and was found to consist of 0.853 g Na, 0.962 g Cr, and 1.185 g O. Determine the empirical formula of the compound.

26. Caffeine is a stimulant found in coffee, tea, cola drinks, and chocolate. Analysis of caffeine shows that it consists of 49.48% carbon, 5.19% hydrogen, 28.85% nitrogen, and 16.48% oxygen by mass. Determine the empirical formula of caffeine.

27. During the operation of a car battery, lead sulphate forms on the battery plates. Analysis of this compound shows that it consists of 68.3% lead, 10.6% sulphur, and 21.1% oxygen by mass. What is the empirical formula of lead sulphate?

28. Lactic acid is the substance responsible for the taste of sour milk. Analysis of a sample of lactic acid shows that its percentage composition by mass is 40.00% carbon, 6.71% hydrogen, and 53.29% oxygen. If the molar mass

of lactic acid is found to be 90 g, determine the molecular formula of lactic acid.

29. A sample of a liquid used in dry-cleaning was found to consist of 10.06% carbon (by mass), 89.10% chlorine, with the remainder being hydrogen. The molar mass of the compound was determined to be 119.6 g. What is the molecular formula of the compound?

30. Sodium hydroxide, NaOH, is used in tomato-canning plants to quickly remove the skins from tomatoes.
    (a) If a plant uses 3.25 kg of sodium hydroxide in a certain time period, how many moles of NaOH is this?
    (b) What is the mass of 6.22 mol of NaOH?
    (c) Determine the percentage by mass of each element in NaOH.

31. The formula of the antibiotic penicillin is $C_{16}H_{18}O_4N_2S$. Determine the percentage by mass of each element in the compound.

32. Analysis of vanillin, the compound responsible for the vanilla favour, showed that the compound consisted of 63.2% carbon, 5.26% hydrogen, and 31.6% oxygen by mass. Determine the empirical formula of vanillin.

33. Analysis of a compound showed that it consisted of 49.0% carbon, 2.72% hydrogen, and 48.3% chlorine by mass. In a separate experiment, the molar mass of the compound was determined to be 147 g. What is the molecular formula of the compound?

CHALLENGE

34. Use the following information, along with what you have learned in this chapter, to answer the three questions below.
    The volume of 20 drops of water = 1.0 cm^3
    The average distance from the earth to the sun
        = 1.5×10^8 km
    The diameter of a water molecule
        = 3.0×10^{-8} cm
    The density of water = 1.0 g/cm^3
    (a) How many water molecules are present in a drop of water?
    (b) If these molecules were lined up end to end, what distance would they cover?
    (c) How many times would this chain of molecules cover the distance between the earth and the sun?

7 FORMULAS AND NAMING OF COMPOUNDS

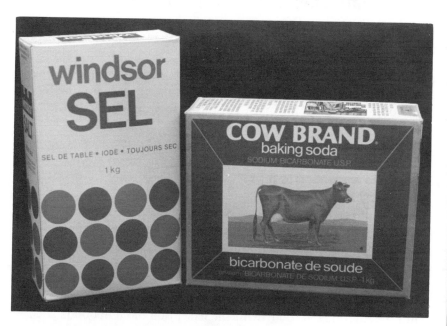

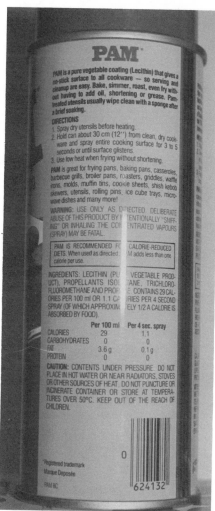

Examine the labels on a number of products around the home. Among the ingredients listed you may see some of these compounds: calcium carbonate, caustic soda, lye, ammonia, aluminum chloride, selenium sulphide, sodium phosphate, monosodium glutamate. What information about the compounds do we get from the names? Some names give more information than others about what is in a compound. For example, the name sodium hydrogen carbonate tells us much more about the constituent elements of the compound than the name baking soda does – yet these are two different names for the same compound.

How is the name of a compound determined? What are the advantages of a common system of *nomenclature* (that is, the naming of compounds)? How do you tell the formula of a compound from its name? The answer to these questions will be found in this chapter.

7.1 Formulas and Compounds: An Introduction

In Chapter 6, you learned that the formula of a compound can be determined only by experiment. Does this mean that once the formula of a compound has been established, we must commit it to memory?

There are over six million known compounds, and several thousand new compounds are prepared each year. Even if you were required to remember the formulas for only 1% of these compounds, you would have to memorize at least sixty thousand formulas! You will be relieved to know that this is not necessary. Chemists have developed a set of rules which allow us to predict the formulas for many compounds. You will have to learn these rules, but you will find that this is far easier than memorizing thousands of formulas.

All compounds can be divided into two categories: **organic compounds** and **inorganic compounds.** Organic compounds are compounds containing carbon, with the exception of compounds such as carbon dioxide, carbon monoxide, and carbonates. Inorganic compounds include all other compounds. There is a much greater number of organic compounds than inorganic compounds. In this chapter, we will deal with the names and formulas of inorganic compounds. This reduces the number of compounds to a more manageable few hundred thousand!

Exercise

1. List the names of ten compounds from the labels of a number of household products. For each compound, write down the names of the elements you think are present.

7.2 Common Names of Compounds

Many compounds, particularly those that have been known for hundreds of years, have been given common names. These names are often derived from a property of the substance. For example, the compound bluestone is obviously named for its colour; laughing gas is named for its effect on people when inhaled in small quantities: and Epsom salts is named after a place. It is even possible for the same substance to have two different common names; for example bluestone is sometimes also known as blue vitriol. The common names of several compounds, along with the corresponding systematic names and formulas, are listed in Table 7.1.

Table 7.1 Common Names, Systematic Names, and Formulas of Some Compounds

Common Name	Systematic Name	Formula	Use
Lime	Calcium oxide	CaO	Fertilizer
Baking soda	Sodium hydrogen carbonate	$NaHCO_3$	Makes dough "rise"
Laughing gas	Dinitrogen oxide	N_2O	General anaesthetic
Limestone, marble	Calcium carbonate	$CaCO_3$	Building material
Lye, caustic soda	Sodium hydroxide	$NaOH$	Drain cleaners
Milk of magnesia	Magnesium hydroxide	$Mg(OH)_2$	Antacid preparation
Muriatic acid	Hydrochloric acid	HCl	Concrete cleaner
Oil of vitriol	Sulphuric acid	H_2SO_4	Making fertilizers
Soda ash	Sodium carbonate	Na_2CO_3	Glass making
Salt	Sodium chloride	$NaCl$	Making chlorine

As more and more compounds became known, the need for a systematic method of nomenclature became greater. Common names were acceptable when the number of compounds known was reasonably small, and when the names were used by a fairly small group of people. These conditions are not true today. Thus, chemists have developed rules which give every chemical compound a unique and descriptive name. Some compounds, however, are so well known by their common names that those names have been retained. For example, water, H_2O, is rarely referred to as hydrogen oxide; ammonia, NH_3, is not called nitrogen trihydride; methane, CH_4, is not usually called carbon tetrahydride!

7.3 Bonding and Chemical Formulas

In Chapter 5 you studied a bonding model which proposed that atoms combine in one of two ways. According to this model, atoms either share electrons, forming a covalent bond, or atoms transfer one or more electrons to other atoms, forming an ionic bond. For example, when hydrogen and oxygen atoms combine to form water, they do so by sharing electrons (covalent bonding). The formula of water determined by experiment is H_2O. This is also the formula predicted by our bonding model. On the other hand, when sodium and oxygen atoms combine to form the compound sodium oxide, the sodium atoms each transfer one electron to the oxygen atom. This results in the formation of oppositely charged ions which attract each other strongly (ionic bonding). The experimentally determined

Figure 7.1
The formulas of sodium oxide and water can be predicted using the bonding model studied earlier. This model assumes that when atoms bond, electrons are either shared by the two atoms (covalent bonding) or are transferred from one atom to the other (ionic bonding).

Na_2O H_2O

formula of sodium oxide is Na_2O. This formula is also predicted by the bonding model. (See Figure 7.1.)

In the examples above, we have made use of a bonding model which produces formulas that have been verified by experiment. Does this mean that to write a chemical formula, we must always draw diagrams showing the electron sharing or transfer, then count the atoms or ions involved? This would give us the correct answer in many cases, but it is too time consuming to be practical.

By experimentally determining the formulas of thousands of compounds, chemists have found that different atoms have different **combining capacities**, or **valences** as they are called in Chapter 5. The combining capacity or valence of an atom is the number of bonds which the atom is able to form. Table 7.2 shows the formulas of the compounds formed when chlorine combines with the elements in the second row of the periodic table (lithium through fluorine) and the valences of those elements. Remember that all of these formulas have been determined by experiment.

Table 7.2 Formulas of Second Row Chlorides and Valences of Second Row Elements

Second Row Element	Li	Be	B	C	N	O	F
Formula of Compound	LiCl	$BeCl_2$	BCl_3	CCl_4	NCl_3	OCl_2	FCl
Valence	1	2	3	4	3	2	1

You will find the same pattern if you consider the formulas of the compounds formed when chlorine combines with the elements in the third row of the periodic table (potassium through chlorine). In fact, the valences of the main group elements seem to follow a pattern. This is shown in Table 7.3.

Table 7.3 Valence of Main Group Elements.

Group No.	IA(1)	IIA(2)	IIIA(13)	IVA(14)	VA(15)	VIA(16)	VIIA(17)	VIIIA(18)
Valence	1	2	3	4	3	2	1	0

How can this pattern be explained? Consider the Lewis symbols of the second row elements.

$$\text{Li}\cdot \quad \text{Be}\cdot \quad \dot{\text{B}}\cdot \quad \cdot\dot{\text{C}}\cdot \quad \cdot\dot{\text{N}}\cdot \quad \cdot\ddot{\text{O}}\colon \quad \cdot\ddot{\text{F}}\colon \quad \colon\ddot{\text{Ne}}\colon$$

The number of bonds formed by each atom is the same as the number of single, or unpaired, *valence electrons* in the atom. The formula of methane is CH_4, because carbon has four unpaired valence electrons and hydrogen has one. Carbon can form four bonds and hydrogen one. The valence of carbon is four and the valence of hydrogen is one.

The formulas of many compounds containing two elements can be correctly predicted using the number of unpaired valence electrons in the atoms of each element. Chemists have devised a faster method of writing formulas, making use of the valences of the elements. An example will make this method clear.

To determine the formula of aluminum chloride (a compound of aluminum and chlorine), the symbols and valences of the elements are written as follows:

valences   3   1

symbols   Al $\diagdown$ Cl   → $AlCl_3$

The valences are "criss-crossed" as shown to produce the formula $AlCl_3$. (Remember that the subscript 1 in a formula is always left out.)

Let us consider another example. What is the formula of magnesium oxide?

2   2

Mg $\diagdown$ O   → Mg_2O_2 →   MgO

With some exceptions, it is customary to reduce the subscripts in a formula to their lowest terms – in this case by dividing both subscripts by two.

This method of formula writing has been devised by chemists to produce a formula for a compound that agrees with the experimentally determined formula.

Exercise

2. Write the formulas for the following compounds.
   - (a) magnesium bromide
   - (b) potassium oxide
   - (c) beryllium fluoride
   - (d) calcium sulphide
   - (e) lithium nitride
   - (f) sodium chloride
   - (g) aluminum oxide
   - (h) barium iodide
   - (i) magnesium hydride
   - (j) strontium chloride

7.4 Naming Binary Compounds

All of the compounds considered so far in this chapter are **binary compounds** – compounds which contain two elements. How do we name these compounds? By agreement amongst chemists, two simple rules apply.

1. The element with the lower electronegativity is written first. If you do not have electronegativity tables to refer to, write the element further to the left in the periodic table first. If the elements are in the same group, write the element which is lower in the group first.

2. Remove the ending from the name of the second element, and add "-ide" in its place.

Keep these rules in mind as you consider the four examples that follow.

1. MgO is magnesium oxide. (Magnesium is a metal and is therefore further to the left in the periodic table.)

2. NaF is sodium fluoride. (Sodium is further to the left in the periodic table.)

3. Al_2S_3 is aluminum sulphide. (Aluminum is further to the left in the periodic table.)

4. SiC is silicon carbide. (Silicon is in the same group as carbon in the periodic table, but is lower.)

Exercise

3. Name the following compounds.

   (a) MgS
   (b) Na_2O
   (c) $CaCl_2$
   (d) K_3N

   (e) $AlBr_3$
   (f) Li_3P
   (g) Al_2S_3
   (h) BaF_2

7.5 Elements with More Than One Valence

In section 7.3, we saw that for many elements, the valence of the element is equal to the number of unpaired valence electrons, as shown in the Lewis symbol of the element. An atom with three unpaired valence electrons has a valence of three and forms three bonds; an atom with four unpaired valence electrons has a valence of four and forms four bonds, and so on.

The valences derived from the number of unpaired electrons are the "normal" or "expected" valences of the elements. Scientists have identified compounds in which some elements display a valence other than the normal valence. For example, when phosphorus and chlorine react, one of two different products may form. The actual product depends on the conditions of the reaction, such as temperature and pressure. According to Table 7.3, phosphorus has a valence of three and PCl_3 is the expected product. The other compound is PCl_5. In this compound, phosphorus forms five bonds and thus has a valence of five. In other words, phosphorus displays a valence of three in some of its compounds and a valence of five in others.

Similarly, other elements, such as iron, copper, tin, nickel, and sulphur, display more than one valence. Some of these are shown in Table 7.4. When we name a compound containing one of these elements, how do we indicate which valence the element is displaying in that compound? How do we distinguish between PCl_3 and PCl_5, for example?

Table 7.4   Compounds Showing Two or More Different Valences of Some Elements

Compound	Number of Bonds Formed by 1st Element	Valence of 1st Element
PCl_3	3	3
PCl_5	5	5
SF_4	4	4
SF_6	6	6
$SnCl_2$	2	2
$SnCl_4$	4	4
$FeCl_2$	2	2
$FeCl_3$	3	3

The Stock System

Over the years, scientists have developed a number of different methods of naming compounds containing elements which may have more than one valence. The method which is now most widely used is the **Stock system.** It is named after Alfred Stock, the originator of the system. The Stock system uses Roman numerals to indicate the valence of the element which may have more than one valence. For example,

PCl_3 is named phosphorus(III) chloride
PCl_5 is named phosphorus(V) chloride

When saying the names of these compounds, the Roman numeral is read as a number – "phosphorus three chloride" and "phosphorus five chloride."

Exercise

4. Name the following compounds using the Stock system.
   - (a) $FeCl_2$
   - (b) $FeCl_3$
   - (c) SnF_2
   - (d) SnF_4
   - (e) SO_2
   - (f) SO_3
   - (g) Cu_2S
   - (h) CuS

5. Write formulas for the following compounds.
   - (a) nickel(II) chloride
   - (b) arsenic(III) sulphide
   - (c) sulphur(IV) fluoride
   - (d) iron(III) phosphide
   - (e) cobalt(II) nitride
   - (f) lead(II) oxide

The Prefix System

A second method of naming compounds where the first element may have more than one valence is the **prefix system.** Generally speaking, this system is used for compounds containing two non-metals, although the Stock system can still be used. A Greek or Latin prefix is attached to the name of an element. It indicates the number of atoms of that element present in a formula unit of the compound. Table 7.5 shows some examples.

Table 7.5 The Prefix System

Number	Prefix	Example
1	Mono-	Carbon monoxide CO
2	Di-	Sulphur dioxide SO_2
3	Tri-	Sulphur trioxide SO_3
4	Tetra-	Carbon tetrachloride CCl_4
5	Penta-	Phosphorus pentachloride PCl_5
6	Hexa-	Sulphur hexafluoride SF_6

Note that if only one atom of the first element is present in the formula unit, the prefix "mono-" is left out. For example, it is not necessary to write monosulphur dioxide for SO_2, but simply sulphur dioxide. A prefix is attached to the name of the first element if more than on atom is present. For example, the compound N_2O_4 is named dinitrogen tetroxide.

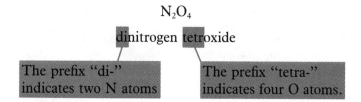

N_2O_4

dinitrogen tetroxide

The prefix "di-" indicates two N atoms

The prefix "tetra-" indicates four O atoms.

When a prefix ending with "o" or "a" is followed by "oxide," the "o" or "a" is omitted. For example, we write tetroxide rather than tetraoxide to indicate the four oxygen atoms.

Exercise

6. Name the following compounds using the prefix system.
   (a) NO_2 (c) P_2O_5
   (b) PCl_3 (d) As_2S_3

7. Write formulas for the following compounds.
   (a) silicon tetrachloride   (c) dinitrogen trioxide
   (b) nitrogen triiodide   (d) uranium hexafluoride

The "ous-ic" system

A third, and the least used system of naming compounds, where the first element may have more than one valence, is the **"ous-ic" system**. This is the oldest system and makes use of the Latin names for many elements. Some examples are shown in Table 7.6.

Table 7.6   Latin Names for Some elements

Element	Latin Name
Copper	Cuprum
Iron	Ferrum
Lead	Plumbum
Tin	Stannum
Gold	Aurum

According to this method, the lower valence of the first element uses the suffix "-ous" and the higher valence uses the suffix "-ic". For example, FeO is named ferrous oxide and Fe_2O_3 is named ferric oxide.

In the first compound, the valence of iron is two and in the second compound the valence is three. Although this system is rarely used by chemists today, you will still see it on many of the labels of bottles of chemicals. (See Figure 7.2.) You should know the meaning of the suffixes. Table 7.7 on page 194 shows the names of some compounds using the "ous-ic" system.

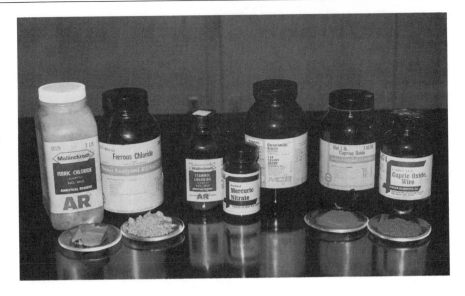

Figure 7.2
Although the "ous-ic" system is not part of the modern system of nomenclature, it is still used by chemical supply companies.

Table 7.7 Names of Selected Compounds Using the "ous-ic" System

Compound	Name
CuO	Cupric oxide
Cu_2O	Cuprous oxide
$SnCl_2$	Stannous chloride
$SnCl_4$	Stannic chloride
$PbBr_2$	Plumbous bromide
$PbBr_4$	Plumbic bromide

Exercise

8. Name the following compounds using the "ous-ic" system.
   (a) $PbCl_2$
   (b) $PbCl_4$
   (c) FeO
   (d) Fe_2O_3

9. For each compound in question 8, give its name using the Stock system.

10. What is the major disadvantage of the "ous-ic" system?

It is important to realize that the three naming systems described above are used only when the first named element has more than one valence. It is not correct, for example, to name the compound Na_2O sodium(I) oxide. The valence of sodium is always one; therefore, the Roman numeral I is redundant.

How do you know which elements have more than one valence? It is difficult to give a rule which can be easily remembered. You may find the following rule of thumb useful. Elements in groups I (1), II (2), and the top part of group III (13) show only one valence. (See Figure 7.3.) Therefore, when the first named element is not in group I (1), II (2), or the top of group III (13), that element probably has more than one valence. There are some exceptions to this rule – for example, fluorine, zinc, and silver – but the rule is nevertheless useful. Remember also that this applies only to the first element in the formula. The valence of the second element is the normal or expected valence.

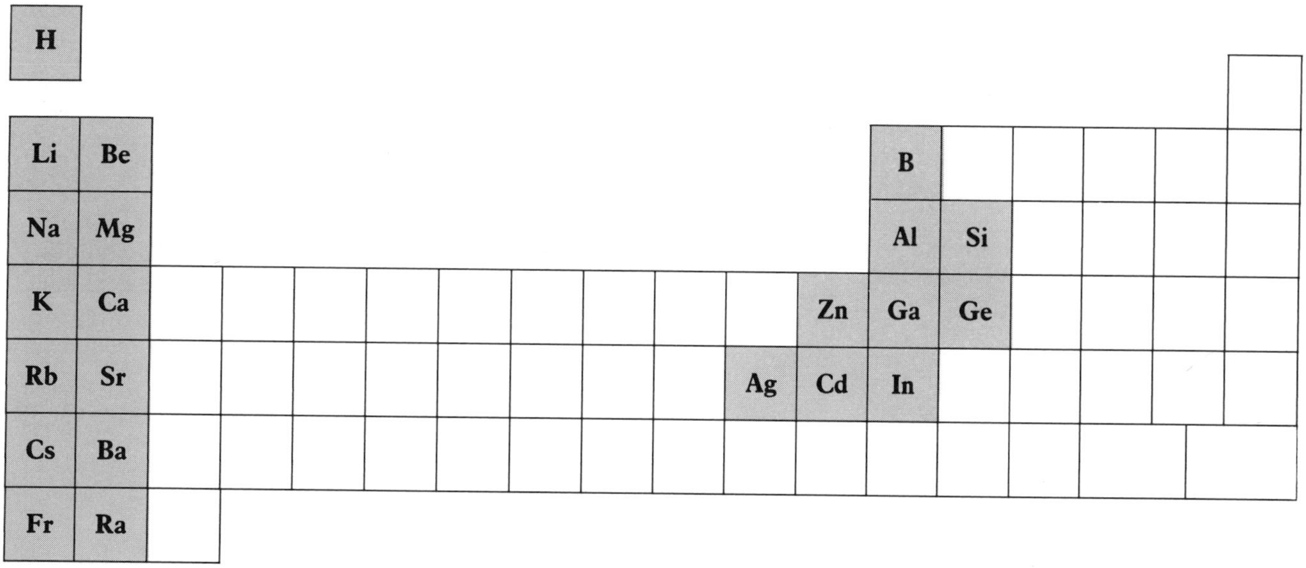

Figure 7.3
The shaded portions of the periodic table represent those elements that display only one valence (combining capacity) when forming compounds. The elements in the unshaded portion of the table display two or more valences when they form compounds.

Exercise

11. Name the following compounds.
    (a) CO_2
    (b) Li_2S
    (c) $NiCl_2$
    (d) N_2O
    (e) SBr_4
    (f) Al_2O_3
    (g) PH_3
    (h) $BeCl_2$
    (i) PbO
    (j) $CaBr_2$

12. Write formulas for the following compounds.
    (a) potassium sulphide
    (b) copper(II) bromide
    (c) dinitrogen pentoxide
    (d) sulphur(IV) fluoride
    (e) ferrous oxide
    (f) magnesium nitride
    (g) iron(III) phosphide
    (h) aluminum carbide
    (i) tin(II) chloride
    (j) calcium fluoride

7.6 Non-binary Compounds with "-ide" Ending

Several compounds that end in "-ide" are not true binary compounds. A few examples are shown below.

NaOH	sodium hydroxide
$Ca(OH)_2$	calcium hydroxide
KCN	potassium cyanide
$Mg(CN)_2$	magnesium cyanide
NH_4Cl	ammonium chloride
$(NH_4)_2S$	ammonium sulphide

Each of these compounds contains more than two elements, yet each name ends in "-ide". The hydroxide ion, OH^-, the cyanide ion, CN^-, and the ammonium ion, NH_4^+, are considered as single units, each with a valence of one. For example, the formula of aluminum hydroxide is derived as shown here.

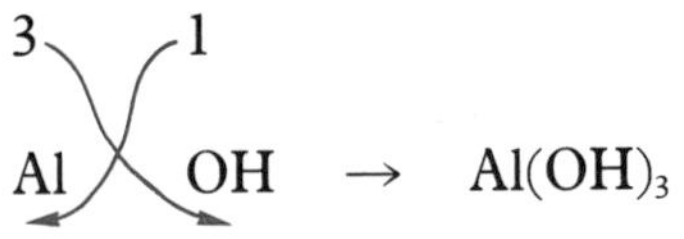

$$\overset{3}{Al} \diagdown \overset{1}{\diagup} OH \quad \rightarrow \quad Al(OH)_3$$

Since three hydroxide ions have to be shown in the formula, brackets are placed around the OH.

Exercise

13. Name the following compounds.
    (a) $Ba(CN)_2$
    (b) NH_4Br
    (c) $Sr(OH)_2$

14. Write formulas for the following.
    (a) lithium hydroxide
    (b) aluminum cyanide
    (c) ammonium phosphide

7.7 Binary Acids

The bond between the hydrogen atom and the chlorine atom in the compound hydrogen chloride, HCl, is a polar covalent bond. The compound hydrogen chloride is a molecular compound. When dissolved in water, hydrogen chloride forms a solution that conducts electricity, turns litmus red, reacts with magnesium to produce hydrogen, and reacts with carbonates to produce carbon dioxide.

Compounds with this set of properties are called **acids**. (You will study acids in greater detail in Chapter 13.) The solution of hydrogen chloride in water is known as hydrochloric acid. (See Figure 7.4.) Other molecular compounds that dissolve in water to produce acidic solutions are shown in Table 7.8.

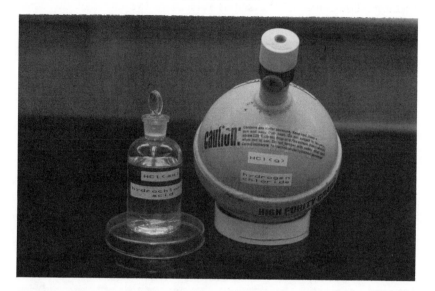

Figure 7.4
Pure hydrogen chloride is a gas at room temperature, and is available as a gas compressed in cylinders. A solution of hydrogen chloride in water is a liquid known as hydrochloric acid.

Table 7.8   Binary Acids and the Compounds from which They Are Derived

Compound	Name	Acid Formed
HF	Hydrogen fluoride	Hydrofluoric acid
HBr	Hydrogen bromide	Hydrobromic acid
HI	Hydrogen iodide	Hydroiodic acid
H_2S	Hydrogen sulphide	Hydrosulphuric acid

The acids listed in Table 7.8 are referred to as **binary acids** – they are acids formed from binary compounds. The formula HCl does not indicate whether it refers to the compound hydrogen chloride or to the acid hydrochloric acid. It is usually clear from the context what the formula refers to. If the HCl is in solution, then you should use the name hydrochloric acid. If the context suggests that the pure substance is being discussed, then the name hydrogen chloride is more appropriate.

7.8   Oxyacids and Polyatomic Ions

Oxyacids are acids containing oxygen in addition to hydrogen and another element. The name of the acid is usually derived from the name of this latter element.

The "-ic" Oxyacids

The first two columns of Table 7.9 give the names and formulas of several oxyacids. When these acids react with other substances, such as metals, bases, or carbonates, compounds called **salts** are formed. The name of the salt is related to the name of the acid. Sulphuric acid, for instance, produces salts called sulphates. The sulphate grouping of atoms is an example of a **polyatomic ion** – a group of atoms covalently bonded and possessing a charge. The single sulphur atom and the four oxygen atoms are bonded together, and the group of atoms possesses a $2-$ charge. The sulphate ion has a valence of two; it can form two bonds. The names, formulas, and valences of the polyatomic ions derived from the acids listed in Table 7.9 are also shown in the table.

Table 7.9 Some "-ic" Oxyacids and Their Corresponding Polyatomic Ions

Acid		Polyatomic Ion		
Name	**Formula**	**Name**	**Formula**	**Valence**
Nitric acid	HNO_3	Nitrate	NO_3^-	1
Chloric acid	$HClO_3$	Chlorate	ClO_3^-	1
Sulphuric acid	H_2SO_4	Sulphate	SO_4^{2-}	2
Carbonic acid	H_2CO_3	Carbonate	CO_3^{2-}	2
Phosphoric acid	H_3PO_4	Phosphate	PO_4^{3-}	3
Acetic acid	CH_3COOH	Acetate	CH_3COO^-	1
Perchloric acid	$HClO_4$	Perchlorate	ClO_4^-	1
_____	_____	Chromate	CrO_4^{2-}	2
_____	_____	Dichromate	$Cr_2O_7^{2-}$	2
_____	_____	Permanganate	MnO_4^-	1

Note: For the last three entries in the table, many compounds containing the polyatomic ions are known. The corresponding acids, however, are not common, and have been omitted.

The acids listed in Table 7.9 are sometimes known are the "-ic" acids. The polyatomic ions derived from these acids are known as the "-ate" ions.

How do we write formulas for compounds containing polyatomic ions? For example, what is the formula of sodium sulphate? You may find it useful to think of the compound as having two parts, in much the same way as a binary compound, with sodium as one part and sulphate as the other. Write the symbols and valences of each part of the compound, then apply the criss-cross rule.

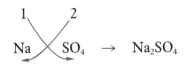

Recall that the sodium ion carries a 1+ charge and the sulphate ion carries a 2− charge. If these charges are included with the formula for sodium sulphate, the formula looks cluttered. It is therefore customary to omit the charges when writing the formula of an ionic compound.

When more than one polyatomic ion is present in a formula, brackets must be placed around the formula of the polyatomic ion. For example, the formula of calcium nitrate is $Ca(NO_3)_2$. The formula shows that two nitrate ions are present for every one calcium ion.

Exercise

15. Write formulas for the following.
    (a) sodium carbonate
    (b) aluminum nitrate
    (c) copper(II) sulphate
    (d) magnesium chlorate
    (e) calcium phosphate
    (f) potassium permanganate
    (g) sodium chromate
    (h) calcium acetate

16. Name the following compounds.
    (a) $HClO_3$
    (b) $CaCrO_4$
    (c) $FePO_4$
    (d) $Mg(NO_3)_2$
    (e) $Al_2(SO_4)_3$
    (f) $Ba(NO_3)_2$
    (g) $K_2Cr_2O_7$
    (h) $(NH_4)_2CO_3$

The "-ous" Oxyacids

Table 7.10 lists the names and formulas of some oxyacids and polyatomic ions which are related to the acids and ions listed in Table 7.9.

Table 7.10 Some "-ous" Oxyacids and Their Corresponding Polyatomic Ions

Acid		Polyatomic Ion		
Name	**Formula**	**Name**	**Formula**	**Valence**
Nitrous acid	HNO_2	Nitrite	NO_2^-	1
Chlorous acid	$HClO_2$	Chlorite	ClO_2^-	1
Hypochlorous acid	$HClO$	Hypochlorite	ClO^-	1
Sulphurous acid	H_2SO_3	Sulphite	SO_3^{2-}	2

The acids shown in Table 7.10 are known are the "-ous" acids. The ions derived from them are known as the "-ite" ions. Notice that the "-ous" acids contain one less oxygen atom than the related "-ic" acid. The same relationship exists between the "-ite" ions and the "-ate" ions. Remember the following rules. (See page 200.)

1. The "-ic" acids produce "-ate" ions.
2. The "-ous" acids produce "-ite" ions.

Table 7.11 summarizes the relationships among a number of related polyatomic ions.

Table 7.11 Relationships Among Related Polyatomic Ions

Ion	Formula	
Perchlorate	ClO_4^-	1 more O atom than "-ate"
Chlorate	ClO_3^-	
Chlorite	ClO_2^-	1 less O atom than "-ate"
Hypochlorite	ClO^-	2 less O atoms than "-ate"

Exercise

17. Write formulas for the following compounds.
    (a) potassium sulphite   (d) barium sulphite
    (b) nitrous acid   (e) magnesium chlorite
    (c) lithium chlorite   (f) sodium hypochlorite

18. Name the following compounds.
    (a) Na_2SO_3 (d) $Fe(NO_2)_2$
    (b) $HClO_2$ (e) $Ca(ClO)_2$
    (c) $KClO_2$ (f) H_2SO_3

7.9 Acid Salts

An **acid salt** is the compound formed when only some of the hydrogen atoms in an acid are replaced by a metal. Take for example, sulphuric acid, H_2SO_4, which has two hydrogen atoms. The "normal" salts of this acid are sulphates, for example, Na_2SO_4, $CuSO_4$ etc., in which both hydrogen atoms are replaced. If only one of the hydrogen atoms is replaced, then the salts are called hydrogen sulphates or bisulphates. Using this method of naming, $NaHSO_4$ is sodium hydrogen sulphate (sodium bisulphate); $Mg(HSO_4)_2$ is magnesium hydrogen sulphate. Examples of other acid salts are listed below.

$NaHCO_3$ is sodium hydrogen carbonate.
$Ca(H_2PO_4)_2$ is calcium dihydrogen phosphate.
K_2HPO_4 is potassium hydrogen phosphate.

The formulas and names of a number of polyatomic ions found in acid salts are shown in Table 7.12

Table 7.12   A Selection of Polyatomic Ions Found in Acid Salts

Name	Formula	Valence
Hydrogen carbonate	HCO_3^-	1
Hydrogen sulphate	HSO_4^-	1
Hydrogen sulphite	HSO_3^-	1
Hydrogen phosphate	HPO_4^{2-}	2
Dihydrogen phosphate	$H_2PO_4^-$	1

Exercise

19. Write formulas for the following.
    (a) potassium hydrogen sulphate
    (b) calcium hydrogen carbonate
    (c) iron(II) dihydrogen phosphate
    (d) magnesium hydrogen phosphate
    (e) sodium hydrogen sulphite

20. Name the following compounds.
    (a) $LiHSO_3$
    (b) $Ba(HSO_4)_2$
    (c) NH_4HCO_3
    (d) Na_2HPO_4

7.10 Hydrates

Normally, when an aqueous solution of a salt is allowed to evaporate, all of the water disappears and crystals of the solute are left. For example, when a solution of potassium nitrate evaporates, crystals of potassium nitrate, KNO_3, remain. There is, however, a group of substances known as **hydrates**, which behave differently. When an aqueous solution of a hydrate evaporates, some of the water remains behind as part of the crystal structure of the substance. The water in the hydrate is referred to as water of hydration or **water of crystallization**. The number of molecules of water of crystallization associated with a formula unit of a particular hydrate is always the same.

To indicate the presence of water of crystallization in a hydrate, the formula of the compound is written first, followed by a dot and the number of water molecules present. For example, the hydrated form of copper(II) sulphate (bluestone) is written as $CuSO_4 \cdot 5H_2O$. This formula indicates that there are five water molecules per formula unit of the compound. The compound is named copper(II) sulphate pentahydrate. Several common hydrates are listed in Table 7.13 on page 203.

The water of crystallization can usually be removed from a hydrate by heating. For example,

$$CuSO_4 \cdot 5H_2O \rightarrow CuSO_4 + 5H_2O$$

Laboratory Manual
How many water molecules are there in a hydrate? You will find out in Experiment 7-1.

HYDRATES

Calcium sulphate dihydrate, $CaSO_4 \cdot 2H_2O$, is a white, chalk-like mineral known as gypsum. In this form, or in a partially dehydrated form, the compound is used to make such diverse things as drywall (a type of plaster board used in construction), moulds, and plaster casts for broken limbs.

When gypsum is heated, some of its water of crystallization is driven off.

$$2(CaSO_4 \cdot 2H_2O) \rightarrow (CaSO_4)_2 \cdot H_2O + 3H_2O$$

The product, which contains a smaller proportion of water of crystallization than the original gypsum, is commonly known as plaster of Paris. This is used to make moulds and casts. When plaster of Paris is mixed with water to form a paste, the mixture hardens quickly. The partially dehydrated compound regains the water that was lost by heating and changes back to the dihydrate. This process is accompanied by the release of heat. If you have ever had a cast applied to a broken arm or leg, you have no doubt experienced this temperature change.

When cobalt(II) chloride hexahydrate is heated, all the water of crystallization is driven off, and the red hydrate changes to a blue anhydrous substance.

$$CoCl_2 \cdot 6H_2O \rightarrow CoCl_2 + 6H_2O$$
$$\text{red} \qquad\qquad \text{blue}$$

Addition of water to the anhydrous substance causes the red colour to reappear. These colour changes can be used to indicate the humidity of the air. If paper soaked with a solution of the hydrate is allowed to dry, it will appear blue and remain blue in low humidity, but will change to red in high humidity.

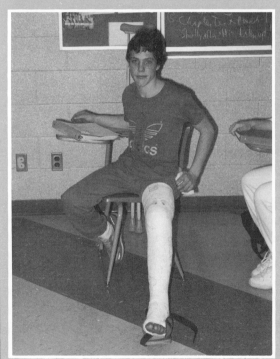

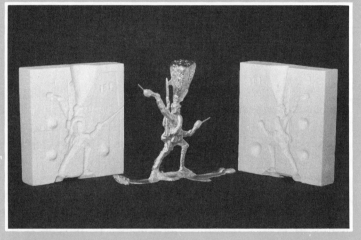

Gypsum has a wide variety of uses. When completely dehydrated, it is used to make drywall. In a partially dehydrated form, it is used to make plaster moulds and casts.

The compound without its water of crystallization is known as an **anhydrous** substance.

Table 7.13 Names and Formulas of Some Hydrates

Name	Formula
Barium hydroxide octahydrate	$Ba(OH)_2 \cdot 8H_2O$
Magnesium sulphate heptahydrate	$MgSO_4 \cdot 7H_2O$
Cobalt(II) chloride hexahydrate	$CoCl_2 \cdot 6H_2O$
Sodium carbonate decahydrate	$Na_2CO_3 \cdot 10H_2O$
Sodium thiosulphate pentahydrate	$Na_2S_2O_3 \cdot 5H_2O$
Lithium chloride tetrahydrate	$LiCl_2 \cdot 4H_2O$

Exercise

21. Write formulas for the following hydrates.
    (a) tin(II) chloride dihydrate
    (b) sodium acetate trihydrate
    (c) sodium sulphate decahydrate

22. Name the following hydrates.
    (a) $CaCl_2 \cdot 2H_2O$
    (b) $ZnSO_4 \cdot 7H_2O$
    (c) $NiCl_2 \cdot 6H_2O$

7.11 Naming Compounds – A Summary

Throughout this chapter, you have been asked to name a number of compounds. The rules for the naming of compounds are summarized below.

1. Write the name of the element with the lowest electronegativity first.

2. If this element has more than one valence, indicate the valence in the compound using the Roman numeral system. If the compound consists of two non-metals, the prefix system may be used instead of the Roman numeral system.

3. Complete the name of the rest of the compound. For example, $CaCl_2$ is calcium chloride; Na_2SO_4 is sodium sulphate; $CuCrO_4$ is copper(II) chromate; SO_3 is sulphur(VI) oxide or sulphur trioxide.

Exercise

23. Name the following compounds.
    (a) AlI_3
    (b) Na_2O
    (c) $Ca(NO_3)_2$
    (d) $Co(NO_3)_2$
    (e) Li_2SO_3
    (f) H_2O
    (g) $BaCO_3$
    (h) HF
    (i) SO_2
    (j) $Al_2(SO_4)_3$

Looking Back – Looking Ahead

In this chapter, you have practised the skills of formula writing and the naming of compounds. In the same way that the study of a second language requires the memorization of some vocabulary, the ability to write formulas and to name compounds requires that you memorize some chemistry vocabulary.

As you go on in the next two chapters to study chemical reactions, chemical equations, and calculations based on equations, you will continue to use the skills developed in this chapter.

Vocabulary Checklist

You should now be able to define or explain the meaning of the following terms.

acid	combining capacity	oxyacid	valence
acid salt	hydrate	polyatomic ion	water of crystallization
anhydrous	inorganic compound	prefix system	
binary acid	organic compound	salt	
binary compound	"ous-ic" system	Stock system	

Chapter Objectives

You should now be able to do the following.

1. Recognize the advantages of a systematic naming system for compounds.

2. Give the name or formula for binary compounds in which the first element has only one common valence.

3. Determine the name or formula for binary compounds in which the first element has more than one common valence, using either the Stock system or the prefix system.

4. State the name or formula for a number of binary acids.

5. Give the name or formula for a number of oxyacids.

6. Derive the name or formula for the compounds of these oxyacids.

Understanding Chapter 7

REVIEW

1. What are the advantages of having a systematic naming system for compounds?

2. How can you tell which elements have more than one valence?

3. The formulas and names of several compounds are given below. Only some names are correct. Identify the incorrect names, state why the name is wrong, and give a correct name for the compound.
   (a) SO_3 sulphur(VI) oxide
   (b) $FeCl_2$ iron(I) chloride
   (c) MgF_2 magnesium difluoride
   (d) P_2S_3 diphosphorus pentasulphide
   (e) CuO copper oxide
   (f) Fe_2O_3 ferrous oxide

APPLICATIONS AND PROBLEMS

4. Write formulas for the following compounds.
   (a) lithium bromide $Li \quad Br \longrightarrow LiBr$
   (b) calcium nitride $Ca^2 \quad N^3 \longrightarrow Ca_3N_2$
   (c) carbon monoxide
   (d) phosphorus(V) fluoride
   (e) barium hydroxide $Ba \ (OH) \longrightarrow Ba(OH)_2$
   (f) iron(II) sulphide

5. Name the following compounds.
   (a) $SiCl_4$ (d) BaS
   (b) MgF_2 (e) Li_3P
   (c) ICl_3 (f) NaI

6. Write formulas for the following compounds.
   (a) magnesium nitrate
   (b) copper(II) oxide
   (c) potassium sulphate
   (d) calcium carbonate
   (e) nitrogen dioxide
   (f) sodium hypochlorite

7. Name the following compounds.
   (a) $Ca(CN)_2$
   (b) $NaClO_3$
   (c) FeO
   (d) NH_4NO_3
   (e) $Al(OH)_3$
   (f) Li_2CO_3

8. Write formulas for the following compounds.
   (a) sulphurous acid
   (b) barium nitrite
   (c) tin(IV) bromide
   (d) ferric oxide
   (e) ammonium sulphate
   (f) magnesium hydride $Mg^2 \quad H^1 \longrightarrow MgH_2$

9. Name the following compounds.
   (a) H_3PO_4 (d) $CaSO_3$
   (b) $Fe(NO_3)_3$ (e) $CuCO_3$
   (c) $KClO_3$ (f) Al_2O_3

10. Write formulas for the following compounds.
    (a) sodium chlorite
    (b) potassium permanganate
    (c) copper(II) acetate
    (d) ammonium perchlorate
    (e) barium hypochlorite
    (f) hydrofluoric acid

11. Name the following compounds.
    (a) HNO_2 (d) Cr_2O_3
    (b) $KClO$ (e) $Fe(OH)_3$
    (c) $(NH_4)_2CO_3$ (f) Na_2CrO_4

12. Write formulas for the following compounds.
    (a) potassium hydrogen sulphite
    (b) magnesium dihydrogen phosphate
    (c) lead(II) nitrate
    (d) sodium permanganate
    (e) ammonium sulphate
    (f) carbon disulphide

13. Name the following compounds.
    (a) $Ba(NO_2)_2$
    (b) H_2S
    (c) $LiHCO_3$
    (d) N_2O_3
    (e) $Sn(OH)_2$
    (f) K_2SO_3

CHALLENGE

14. Imagine that chemists have just made for the first time a compound of strontium, Sr, and nitrogen, N. How would they decide what the formula of the commpound is? Predict what you think the formula would be.

8 CHEMICAL REACTIONS

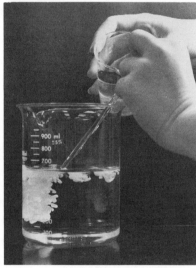

Chemical changes or reactions range from the dramatic explosion of dynamite to the quiet burning of a piece of paper; from the rapid formation of a precipitate to the slow decomposition of a rotting piece of wood. The changes that occur during a chemical reaction can be summarized by means of a chemical equation. In this chapter, you will find out about different types of chemical equations. You will also learn how to balance equations, how to classify chemical reactions, and how to predict whether or not a chemical reaction will occur when substances are mixed together.

8.1 Chemical Reactions – A Review

Despite the great variety of chemical reactions, one characteristic is common to all of them. Every chemical reaction produces at least one new substance. Recall the evidence that shows a chemical reaction has occurred. There may be (a) a change in colour, (b) the formation of a gas, (c) the formation of a precipitate and/or (d) the release or absorption of a large amount of energy. You probably would not observe all these changes in a single chemical reaction. Most chemical reactions, though, are accompanied by one or more of these four changes.

Exercise

1. Consider each of the four changes which give evidence for a chemical reaction. For each one, give an example of a chemical reaction where that change is observed.

8.2 Chemical Equations

A chemical reaction can be described by three different types of equations: **word equations**, unbalanced or **skeleton equations**, and **balanced equations**. Let us review how these terms apply to the reaction between magnesium and oxygen.

Word Equations

Magnesium burns brightly in oxygen to form a white powder. (See Figure 8.1.) The white powder has been shown experimentally to consist only of magnesium and oxygen. This compound is named magnesium oxide. The word equation for the reaction is

$$\text{magnesium} + \text{oxygen} \rightarrow \text{magnesium oxide}$$

The "+" means "reacts with", and the arrow means "to produce". The word equation is read as follows: "Magnesium reacts with oxygen to produce magnesium oxide." Magnesium and oxygen are the **reactants**. Magnesium oxide is the **product**.

A word equation gives limited information about a chemical reaction. The word equation shown above only identifies the reactants and the product. There is no indication of the masses of magnesium and oxygen that react. Neither is there any indication of the mass of magnesium oxide produced. It is often important to be able to determine the masses of the reactants and products in a chemical reaction. You will learn how this is done in Chapter 9.

Figure 8.1
Magnesium burning in air with a very bright flame to form magnesium oxide

Skeleton (Unbalanced) Equations

In a skeleton equation, chemical symbols and formulas are substituted for the reactants and products in the word equation. Here is the skeleton equation for the reaction between magnesium and oxygen:

$$Mg + O_2 \rightarrow MgO$$

Each symbol or formula is written independently of the others. The formulas of oxygen and magnesium oxide have been found by experiment to be O_2 and MgO respectively.

Exercise

2. Write skeleton equations for the following reactions.
   (a) Calcium reacts with water to produce calcium hydroxide and hydrogen.
   (b) Propane (C_3H_8) burns in air (oxygen) to produce carbon dioxide and water vapour.

Balanced Equations

The skeleton equation for the reaction between magnesium and oxygen is a useful summary of the changes that occur during the reaction. However, this summary still has limitations. The reactant (left) side of the equation shows one oxygen molecule consisting of two oxygen atoms. On the product (right) side of the equation there is only one oxygen atom. Can an oxygen atom simply disappear? According to the law of conservation of mass, the answer is obviously no. (Recall that, according to this law, the total mass of the reactants in a chemical reaction is equal to the total mass of the products. Dalton explained this law in terms of atoms. He stated that, in a chemical reaction, atoms are neither created nor destroyed, but are rearranged to form new substances.) Thus, the skeleton equation

$$Mg + O_2 \rightarrow MgO$$

is said to be **unbalanced**. The number of oxygen atoms on each side of the equation is different.

Balancing Simple Equations

A chemical equation must be **balanced**. In other words, the number of atoms of each element must be the same on both sides of the equation. This cannot be accomplished by changing the formula of any substance in the equation. Remember that the formula of a compound is found by experiment and cannot be changed to suit the wishes of chemistry students or teachers! They only way to balance an equation is to place numbers, called **coefficients**, in front of whole formulas. A coefficient applies to the whole formula that follows it. (See Figure 8.2.) The skeleton equation

$$Mg + O_2 \rightarrow MgO$$

can be balanced by, first of all, placing a coefficient 2 in front of MgO. This will balance the oxygen atoms.

$$Mg + O_2 \rightarrow 2MgO$$

To balance the magnesium atoms, a coefficient 2 must be placed in front of Mg.

$$2Mg + O_2 \rightarrow 2MgO$$

Figure 8.2
A coefficient in a balanced equation applies to the whole formula following it

NH_3 means one molecule of ammonia consisting of 1 N atom and 3 H atoms

$2 NH_3$ means two molecules of ammonia consisting of 2 N atoms and 6 H atoms

Check the number of atoms of each element on each side of the equation. You can see that the equation is now balanced. You can also see that all of the atoms that are present in the reactants are present in the product. There is therefore no change in mass during the reaction, in accordance with the law of conservation of mass. (See Figure 8.3.)

Many simple equations can be balanced using the method described above. This is sometimes known as **balancing by inspection**.

Figure 8.3
The unbalanced equation for the reaction between magnesium and oxygen suggests that the mass of the reactants is greater than the mass of the products. The balanced equation for the same reaction indicates that the mass of the reactants is equal to the mass of the products.

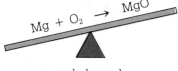

unbalanced equation

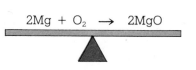

balanced equation

Exercise

3. Copy the following equations into your notebook. Balance the equations by inspection. *Do not write in your textbook.*
   (a) Na + Cl$_2$ → NaCl
   (b) H$_2$ + O$_2$ → H$_2$O
   (c) K + O$_2$ → K$_2$O
   (d) H$_2$ + Cl$_2$ → HCl
   (e) N$_2$ + H$_2$ → NH$_3$
   (f) CO + O$_2$ → CO$_2$

Balancing More Complex Equations

Many chemical reactions involve more than a total of three reactants and products. Balancing these more complex equations involves the same principles used in balancing the relatively simple equations. Sample Problem 8.1 shows how a more complex equation is balanced.

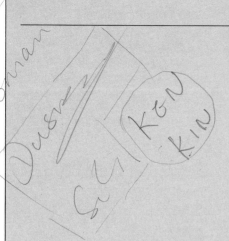

Sample Problem 8.1

Methane, CH_4, burns in air to produce carbon dioxide and water. Write a balanced equation for the reaction.

Solution

The word equation for the reaction is usually written first.

methane + oxygen → carbon dioxide + water

The skeleton equation is then written.

$$CH_4 + O_2 \rightarrow CO_2 + H_2O$$

Now you can begin to balance the equation. The carbon atoms are already balanced since there is one on each side of the equation. There are four hydrogen atoms on the left side of the equation, and only two on the right side. To balance the hydrogen atoms, a 2 must be placed in front of H_2O to give

$$CH_4 + O_2 \rightarrow CO_2 + 2H_2O$$

There are two oxygen atoms on the left side of the equation, but there are four oxygen atoms on the right side. To balance the oxygen atoms, a 2 is placed in front of O_2 on the left. The balanced equation is

$$CH_4 + 2O_2 \rightarrow CO_2 + 2H_2O$$

Check the number of atoms on each side of the equation.

Left side:
1 carbon atom
4 hydrogen atoms
4 oxygen atoms

Right side:
1 carbon atom
4 hydrogen atoms
4 oxygen atoms

The number of atoms of each element on each side of the equation is the same. The equation is therefore balanced.

Guidelines for Balancing Equations

When asked to balance more complex equations, some students may have difficulty knowing where to start. Although there are no set rules for balancing equations, you may find the following guidelines useful. They are not strict rules to be followed but are suggestions that you may find helpful.

1. Balance the element, other than hydrogen and oxygen, that has the greatest number of atoms in any reactant or product.
2. Balance other elements, other than hydrogen and oxygen.
3. Balance oxygen or hydrogen, whichever one is present in the

combined state. Leave until last whichever one is present in the uncombined state.

4. Check that the equation is balanced by counting the number of atoms of each element on each side of the equation.

When the equation is balanced, the coefficients should be in their lowest terms. For example, the balanced equation for the reaction between hydrogen and oxygen is

$$2H_2 + O_2 \rightarrow 2H_2O$$
$$not\ 4H_2 + 2O_2 \rightarrow 4H_2O$$

It is also customary to clear any fractions by multiplying all the coefficients by the common denominator of the fractions. For example,

$$C_3H_6 + \frac{9}{2}O_2 \rightarrow 3CO_2 + 3H_2O$$

is written as

$$2C_3H_6 + 9O_2 \rightarrow 6CO_2 + 6H_2O$$

Exercise

4. Copy the following equations into your notebook, then balance them. *Do not write in your textbook.*
   (a) $Na + H_2O \rightarrow NaOH + H_2$
   (b) $CaCO_3 + HCl \rightarrow CaCl_2 + H_2O + CO_2$
   (c) $Mg + HCl \rightarrow MgCl_2 + H_2$
   (d) $KClO_3 \rightarrow KCl + O_2$
   (e) $C_3H_8 + O_2 \rightarrow CO_2 + H_2O$
   (f) $Al + H_2SO_4 \rightarrow Al_2(SO_4)_3 + H_2$
   (g) $Fe_2O_3 + C \rightarrow Fe + CO_2$
   (h) $NH_3 + O_2 \rightarrow NO + H_2O$
   (i) $Fe(OH)_3 + H_2SO_4 \rightarrow Fe_2(SO_4)_3 + H_2O$
   (j) $CaC_2 + H_2O \rightarrow Ca(OH)_2 + C_2H_2$
   (k) $CH_3OH + O_2 \rightarrow CO + H_2O$
   (l) $FeCl_2 + KMnO_4 + HCl \rightarrow FeCl_3 + KCl + MnCl_2 + H_2O$
   (m) $Cu + HNO_3 \rightarrow Cu(NO_3)_2 + NO + H_2O$

Types of Chemical Reaction

Many thousands of chemical reactions take place in nature, in industry, and in laboratories all over the world. By searching for patterns in many, many reactions, chemists have found that there are four main types of chemical reaction. These are discussed in detail in the following sections. The four types of reaction are *synthesis*, *decomposition*, *single replacement*, and *double replacement*. This classification scheme is useful because it enables us to predict the outcome of reactions. It is also useful for writing equations for reactions.

8.3 Synthesis (Combination) Reactions

Laboratory Manual
In Experiment 8-1, you will observe the formation of new substances from synthesis reactions.

A **synthesis reaction** involves the direct combination of two substances to produce one new substance. We can write the following general equation to represent a synthesis reaction:

$$X + Y \rightarrow Z$$

In this general equation, X and Y may be elements or compounds, and Z is a compound.

The reactions shown below, which take place during the formation of acid rain, are examples of synthesis reactions.

$S + O_2 \rightarrow SO_2$ (two elements combining)
$2SO_2 + O_2 \rightarrow 2SO_3$ (an element and a compound combining)
$SO_3 + H_2O \rightarrow H_2SO_4$ (two compounds combining)

Exercise

5. Name the four compounds whose formulas are given in the equations of the synthesis reactions shown above.

You can often predict the formula of the product of a synthesis reaction involving elements if you know the valences of the reacting elements. For example, when sodium reacts with oxygen, we can predict that the product is sodium oxide and that the formula of the product is Na_2O. The balanced equation for the reaction is

$$4Na + O_2 \rightarrow 2Na_2O$$

How can you tell that a reaction between two substances will be a synthesis reaction? In some cases it should be obvious. For example, two elements reacting together can only combine to form a compound. The situation is more complex if either of the reactants is a compound. Some generalizations can be made, however. The equations shown below summarize several different types of synthesis reactions.

1. Metal or non-metal + oxygen → an oxide

$$2Mg + O_2 \rightarrow 2MgO$$
$$C + O_2 \rightarrow CO_2$$
$$4Al + 3O_2 \rightarrow 2Al_2O_3$$

A reaction involving the combination of an element with oxygen is known as a **simple oxidation reaction**.

2. Metal + non-metal → a binary compound

$$Ca + Cl_2 \rightarrow CaCl_2$$
$$2K + S \rightarrow K_2S$$

ionic bond
Binary-2
compound

3. Metal oxide + water → a base

$$MgO + H_2O \rightarrow Mg(OH)_2$$

4. Non-metal oxide + water → an acid

$$CO_2 + H_2O \rightarrow H_2CO_3$$

Two important examples of synthesis reactions do not fit into the categories above. The first is the synthesis of ammonia. Ammonia, NH_3, is one of the most important chemicals of the 20th century. Enormous amounts of ammonia are used in the production of fertilizers and other chemicals. (See Figure 8.4.) Ammonia is prepared by reacting hydrogen and nitrogen at a relatively high temperature and pressure in the presence of a catalyst. (A **catalyst** is a substance that speeds up the rate of a reaction without being permanently changed by the reaction.) The balanced equation for the reaction is

$$N_2 + 3H_2 \rightarrow 2NH_3$$

The second example is the synthesis of ammonium sulphate, a compound widely used as a solid fertilizer. It is prepared using the reaction represented by the following equation.

$$NH_3 + H_2SO_4 \rightarrow (NH_4)_2SO_4$$

Figure 8.4
This plant produces ammonia by combining nitrogen and hydrogen at high temperature and pressure.

Exercise

6. Copy the following incomplete equations into your notebook. Predict the formula of the product in each case, then balance the equation. *Do not write in your textbook.*
   (a) $Li + O_2 \rightarrow$
   (b) $P_4 + O_2 \rightarrow$
   (c) $Mg + N_2 \rightarrow$
   (d) $K_2O + H_2O \rightarrow$
   (e) $SO_2 + H_2O \rightarrow$

PUTTING CHEMISTRY TO WORK

BRENDA WANG / CLINICAL PHARMACIST

When a doctor in a hospital prescribes a drug for a patient, the prescription is reviewed by a clinical pharmacist, an expert in the chemistry of medicine. Brenda Wang is a clinical pharmacist in a hospital which specializes in treating cancer victims.

"I look for potential problems with a prescription that the prescribing doctor may be unaware of," Brenda says seriously. "There could be dangerous interactions among different drugs given to one patient; a patient could have an allergy to a certain kind of drug; or there could be a problem with the dose or frequency." Brenda's specialized knowledge allows her to recommend changes or improvements based on her review. "It would be very difficult for any physician to keep up with all the different aspects of pharmaceuticals as well as their own specialty. That is my role in the hospital — to provide them with that expertise."

A clinical pharmacist also reviews the medications given to a particular patient each week. "I check that the proper dose has been given in the correct manner.

Sometimes I need to suggest a different way of administering a drug so that the patient would tolerate it better."

To become a pharmacist, Brenda studied at a university faculty of pharmacy for four years followed by a nine-month apprenticeship. She chose this field because it combined her interest in science and math with a natural talent. "I was always good at memorizing information," Brenda grins. "There is a lot of that in pharmacy courses."

Brenda began her career as a retail pharmacist, dispensing drugs over the counter in a drugstore, but she wasn't satisfied. "I found being a retail pharmacist very monotonous. There was very little opportunity to follow a patient's progress after they left the counter."

Brenda is much happier, and busier, working in the hospital environment. "I deal with patients over a long term," she notes, "and I'm constantly in contact with very highly trained professional staff as well. It's a challenging and satisfying combination."

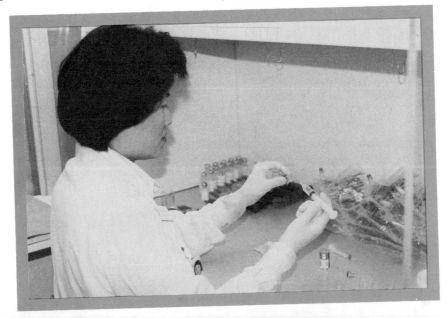

8.4 Decomposition Reactions

A **decomposition reaction** is a reaction involving the breaking up of a compound into simpler substances. The general equation for a decomposition reaction is

$$XY \rightarrow X + Y$$

XY is a compound; X and Y may be either elements or compounds. A decomposition reaction is exactly the reverse of a synthesis reaction.

An example of a decomposition reaction is the roasting of limestone (calcium carbonate) to produce lime (calcium oxide) and carbon dioxide.

$$CaCO_3 \rightarrow CaO + CO_2$$

The calcium oxide produced is reacted with water to form calcium hydroxide, which is used to make mortar and cement.

Sodium hydrogen carbonate (baking soda) also decomposes when heated.

$$2NaHCO_3 \rightarrow Na_2CO_3 + H_2O + CO_2$$

The carbon dioxide released is responsible for the rising of cakes and other baked goods. The same reaction, on an industrial scale, is one of the steps in the produciton of sodium carbonate, or soda ash. This is one of the raw materials in the manufacture of glass.

It is much more difficult to predict the products of a decomposition reaction than it is to predict those of a synthesis reaction. If the compound undergoing decomposition is a binary compound, the products are often the two elements in the compound. For example,

$$2H_2O \rightarrow 2H_2 + O_2$$
$$2HgO \rightarrow 2Hg + O_2$$

Many carbonates and hydrogen carbonates decompose on heating as shown in the earlier examples.

Laboratory Manual

You will observe the decomposition of magnesium carbonate and of water in Experiment 8-2.

Exercise

7. Predict the products of, and write balanced equations for, the following decomposition reactions. *Do not write in your textbook.*
   (a) $Ag_2O \rightarrow$
   (b) $CuCO_3 \rightarrow$
   (c) $KHCO_3 \rightarrow$

HYDROGEN FROM THE DECOMPOSITION OF WATER

Hydrogen is sometimes described as "the fuel of the future". This suggests that it is not presently used as a fuel. In fact, hydrogen has been used in the United States' space program for many years, and is the fuel used in the main engines of the space shuttle. Except in research vehicles, however, limited use is made of hydrogen as a fuel for conventional transportation.

Hydrogen is a clean fuel, since the only product of its combustion is water. There are some problems associated with the transportation and storage of the gas, but a great deal of

research is presently underway. It may only be a few years until the use of hydrogen as a fuel is more widespread. At the present time, there is no economical method of producing hydrogen in large quantities. Since water is a readily available cheap starting material, scientists are trying to find ways of decomposing water into hydrogen and oxygen economically. The decomposition can be carried out by passing electricity through water, but this is a very expensive process. One of the areas of current research involves the use of solar energy to generate electricity to decompose water.

8.5 Single Replacement Reactions

Figure 8.5
In this reaction, copper has replaced silver in silver nitrate solution. Beautiful crystals of silver are formed.

Laboratory Manual
Single replacement reactions are the subject of Experiment 8-3

In a **single replacement reaction**, one element replaces another element in a compound. (This type of reaction is also referred to as a **substitution reaction**.) (See Figure 8.5.) The general equation for a single replacement reaction is

$$X + YZ \rightarrow Y + XZ$$

X and Y are elements; YZ and XZ are compounds.

Some examples of single replacement reactions are

(a) $Fe + CuSO_4 \rightarrow Cu + FeSO_4$
(b) $Cu + 2AgNO_3 \rightarrow 2Ag + Cu(NO_3)_2$
(c) $Mg + 2HCl \rightarrow MgCl_2 + H_2$
(d) $2Li + H_2O \rightarrow 2LiOH + H_2$

Whether or not one element can replace another element in a compound depends on the relative reactivities of the two elements. For example, iron is able to replace copper in reaction (a) above because iron is more reactive than copper. As discussed in Chapter 3, chemists, by observing many reactions of metals, have been able to produce a list of metals in order of their chemical reactivity. This list is known as the **activity series** of the metals. (See Table 8.1.) The most reactive metals are at the top of the list. Hydrogen is usually included in the activity series even though it is a non-metal. Like metals, however, hydrogen forms positive ions. Some metals can replace hydrogen in compounds such as water or acids, for example, reactions (c) and (d) above.

Table 8.1   Activity Series of Metals

Potassium Sodium Barium Strontium Calcium	These metals react with cold water to produce hydrogen.
Magnesium Aluminum Zinc Iron Nickel Tin Lead	These metals react with acids to produce hydrogen.
Hydrogen	
Copper Mercury Silver Platinum Gold	These metals do not react with water or with acids.

uban

Figure 8.6
A spectacular reaction occurs when aluminum powder reacts with iron (III) oxide. The reaction is usually started with a length of burning magnesium. A large amount of energy is released as the reaction takes place. This reaction is known as the thermite reaction. At one time, this reaction was used to weld railroad rails in remote areas. The equation for the reaction is

$$2A\ell + Fe_2O_3 \longrightarrow 2Fe + A\ell_2O_3$$

Figure 8.7
Iron is prepared from iron ore (impure iron(III) oxide) by reacting the ore with carbon in a blast furnace. The equation for the reaction is

$$3C + 2Fe_2O_3 \longrightarrow 4Fe + 3CO_2$$

217

The activity series can help you predict whether or not a single replacement reaction takes place between an element and a compound. Any metal will replace in a compound a metal below it in the activity series. For example, magnesium will replace nickel in nickel compounds. Lead, however, will not replace aluminum in aluminum compounds since lead is listed below aluminum in the activity series. Figures 8.6 and 8.7 show examples of single replacement reactions which are important in industry.

Exercise

8. Copy the following incomplete equations into your notebook. Use the activity series to predict whether or not a reaction will occur in each case. Write a balanced equation for any reaction that does take place. *Do not write in your textbook.*
   (a) $Zn + Pb(NO_3)_2 \rightarrow$ (c) $Al + NiCl_2 \rightarrow$
   (b) $Ag + CuSO_4 \rightarrow$ (d) $Mg + H_2SO_4 \rightarrow$

An Activity Series for the Halogens

In Chapter 3 you learned that the halogens – fluorine, chlorine, bromine, and iodine – have different reactivities. Fluorine is the most reactive halogen. In fact, it is the most reactive element. The activity series for the halogens is shown in Table 8.2

Table 8.2   Activity Series for the Halogens

Fluorine	Most reactive
Chlorine	
Bromine	
Iodine	Least reactive

Any halogen can replace in a compound a halogen below it in the activity series. For example, chlorine can replace bromine in sodium bromide.

$$Cl_2 + 2NaBr \rightarrow 2NaCl + Br_2$$

This reaction is used to produce bromine from sea water.

Exercise

9. Predict whether or not a reaction will occur in each of the following cases. Write a balanced equation for any reaction that does occur. *Do not write in your textbook.*
   (a) $Cl_2 + KI \rightarrow$ (c) $Br_2 + NaI \rightarrow$
   (b) $I_2 + NH_4Cl \rightarrow$

8.6 Double Replacement Reactions

A **double replacement reaction** involves the reaction of two compounds to form two new compounds. The general equation for such a reaction is

$$WX + YZ \rightarrow WZ + YX$$

Reactions which result in the formation of a precipitate are examples of double replacement reactions. (See Figure 8.8.)

When solutions of potassium iodide and lead nitrate are mixed, a precipitate of lead iodide forms. The balanced equation for the reaction is

$$2KI + Pb(NO_3)_2 \rightarrow PbI_2 + 2KNO_3$$

Both potassium iodide and lead nitrate are ionic compounds. Potassium iodide solution contains K^+ ions and I^- ions. Lead nitrate solution contains Pb^{2+} ions and NO_3^- ions. Lead iodide is insoluble in water. When the Pb^{2+} ions from the lead nitrate solution and the I^- ions from the potassium iodide solution come in contact, the ions combine to form a precipitate of lead iodide, PbI_2. Potassium nitrate, on the other hand, is a soluble compound. Thus, K^+ ions and NO_3^- ions remain in solution.

A precipitate forms because the combination of a positive ion from one solution and a negative ion from the other solution forms a compound which is insoluble in water. To correctly predict the formation of a precipitate, you must know which substances are soluble in water and which substances are insoluble. You will learn some solubility rules in a later chapter. For the purposes of this chapter, you will be told whether a substance is soluble or insoluble.

Laboratory Manual
In Experiment 8-4, you will observe double replacement reactions.

Figure 8.8
Potassium iodide solution and lead nitrate solution react to form a precipitate of lead iodide.

Exercise

10. Carbonates, sulphides, and hydroxides are generally insoluble, except those of the group I metals and ammonium. All compounds of the group I metals and of ammonium are soluble.

    Use this information to write balanced equations for the following reactions. Indicate which product is the precipitate.

    (a) $Na_2CO_3 + CuSO_4 \rightarrow$ (c) $KOH + Co(NO_3)_2 \rightarrow$

    (b) $(NH_4)_2SO_4 + CaCl_2 \rightarrow$ (d) $Na_2S + Pb(NO_3)_2 \rightarrow$

Neutralization Reactions

Another important example of a double replacement reaction is a **neutralization reaction**. This involves the reaction between an acid and a base to form a salt and water. For example, hydrochloric acid and sodium hydroxide react to form sodium chloride and water.

$$HCl + NaOH \rightarrow NaCl + H_2O$$

Laboratory Manual
You will study the neutralization of an acid in Experiment 8-5.

This is similar to a precipitation reaction in that an ion from one solution (the H^+ ion from the acid) combines with an ion from the other solution (the OH^- ion from the base). The product of this combination is water, which is a molecular compound. (Recall that a precipitate is an ionic compound.) The first part of the name of the salt produced is derived from the base (and is always a metal or ammonium). The second part of the name of the salt is derived from the acid.

Exercise

11. Magnesium hydroxide, $Mg(OH)_2$, commonly called milk of magnesia, is often used to neutralize stomach acid, HCl. Write a balanced equation for the reaction.

12. Write balanced equations for the following neutralization reactions. Name the salt produced in each case. *Do not write in your textbook.*
    (a) $Cu(OH)_2 + HNO_3 \rightarrow$
    (b) $LiOH + H_3PO_4 \rightarrow$
    (c) $Fe(OH)_3 + H_2SO_3 \rightarrow$
    (d) $Ca(OH)_2 + HBr \rightarrow$

8.7 Chemical Reactions and Heat

Many chemical reactions are accompanied by the release of heat to their surroundings. Think, for example, of the burning of propane in a gas barbecue. Such reactions are said to be **exothermic**. (See Figure 8.9.) Much of the energy used in industry, in transportation, and in the home is supplied by exothermic chemical reactions. Some reactions used for this purpose include the combustion of oil, coal, gasoline, and natural gas.

To show that a reaction is exothermic, we indicate the release of heat or energy with the products of the reaction in the equation. The heat term, as it is called, is added to the right side of the equation. For example, the equation for the combustion of methane, the major constituent of natural gas, is

$$CH_4 + 2O_2 \rightarrow CO_2 + 2H_2O + energy$$

Laboratory Manual
In Experiment 8-6, you will examine and classify exothermic and endothermic reactions.

In some exothermic reactions, heat must be supplied to start the reaction. Once started, the reaction proceeds with the release of energy. The burning of a match is a good example. Before the match can burn, it must be given some energy. Striking a match generates the heat required to start the burning. Once the match has started to burn, however, it continues to do so until there is

Figure 8.9
The reaction between zinc and sulphur is, quite obviously, an example of an exothermic reaction.

nothing left to burn. The amount of energy released during the burning is greater than the small amount of energy needed to light the match.

Other chemical reactions take place only when heat is continuously supplied to the reactants. Such reactions are said to be **endothermic**. The decomposition of limestone, $CaCO_3$, discussed in section 8.4, is an example of an endothermic reaction. The heat term for an endothermic reaction is written on the left side of the equation. This indicates the supply of heat with the reactants. For the decomposition of limestone, the equation is

$$CaCO_3 + energy \rightarrow CaO + CO_2$$

The heat is much like a reactant – without it, the reaction cannot take place. If the supply of heat is removed, the reaction stops. If an endothermic reaction takes place at room temperature, the reaction absorbs heat from the surroundings and the temperature of the surroundings drops. The dissolving of ammonium nitrate in water is an example of such an endothermic reaction. Figure 8.10 shows a practical application of this reaction.

Figure 8.10
An endothermic reaction absorbs energy from its surroundings. In an athletic cold pack, ammonium nitrate and water are contained in separate compartments. When the barrier between them is broken, the substances mix, an endothermic reaction occurs, and the temperature of the water drops dramatically.

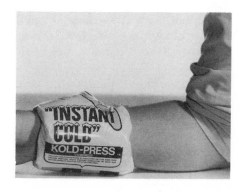

JOHN C. POLANYI AND CHEMICAL REACTIONS

Note: This autobiographical sketch of Dr. Polanyi was written approximately one and a half years before he, and two colleagues from the United States, were awarded the Nobel Prize for chemistry on October 15, 1986.

"At school my interests were in history, politics, and science. Since my older brother went on to study history and politics at the university, I decided that I would study chemistry. I never felt ordained for a career in chemistry, and I never abandoned my keen interest in history and politics.

"At first, science at the university was a disappointment to me. In the laboratory I found that it was necessary to follow procedures that had not been fully explained (if, indeed, the explanations were known) in order to obtain the 'right' result. Out of curiosity I would vary the method from that given in the laboratory manual, with the consequence that I routinely got the 'wrong' result. All this was symptomatic of the fact that I lacked the discipline to learn — or at any rate to learn with any degree of pleasure — the large number of rules that one must master before one can play the game of science.

"The rules of this magnificent game are a hundred times more complex than those of chess. But who would choose to play such a game, I wondered. Without any great enthusiasm, I persisted through some years of training, often reaching outside of science for the intellectual stimulus, excitement, and relevance to the world's great problems, which science seemed to me to lack.

"Slowly, however, by doing science, I came to be captivated and engrossed by the beauty of the subject matter and the thrill of the challenge that it offered.

"Today, I retain my interest in history and particularly my concern for the grave political questions with which late twentieth century humans are faced, but now I begrudge the time spent away from the excitement of the laboratory and the fellowship of the international community of scientists to which I belong.

"This is a peculiar story for a scientist to give of the birth of his interest in his subject. It happens to be my story, and may serve to illustrate that in the choice of a career, as in other major decisions, a lifelong commitment need not start with a love affair.

"I have given an account of how I became a scientist, but have said nothing about my choice of speciality. Here my story may be more commonplace. Chemistry was the field in which I felt at home. My father was for a major part of his career a chemist, and so the sight and smell of a chemistry laboratory was lodged in my subconscious. At a more rational level, I found in the course of my studies that physics was too arid for my tastes — too close to the abstractions of mathematics — whereas biology, for all the appeal of living systems, seemed lacking in quantitative structure. This was a personal assessment, based on a wealth of ignorance. I would not care to defend that judgement today, but it turned out to be a suitable one for me several decades ago.

"As I advanced through my undergraduate years, I had more decisions to make. I chose to concentrate on that part of chemistry that makes the greatest use of the methods of physics: 'chemical physics'. (By convention this is somewhat to the physics side of 'physical chemistry'.)

"Chemistry in the first half of this century tended to concentrate on the bulk properties of matter. In particular, the rates of chemical reactions (which I studied as a graduate student from 1949-1952, working toward a doctorate at Manchester University in England) were studied as a function of concentration and temperature. The reacting species — the individual atoms and molecules — by contrast have no such properties as concentration and temperature; they are described by their energy states of translation, vibration, rotation, and so forth.

"It was as if the chemists were sociologists, dealing with the behaviour of societies, and the physicists were psychologists recognizing only the rules of behaviour for individuals. It was evident in the early 1950s that the time was ripe for a melding of these viewpoints; if each constituted such a powerful system of thought alone, they would be still more powerful in conjunction.

"All this is clearer in retrospect than it was at the time. Nonetheless, hints of this new direction in chemical physics, and particularly in the treatment of chemical reaction rates, were readily apparent in many discussions I heard as a graduate student. These discussions provided me with the underlying direction of my interests for my entire time in science.

"Subsequently, as a Postdoctoral Fellow at the National Research Council in Ottawa and at Princeton University, in my reading, in the calculations that I attempted, and the experiments on which I embarked, I was looking for ways of relating the rates of chemical processes to the states of motion of the colliding molecules. It turned out to be easier initially to look at the motions of the newly born products of chemical reactions, than to control those in the reagents. Fortunately, these types of information are complementary since one can, so to speak, run the movie of a reactive collision backwards, whereupon the products become the reagents.

"On arriving at the University of Toronto in 1956, I suggested to my first graduate student,

Ken Cashion, that he look for the motion of hydrogen chloride molecules newly born in the reaction of atomic hydrogen with molecular chlorine. The reaction

$$H + Cl_2 \rightarrow HCl + Cl$$

was known to be strongly exothermic (200 kJ/mol of energy is liberated when the strong HCl bond is formed). It followed that there must be vibration and rotation in the new-born HCl molecule, with the balance of the energy present as relative motion (that is, translation) of the HCl and the Cl. What was entirely unknown was the apportionment of the reaction energy among these three types of motion. If this information could be obtained, one could start to picture the pattern of molecular motion involved in the act of chemical reaction.

"The apparatus that Ken Cashion constructed consisted of an electrical discharge in a tube containing H_2; the H_2 dissociated in the discharge to give hydrogen atoms. The H atoms then flowed down a short tube to a point where they encountered a jet of molecular chlorine. The reaction $H + Cl_2$ ensued, at low pressure so that the newly formed HCl product would, we hoped, emit radiation prior to being robbed of its energy in collisions.

"The emission, we knew from molecular physics, would be in the infrared (at a wavelength ~3 μm). In the first experiments the reaction zone was viewed by means of an infrared spectrometer through a sodium chloride prism, using a sensitive thermocouple as detector.

"With the reagents flowing, there was no visible emission whatever from the reaction zone. However, when Ken Cashion scanned the prism through the angles corresponding to the range of wavelengths from 2 μm to 4 μm, a broad peak was traced on the chart recorder attached to the thermocouple; HCl was being formed by the reaction in a range of vibrational energy states indicative of a highly specific motion in this new-born molecule.

"As the recorder traced the infrared, chemiluminescent spectrum of HCl, Ken Cashion, ⟶

having been recently ordained as a priest, felt that he should keep his excitement within the bounds of propriety; he merely beat on the top of the spectrometer with his fists, while shouting, 'Holy crowbar!'

"These experiments were reported in 1958 (*Journal of Chemical Physics*, Vol.29, p.455). The paper concluded with the sentence: "The method promises to provide for the first time information concerning the distribution of vibrational and possibly rotational energy among the products of a three-centre reaction."

"This work took its place in the early stages of development in laboratories around the world that led to the establishment of a new field called 'reaction dynamics' – the study of the motions of molecules in the course of the molecular collisions that we call *chemical reaction*."

Exercise

13. Identify each of the following reactions as either exothermic or endothermic.
    (a) $C + O_2 \rightarrow CO_2 +$ energy
    (b) $H_2 + I_2 +$ energy $\rightarrow 2HI$
    (c) $2C_4H_{10} + 13O_2 \rightarrow 8CO_2 + 10H_2O +$ energy
    (d) $NaOH + HCl \rightarrow NaCl + H_2O +$ energy
    (e) $2Pb(NO_3)_2 +$ energy $\rightarrow 2PbO + 4NO_2 + O_2$

14. Write balanced equations for the following reactions. Include the heat term with the equation.
    (a) Carbon monoxide burns in oxygen to produce carbon dioxide, releasing energy in the process.
    (b) When potassium chlorate, $KClO_3$, is heated strongly, it decomposes into potassium chloride and oxygen.

Looking Back – Looking Ahead

In this chapter we have seen that many chemical reactions can be classified as one of the following types.

1. Synthesis – a combining of two simple substances (elements or compounds) to form a single compound.
2. Decomposition – the breaking up of a compound to form simpler substances (elements or compounds).
3. Single replacement – one element takes the place of another element in a compound.
4. Double replacement – an ion from one compound combines with an ion from another compound to form a precipitate or water.

If we recognize a reaction as being of a certain type, we can often predict the products of the reaction. Not every chemical reaction can be classified as one of the four types described in this chapter.

You will study other types of reactions in a later chemistry course. It is a worthwhile practice during this course for you to attempt to classify the chemical reactions that you carry out as one of the four types discussed.

Vocabulary Checklist

You should now be able to define or explain the following terms.

activity series	double replacement	product	skeleton equation
balanced equation	reaction	reactant	substitution reaction
catalyst	endothermic	simple oxidation	synthesis reaction
coefficient	exothermic	reaction	unbalanced equation
decomposition	neutralization	single replacement	word equation
reaction	reaction	reaction	

Chapter Objectives

You should now be able to do the following.

1. Appreciate the wide variety of chemical reactions.

2. Recognize evidence that indicates a chemical reaction has occurred.

3. Explain why a chemical equation should be balanced.

4. Balance a variety of chemical equations.

5. Classify chemical reactions as synthesis, decomposition, single replacement or double replacement reactions, given the equation for the reaction.

6. Complete and balance equations for the four reaction types listed in objective 5, given the reactants.

7. Identify a reaction as exothermic or endothermic, given the equation for the reaction, including the heat term.

8. Write balanced equations, including the heat term, for exothermic and endothermic reactions.

Understanding Chapter 8

REVIEW

1. Indicate which of the following statements are incorrect. Rewrite them to make them correct.
   (a) An exothermic reaction is one that releases energy to the surroundings.
   (b) The substances formed during a reaction are the reactants.
   (c) A synthesis reaction involves the breaking down of a substance into simpler substances.
   (d) A substance that speeds up a chemical reaction without being permanently changed during the reaction is a catalyst.
   (e) The equation $N_2 + 3H_2 \rightarrow 2NH_3$ indicates that 1 g of nitrogen reacts with 3 g of hydrogen.
   (f) In a chemical reaction, the relative number of moles of each substance is given by the coefficient in front of the formulas in the balanced equation.

(g) The equation $Bi_2S_3 + 6HCl \rightarrow 2BiCl_3 + 3H_2S$ represents a single replacement reaction.

(h) The reaction between an acid and a base can be classified as both a double replacement and a neutralization.

(i) In a balanced equation, the number of moles of reactants must equal the number of moles of products.

(j) One mole of sucrose, $C_{12}H_{22}O_{11}$, contains 22 hydrogen atoms.

2. Identify each of the following reactions as either exothermic or endothermic.

(a) $2Li + 2H_2O \rightarrow 2LiOH + H_2 + energy$

(b) $3C + 2Fe_2O_3 + energy \rightarrow 4Fe + 3CO_2$

(c) $2NaHCO_3 + energy \rightarrow$
$$Na_2CO_3 + CO_2 + H_2O$$

APPLICATIONS AND PROBLEMS

3. Copy the following equations into your notebook, then balance them.

(a) $C_4H_{10} + O_2 \rightarrow CO_2 + H_2O$

(b) $Al_4C_3 + H_2O \rightarrow CH_4 + Al(OH)_3$

(c) $Ca_3(PO_4)_2 + SiO_2 + C \rightarrow$
$$P_4 + CaSiO_3 + CO$$

(d) $(NH_4)_2Cr_2O_7 \rightarrow N_2 + Cr_2O_3 + H_2O$

(e) $FeS_2 + O_2 \rightarrow Fe_2O_3 + SO_2$

(f) $NH_4NO_3 \rightarrow N_2O + H_2O$

(g) $Bi_2O_3 + H_2 \rightarrow Bi + H_2O$

(h) $FeCl_3 + (NH_4)_2S \rightarrow Fe_2S_3 + NH_4Cl$

(i) $C_{10}H_{16} + Cl_2 \rightarrow HCl + C$

(j) $NaI + MnO_2 + H_2SO_4 \rightarrow$
$$Na_2SO_4 + MnSO_4 + H_2O + I_2$$

4. The following reactions are examples of synthesis reactions. Complete the equations by writing the formulas of the products, then balance the equations.

(a) $Ca + N_2 \rightarrow$

(b) $CaO + CO_2 \rightarrow$

(c) $Al + Br_2 \rightarrow$

(d) $Li + O_2 \rightarrow$

(e) $P_4 + I_2 \rightarrow$

(f) $BaO + H_2O \rightarrow$

5. Complete the following equations for decomposition reactions, then balance the equations.

(a) $HgO \rightarrow$

(b) $MgCO_3 \rightarrow$

(c) $HCl \rightarrow$

(d) $H_2O \rightarrow$

6. Predict whether a reaction will take place for each of the following. Write a balanced equation for any reaction that does occur.

(a) $Mg + Pb(NO_3)_2 \rightarrow$

(b) $Cu + ZnCl_2 \rightarrow$

(c) $Cl_2 + LiI \rightarrow$

(d) $Sn + AgNO_3 \rightarrow$

7. In each of the following reactions, a precipitate forms. Noting that all compounds of group I metals are soluble, write a balanced equation for each reaction. Identify the precipitate in each case.

(a) $AgNO_3 + K_2S \rightarrow$

(b) $BaCl_2 + Na_2CO_3 \rightarrow$

(c) $Ca(NO_3)_2 + Li_2SO_4 \rightarrow$

(d) $CoCl_2 + NaOH \rightarrow$

8. Identify each of the following reactions as a synthesis, decomposition, single replacement, or double replacement reaction.

(a) $Mg(OH)_2 + 2HNO_3 \rightarrow Mg(NO_3)_2 + 2H_2O$

(b) $H_2O + SO_3 \rightarrow H_2SO_4$

(c) $FeCl_3 + 3NaOH \rightarrow Fe(OH)_3 + 3NaCl$

(d) $Cl_2 + ZnI_2 \rightarrow ZnCl_2 + I_2$

(e) $H_2SO_4 + Mg \rightarrow MgSO_4 + H_2$

(f) $K_2O + H_2O \rightarrow 2KOH$

(g) $2NaClO_3 \rightarrow 2NaCl + 3O_2$

(h) $2AsCl_3 + 3H_2S \rightarrow As_2S_3 + 6HCl$

(i) $2Pb(NO_3)_2 \rightarrow 2PbO + 4NO_2 + O_2$

(j) $3NaOH + H_3PO_4 \rightarrow Na_3PO_4 + 3H_2O$

9. The presence of small amounts of calcium chloride dissolved in water is responsible for one type of water hardness. Hard water does not form a lather with soap, but forms a scum instead. The calcium ions can be removed from the water by the addition of sodium carbonate. A precipitate of calcium carbonate forms, and this can be removed by filtration. Write a balanced equation for the reaction between calcium chloride and sodium carbonate.

10. Each of the following is a neutralization reaction. Complete and balance each equation.

(a) $KOH + HBr \rightarrow$

(b) $Ba(OH)_2 + HCl \rightarrow$

(c) $Mg(OH)_2 + H_2SO_4 \rightarrow$

(d) $Al(OH)_3 + HNO_3 \rightarrow$

(e) $Ca(OH)_2 + HClO_4 \rightarrow$

11. Write balanced equations for the following reactions, including the heat term in the equation.
    (a) Hydrogen and chlorine react to form hydrogen chloride, releasing energy in the process.
    (b) When sodium chlorate is heated, it decomposes into sodium chloride and oxygen.
    (c) Calcium reacts with water, releasing hydrogen gas and forming a solution of calcium hydroxide. The temperature of the water increases during the reaction.

12. Write balanced equations for the following reactions. Include the heat term with the equation.
    (a) When calcium oxide is added to water, calcium hydroxide is formed. The temperature of the water increases during the reaction.
    (b) The process of photosynthesis in plants produces glucose, $C_6H_{12}O_6$, and oxygen, from the raw materials carbon dioxide and water. Energy, usually from the sun, is absorbed during photosynthesis.
    (c) Magnesium reacts with sulphuric acid, forming magnesium sulphate and releasing hydrogen gas. The solution becomes warmer as the reaction proceeds.

CHALLENGE

13. For each of the following reactions, write a balanced equation. Indicate to which of the four types of reaction studied in this chapter each reaction belongs.
    (a) Ammonia gas and hydrogen chloride gas react to form ammonium chloride, a white solid.
    (b) Sulphur dioxide, formed during the burning of sulphur-containing coal, may be removed from smokestack gases by passing the gases over solid calcium oxide. Calcium sulphite is formed by this reaction.
    (c) If a bottle of hydrogen peroxide solution, H_2O_2, is left to stand at room temperature, oxygen gas is slowly released. After a period of time, the bottle contains only water.
    (d) In some water treatment treatment plants, solutions of aluminum sulphate and calcium hydroxide are added to the water. A "sticky" precipitate of aluminum hydroxide forms. This removes some of the small particles in the water as it settles to the bottom.

11 (c)

calcium + water $\longrightarrow$ hydrogen + calcium hydroxide + heat energy

$Ca + 2H_2O \longrightarrow H_2 + Ca(OH)_2 + energy$

12 (c) $Mg + H_2SO_4 \longrightarrow MgSO_4 + H_2 + energy$

13 (d) $Al \overset{3}{(SO_4)} \overset{2}{} \qquad Al_2(SO_4)_3$

$Al_2(SO_4)_3 + Ca(OH)_2 \longrightarrow Al(OH)_3 + CaSO_4$

IV CHEMICAL REACTION CALCULATIONS

9 CALCULATIONS USING BALANCED EQUATIONS

A fireworks manufacturer, a designer of car engines, a chemist in charge of a chemical manufacturing plant – what do these people have in common? Well, at least one common feature is that they all need to know the relative amounts of materials that react in some reactions in their specific area of work. For example, the fireworks manufacturer must mix together the correct amounts of chemicals; the wrong amounts could result in a disappointing failure or in an uncontrolled, violent explosion. The engine designer must know the relative amounts of fuel and oxygen needed for safe and efficient combustion. The chemist must know the relative amounts of reactants required to produce a certain product. In addition, the chemist is interested in knowing how much product should theoretically be formed from specific quantities of starting materials.

The study of the relative amounts of substances involved in a chemical reaction is known as **stoichiometry**. (The word stoichiometry comes from two Greek words meaning element and measure.) A knowledge of stoichiometry is essential whenever quantitative information about a chemical reaction is required.

9.1 Interpreting Balanced Equations

Laboratory Manual
In Experiment 9-1, you will determine the number of moles of copper obtained when a given number of moles of iron react with copper(II) sulphate.

The starting point for any problem involving quantities of reactants and products in a chemical reaction is the balanced equation for the reaction. Recall that in a balanced equation, the number of atoms of each element is the same on both sides of the equation. It is important that you understand exactly what information is conveyed by such an equation. Consider, for example, the main reaction that takes place in a gas barbecue. This is the combustion of propane, C_3H_8, to form carbon dioxide and water. The balanced equation for the reaction is

$$C_3H_8 + 5O_2 \rightarrow 3CO_2 + 4H_2O$$

You can interpret this equation in terms of the number of molecules of the reactants and products. The equation states that one molecule of propane reacts with five molecules of oxygen to produce three molecules of carbon dioxide and four molecules of water.

Although this is a perfectly correct interpretation, it is not very practical. Chemical reactions usually involve much larger numbers of molecules than those numbers suggested by the equation. For this reason, it is more common to interpret balanced equations in terms of moles rather than molecules. (See Figure 9.1.) Recall that 1 mol of a substance is the amount of substance that contains 6.02×10^{23} particles (atoms, ions, or molecules) of which the substance is composed. Thus, the equation tells us that 1 mol of propane reacts with 5 mol of oxygen to produce 3 mol of carbon dioxide and 4 mol of water vapour. The equation actually indicates the **mole ratio** of reactants and products. When propane burns in a plentiful supply of oxygen, the propane reacts with oxygen in the ratio of 1 mol of propane to 5 mol of oxygen. The quantities of products formed are always in the ratio of 3 mol of carbon dioxide to 4 mol of water vapour. The amount of propane used will not affect these ratios.

Figure 9.1
In a gas barbecue, the propane and oxygen react in the mole ratio 1 mol of propane to 5 mol of oxygen. The products are formed in the mole ratio 3 mol of carbon dioxide to 4 mol of water.

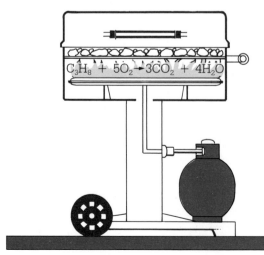

Sample Problem 9.1

Methanol, CH_3OH, is sometimes added to gasoline to enhance its burning properties. It may also be used as a fuel on its own. Methanol burns according to the following equation.

$$2CH_3OH + 3O_2 \rightarrow 2CO_2 + 4H_2O$$

If 3.50 mol of methanol are burned in a plentiful supply of oxygen,
(a) how many moles of oxygen are used?
(b) how many moles of water are produced?

Solution

We can solve this problem by using conversion factors as we did in Chapter 6. The correct conversion factors are obtained from the balanced equation for the reaction.

(a) Number of moles of O_2 consumed $= 3.50 \text{ mol } CH_3OH \times \dfrac{3 \text{ mol } O_2}{2 \text{ mol } CH_3OH}$

$= 5.25 \text{ mol } O_2$

Note that the conversion factor $\dfrac{3 \text{ mol } O_2}{2 \text{ mol } CH_3OH}$

is, in fact, the mole ratio relating oxygen and methanol obtained from the balanced equation for the reaction.

(b) Number of moles of H_2O formed $= 3.50 \text{ mol } CH_3OH \times \dfrac{4 \text{ mol } H_2O}{2 \text{ mol } CH_3OH}$

$= 7.00 \text{ mol } H_2O$

Again, the conversion factor $\dfrac{4 \text{ mol } H_2O}{2 \text{ mol } CH_3OH}$

is the mole ratio relating water and methanol obtained from the balanced equation.

Exercise

1. Ammonia, NH_3, is an important raw material in the manufacture of fertilizers. It is produced from a reaction shown by the following balanced equation.

$$N_2(g) + 3H_2(g) \rightarrow 2NH_3(g)$$

How many moles of the following are required to manufacture 5.0 mol of ammonia?
(a) nitrogen
(b) hydrogen

2. Refer again to the balanced equation for the combustion of propane.

$$C_3H_8(g) + 5O_2(g) \rightarrow 3CO_2(g) + 4H_2O(g)$$

(a) How many moles of oxygen are required to react with
   (i) 3.0 mol of propane? 15
   (ii) 20.0 mol of propane? 100
   (iii) 0.20 mol of propane? 1

(b) If 50.0 mol of oxygen are available, what is the maximum number of moles of propane that can be burned? 10

(c) If 200.0 mol of propane and 200.0 mol of oxygen are available, how many moles of each will be used? How many moles of carbon dioxide and of water will be formed?
   40   200   120   160

3. The combustion of octane, a major constituent of gasoline, is represented by the following equation.

$$2C_8H_{18}(g) + 25O_2(g) \rightarrow 16CO_2(g) + 18H_2O(g)$$

(a) How many moles of oxygen are required to react with 5.00 mol of octane?

(b) How many moles of octane must be burned to produce 7.2 mol of water?

(c) If 2.2 mol of carbon dioxide were produced from burning octane, how many moles of water were produced at the same time? How many moles of octane must have burned?

What an Equation Does *Not* Tell Us

A balanced equation for a reaction does not show us everything that may be known about a chemical reaction. For example, an equation conveys no information about the rate at which the reaction takes place. The reaction may be extremely slow, such as the formation of rust, or it may be very rapid, such as the explosion of a hydrogen-oxygen mixture.

In addition, a balanced equation does not indicate whether a reaction is exothermic (gives out heat) or endothermic (takes in heat). Recall from Chapter 7 that a heat term can be included with an equation to give us this information. A balanced equation does not show the conditions under which the reaction occurs. There is no indication, for example, of the temperature or pressure needed for the reaction to occur.

Finally, a balanced equation gives no information about the processes that occur at the molecular level. The equation indicates a starting point (the reactants) and a finishing point (the products). Nothing is suggested, however, about the processes that transform the individual atoms and molecules of the reactants into products. This information is essential to fully understand exactly how a chemical reaction takes place. The balanced equation is not the source of this information.

9.2 Solving "Mass-to-Mass" Problems

A "mass-to-mass" problem is one in which the mass of a reactant or product is given. You are then asked to calculate the mass of another reactant required or the mass of another product formed.

Solving stoichiometry problems is particularly important to chemists in fields such as food chemistry, drug chemistry, forensic chemistry, and any industry where chemicals are manufactured. Regardless of the type of chemical reaction involved, the method of solving the problem consists of the same series of steps. Let us consider an example.

Laboratory Manual
What product forms when sodium hydrogen carbonate decomposes by heating? Find out in Experiment 9-2.

Sample Problem 9.2

Iron can be produced form iron ore, Fe_2O_3, by reacting the ore with carbon monoxide, CO (Figure 9.2). Carbon dioxide is also produced. What mass of iron can be formed from 425 g of iron ore?

Solution

Step 1

Write the balanced equation for the reaction and identify the given and required substances.

$$Fe_2O_3 + 3CO \rightarrow 2Fe + 3CO_2$$

mass   mass
given   required

Step 2

Convert the given mass of iron ore to the corresponding number of moles of iron ore.

$$\text{Number of moles of } Fe_2O_3 = 425 \text{ g } Fe_2O_3 \times \frac{1 \text{ mol } Fe_2O_3}{159.6 \text{ g } Fe_2O_3}$$

$$= 2.66 \text{ mol } Fe_2O_3$$

Step 3

Use the mole ratio in the balanced equation for the conversion factor to find the number of moles of iron produced.

$$\text{Number of moles of iron produced} = 2.66 \text{ mol } Fe_2O_3 \times \frac{2 \text{ mol Fe}}{1 \text{ mol } Fe_2O_3}$$

$$= 5.32 \text{ mol Fe}$$

Step 4

Convert number of moles of iron to mass of iron in grams.

$$\text{Mass of iron produced} = 5.32 \text{ mol Fe} \times \frac{55.9 \text{ g Fe}}{1 \text{ mol Fe}}$$

$$= 297 \text{ g Fe}$$

Figure 9.2
Producing iron from iron ore in a blast furnace

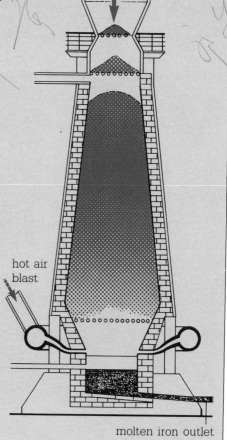

iron ore (Fe_2O_3)
coke (C)
calcium carbonate ($CaCO_3$)

hot air blast

molten iron outlet

Study the steps in Sample Problem 9.2 carefully. Make sure that you understand them. Figure 9.3 summarizes the steps involved in solving a mass-to-mass problem.

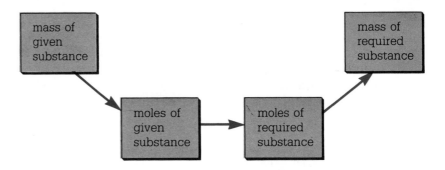

Figure 9.3
A stoichiometry problem involving only masses can be solved by following the steps shown.

Exercise

4. The main engines of the U.S. Space Shuttle are powered by liquid hydrogen and liquid oxygen. If 1.02×10^5 kg of liquid hydrogen are carried on a particular launch, what mass of liquid oxygen is necessary for all the hydrogen to burn? The equation for the reaction is

$$2H_2(g) + O_2(g) \rightarrow 2H_2O(g)$$

5. In some countries (for example, Brazil) ethanol, C_2H_5OH, is added to gasoline to produce a fuel called gasohol. Some race car engines are actually designed to operate on pure ethanol as fuel. Ethanol burns according to the following equation.

$$C_2H_5OH(aq) + 3O_2(g) \rightarrow 2CO_2(g) + 3H_2O(g)$$

(a) What mass of oxygen is required to react with 1200 g of ethanol?
(b) If 655 g of water is produced, what mass of ethanol is burned?

6. Sodium hydroxide or caustic soda, NaOH, used in many household drain cleaners, can be prepared by the reaction of calcium hydroxide, $Ca(OH)_2$, with sodium carbonate or soda ash, Na_2CO_3. Calcium carbonate, $CaCO_3$, is also formed in the reaction.
(a) Write the balanced equation for the reaction.
(b) What mass of sodium hydroxide can be prepared by the reaction of 225 g of calcium hydroxide with excess sodium carbonate?

7. One of the steps involved in obtaining a metal from its sulphide ore is the "roasting" of the sulphide. This step involves changing the sulphide into an oxide by heating it strongly in air. The balanced equation for the roasting of iron(II) sulphide is as follows:

$$4FeS(s) + 7O_2(g) \rightarrow 2Fe_2O_3(s) + 4SO_2(g)$$

The sulphur dioxide produced in this reaction, and in other reactions similar to it, is responsible for much of the acid rain that falls on North America. What mass of iron(III) oxide can be obtained by the roasting of 774 g of the sulphide ore? What mass of sulphur dioxide is produced?

9.3   Limiting Reactant Calculations

Recall that a balanced equation indicates the mole ratios in which the reactants are used up and products are formed. For example, methane, CH_4, burns in oxygen as follows.

$$CH_4(g) + 2O_2(g) \rightarrow CO_2(g) + 2H_2O(l)$$

According to this equation, methane and oxygen react in the ratio of 1 mol of methane to 2 mol of oxygen.

If 1 mol of methane and 3 mol of oxygen are mixed and allowed to react, then all of the methane and 2 mol of oxygen are used up. There is 1 mol of oxygen left over. The reactant that is completely used up, in this case methane, is called the **limiting reactant**. Any other reactants, oxygen in this example, are called **excess reactants**. The amount of product formed by the reaction is limited by the amount of the reactant that is completely used up – hence the term limiting reactant. In the methane-oxygen reaction being discussed, the addition of more oxygen to the reaction mixture has no effect on the amount of product formed. Sample Problems 9.3 and 9.4 illustrate how to determine which is the limiting reactant in a reaction.

Sample Problem 9.3

Zinc and sulphur react to form zinc sulphide according to the following balanced equation.

$$Zn + S \rightarrow ZnS$$

If 6.00 g of zinc and 4.00 g of sulphur are available for reaction, determine
(a) the limiting reactant;
(b) the mass of zinc sulphide produced.

Solution

(a) To determine which reactant is the limiting reactant, we first calculate the mole ratio of the given reactants in the problem. We then compare this ratio with the mole ratio in the balanced equation for the reaction.

   The given mass of each reactant must first be converted to the corresponding numbers of moles.

$$\text{Number of moles of Zn} = 6.00 \text{ g Zn} \times \frac{1 \text{ mol Zn}}{65.4 \text{ g Zn}}$$

$$= 0.0917 \text{ mol Zn}$$

$$\text{Number of moles of S} = 4.00 \text{ g S} \times \frac{1 \text{ mol S}}{32.1 \text{ g S}}$$

$$= 0.125 \text{ mol S}$$

According to the balanced equation for the reaction, 1 mol of Zn reacts with 1 mol of S. Therefore, 0.0917 mol of Zn requires 0.0917 mol of S. Since our calculations show that the mole ratio is actually 0.0917 mol of Zn to 0.125 mol of S, sulphur is the excess reactant. Zinc is the limiting reactant, since all of it is used up.

(b) The amount of product formed is determined by the amount of the limiting reactant. We found in part (a) that 0.0917 mol of Zn is used up. The balanced equation shows that 1 mol of Zn produces 1 mol of ZnS. Thus 0.0917 mol of Zn produces 0.0917 mol ZnS.

$$\text{Mass of ZnS produced} = 0.0917 \text{ mol ZnS} \times \frac{97.5 \text{ g ZnS}}{1 \text{ mol ZnS}}$$

$$= 8.94 \text{ g ZnS}$$

Sample Problem 9.4

Aluminum reacts with bromine to form aluminum bromide, as shown by the following balanced equation.

$$2Al + 3Br_2 \rightarrow 2AlBr_3$$

If 15.8 g of Al and 55.6 g of Br_2 are available for reaction, determine
(a) the limiting reactant;
(b) the mass of $AlBr_3$ produced.

Solution

(a) Number of moles of Al $= 15.8 \text{ g Al} \times \dfrac{1 \text{ mol Al}}{27.0 \text{ g Al}}$

$$= 0.585 \text{ mol Al}$$

$$\text{Number of moles of Br}_2 = 55.6 \text{ g Br}_2 \times \frac{1 \text{ mol Br}_2}{159.8 \text{ g Br}_2}$$

$$= 0.348 \text{ mol Br}_2$$

We see from the balanced equation that 2 mol of Al reacts with 3 mol of Br_2. Therefore, 0.585 mol Al reacts with

$$0.585 \text{ mol Al} \times \frac{3 \text{ mol Br}_2}{2 \text{ mol Al}}$$

$$= 0.878 \text{ mol Br}_2$$

Since only 0.348 mol of Br_2 is available, Br_2 is obviously the limiting reactant.

(b) We use the number of moles of the limiting reactant, that is, 0.348 mol of Br_2, to calculate the amount of product formed.

$$\text{Number of moles of AlBr}_3 \text{ formed} = 0.348 \text{ mol Br}_2 \times \frac{2 \text{ mol AlBr}_3}{3 \text{ mol Br}_2}$$

$$= 0.232 \text{ mol AlBr}_3$$

$$\text{Mass of AlBr}_3 \text{ formed} = 0.232 \text{ mol AlBr}_3 \times \frac{266.7 \text{ g AlBr}_3}{1 \text{ mol AlBr}_3}$$

$$= 61.9 \text{ g AlBr}_3$$

Exercise

8. Sodium and chlorine react according to the following equation.

$$2Na + Cl_2 \rightarrow 2NaCl$$

If 12.5 g of sodium and 25.5 g of chlorine are available for reaction, determine
(a) the limiting reactant;
(b) the mass of NaCl produced.

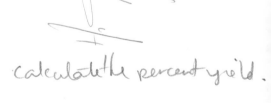 *(handwritten) If the actual yield of NaCl is 20.0g calculate the percent yield.*

9. One type of stomach antacid consists of magnesium hydroxide, $Mg(OH)_2$. This reacts with stomach acid, which is mainly hydrochloric acid, as follows.

$$Mg(OH)_2 + 2HCl \rightarrow MgCl_2 + 2H_2O$$

If 2.86 g of $Mg(OH)_2$ and 5.15 g of HCl are available for reaction, what mass of $MgCl_2$ is produced?

10. Hydrazine, N_2H_4, is used as a rocket fuel. It reacts with oxygen as follows.

$$N_2H_4 + O_2 \rightarrow N_2 + 2H_2O$$

In a particular rocket engine, 2.29 kg of hydrazine and 3.14 kg of oxygen are available for reaction.
(a) Show by calculation which reactant is in excess.
(b) Calculate the mass of water produced.

9.4 Percentage Yield

In all of the problems we have dealt with so far in this chapter, we have carried out calculations based on balanced equations for the reactions. It is important to realize that the quantities we calculated are theoretical quantities. They are not necessarily those that would actually be obtained by experiment or in an industrial process.

As you have seen, it is relatively easy to calculate the mass of product that *should* be produced from a known mass of reactant. This is called the **theoretical yield**. For many chemical reactions the **experimental yield**, which is the mass of product *actually* obtained, is less than the theoretical yield. There are a number of possible reasons for this. Side reactions may occur, producing other products, thus reducing the quantity of the desired product. The experimental technique may result in some of the product being lost. Or the reaction may not go to completion. Whatever the reason, the experimental yield is usually less than the theoretical yield.

The relationship between experimental yield and theoretical yield is expressed in a quantity called **percentage yield**.

$$\text{Percentage yield} = \frac{\text{experimental yield}}{\text{theoretical yield}} \times 100\%$$

Both the experimental yield and the theoretical yield must be expressed in the same units, usually in grams or kilograms.

In the chemical industry it is important for a manufacturer of a chemical to know the efficiency of the production process. To determine this, a chemist or chemical engineer must calculate the theoretical yield of product and compare it with the experimental yield. If the percentage yield is low, the chemist will want to find ways to improve it. These may include changing the conditions of the

Laboratory Manual
You will predict and determine the percentage yield of a precipitate in Experiment 9-3.

Sample Problem 9.5

The overall balanced equation for the production of ethanol, C_2H_5OH, from sugar is as follows.

$$C_6H_{12}O_6(aq) \rightarrow 2C_2H_5OH(aq) + 2CO_2(g)$$

(a) What is the theoretical yield of ethanol available from 10.0 g of sugar?
(b) If in a particular experiment, 10.0 g of sugar produces 0.664 g of ethanol, what is the percentage yield?

Solution

(a) Number of moles of sugar used $= 10.0 \text{ g sugar} \times \dfrac{1 \text{ mol sugar}}{180 \text{ g sugar}}$

$= 0.0555 \text{ mol sugar}$

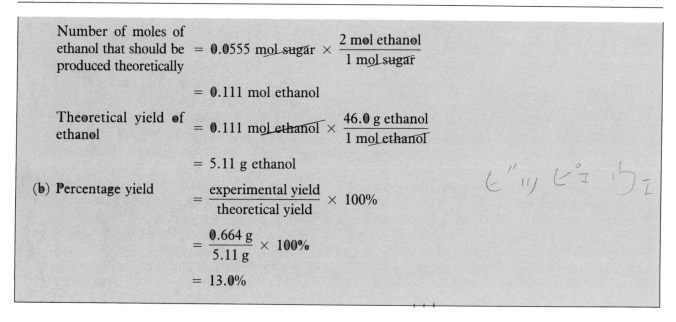

Number of moles of ethanol that should be produced theoretically	$= 0.0555 \text{ mol sugar} \times \dfrac{2 \text{ mol ethanol}}{1 \text{ mol sugar}}$
	$= 0.111 \text{ mol ethanol}$
Theoretical yield of ethanol	$= 0.111 \text{ mol ethanol} \times \dfrac{46.0 \text{ g ethanol}}{1 \text{ mol ethanol}}$
	$= 5.11 \text{ g ethanol}$
(b) Percentage yield	$= \dfrac{\text{experimental yield}}{\text{theoretical yield}} \times 100\%$
	$= \dfrac{0.664 \text{ g}}{5.11 \text{ g}} \times 100\%$
	$= 13.0\%$

reaction such as temperature and pressure, or changing the experimental procedures. For obvious reasons, a manufacturer of a chemical would like to have a percentage yield as close to 100% as possible. In practice this may be difficult or even impossible. Some industrial chemical reactions even have a percentage yield of less than 50%. For example, ammonia is produced industrially by reacting nitrogen and hydrogen gases at about 500°C and at a pressure approximately 400 times atmospheric pressure. (See Figure 9.4.) Under these conditions, the reaction does not go to completion. The percentage yield is about 30%. In situations like this, where the percentage yield is low, it is common practice to recycle unused reactants. This reduces wastage of the starting materials, and obviously reduces the manufacturer's costs.

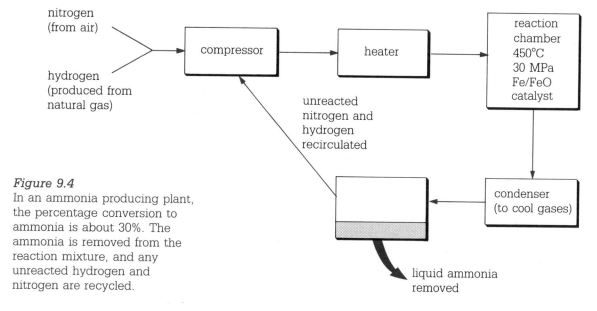

Figure 9.4
In an ammonia producing plant, the percentage conversion to ammonia is about 30%. The ammonia is removed from the reaction mixture, and any unreacted hydrogen and nitrogen are recycled.

ETHANOL PRODUCTION

The production of ethanol, C_2H_5OH, is one of the oldest chemical processes known. When and how the process was discovered are not known, but the discovery is believed to have been accidental. Today, ethanol is produced by the fermentation of organic materials such as grains. In fermentation, the sugars in grains are converted to alcohol and carbon dioxide by the enzymes in yeast.

As fermentation occurs, the concentration of ethanol in the reaction mixture increases. When the concentration of ethanol reaches about 15% by volume, the enzymes are unable to continue their work and the reaction stops. This is the most concentrated ethanol solution attainable from the process unless further steps are taken. For example, distillation is widely used to produce more concentrated ethanol solutions.

When the ethanol solution is distilled, the ethanol distills first. (Ethanol has a lower boiling point than water.) The distillate has a very high concentration of ethanol – about 93%. This highly concentrated ethanol solution is extremely dangerous. It can prove fatal if consumed in this form. In diluted form, ethanol is used in alcoholic beverages. Alcoholic beverages consist of ethanol solutions in various concentrations, ranging from 4%–5% ethanol in beer, to 9%–12% ethanol in wine, 20% in sherry, and 35%–40% in distilled spirits such as rye, rum, gin, scotch, etc. In the production of beer and wine, no distillation is necessary. Since the other beverages contain more than 15% ethanol, the solution produced by fermentation must be distilled.

In addition to differences in ethanol concentration, alcoholic beverages also differ in the source of the sugar or starch which is fermented. Beer originates from barley, most wine is produced from grapes, and distilled spirits are produced from various sources, for example, vodka comes from rye and rum from molasses.

Even in diluted form, ethanol cannot be considered to be a "safe" drink. Although it has the properties of a mild sedative when taken in small quantities, there are serious problems associated with the misuse and abuse of alcohol. Consumption of ethanol in large quantities, particularly over a long period of time, can cause severe damage to the liver and brain. In addition, it has been shown that approximately 50% of all fatal automobile accidents in North America involve at least one driver who has earlier consumed alcohol. Many experts in the area of drug counselling consider that alcohol abuse is the major drug problem that our society faces today.

Producing ethanol by the fermentation of the sugar in grain

Exercise

11. Solid carbon dioxide (dry ice) may be used for refrigeration. Some of this carbon dioxide is obtained as a by-product when hydrogen is produced from methane in the following reaction.

$$CH_4(g) + 2H_2O(g) \rightarrow CO_2(g) + 4H_2(g)$$

(a) What mass of carbon dioxide should be obtained from the complete reaction of 1250 g of methane?

(b) If the actual yield obtained is 3000 g, what is the percentage yield?

12. Ammonium nitrate is an important compound used both as a fertilizer and as an explosive. It is produced by reacting ammonia with concentrated nitric acid.

$$NH_3(s) + HNO_3(ag) \rightarrow NH_4NO_3(s)$$

(a) What mass of ammonium nitrate can theoretically be produced from the reaction of 375.0 g of ammonia with excess nitric acid?

(b) If the percentage yield is 88.50%, what mass of ammonium nitrate is actually obtained?

Looking Back – Looking Ahead

In this chapter, you have studied the meaning of balanced chemical equations, and used the mole concept in solving problems involving chemical equations. You have also learned to calculate the percentage yield of chemical reactions. All of the problems you solved dealt only with the masses of reactants and products.

In the next unit, you will investigate gases. In particular, you will extend your study of stoichiometry to include gas volumes in Chapter 12. Then, in Chapter 14, you will apply your problem solving skills to chemical reactions which take place in solution.

Vocabulary Checklist

You should now be able to define or explain the meaning of the following terms.

excess reactant	limiting reactant	percentage yield	theoretical yield
experimental yield	mole ratio	stoichiometry	

PUTTING CHEMISTRY TO WORK

MARY BROWN / QUALITY ASSURANCE INSPECTOR

The preparation of food is quite correctly part of chemistry. In order for a recipe to work consistently and well, especially in large quantities, food chemists must be aware of the characteristics of compounds and their action in mixtures and solutions. They must also know the effects of heating, pH, and many other factors.

Mary Brown is a quality assurance inspector for a large bakery. "I am responsible for overseeing the entire baking process as well as monitoring the quality of the finished products — the hamburger and hot dog buns which we prepare for a popular restaurant chain in Ontario." The key is to maintain consistency in the product, so that consumers will continue to be satisfied, whether in restaurants in Cornwall, Toronto, Kenora or any other Ontario city.

The bakery operates around the clock. "Once per shift, I determine the acidity of the brew — a mixture of yeast, water, sugar, and additives that will be mixed with the flour to make the dough. I also do some laboratory testing of incoming ingredients." Once every hour, Mary also performs spot checks of the entire baking process. "I make sure everything is consistent — from the mass of the dough pieces to the temperatures and humidities in the ovens." Each half hour, Mary inspects and rates the finished buns as they are packaged.

The many people who work at other jobs in the plant depend upon the quality of Mary's work. "It's a responsibility that gives me a great deal of satisfaction. On my shift, I'm the only person testing the chemistry of the products. I have to detect any problem and trace the cause so it can be corrected." Mary smiles. "Consumers can take for granted that the food they buy from one day to the next will be up to the standard they expect because of the work of inspectors such as myself. It's a good feeling."

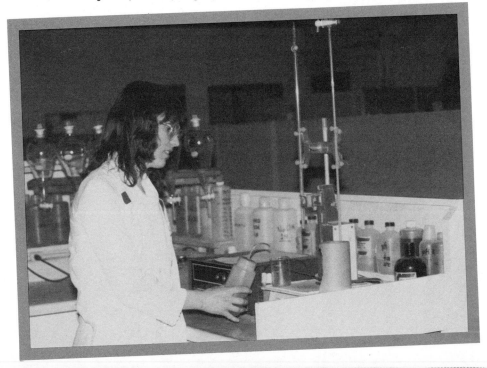

Chapter Objectives

You should now be able to do the following.

1. Interpret a balanced chemical equation in terms of mole ratios of reactants and products.

2. Appreciate the need for quantitative information about chemical reactions.

3. Appreciate that a chemical equation gives limited information about a chemical reaction.

4. Calculate the number of moles of a reactant consumed or product formed, given a balanced equation and the number of moles of another reactant consumed or product formed.

5. Calculate the mass of a reactant consumed or product formed, given a balanced equation and the mass of another reactant consumed or product formed.

6. Determine the excess reactant and the limiting reactant, given a balanced equation and the mass of each reactant present.

7. Calculate the percentage yield of any product in a chemical reaction, given the experimental yield and the balanced equation.

Understanding Chapter 9

REVIEW

1. What is stoichiometry and why is it important?

2. Consider the following balanced equation.

$$4FeS(s) + 7O_2(g) \rightarrow 2Fe_2O_3(s) + 4SO_2(g)$$

State three important pieces of information about this reaction which are not given by the equation.

APPLICATIONS AND PROBLEMS

3. The reaction between hydrazine, N_2H_4, and hydrogen peroxide, H_2O_2, is used to power some rocket engines. The balanced equation for the reaction is as follows.

$$N_2H_4 + 7H_2O_2 \rightarrow 2HNO_3 + 8H_2O$$

2.68 mol of hydrazine were completely consumed in a rocket engine.
(a) How many moles of H_2O_2 were required?
(b) How many moles of HNO_3 were produced?
(c) How many moles of water were formed?

4. The metal tungsten, which is used to make filaments for incandescent light bulbs, can be obtained from its oxide by heating it with hydrogen.

$$WO_3(s) + H_2(g) \rightarrow W(s) + 3H_2O(g)$$

(a) What mass of tungsten can be obtained from 250 g of tungsten(VI) oxide?
(b) What mass of hydrogen is required in part (a)?

5. Oxygen can be prepared in the laboratory by adding manganese dioxide, MnO_2, to hydrogen peroxide, H_2O_2. The hydrogen peroxide decomposes according to the equation

$$2H_2O_2(l) \rightarrow 2H_2O(l) + O_2(g)$$

The manganese dioxide is a catalyst and is unchanged at the end of the reaction. Calculate the mass of oxygen produced by the decomposition of 4.86 g of hydrogen peroxide.

6. The electrical energy requirements of the Space Shuttle are supplied by three fuel cells. These are devices that allow hydrogen and oxygen to react, producing electricity in the process. Water is formed during the reaction. In fact, the total water needs of the Shuttle are met in this way. If each crew member requires 3.0 kg of water per day, calculate the mass of oxygen needed to produce the water requirements for a six-person crew on a seven-day mission.

7. Our bodies use glucose, $C_6H_{12}O_6$, as a source of energy. The glucose reacts with oxygen through a complex series of reactions. The overall reaction can be represented by the following equation.

$$C_6H_{12}O_6(aq) + 6O_2(g) \rightarrow 6CO_2(g) + 6H_2O(l)$$

(a) What mass of oxygen is needed to completely react with 12.4 g of glucose?
(b) What mass of water is produced at the same time?

THE TARNISHING OF SILVER

Sodium thiosulphate reacts with hydrogen sulphide more readily than silver reacts. This property can be used to protect silver from tarnishing. If some sodium thiosulphate crystals in a porous container – a cloth, for example – are placed in a silverware drawer or chest, the sodium thiosulphate will react with any traces of hydrogen sulphide. The silverware will not tarnish.

8. You may have seen the thick haze commonly found over highly industrial areas. One of the substances responsible for this is ammonium sulphate, $(NH_4)_2SO_4$, which forms in the air by the reaction between ammonia, NH_3, and sulphuric acid, H_2SO_4.
    (a) Write the balanced equation for the reaction.
    (b) What mass of ammonium sulphate would be formed by the complete reaction of 725 g of ammonia?

9. Silver is widely used for jewellery and for tableware. Although the metal is highly resistant to oxidation by air at ordinary temperatures, it becomes tarnished when exposed to air containing small amounts of hydrogen sulphide. The tarnish is a layer of silver sulphide. The overall reaction is:

    $$4Ag(s) + 2H_2S(g) + O_2(g) \rightarrow 2Ag_2S(s) + 2H_2O(g)$$

    What mass of silver sulphide would form from the reaction of 0.015 g of silver?

10. You may have seen the reaction between sugar and concentrated sulphuric acid to form carbon. The sulphuric acid acts as a dehydrating agent and the equation for the reaction is as follows.

    $$C_6H_{12}O_6(s) + H_2SO_4(l) \rightarrow 6C(s) + 6H_2O(l) + H_2SO_4(aq)$$

    What mass of carbon can be obtained from the dehydration of 20.8 g of sugar?

11. Sodium phosphate, Na_3PO_4, is sometimes referred to as trisodium phosphate or TSP. It is often used for cleaning purposes. (See Figure 9.5.) The substance must be handled with extreme care since it is corrosive. Sodium phosphate may be prepared by the reaction shown below.

    $$3NaOH(aq) + H_3PO_4(aq) \rightarrow Na_3PO_4(aq) + 3H_2O(l)$$

    Calculate the mass of sodium phosphate that can be produced from 12.4 g of sodium hydroxide reacting with excess phosphoric acid.

Figure 9.5
TSP, trisodium phosphate, is used as an all-purpose cleaning material around the home. It is a corrosive substance and must be handled with care.

SPECIAL TOPIC 9.3

NITROGLYCERINE

Nitroglycerine is such an unstable substance that it may explode even with movement. Alfred Nobel, a Swedish chemist and inventor, found that when nitroglycerine was mixed with a clay-like material called kieselguhr or diatomaceous earth, it became much less sensitive. This allowed nitroglycerine to be transported much more safely.

The mixture of kieselguhr and nitroglycerine is more commonly known as dynamite. When Nobel died in 1896, he left a fortune worth several millions of dollars, largely made in manufacturing explosives. This money was used to establish the Nobel Prizes, awarded annually for outstanding achievement in the areas of science, literature, and peace.

12. In terms of annual production, nitric acid, HNO_3, is one of the top ten industrial chemicals in North America. It is used in the production of fertilizers such as ammonium nitrate, and explosives such as nitroglycerin. The first step in the manufacture of nitric acid by the Ostwald process is the reaction between ammonia and oxygen, in the presence of a catalyst. The equation for the reaction is as follows.

$$4NH_3(g) + 5O_2(g) \rightarrow 4NO_2(g) + 6H_2O(g)$$

(a) What mass of oxygen is required to completely react with 1.22 kg of ammonia?
(b) What mass of NO_2 is produced at the same time?

13. Aspirin, or acetylsalicylic acid, $C_9H_8O_4$, is the major constituent of many headache relief compounds. It is produced from salicylic acid, $C_7H_6O_3$, and acetic anhydride, $C_4H_6O_3$, according to the following equation.

$$C_7H_6O_3 + C_4H_6O_3 \rightarrow C_9H_8O_4 + C_2H_4O_2$$

What mass of salicylic acid is needed to produce 200 kg of aspirin?

14. Lead(II) chloride solution reacts with sodium chromate solution to form a precipitate of lead(II) chromate.

$$PbCl_2(aq) + Na_2CrO_4(aq) \rightarrow$$
$$PbCrO_4(s) + 2NaCl(aq)$$

In an experiment, a student mixed a solution containing 20.0 g of lead(II) chloride with another solution containing sodium chromate.

(a) What mass of sodium chromate is required to react with all of the lead(II) chloride?
(b) What is the theoretical yield of lead(II) chromate?
(c) Calculate the percentage yield if the student obtained 22.0 g of lead chromate in an experiment.

15. Phosphorus is an essential element for plant growth. One way of getting the phosphorus into the soil is to use a water-soluble phosphate fertilizer. One such fertilizer is prepared by reacting calcium phosphate, an insoluble compound, with sulphuric acid. The equation of the reaction is as follows.

$$Ca_3(PO_4)_2(s) + 2H_2SO_4(aq) \rightarrow$$
$$Ca(H_2PO_4)_2(aq) + 2CaSO_4$$

The dried mixture of products is known as superphosphate. The important component of the mixture is the calcium dihydrogen phosphate. Calculate the mass of calcium dihydrogen phosphate obtained from 1000 kg of calcium phosphate.

16. It has been estimated that approximately 4×10^9 t of oxygen are produced annually by the process of photosynthesis. The equation for this reaction is

$$6CO_2 + 6H_2O \rightarrow C_6H_{12}O_6 + 6O_2$$

(a) What mass of carbon dioxide is consumed in producing this mass of oxygen?
(b) What mass of carbohydrate, $C_6H_{12}O_6$, is formed?

17. Ethane, C_2H_6, burns in oxygen as follows.

$$2C_2H_6 + 7O_2 \rightarrow 4CO_2 + 6H_2O$$

What mass of carbon dioxide is produced if 10.2 g of ethane and 44.6 g of oxygen are mixed and ignited?

18. Sodium phosphate, Na_3PO_4, can be prepared by the following reaction.

$$3NaOH + H_3PO_4 \rightarrow Na_3PO_4 + 3H_2O$$

A student examines this equation and decides that if 36.0 g of NaOH reacts with 12.0 g of H_3PO_4, 12.0 g of Na_3PO_4 will be produced.
(a) Determine the limiting reactant.
(b) What mass of sodium phosphate should be produced?
(c) What is the flaw in the student's reasoning?

19. One of the steps in the manufacture of sulphuric acid (and, incidentally, in the formation of acid rain) is the oxidation of sulphur dioxide to sulphur trioxide.

$$2SO_2(g) + O_2(g) \rightarrow 2SO_3(g)$$

If 3.55 kg of sulphur dioxide is burned in excess oxygen and 4.20 kg of sulphur trioxide is produced, calculate the percentage yield.

20. Acetylene, C_2H_2, used in welding, can be produced by the reaction of calcium carbide, CaC_2, with water. The equation for the reaction is as follows.

$$CaC_2(s) + 2H_2O(l) \rightarrow Ca(OH)_2(aq) + C_2H_2(g)$$

(a) What is the theoretical yield of acetylene if 2.38 g of calcium carbide is reacted with water?
(b) If 0.77 g of acetylene is actually produced, calculate the percentage yield.

21. Esters are compounds which usually have very pleasant aromas. A number of esters are used as artificial flavouring agents in foodstuffs. Ethyl butanoate is one such ester – it has the flavour of pineapple. It is prepared according to the following equation.

$$\underset{\text{butanoic acid}}{C_3H_8COOH(l)} + \underset{\text{ethanol}}{C_2H_5OH(l)} \rightarrow$$

$$C_3H_8COOC_2H_5(l) + H_2O(l)$$

If the percentage yield in this reaction is 35.5%, calculate the mass of of ester produced when 12.5 g of butanoic acid reacts with excess ethanol.

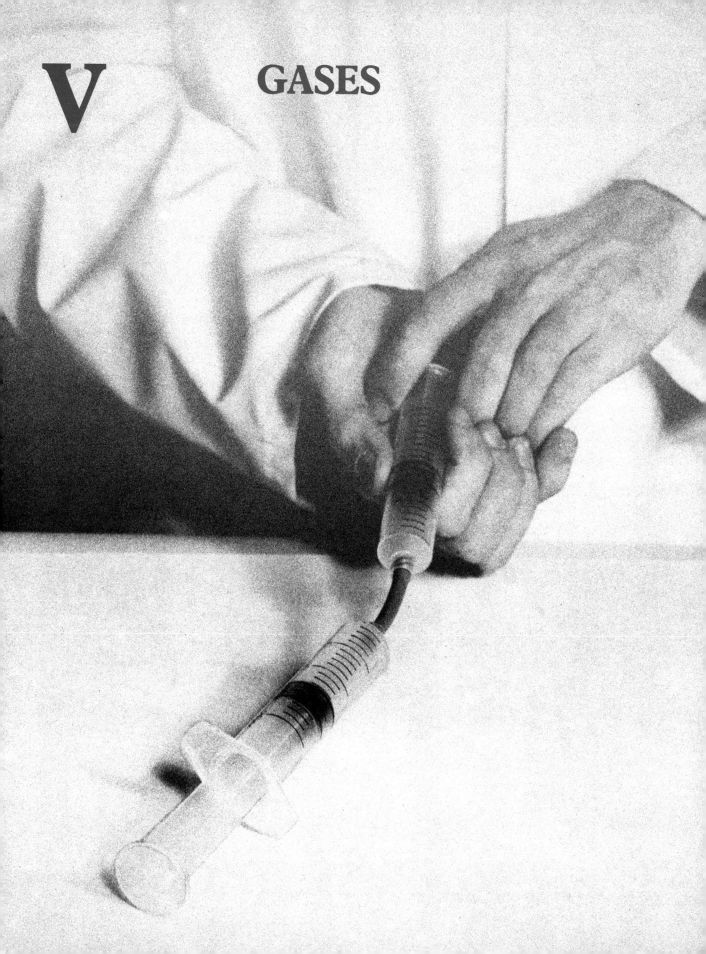

V GASES

10 STATES OF MATTER – THE PARTICLE THEORY

So far in this textbook you have studied the reactions of elements and compounds. You have used scientific models to help explain why those reactions take place. Using the mole concept you can also predict the quantity of reactants and products in a chemical reaction. In order for chemical reactions to take place, however, the atoms of the reacting substances must come into contact. How does this happen and how do chemists control these reacting substances?

Matter exists in a bewildering variety of forms. These forms can be classified into three states, namely, solids, liquids, and gases. Imagine the states of matter to be like a collection of small plastic pieces. The particles that make up matter can be joined together in rigid shapes (solid), or they can be poured as a collection of separate particles from container to container (liquid), or the separate pieces scattered about (gas). In this chapter, we will review the kinetic molecular theory of matter. This important model for matter will be extended so that a detailed study of the physical and chemical behaviour of gases can be carried out in Chapter 11.

10.1 The Kinetic Molecular Theory

The **kinetic molecular theory** or particle theory of matter is based on a considerable amount of experimental data. There are two main ideas in this theory.

1. Matter is made up of particles (atoms, ions, or molecules).
2. These particles are in constant motion. (The word kinetic comes from the Greek word *kinein*, meaning to move.)

The kinetic molecular theory also postulates that particles interact with each other. They exhibit attractive forces over short distances and repulsive forces when they are in contact with each other, as shown in Figure 10.1.

Figure 10.1
The forces between molecules. Molecules attract more strongly the closer they approach each other but a very strong repulsion arises when they come into contact.

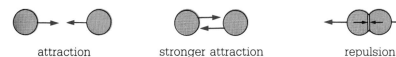

attraction   stronger attraction   repulsion

10.2 Matter and Its States

If you perform a few simple experiments on a solid, a liquid, and a gas, you will discover that only two properties, shape and volume, are needed to characterize each state. (See Table 10.1.)

Table 10.1   Properties of the Three States of Matter

State	Properties
Solid	Rigid fixed shape (constant shape) Incompressible (constant volume)
Liquid	No fixed shape (variable shape) Incompressible (constant volume)
Gas	No fixed shape (variable shape) Compressible (variable volume)

How can the properties in Table 10.1 be explained? You can use the kinetic molecular theory to follow the behaviour of the particles which make up matter. Think about water in its solid state, ice. You can explain the rigid shape of ice by considering the attractive forces between its molecules. These attractive forces hold the molecules in specific positions in an ordered structure. The molecules have very low kinetic energy; the only motion they can undergo is vibration. A solid like ice is incompressible because the molecules are close enough together to cause a dramatic increase in repulsive forces if an attempt is made to force the molecules any closer.

The molecules of a liquid such as water must also be close together since a liquid resists compression. The particles still attract each other, but the attractive forces do not lock the molecules in fixed positions. This is why liquids do not have a fixed shape. The attraction between water molecules is much less effective in liquid than in solid because of increased molecular kinetic energy in the liquid state. The faster moving molecules of liquid water make the forces of attraction between them less effective. (See Figure 10.2.) Thus, these higher energy molecules are able to move past each other.

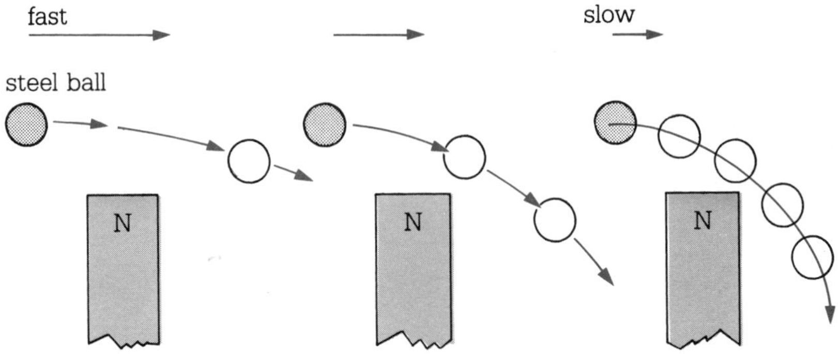

Figure 10.2
Particle speed and force effectiveness. A magnet causes very little deflection on a fast-moving steel ball. As the ball's speed drops the deflection becomes greater.

The kinetic energy of water molecules in the gaseous state is so great that it completely overwhelms the force of attraction between the molecules. According to the kinetic molecular theory, gas particles move very quickly. They tend to move far away from each other, filling any container in which they are placed.

Gases and the Kinetic Molecular Theory

The behaviour of gases can be used to deduce more features of the kinetic molecular theory. Gases in a sealed container maintain their gaseous state and initial pressure indefinitely. Therefore, gas particles undergo collisions – collisions where there is no energy being given off. Even though gas particles are colliding with each other and with the walls of their container, they must rebound without releasing any energy. If there were energy released on collision, the gas particles would slow down and eventually condense into a liquid. This is not what is observed to happen.

TYPES OF MOLECULAR MOTION

Kinetic energy, the energy of motion, can be absorbed by molecules in three different ways as follows.

1. Translational motion in which the entire molecule moves from place to place.

2. Rotational motion in which the molecule spins around like a propeller.

3. Vibrational motion in which the atoms in a molecule or the molecules in a solid vibrate back and forth about the same fixed location.

Molecules can absorb kinetic energy as motion in three different ways.

(b) Rotational motion

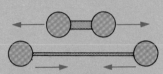
(c) Vibrational motion

(a) Translational motion

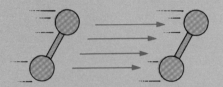

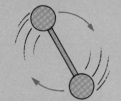

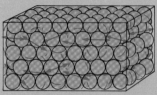
in a molecule

in a solid

GAS VOLUME AND MOLECULAR VOLUME

Liquid water occupies much less volume than steam does. If we assume that water molecules in the liquid state are in contact with one another, then the volume of 1 g of water molecules is 1 mL. When 1 g of water molecules is turned into a gas (steam) at 100°C and at atmospheric pressure, the gas occupies a volume of 1700 mL. Since the volume of the gas molecules is still 1 mL, then the remaining 1699 mL of gas volume must be "empty." The gas molecules are very far apart from one another. In this example 99.94% of the volume occupied by the gas is, in fact, "free" space.

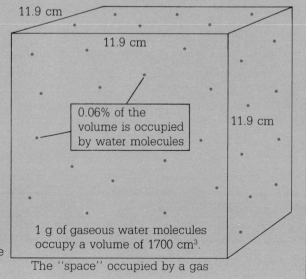

11.9 cm

11.9 cm

0.06% of the volume is occupied by water molecules

11.9 cm

1 g of liquid water molecules occupy a volume of 1 cm³

1 cm
1 cm
1 cm

99.94% of the volume is free space

1 g of gaseous water molecules occupy a volume of 1700 cm³.

The "space" occupied by a gas

The fact that gases can be easily compressed can be explained if we hypothesize that there are large spaces between their particles. (See Figure 10.3.) This hypothesis is supported by the fact that gases have lower densities compared to their corresponding liquids or solids. (See Table 10.2.)

Table 10.2 Comparing Solid, Liquid, and Gas Densities

Substance	Density (kg/m³)		
	Solid	Liquid (25°C)	Gas (atmospheric pressure)
Water	917 (0°C)	997	0.59 (100°C)
Mercury	14 193 (−40°C)	13 534	3.63 (400°C)

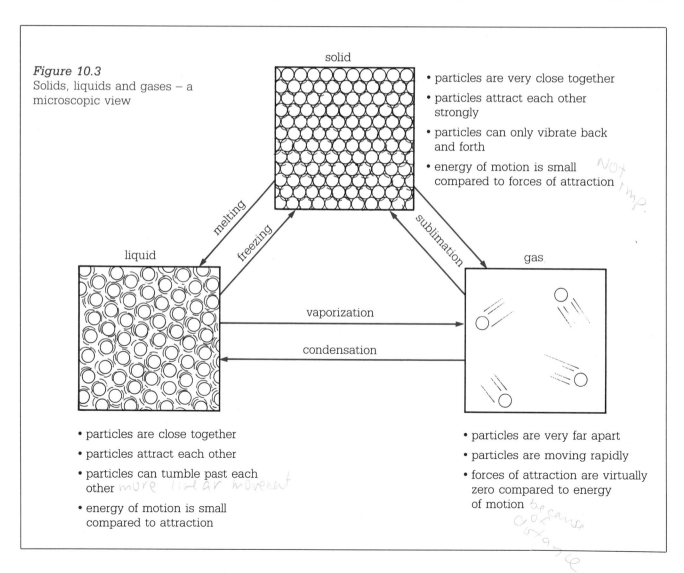

Figure 10.3
Solids, liquids and gases – a microscopic view

solid
• particles are very close together
• particles attract each other strongly
• particles can only vibrate back and forth
• energy of motion is small compared to forces of attraction   *Not imp.*

melting freezing sublimation

liquid

vaporization

condensation

gas

• particles are close together
• particles attract each other
• particles can tumble past each other *more linear movement*
• energy of motion is small compared to attraction

• particles are very far apart
• particles are moving rapidly
• forces of attraction are virtually zero compared to energy of motion *because of distance*

Thus, we can make a more complete statement of the kinetic molecular theory for gases as follows.

1. All matter is made up of tiny particles.
2. The particles are in constant motion. The higher the temperature, the more energy of motion (kinetic energy) they have and the faster they move.
3. There are forces of attraction and repulsion between particles. The closer they approach each other, the stronger this attraction becomes. When particles come into contact, they repel each other strongly.
4. In the gaseous state, particles are far apart. The volume of the particles is very small compared to the volume the gas "occupies."
5. The forces of attraction between gas particles are very small.
6. Particles are elastic; no energy is released or absorbed upon collison.

10.3 Matter and Its Physical Changes

When matter changes state, energy is always involved. Melting, vaporization, and sublimation from solid to gas all involve the absorption of thermal energy. Ice must be heated before it melts; water must be heated before it boils. Freezing, condensation, and sublimation from gas to solid all involve the release of thermal energy. (See Figure 10.4.)

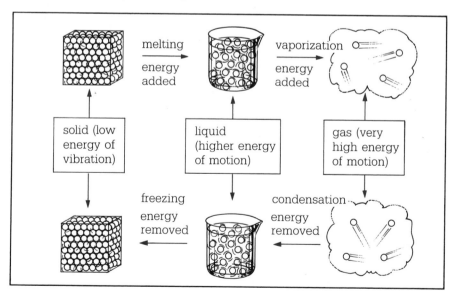

Figure 10.4
Energy and change of state. Melting, vaporizing, and sublimation (solid to gas) absorb energy. Freezing, condensation, and sublimation (gas to solid) release energy.

The type of particles involved, the forces between them, and the kind of energy (kinetic or potential) released or absorbed, govern the physical behaviour of matter. The thermal expansion of solids, liquids, and gases, and the changes of state you studied in earlier science courses are examples of physical behaviour of matter. The

FORMS OF ENERGY

Energy, the ability to do work, comes in many forms. Two of these energy forms are important for discussing the kinetic molecular theory: kinetic and potential energy.

KINETIC ENERGY

Kinetic energy is the energy an object has due to its motion. The faster an object is moving, the more kinetic energy it has. In fact, if the speed of an object doubles, its kinetic energy increases four times. The equation for calculating the kinetic energy, E_k, of an object of mass m and moving at a speed v is

$$E_k = \frac{1}{2}mv^2$$

POTENTIAL ENERGY

Potential energy is the energy an object possesses due to its position relative to a body that either attracts or repels the object. A brick held above the earth has potential energy both because it is separated from the earth's surface (position) and because of the gravitational attraction between the earth and the brick. The potential energy of an object depends on the strength of the attractive (or repulsive) force and on the distance between the object and the attracting (or repelling) body. For example, the stronger the attractive force and the farther away two objects are, the more potential energy they possess.

CONSERVATION OF ENERGY

Energy can be transformed from one form to another. When this happens the total energy after the transformation is the same as the total energy before – no energy is lost or gained. This is a statement of the Law of Conservation of Energy. Much experimental evidence, some of which you may have studied in earlier science courses, has verified this law.

Kinetic and potential energy exchange. Energy can be transformed from one form to another. If no energy is changed into heat due to friction, this exchange can continue indefinitely.

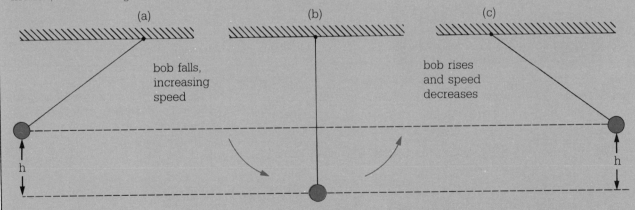

(a) pendulum bob not moving, but is above its lowest point (no kinetic energy, but some potential energy due to attraction between earth and the bob)

(b) pendulum bob at its lowest point and moving at its maximum speed (no potential energy, but same amount of kinetic energy as there was potential energy in (a))

bob falls, increasing speed

bob rises and speed decreases

(c) pendulum bob back up to same height as in (a)

potential energy at (a) = kinetic energy at (b) = potential energy at (c)

temperature of an object is a measure of the average kinetic energy of the particles making up the object. The higher the average kinetic energy and speed of motion of the particles, the higher the temperature. (See Figure 10.5.) Thermal energy, on the other hand, is the total kinetic energy of all of the particles in the object.

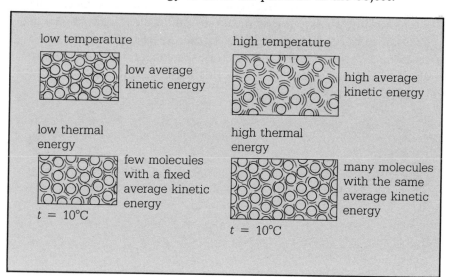

low temperature

low average kinetic energy

high temperature

high average kinetic energy

low thermal energy

few molecules with a fixed average kinetic energy

$t = 10°C$

high thermal energy

many molecules with the same average kinetic energy

$t = 10°C$

Figure 10.5
Temperature and thermal energy. The temperature of an object is a measure of the average kinetic energy of its molecules. The thermal energy an object contains is a measure of the total kinetic energy of its molecules.

States of Matter

The state of a substance indicates its relative amount of potential energy; the solid state has the least potential energy and the gas state the most. Recall that attracting objects that are separated have potential energy – they tend to pull themselves together, thus releasing their energy in some other form. The particles of a gas, which are separated far apart, have large amounts of potential energy. The particles of a liquid are relatively closer together and possess less potential energy. The particles of a solid, being very close together, have very little potential energy.

Let us use the kinetic molecular theory to explain what happens when a solid, for example a block of copper, is heated. As energy is added to the block, the copper atoms absorb it as kinetic energy. They begin to vibrate faster and faster. The temperature of the block (the average kinetic energy of its particles) must increase. Since the atoms are vibrating more, the average space each takes up increases. Thus, the block of copper expands in size as its temperature rises. (See Figure 10.6.)

low temperature

low kinetic energy

high temperature

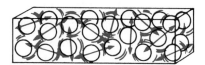

high kinetic energy

Figure 10.6
The thermal expansion of a solid. As the temperature of a solid rises, the average kinetic energy (and speed) of its particles increases. This more vigorous motion results in the particles moving farther apart.

Change of State

Imagine that the copper block shown in Figure 10.6 is heated still further. We can use the kinetic molecular theory to predict that the increased kinetic energy (and vibrational motion) of the copper atoms will decrease the effectiveness of the attraction between the atoms to such an extent that the atoms no longer remain locked in position. They will move slightly farther apart and start to move past one another as in a liquid, that is, the copper will melt (Figure 10.7). The temperature at which this occurs is the **melting point** (1084°C) of copper.

Figure 10.7
Copper melting. Thermal energy added to atoms locked in a solid can increase their vibrational kinetic energy to a level where they can break out of their fixed positions and begin to move past each other (translational motion). A liquid forms.

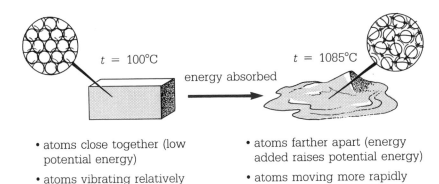

$t = 100°C$ energy absorbed $t = 1085°C$

- atoms close together (low potential energy)
- atoms vibrating relatively slowly (low kinetic energy)

- atoms farther apart (energy added raises potential energy)
- atoms moving more rapidly (higher kinetic energy)

Energy and Change of State

When a sample of crushed ice from a deep freezer is heated at a fixed rate and the temperature monitored, a graph similar to the one shown in Figure 10.8 may be drawn. The various sections of the graph are associated with temperature increases and changes of state. How can the kinetic molecular theory be used to explain these observations? To increase the separation between the attracting particles requires energy. The potential energy of the molecules in liquid water is therefore greater than the potential energy of the molecules in ice. The energy being supplied to melt ice goes into pulling molecules slightly farther apart instead of increasing their speed or kinetic energy. The kinetic molecular theory therefore explains why the temperature of a melting substance remains constant until all of the solid has melted. The energy needed to cause the extra particle separation in liquid form and the consequent melting is called the substance's **heat of fusion**. Similarly, the potential energy needed to completely separate the particles as they change from a liquid to a gas is called the **heat of vaporization**.

Recall that temperature is a measure of the average kinetic energy of the particles in a substance. The sections of the graph in Figure 10.8 where the temperature is rising indicate that the incoming heat energy is being used to increase the speed of motion of the particles.

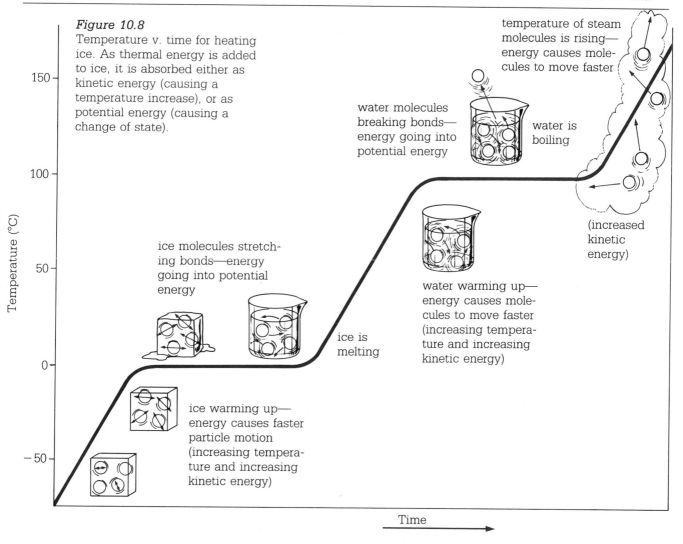

Figure 10.8
Temperature v. time for heating ice. As thermal energy is added to ice, it is absorbed either as kinetic energy (causing a temperature increase), or as potential energy (causing a change of state).

Thus when energy is added to matter it can be absorbed in two ways:

1. By changing the potential energy of the particles, resulting in a change of state.
2. By changing the kinetic energy of the particles, resulting in a change in temperature.

Using the Potential Energy of a Change of State

The heat of fusion of ice is used to keep drinks cold. Ice absorbs 333 J of energy for every gram of ice melted. For example, if 33.3 kJ of energy from the surroundings is absorbed by each of the drinks shown in Figure 10.9 (page 258), the temperature of the drink without ice would rise from 0°C to 20°C. The drink containing ice would remain at 0°C, but 100 g of ice would melt.

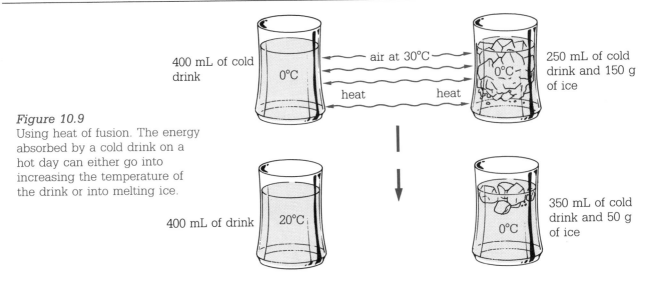

Figure 10.9
Using heat of fusion. The energy absorbed by a cold drink on a hot day can either go into increasing the temperature of the drink or into melting ice.

400 mL of cold drink

0°C

air at 30°C

heat

heat

250 mL of cold drink and 150 g of ice

0°C

400 mL of drink

20°C

350 mL of cold drink and 50 g of ice

0°C

A similar phenomenon occurs when a liquid changes to a gas. For example, vegetables cooking in boiling water at 100°C will not cook any faster if the heat input is increased. Any extra energy converts water into steam, but does not increase the boiling temperature of the water. Boiling vegetables with a medium heat setting on the stove is just as effective as rapidly boiling them at a high heat setting, and furthermore conserves energy. (See Figure 10.10.)

Figure 10.10
Boiling water will not exceed 100°C. Any extra energy provided will simply convert water into steam at a faster rate.

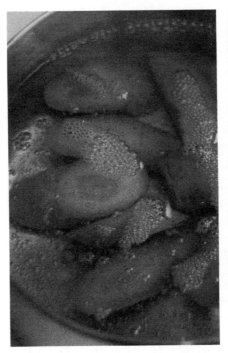

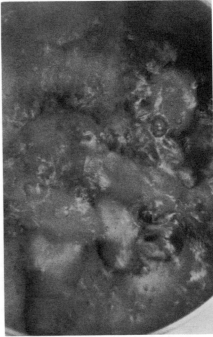

(a) Low heat setting. The water temperature is 100°C. Water is turning into steam at a slow rate. Cooking time is 35 min.

(b) High heat setting. The water temperature is 100°C. Water is turning into steam at a faster rate. Cooking time is still 35 min.

Steam is a very important energy transfer medium. Water molecules have a relatively large attraction for each other. A considerable amount of potential energy is required to separate the molecules, that is, water has a high heat of vaporization. Therefore, steam contains a large amount of energy. It is used in a variety of ways. For example, many large institutions, such as universities and governments, that have several buildings located in the same vicinity often have just one large boiler (furnace) which produces and pumps steam to other locations for heating purposes. Since each building does not need its own heating plant to build and maintain, much money can be saved on construction and operating costs.

Exercises

1. Use sketches and the ideas of the kinetic molecular theory to explain why liquid mercury or alcohol used in thermometers expands when it is heated.

2. Steaming is sometimes used to cook vegetables. In this method of cooking the vegetables are suspended in a metal basket above boiling water. Explain how the heat required to cook the vegetables is obtained from the water.

10.4 Gas Pressure

When you blow into a balloon, its walls expand. Like a stretched elastic, the balloon walls tend to collapse, squeezing on the air trapped inside. Why doesn't the balloon collapse completely under this tension? The squeezing force exerted by the stretched balloon walls must be matched by a **force** exerted by the trapped gas.

A force being exerted over a unit area is called **pressure** (that is, pressure = force/area). How can you explain the gas pressure that keeps a balloon inflated? You can use a simple model to answer this question. Imagine that the particles which make up air are tennis balls. According to the kinetic molecular theory, these tennis balls (air particles) are moving very quickly and are elastic. Now, picture one of these moving particles striking a stationary surface such as a tennis racquet. As you would expect, the ball is reflected; the person holding the racquet feels a momentary push. If the racquet is repeatedly struck by many tennis balls, these momentary pushes would be felt as a single, continuous push – pressure. (See Figure 10.11 on page 260.) This is precisely what happens in the case of the balloon. The continuous collisions of air particles with the balloon walls exert pressure on these walls. The balloon remains inflated.

Laboratory Manual
In Experiment 10-1, you will measure air pressure both qualitatively and quantitatively.

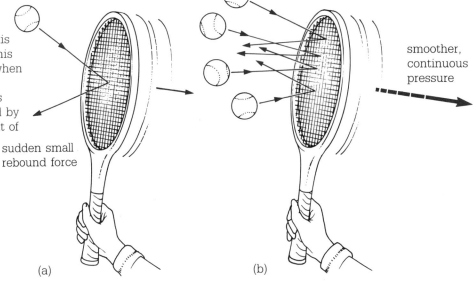

Figure 10.11
A gas pressure model – tennis balls striking a racquet. Tennis balls exert a rebound force when they strike a racquet. Gas pressure can be explained as the rebound pressure caused by the continuous bombardment of gas molecules.

sudden small rebound force

smoother, continuous pressure

(a)   (b)

You can use the kinetic molecular theory to explain the pressure exerted by any gas. **Gas Pressure** is the force caused by the continual bombardment of a surface by the tremendous number of perfectly elastic particles that make up a gas.

In SI, pressure is measured in units called **pascals** (Pa). One pascal is the force of 1 N (newton) spread over an area of 1 m² (1 Pa = 1 N/m²). This is equivalent to the force of gravity on 100 g of sand spread over 1 m². (See Figure 10.12.) Normal average air pressure at sea level is equivalent to the force of gravity on 1.013×10^4 kg of sand spread over the same 1 m² (1.013×10^5 Pa). Air pressure is usually measured in units of kilopascals, kPa. Normal average air pressure at sea level is 101.3 kPa.

Figure 10.12
Standard air pressure. A pressure of 1 Pa is equivalent to the pressure exerted by 100 g (0.1 kg) of sand spread over an area of 1 m², a layer less than 1 mm thick. Average atmospheric pressure at sea level is equivalent to the pressure exerted by 10 130 kg of sand spread over an area of 1 m², a pile about 6 m high.

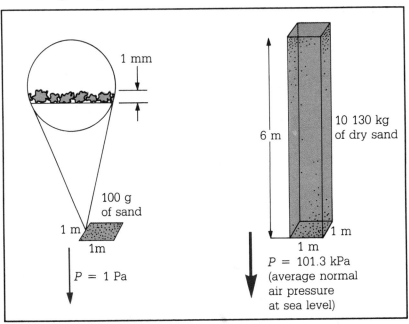

1 mm

100 g of sand

1 m
1m

P = 1 Pa

6 m

10 130 kg of dry sand

1 m
1 m

P = 101.3 kPa (average normal air pressure at sea level)

Sample Problem 10.1

Explain how a drinking straw works when you use it to "draw up" a drink.

Solution

When you draw on a straw, you reduce the air pressure inside the straw. Air pressure on the surface of the drink is greater than the pressure inside the straw. Since a liquid is not compressible, air pressure on the surface of the drink is transmitted throughout the liquid in all directions, including the upward direction. The excess air pressure therefore pushes the liquid up the straw. (See Figure 10.13.)

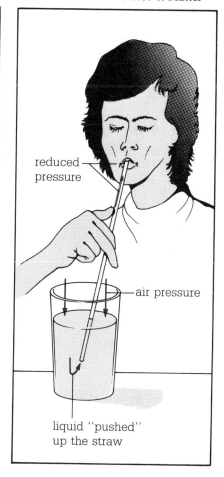

reduced pressure

air pressure

liquid "pushed" up the straw

Figure 10.13
Air pressure pushing a liquid up a straw. Reducing the pressure in a straw allows the atmospheric pressure to push the liquid up the straw.

Exercise

3. Could a drinking straw be of indefinite length vertically and still work? Explain why or why not.

Gas Pressure and Area

The total force exerted by a particular gas pressure depends on the area exposed to the pressure. The larger the area that experiences the pressure, the greater the total force. Relatively low pressure gas can be used to support very large but light fabric and plastic roofs. Air blowers, for example, supply enough pressure to support the rubberized fabric roofs on domed structures. (See Figure 10.14.) Propeller type air blowers supply enough air pressure to suspend hovercraft a few decimetres above the ground.

Figure 10.14
Relatively low air pressures can support large roof areas. This cable-reinforced Airdome houses 25 000 t of chemical fertilizer at the Port of Hamilton, Ontario. The fertilizer can be unloaded continuously from ships, and can be fed into the Airdome via special loading inlets in the roof.

261

Gas Pressure and Number of Particles

According to the kinetic molecular theory, the more gas particles there are in a given volume, the more collisions per second on a unit area occur and the higher the gas pressure. Thus, if more gas particles are pumped into a container with a given volume, the gas pressure will increase. A very common example you may have experienced is pumping air into a car tire to increase the pressure.

Seal the end of a syringe with its plunger pulled out as far as possible, and then push on it. Can you push it in all the way? This would prove to be an impossible task! Why? You can use the kinetic molecular theory to explain this observation. As you push in the plunger, the number of trapped gas particles per unit volume rises, increasing the pressure on the plunger's surface. Eventually the increasing pressure on the plunger's surface results in a force greater than you are able to exert to push the plunger further. (See Figure 10.15.)

Figure 10.15
Force and gas pressure. The pressure exerted by a gas depends on the number of gas molecules that strike a given area per second.

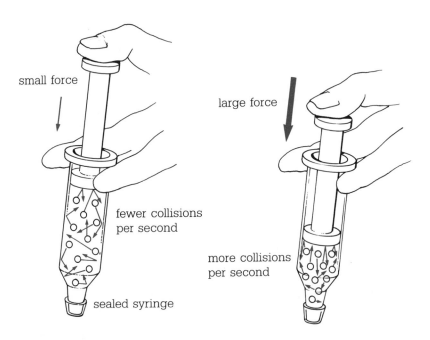

small force

large force

fewer collisions
per second

more collisions
per second

sealed syringe

Gas Pressure and Temperature

Laboratory Manual
You will observe temperature changes during rapid evaporation of a liquid in Experiment 10-2.

If you place a sealed syringe containing some air in hot water, you will notice that the plunger moves out slightly. Can we use the kinetic molecular theory to explain this behaviour? As the hot water increases the temperature of the gas in the syringe, the gas particles gain more kinetic energy and move faster. Each particle that collides with the plunger exerts a greater momentary push and the pressure increases slightly. (See Figure 10.16.) In addition, since the particles are moving faster, each one collides more frequently with the plunger. This also causes the pressure to rise.

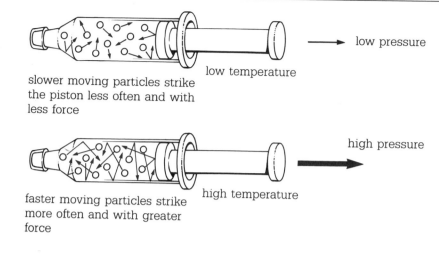

slower moving particles strike the piston less often and with less force

low pressure

low temperature

faster moving particles strike more often and with greater force

high temperature

high pressure

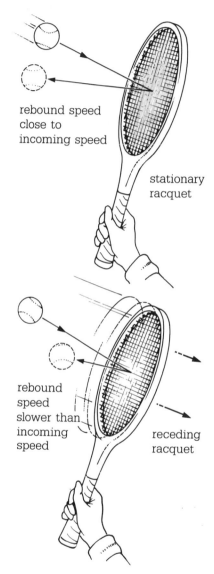

◀ *Figure 10.16*
Gas pressure and temperature. The pressure exerted by a gas depends on the speed (temperature) of its gas molecules.

rebound speed close to incoming speed

stationary racquet

rebound speed slower than incoming speed

receding racquet

Figure 10.17
How energy can be extracted from a "gas" particle – the tennis ball model. A tennis ball rebounds from a receding tennis racquet slower than it does from a stationary one. If the tennis ball causes the racquet to recede, some of the ball's kinetic energy must have been transferred to the racquet.

Many of our sources of energy make use of gases. Chemical combustion reactions produce hot gases which can be used for many purposes. For example, combustion reactions in furnaces provide heat; in automobile engines or steam turbines used to run electrical generators, mechanical energy is extracted from hot gases from combustion reactions. How is this done?

If compressed air such as you might find in a service station air hose is allowed to expand freely to normal air pressure, the nozzle where the gas is expanding becomes cold. If you pull out the plunger of a sealed syringe, allowing the air inside to expand, the temperature drops. Why does the temperature of a gas decrease when the gas expands? Let us use the tennis ball and racquet model of the kinetic molecular theory used earlier. A perfectly elastic tennis ball striking a perfectly elastic, stationary racquet would rebound with the same speed with which it approaches the racquet. What happens if the tennis racquet is receding from the approaching ball? If you try this experiment with a real ball you will observe that the ball rebounds with much less speed than it does from a stationary racquet. (See Figure 10.17.) The plunger of the sealed syringe being pulled out is similar to the receding tennis racquet. Air particles striking the receding plunger rebound at a slower speed. Some of the kinetic energy possessed by the particles is lost to the plunger, pushing it farther out of the syringe. The temperature of the particles drops.

Exercise

4. When you pump up a bicycle tire using a hand pump you may have noticed that the end of the pump where the compressed gas is expelled becomes hot. Use the tennis ball and racquet model to explain why this happens.

Gas Pressure and Boiling Point

Changing the pressure surrounding a liquid changes its boiling point. Water, for example, boils at 100°C at normal atmospheric pressure at sea level (101.3 kPa). The boiling point of water is lower at higher altitudes where atmospheric pressure is lower. The boiling point of water at the top of Mount Everest (8850 m above sea level where the air pressure is 34 kPa) is 72°C. Once again, let us use the kinetic molecular theory to explain this phenomenon.

In order for a substance to boil and form bubbles in the liquid, the kinetic energy of the vaporizing particles must be high enough to be able to push out of the liquid and form vapour bubbles. The greater the pressure pushing down on the top of the liquid, the higher the kinetic energy the particles need to do this, that is, the higher the temperature required. A particle "trying" to boil might be compared to a person trying to get through a trap door. (See Figure 10.18.) Enough push (pressure) must be exerted to lift the door before the person can get out. If a heavy object is placed on the door (higher "air" pressure), the person will have to push harder (have a higher kinetic energy level) in order to lift the trap door and get out (form a gas bubble). Thus, the kinetic molecular theory agrees with experimental observations: a liquid will boil at a lower temperature when the pressure is low and at a higher temperature when the pressure is high. The **standard boiling point** of a substance is its boiling temperature at normal atmospheric pressure at sea level, that is, at 101.3 kPa. It is the temperature at which the pressure of the gas particles (coming from the boiling liquid) is equal to the normal atmospheric pressure at sea level.

Figure 10.18
Boiling point and pressure. As the air pressure above a liquid rises, the boiling point of the liquid also rises.

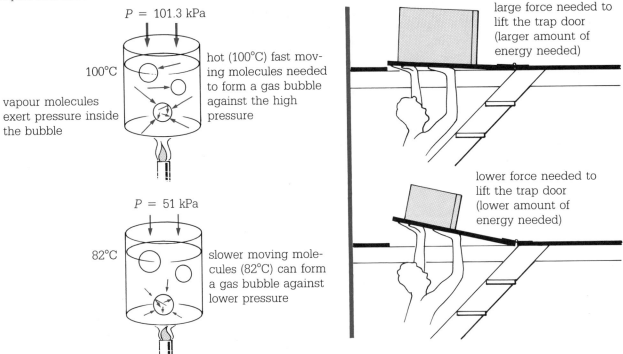

P = 101.3 kPa

100°C

vapour molecules exert pressure inside the bubble

hot (100°C) fast moving molecules needed to form a gas bubble against the high pressure

P = 51 kPa

82°C

slower moving molecules (82°C) can form a gas bubble against lower pressure

large force needed to lift the trap door (larger amount of energy needed)

lower force needed to lift the trap door (lower amount of energy needed)

SPECIAL TOPIC 10.4

"BOILING WATER UNTIL IT FREEZES"

Water can be made to boil until it actually freezes into ice! If a small amount of water is placed in an insulated container and placed in a jar attached to a vacuum pump, the water will start to boil as the pump reduces the pressure above the water. The water molecules which vaporize draw heat (heat of vaporization) from other water molecules in the container, thereby lowering the temperature of the remaining water. This energy drain can be so great that the temperature of the remaining water can drop to 0°C. Some water can provide even more energy by giving up its heat of fusion by freezing.

A similar effect can be observed when a carbon dioxide fire extinguisher is used. The carbon dioxide in the fire extinguisher's cylinder is under enough pressure to keep it liquid at room temperature. When the liquid carbon dioxide squirts out into the comparatively low (atmospheric) pressure in a room, some of it instantly vaporizes. The heat needed for this vaporization comes from other carbon dioxide molecules, which drop in temperature and freeze into a snow of dry ice with a temperature of −78.5°C.

This scientific principle has been used to develop a self-cooling pop can. A small container holding liquid carbon dioxide is built right into the can. When the can is opened, the liquid carbon dioxide vaporizes through a cylinder in the pop and escapes out of the top of the can. The heat absorbed by the vaporizing carbon dioxide can lower the temperature of the pop by about 16°C in a few seconds. This new product has been developed but its actual production will depend on its cost (estimated at about 10¢ per can), its safety, and consumer acceptance.

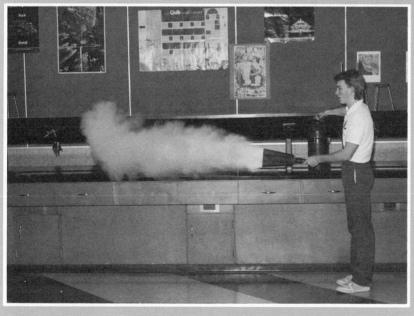

Dry ice snow from a carbon dioxide fire extinguisher. Part of the liquid carbon dioxide ejected from the fire extinguisher vaporizes, extracting enough energy from the remaining carbon dioxide liquid to freeze it to solid dry ice at −79.5°C.

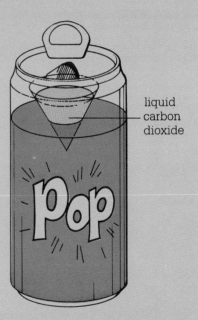

liquid carbon dioxide

A self-cooling pop can. The beverage can cools itself when the liquid carbon dioxide contained in the cone shaped compartment under the flip-top vaporizes, taking energy from the can's contents.

Gases and Energy Transfer

Appliances such as refrigerators, heat pumps, and air conditioners all work on the principle that energy is transferred when a fluid changes state and when a vapour expands or is compressed. For example, a compressor in a refrigerator compresses a gas, such as freon, increasing its temperature. (See Figure 10.19.) The hot freon gas moves to a condenser on the outside of the refrigerator. Air at room temperature cools the gas down and it condenses into a liquid, releasing even more energy (the heat of vaporization). The liquid freon, now at room temperature, is then passed through a small hole in a restriction valve into the low pressure tubes which are inside the refrigerator. There, the freon liquid vaporizes rapidly. The heat of vaporization needed to do this is drawn from the kinetic energy of the freon molecules which become very cold. This cold gas absorbs heat from the refrigerator and its contents, thereby cooling them. The freon gas is once again fed into the compressor and the cycle starts over again.

Figure 10.19
A refrigerator makes use of the heat of vaporization and condensation.

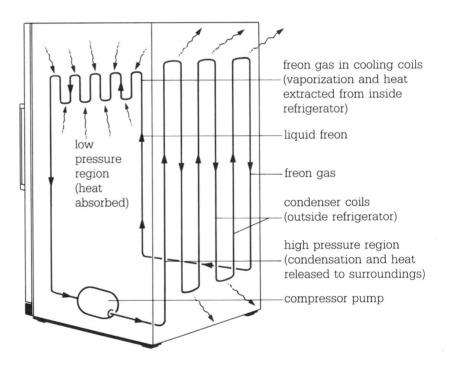

freon gas in cooling coils (vaporization and heat extracted from inside refrigerator)

liquid freon

freon gas

condenser coils (outside refrigerator)

high pressure region (condensation and heat released to surroundings)

compressor pump

low pressure region (heat absorbed)

Exercise

5. In the winter a heat pump is used to take heat from the air or ground outside and "pump" it into a building. In the summer the operation of the heat pump is reversed, taking heat from a building and "pumping" it outside. Explain how the basic refrigerator principles are applied in the heat pump.

PUTTING CHEMISTRY TO WORK

DR. GLORIA ELLENTON / METEOROLOGIST AND ATMOSPHERIC MODELLER

A stack belching clouds of smoke into the atmosphere has become the classic example of how our society pollutes the environment. But learning exactly what is happening requires the answers to several questions. How far does the pollution spread? How high in the atmosphere does it reach? What is the effect of wind or temperature? Are neighbouring cities being affected?

The answers, or at least the best possible predictions of these answers, are obtained through the work of scientists such as Gloria Ellenton. "My work consists of analyzing weather data and then designing mathematical models on a computer which use that data to make predictions." Among the things which Gloria can predict are wind flow patterns or temperature changes in the atmosphere. "Such models can be used in atmospheric pollution studies even quite close to the source of pollution, for example, the quality of air in communities immediately around an incinerator."

Gloria's favorite subjects in high school were physical education, mathematics, and physics. "What to do with that combination was a puzzle," she remembers. "I took math and physics in university and ended up as a weather forecaster so that I could work outdoors." After spending a few years at home to raise a family, Gloria was ready to settle on a career. "I went back to university and studied weather prediction modelling and atmospheric pollution."

Gloria's business operates from an office in her home. "I supply expertise to government agencies and industry." The information from models is used to design the regulations which control emissions from industries. "It is rewarding to think of the important decisions that result from my models, especially when I'm doing the more tedious, number-crunching part of my work."

One of the most important models Gloria helped develop concerns acid precipitation. "This model calculates the transport of air-borne contaminants over hundreds of kilometres. It also calculates how these pollutants change chemically as they travel and fall to earth in wet or dry weather." When Gloria looks out her window at the rain, she is aware that her work crosses the boundaries of Canada and is involved in continuing negotiations between countries to reduce air pollution.

10.5 The Absolute or Kelvin Temperature Scale

Laboratory Manual
What is the relationship between the temperature of a gas and its pressure? In Experiment 10-3, you will find out.

Recall that according to the kinetic molecular theory, the temperature of an object is a measure of the average kinetic energy of the particles which make up the object. Using this theory you can predict that as the temperature of an object decreases, the particles, which are in constant motion, will slow down. How long can this continue? Obviously until the particles come to a complete stop. When the particles in an object are not in motion, they do not possess kinetic energy. Thus, you predict that the object would be as cold as it can possibly be. It seems reasonable to predict that there is a temperature below which it is impossible to cool any object further! This coldest possible temperature is called **absolute zero**. Is there any experimental evidence for such an ultimate cold temperature?

Experimentally, as the temperature of a gas trapped in a container of fixed volume drops, so does the pressure. (See Figure 10.20.) If a pressure versus temperature graph of experimental values is plotted, you will find that the data points lie on a straight line (with some slight variation due to experimental error). What would happen to the pressure if this regularity persists as the temperature continues to decrease? You can make this prediction by extrapolating the straight line on the graph below the temperatures for which you have data. Assuming that the regularity observed at high temperatures is also true at low temperatures, then the approximate temperature at which the pressure of the gas is zero can be found as shown in Figure 10.20.

SPECIAL TOPIC 10.5

ABSOLUTE ZERO AND HEISENBERG'S UNCERTAINTY PRINCIPLE

It is generally accepted that particles can never be totally still, even at absolute zero. A theory from particle physics, Heisenberg's uncertainty principle, states that both the position and energy of any particle cannot be measured at the same time beyond a certain level of accuracy. Much experimental evidence has supported Heisenberg's uncertainty principle. If a particle was actually at rest, both its position and energy would be known exactly. Thus, scientists think that particles at absolute zero still have some small, irreducible amount of kinetic energy.

Heisenberg's uncertainty principle

HEISENBERG MAY HAVE SLEPT HERE

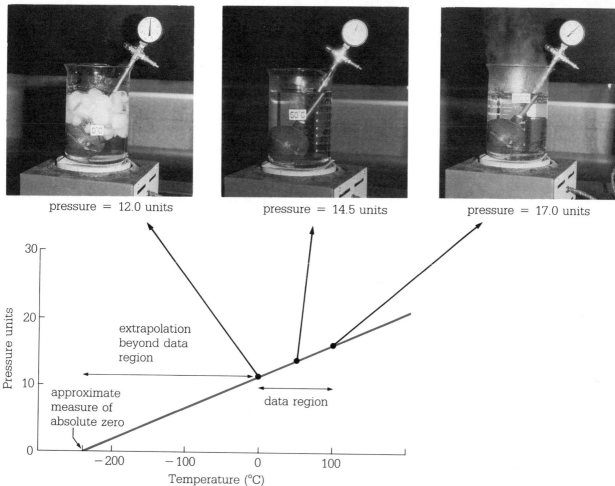

pressure = 12.0 units   pressure = 14.5 units   pressure = 17.0 units

Figure 10.20
An experiment to determine absolute zero. As the temperature of a gas decreases its pressure and speed of its molecules also decrease. At a pressure of zero its molecules must have stopped moving. The temperature at which this occurs is absolute zero.

According to the kinetic molecular theory, a sample of gas would exert zero pressure only when the molecules are not in motion. This would be the case when the temperature is at absolute zero. The accepted value for absolute zero is $-273.18°C$. This value is rounded to $-273°C$ for the purposes of this textbook.

Temperature Scales

On the Celsius temperature scale, zero is set at the melting point of water. There is a temperature scale that has its zero set to coincide with absolute zero. This is the absolute or **Kelvin scale**. A degree on the Kelvin scale is called a **kelvin** (symbol K). It is exactly the same as a degree on the more familiar Celsius scale. The only difference between the two scales is that $0°C$ is the melting point of water whereas 0 K is absolute zero.

In order to convert temperatures in degrees Celsius to kelvins, 273 is added to the Celsius reading. For example, 0°C becomes 273 K and 100°C becomes 373 K. Subtracting 273 from a Kelvin reading will convert it to degrees Celsius. For example, 0 K becomes −273°C.

The Celsius scale is used in everyday life. The normal range of weather tempeatures we experience is relatively small, usually between −40°C and +40°C. The Kelvin scale relates directly to particle motion and is a measure of the average kinetic energy of particles in a sample of matter. (Zero on the Kelvin scale measures zero kinetic energy.) It is, therefore, the scale used in many equations which relate particle motion and temperature. (Some of these equations will be studied in Chapter 11.) For example, if the Kelvin temperature of an object doubles from 200 K to 400 K, then the average kinetic energy of the particles in the object also doubles.

Exercise

6. Copy and complete the following chart in your notebook. *Do not write in your textbook.*

	Temperature (°C)	Temperature (K)
Freezing point of water	0	
A very hot day	40	
The hottest day on record (1922, in Libya)		331
A very cold day		233
The coldest day on record (1960, in Antarctica)	−88	
An intense, life threatening fever		314
Absolute zero		
Oil temperature for cooking french fries		435
Coldest temperature possible by mixing sodium chloride and ice	−18	

Exercise

7. If the outside air temperature increases from 10°C to 20°C, can we say that the temperature has doubled? Explain your answer.

10.6 Molecular Mass and Rate of Diffusion: Graham's Law

You may have observed that a helium filled rubber balloon slowly collapses over a day or so while a similar air filled balloon stays inflated for a much longer period of time. (See Figure 10.21.) What is the difference between air molecules (N_2 and O_2) and helium atoms that causes helium to escape from a balloon so much faster than air molecules do? As you learned in the last section, one of the assumptions of the kinetic molecular theory is that the kinetic energy of particles varies directly as the absolute temperature.

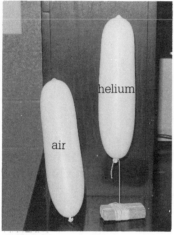
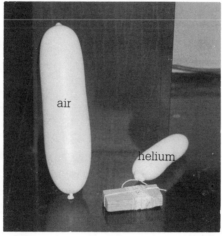

time 0 h   time 24 h

Figure 10.21
Helium versus air. A helium-filled balloon collapses over a 24 h period while an air-filled balloon maintains its pressure.

The kinetic energy, E_k, of a molecule with mass m and speed v is given by the equation

$$E_k = \frac{1}{2}mv^2$$

By rearranging the equation, you can express the speed of molecules in terms of their kinetic energy.

$$v^2 = \frac{2E_k}{m}$$

$$v = \sqrt{\frac{2E_k}{m}}$$

In other words, the kinetic molecular theory predicts that the average speed of molecules in a sample of gas depends on molecular mass. The last equation suggests that average molecular speed varies inversely as the square root of molecular mass. Thus, you can predict that at the same temperature (same kinetic energy), atoms of the gas argon ($m = 40$ u) would have an average speed double that of molecules of bromine gas ($m = 160$ u) since the mass of bromine molecules is four times the mass of argon atoms. (See

Laboratory Manual
In Experiment 10-4, you will find the relative rates of diffusion for two gases. Then, in Experiment 10-5, you will determine the relative velocities of gas molecules and derive a mathematical relationship for the velocities of gases. Finally, you will examine one gas, hydrogen, in more detail in Experiment 10-6.

Figure 10.22

The speed of different gas molecules. Argon atoms move more quickly than bromine molecules, because they have the same average kinetic energy at the same temperature, but argon atoms have a smaller atomic mass than the molecular mass of bromine molecules.

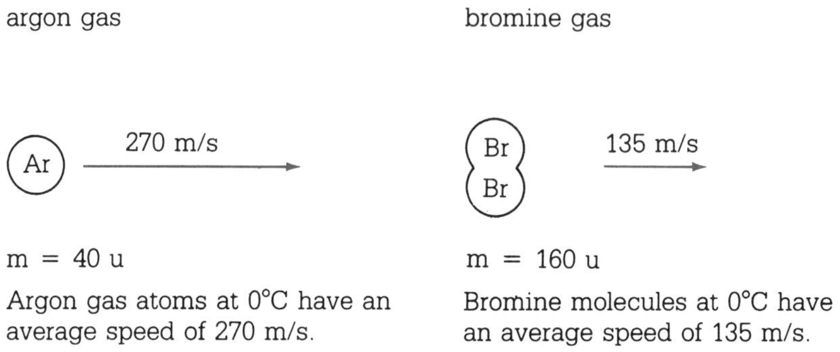

argon gas   bromine gas

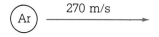

m = 40 u   m = 160 u

Argon gas atoms at 0°C have an average speed of 270 m/s.

Bromine molecules at 0°C have an average speed of 135 m/s.

$$E_{K_{Ar}} = \frac{1}{2} \times 40 \times (270)^2$$
$$= 1.46 \times 10^6 \text{ energy units}$$

$$E_{K_{Br}} = \frac{1}{2} \times 160 \times (135)^2$$
$$= 1.46 \times 10^6 \text{ energy units}$$

Figure 10.22.) In order to have the same kinetic energy (same temperature) lighter gas molecules must, on average, move faster than heavier gas molecules.

This predicton that average molecular speed varies as the square root of molecular mass can be experimentally verified by comparing the rate at which molecules of different masses diffuse through air. (Diffusion refers to the movement of one kind of molecule through a second kind.)

The relationship between molecular mass and rate of diffusion was first experimentally determined by the English scientist, Thomas Graham, in 1883. He measured the rates at which equal volumes of various gases escaped from an apparatus at the same temperature and pressure. Graham found that the less dense a gas was (that is, the lighter its particles), the faster the gas escaped. **Graham's Law** states that the *rate of diffusion of a gas is inversely proportional to the square root of the mass of its molecules.* This law is direct experimental evidence for the kinetic molecular theory.

Application of Graham's Law

Hydrogen is the lightest gas. Thus, hydrogen gas molecules can move faster than the molecules of any other gas at the same temperature. Because of this high speed, hydrogen gas is able to transfer large amounts of heat energy quickly. This is one of the reasons why large electrical generators are sealed and filled with hydrogen gas. Hydrogen gas transfers heat (kinetic energy) out of the core of a generator very efficiently.

Looking Back – Looking Ahead

The kinetic molecular theory provides an excellent model that you can use to explain a great deal of the physical behaviour of the three states of matter. It is especially useful in explaining the behaviour of gases. In the next chapter this theory will be extended even further in a more detailed and quantitative study of gases.

Vocabulary Checklist

You should now be able to define or explain the meaning of the following terms.

absolute zero	heat of fusion	kinetic molecular theory	standard boiling point
force	heat of vaporization	melting point	temperature
gas pressure	kelvin	pascal	
Graham's Law	Kelvin scale	pressure	

Chapter Objectives

You should now be able to do the following.

1. Describe the three states of matter in terms of their large-scale, directly observable properties, and explain the differences among the three states using the kinetic molecular theory.

2. Explain changes of state in terms of potential and kinetic energy and the kinetic molecular theory.

3. List the main points of the kinetic molecular theory for gases.

4. Use the kinetic molecular theory to explain gas compressibility, gas pressure, and the dependence of gas pressure on temperature.

5. Relate temperature to the average kinetic energy of the particles in a sample, explain the existence of an absolute zero temperature, and relate the absolute (Kelvin) temperature scale to the Celsius scale.

6. Use the kinetic molecular theory and the concept of particle kinetic energy to explain why lighter particles diffuse faster than heavier particles at the same temperature.

7. Identify three uses or applications of the kinetic molecular theory.

Understanding Chapter 10

REVIEW

1. What does the kinetic molecular theory say about each of the following for each of the three states of matter?
   (a) spaces between particles
   (b) particle movement
   (c) forces between particles

2. Copy and complete the chart on page 274 in your notebook. *Do not write in your textbook.*

State		Gas (steam)	Liquid (water)	Solid (ice)
Properties	Shape			
	Volume			
	Relative density			
	Compressibility			
Particle Characteristics	Distance between			
	Relative force effectiveness			
	Relative kinetic energy			
	Type of motion possible			
	Microscopic sketch			

3. List the points of the kinetic molecular theory for gases.

4. Explain the difference between thermal energy and temperature.

5. Use sketches in defining or explaining the following terms.
   (a) gas pressure
   (b) boiling point
   (c) standard boiling point

6. Copy and complete the following table in your notebook.

Temperature (°C)	Temperature (K)	Significance of the Temperature
0		
	0	
		Boiling point of water
	310	Normal body temperature
−78		Sublimating dry ice
	600	Melting point of lead
−183		Boiling point of oxygen
	234	Freezing point of mercury
78		Boiling point of ethanol

APPLICATIONS AND PROBLEMS

7. Explain why the particles in solid ice stick together while those of steam do not (even when they get very close in a collision).

8. Identify the type of energy involved in each of the following situations or changes.
   (a) Two water molecules are a short distance apart.
   (b) The temperature of a glass of water increases from 5°C to 35°C.
   (c) A gas particle moves with a speed of 1200 m/s.
   (d) Two colliding oxygen molecules have their nuclei very close together.
   (e) Water at 100°C is being changed to steam at 100°C.

9. Use the kinetic molecular theory to explain how each of the following changes of state occur. Indicate whether energy (heat) is being added or removed in each case.
   (a) melting
   (b) boiling
   (c) condensation
   (d) sublimation (gas to solid)

10. When long bridges are constructed, the road bed is built in sections with spaces between the sections. Why must this be done?

11. (a) What is significant about a temperature of 0 K?

(b) What is significant about a temperature of 0°C?

12. Explain each of the following situations in terms of how energy is produced or used.
    (a) You feel cold immediately after getting out of a warm swimming pool, but you feel comfortably warm after you have dried yourself.
    (b) Skin exposed for just a short time to steam at 100°C can suffer severe burns. Skin exposed to air at 100°C (for example, in a hot oven) for the same length of time suffers only a mild burn or none at all.

13. (a) Use Graham's law of diffusion to explain why a rubber balloon filled with helium takes only a day or so to lose pressure and deflate while a rubber balloon filled with air stays the same or deflates very little.
    (b) Why does a metallized balloon filled with helium not lose pressure and deflate?

CHALLENGE

14. You wish to have a "five minute" boiled egg for breakfast. For each of the following locations or situations, would you cook your egg less than or more than five minutes to produce your "five minute" boiled egg? Explain your answers.
    (a) You are on a skiing trip at the top of Whistler Mountain in British Columbia.
    (b) You have breakfast just before you start work 2000 m underground in a gold mine in Timmins, Ontario.
    (c) You have breakfast on a very clear and bright sunny day.
    (d) You have breakfast on a very gloomy, rainy day.

15. If the pressure inside a 1 L container of oxygen gas is 100 kPa, what would the new pressure be in each of the following situations?
    (a) The Kelvin temperature of the container and the oxygen gas is doubled.
    (b) The 1 L container is collapsed to 1/2 L with the temperature held constant.
    (c) Each of the oxygen molecules (molecular mass 32 u) is replaced by a SF_6 molecule (molecular mass 146 u). The temperature remains the same.
    (d) The average kinetic energy of the oxygen molecules is doubled.
    (e) The volume of the container is expanded to 10 L. The temperature is held constant.
    (f) The average speed of the oxygen molecules is doubled.

16. Large electrical generators are often driven by steam turbines. The steam enters the turbine at a high temperature and pressure and leaves it at a much lower temperature and pressure (still as steam). Explain, in terms of moving steam molecules, how the steam delivers the energy that the turbine provides to the generator.

17. The $^{235}_{92}U$ isotope is capable of undergoing a nuclear reaction (fission) producing heat. The $^{238}_{92}U$ isotope does not undergo fission efficiently in nuclear reactors. Natural uranium is a mixture of these two isotopes. One method of separating these isotopes is to combine uranium with fluorine to make the gaseous compound, uranium hexafluoride, UF_6, and allow it to pass repeatedly through plates containing microscopic tunnels. After this diffusion is repeated thousands of times, a significant separation of the two isotopes is obtained. Identify the principle behind this separation technique and explain how it works.

HANDLING GASES – THE GAS LAWS

Handling gases properly can be a life or death proposition. As humans venture into the hostile environments of water and space, they must take life-giving gases with them.

A scuba diver operating at moderate depths uses compressed air to breathe. Returning to the surface can be quite hazardous if not done properly. As the diver comes up, the pressure of the surrounding water drops, and the compressed air in the lungs expands.

Surfacing must be done very slowly to allow the decompressing air to escape the lungs without causing damage to them.

There is an additional problem at greater depths. Like all gases, the solubility of nitrogen gas, which makes up almost 80% of normal air, increases with pressure. A deep sea diver's blood can become saturated with dissolved nitrogen during a dive. If the diver returns too quickly to the relatively low pressures at the surface, nitrogen gas bubbles out of the blood, causing a painful and life threatening condition called "the bends." Deep sea divers must either use a different solution of gases (such as the less soluble gas helium in oxygen), or spend many hours in a decompression chamber after a dive. The pressure in a decompression chamber is slowly lowered over many hours. This allows nitrogen gas to come out of the blood slowly enough to be safely disposed of by normal breathing.

As many gases are used in everyday life, it is important to handle them properly. Natural gas, propane, pressurized spray cans, and even compressed air in car tires and fire extinguishers, can be hazardous if they are not handled carefully. In this chapter you will investigate the behaviour of gases. You will use the information obtained in these investigations to develop quantitative models (mathematical equations) to describe this behaviour. Then in Chapter 12, you will be able to use these relationships to solve stoichiometric problems involving gases.

11.1 Measuring Gases

Matter can be measured in different ways. Solids are very easily measured using either mass or volume. We can use a scale to measure 5 kg of potatoes or a tape measure to find the volume of a block of marble. Liquids can be measured by pouring them into containers such as measuring cups or graduated cylinders. Liquids can also be measured by mass. Propane, the gas used in many gas barbecues, is a liquid when under pressure. It is often sold by determining the mass of propane added to an empty cylinder. Gases, however, do not have a fixed volume. The same mass of a gas could occupy a very large or a much smaller volume depending on its pressure. Determining the mass of a gas can pose problems. For example, measuring the mass of gases such as hydrogen and helium can produce a negative result due to the buoyant force of air. (See Figure 11.1 on page 278.)

When measuring the quantity of substances in the gas state, other factors have to be considered. The pressure under which the gas is kept determines how much of it is squeezed into a particular volume. In addition, the temperature of a gas has an effect on the gas pressure. For example, you may have experienced the increase in air pressure of a volleyball as it gets harder when it is taken from

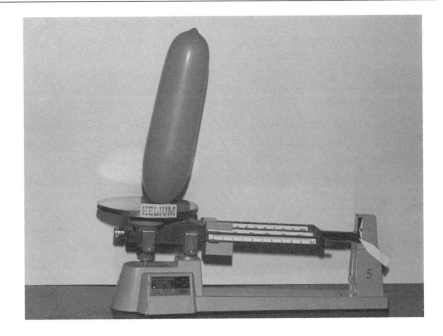

Figure 11.1
Measuring the mass of gases is not a straightforward procedure. This sample of helium gas gives a *negative* mass reading when it is measured directly. This effect is due to the buoyant force of the surrounding air.

a colder to a warmer location. To indicate a quantity of a gaseous substance, its volume at a known temperature and pressure must be specified. The natural gas supplied to the home is sold in units of cubic metres delivered through the gas meter at a pressure of 103 kPa at an effective temperature of 16°C. (See Figure 11.2.) The gas meter contains a mechanism that compensates for actual temperatures above or below 16°C.

Gas cylinders containing oxygen have a range of standard volumes. They are sold with the gas compressed to 1.65×10^4 kPa at 20°C at the time of filling. These are standards agreed upon by the gas suppliers and their consumers such as welders.

Figure 11.2
A natural gas meter reduces the incoming gas pressure from 500 kPa to an outlet pressure of 103 kPa. The volume of gas at this pressure is recorded by the meter. Since temperature also affects the volume of a gas, a compensating mechanism in the meter adjusts the volume measurement to what would be read at 16°C.

11.2 Volume and Temperature

When a given mass of air is heated, it expands taking up more volume. The exciting and colourful sport of hot air ballooning (see Figure 11.3) depends on the volume-temperature relationship of air. Since the same mass of cold air occupies a larger volume when hot, the density of hot air trapped in the balloon is less than that of the colder outside air. Just as less dense wood floats in denser water, the hot air in the balloon floats in the cold outside air.

Laboratory Manual
In Experiment 11-1, you will investigate the quantitative relationship between the volume and temperature of a gas.

Figure 11.3
Using the gas volume-temperature relationship. A hot air balloon floats because the volume occupied by a given mass of air is greater when the temperature is hotter.

Determining The Volume-Temperature Relationship

The results of an experiment done to determine the volume-temperature relationship for the gas, air, are listed in Table 11.1 on page 282. In this experiment temperature was the independent variable and volume the dependent variable. The pressure and amount of gas (number of molecules) were held constant; these were the controlled variables.

EXPERIMENT DESIGN

What are the factors that affect the mass of a gas that can be contained in a given volume? As discussed in Chapter 10, temperature, pressure, and the number of molecules of gas (number of moles) all have an effect on determining the volume of a gas. These factors are all related. Changing one factor results in a change in at least one of the others. A factor or condition which can be measured, altered, or controlled is called a *variable*. The temperature, pressure, volume, and number of moles are all variables to consider when investigating gases.

In order to obtain useful experimental information concerning the behaviour of gases, one of these variables, such as the temperature, is changed. Any effect on *one* of the other variables, such as the volume, is measured. The remaining variables, pressure and amount of gas (number of moles), are held constant.

The factor that is manipulated or changed by the experimenter is called the *independent variable*. The factor that changes as a result of varying the independent variable is called the *dependent variable*. The factors that are held constant and are not allowed to change are called the *controlled variables*. Many experiments may have to be performed to determine the effect of several variables on each other.

CONTROL OF VARIABLES

A critical part of any experiment is the control of variables. The results of many experiments are difficult to interpret because variables either have not been, or could not be, properly controlled. Human diet and health experiments, for example, are often surrounded by uncertainty and controversy. It is not easy to come up with valid conclusions following these kinds of experiments because of the large number of variables associated with humans. Variables such as genetic inheritance, smoking and exercise habits, job or school tensions, emotional state, and many others like these all influence health. Each one of these variables must be considered, and if possible, controlled, if proper conclusions are to be drawn.

Listing these variables is hard enough. Controlling them is a considerable challenge! Some areas of study, such as economics and meteorology (weather prediction), are considered inexact sciences due to the extreme difficulty of identifying and controlling all of the variables that can affect their outcomes.

Gas volume v. temperature

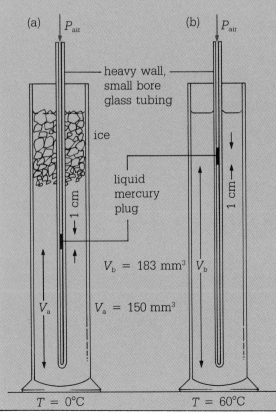

Independent Variable:	temperature. The experimenter changes the temperature.
Dependent Variable:	volume. The volume of the gas changes in response to the change in temperature.
Controlled Variable:	pressure. The pressure on the gas remains constant during the experiment (atmospheric pressure and the pressure caused by the mercury in the plug).
	number of moles. The number of gas molecules trapped under the mercury droplet stays the same.

Gas volume v. pressure

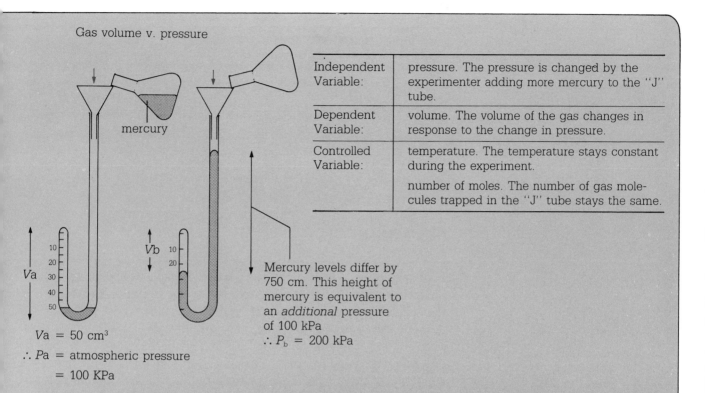

mercury

Va

Vb

| 10 |
| 20 |
| 30 |
| 40 |
| 50 |

| 10 |
| 20 |

$Va = 50 \text{ cm}^3$

$\therefore Pa = $ atmospheric pressure

$= 100 \text{ KPa}$

Mercury levels differ by 750 cm. This height of mercury is equivalent to an *additional* pressure of 100 kPa

$\therefore P_b = 200 \text{ kPa}$

Independent Variable:	pressure. The pressure is changed by the experimenter adding more mercury to the "J" tube.
Dependent Variable:	volume. The volume of the gas changes in response to the change in pressure.
Controlled Variable:	temperature. The temperature stays constant during the experiment.
	number of moles. The number of gas molecules trapped in the "J" tube stays the same.

Gas pressure v. temperature

Independent Variable:	temperature. The temperature is changed by the experimenter.
Dependent Variable:	pressure. The pressure changes in response to the temperature change.
Controlled Variable:	volume. The volume of the solid container stays relatively constant.
	number of moles. The system is sealed, and no gas molecules are added or removed during the experiment.

Table 11.1 Volume and Temperature of a Sample of Air

Temperature (°C) (Independent Variable)	Volume (mm³) (Dependent Variable)	Controlled Variables
100	126	(a) pressure
75	119	(b) number of moles of gas
50	109	
25	102	
0	92	

Data Analysis

One of the most useful ways of organizing and seeking regularities in quantitative data is to plot the data on a graph. Conventionally, the independent variable is plotted on the horizontal or x axis and the dependent variable is plotted on the vertical or y axis, as shown in Figure 11.4. Since mathematical relationships are usually expressed as y being a function of x, this allows you to express how the dependent variable (on the y axis) varies with changes in the independent variable (on the x axis). For example, in the volume-temperature experiment being discussed (Table 11.1), the gas volume varied as the temperature changed (volume versus temperature). The reverse, temperature versus volume, would imply that the temperature was measured as the volume was changed. This would change the controlled variable, pressure. Thus the experiment cannot be carried out in this way.

Figure 11.4
Data is usually graphed to reflect how an experiment has been carried out. The dependent variable is said to vary as the independent variable is changed. Mathematical equations derived from these graphs also express this relationship.

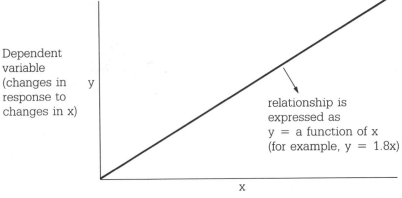

Dependent variable (changes in response to changes in x)   y

relationship is expressed as y = a function of x (for example, y = 1.8x)

x

Independent variable (changed by experimenter)

The data listed in Table 11.1 have been graphed in Figure 11.5. A regularity is immediately apparent. Within experimental uncertainty, the data points lie on a straight line, indicating a linear relationship. You can apply straight line graph theory from mathematics to determine an equation which describes how the volume of the sample of air used in the experiment varies with temperature.

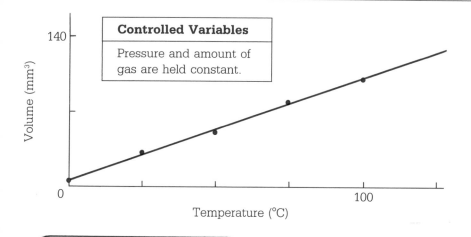

Volume (mm³) — Temperature (°C)

Controlled Variables

Pressure and amount of gas are held constant.

140

0   100

Figure 11.5
Analyzing experimental
data – graphing

SPECIAL TOPIC 11.2

GRAPHICAL ANALYSIS

The mathematical expression for a straight line on a graph is $y = mx + b$. The variable y stands for the vertical or y axis value of a point; x represents the horizontal or x axis value; m represents the slope of the line; and b stands for the y intercept value (where the line crosses the y axis).

The slope of a line is determined by first measuring the rise (on the y axis) and the run (on the x axis) between any two points on the line. The slope is then calculated by dividing the rise by the run.

In plotting experimental results, the x and y axes have identified variables. For example, in an experiment to find the relationship between volume and temperature, the x axis (the independent variable) measures temperature, T, and the y axis (the dependent variable) measures volume, V. Thus, the straight line equation

$$y = mx + b$$

becomes

$$V = mT + b$$

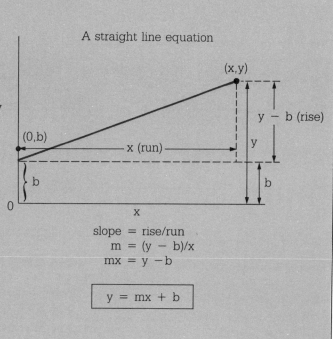

A straight line equation

(x,y)

$y - b$ (rise)

$(0,b)$

x (run)

y

b

b

0   x

slope = rise/run
$$m = (y - b)/x$$
$$mx = y - b$$

$$y = mx + b$$

Determining the Equation

The graph in Figure 11.5 has been replotted in Figure 11.6 on page 284. The straight line has been extrapolated to the x intercept ($y = 0$). The slope of this line, m, as determined from the graph, is 0.34 mm³/°C. The x intercept is about -273°C (absolute zero!). A gas volume versus temperature experiment provides a means of measuring the value of absolute zero.

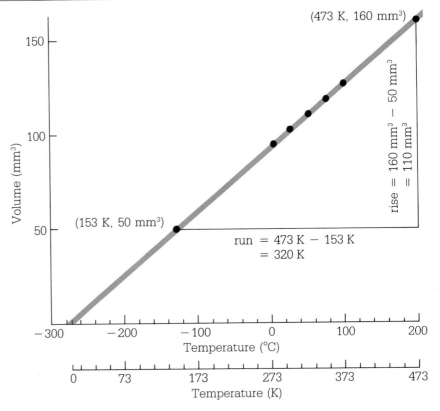

Figure 11.6
Determining a straight line equation. The slope of the line is obtained by determining the rise and run between two points on the graph. The y intercept (b) is set to zero by using the Kelvin scale (that is, at zero volume the temperature is 0 K).

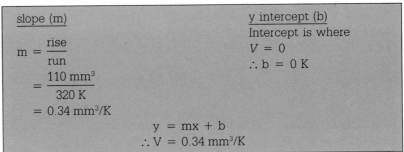

In the graph in Figure 11.6 the y axis represents volume V. If you use the Kelvin temperature scale for the x axis, as shown by the lower horizontal scale in Figure 11.6, then x represents the temperature T in kelvins and the y intercept b becomes zero. Thus, the general equation for a straight line

$$y = mx + b$$

becomes

$$V = mT + 0$$

The slope, m, from the graph in Figure 11.6, is 0.34 mm³/K. Substituting this value in the last equation, the equation becomes

$$V = 0.34\ T \text{ (volume in mm}^3\text{)}$$

This equation describes the volume-temperature relationship at constant pressure for the specific amount of gas used in the experiment.

EXTRAPOLATION AND REALITY

The graph in Figure 11.6 has been extrapolated (that is, extended beyond the region for which we have data points) to determine the x intercept. This was found to be about −273°C or absolute zero. What does this mathematical extrapolation mean physically?

The straight line graph indicates that as the temperature of a gas drops, so does its volume.

Volume of a gas as it cools.
A real gas such as ammonia has molecules of definite volume. A real gas will condense into a liquid having a virtually fixed volume before absolute zero temperature is reached.

The kinetic molecular theory predicts that if the temperature is lowered to absolute zero, −273°C, then the gas particles would not be moving and their volume would be at a minimum. The extrapolation of this graph would provide an experimental measurement of absolute zero. But how can a gas shrink to zero volume? Gas molecules, like any other molecules, do occupy some space — they would condense to a liquid or solid of a virtually fixed volume before absolute zero is reached, as shown in the accompanying diagrams.

Predicting behaviour by extrapolating beyond the region for which data has been gathered can produce unreliable results. The concept of an imaginary "ideal gas" whose molecules have zero volume is used to overcome these "real gas" difficulties. This concept is discussed later in the chapter.

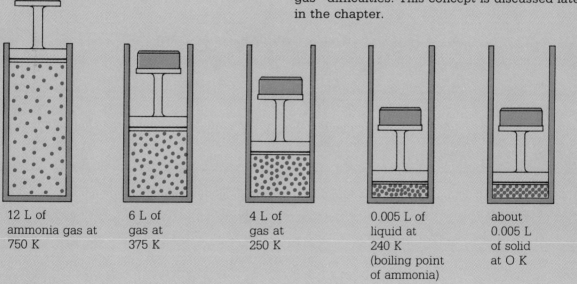

12 L of ammonia gas at 750 K

6 L of gas at 375 K

4 L of gas at 250 K

0.005 L of liquid at 240 K (boiling point of ammonia)

about 0.005 L of solid at 0 K

The General Volume-Temperature Relationship: Charles' Law

For any quantity of a gas kept at a fixed pressure, experiments show that $V = mT$, where m is a constant (the slope). The value of the constant depends on two factors: the amount or number of moles of gas there are in the sample, and the pressure under which it is kept. The volume of a gas varies directly as its Kelvin temperature. As the Kelvin temperature rises or falls, gas volume also rises or falls in the same proportion.

This linear relationship was first determined experimentally and stated as a law in 1787 by Jacques Charles, a French scientist. **Charles' law** states that *the volume of a fixed quantity of gas kept at constant pressure is directly proportional to the Kelvin temperature.*

In mathematical terms Charles' law can be written as

$$V \propto T$$

The symbol $\propto$ means "proportional to" or "varies directly as."

Sample Problem 11.1

For the gas sample discussed in Figure 11.6, determine the volume if the temperature is increased to 230°C while the pressure remains constant.

Solution

Step 1
Determine the Kelvin temperature.

$$T = 230°C + 273$$
$$T = 503 \text{ K}$$

Step 2
Substitute this temperature into the straight line equation derived from the experiment ($V = 0.34\ T$).

$$V = 0.34\ \frac{mm^3}{K} \times T$$

$$V = 0.34\ \frac{mm^3}{K} \times 503\ K$$

$$V = 171\ mm^3$$

Therefore, the volume of the gas sample is 171 mm³ at the conditions specified.

Exercise

1. A volume-temperature experiment which was performed on a sample of hydrogen gas produced the Charles' law relationship: $V = 0.167\ T$ (V is in litres). At what temperature, in degrees Celsius, would this sample of gas occupy 50 L if the pressure was held constant?

Using Charles' Law

Charles' law can be used to solve problems involving gas volume and temperature. The expression we have developed earlier, $V \propto T$, (or $V = mT$), means that at constant pressure, if the Kelvin temperature of any sample of gas is doubled, the volume of the gas will also double. In Table 11.2, for example, if the temperature of 20 L of a gas at 100 K is doubled to 200 K, then the volume also doubles to 40 L. Likewise, if the temperature increases by a factor of three (to 300 K), then the volume increases by the same factor to 60 L. Whatever factor change the absolute temperature undergoes, the volume of the gas changes by the same factor (provided the pressure remains constant).

Table 11.2 Gas Volume Change as Temperature Changes (at constant pressure)

Temperature (K)	Temperature Change Factor	Volume (L)
100	–	20
200	200/100 = 2	40
300	300/100 = 3	60
50	50/100 = 0.5	10
25	25/100 = 0.25	5
231	231/100 = 2.31	?

What is the volume of the gas sample discussed in Table 11.2 if the tempreature changes to 231 K? As indicated, the temperature change factor is 2.31. Thus, in accordance with Charles' law, the volume also changes by a factor of 2.31. The conversion factor for determining the new volume is the temperature change factor. The volume becomes 2.31 × 20 L or 46.2 L. This procedure can be summarized as follows using V_1 and T_1 as the original volume and temperature, and V_2 and T_2 as the new volume and temperature.

new volume = original volume × conversion factor

$$V_2 \quad = \quad V_1 \quad \times \quad \frac{T_2}{T_1}$$

Sample Problem 11.2

If 50 cm³ of gas in a syringe at 15°C is heated to 50°C and the syringe's piston is allowed to move outward against (constant) atmospheric pressure, calculate the new volume of the hot gas. ⟶

Solution

Step 1
List the given information, identify the unknown, and convert temperatures to kelvins.

$$V_1 = 50 \text{ cm}^3 \qquad V_2 = ?$$

$$T_1 = 15°C \qquad T_2 = 50°C$$

$$= 288 \text{ K} \qquad = 323 \text{ K}$$

Step 2
Identify the conversion factor, substitute, and calculate the answer.

$$V_2 = V_1 \qquad \times \qquad \frac{T_2}{T_1}$$

$$V_2 = 50 \text{ cm}^3 \times \frac{323 \text{ K}}{288 \text{ K}}$$

(Remember, if the temperature increases, so must the volume. The conversion factor must therefore be greater than one.)

$$V_2 = 56 \text{ cm}^3$$

Therefore, the new volume of the hot gas is 56 cm³.

Exercise

2. What is the final volume if 3.4 L of nitrogen gas at 400 K is cooled to 200 K and kept at the same pressure?

3. Determine the final volume of 20 L of a gas whose temperature changes from $-73°C$ to $327°$ C if the pressure remains constant.

4. A partially filled plastic balloon contains 3.4×10^3 m³ of helium gas at 5°C. The noon day sun heats this gas to 37°C. What is the volume of the balloon if atmospheric pressure remains constant?

11.3 Volume and Pressure

Laboratory Manual
You will find a quantitative relationship between the pressure and volume of a gas in Experiment 11-2.

In order to determine the behaviour of gases completely, other experiments with different independent and/or dependent variables must be performed. A British scientist, Robert Boyle, in the mid 1600s performed a series of experiments on gases using pressure as his independent variable and gas volume as the dependent variable. Boyle kept constant the temperature and amount of gas, which are therefore the controlled variables. Figure 11.7 illustrates the apparatus he used. Sample results are given in Table 11.3.

Figure 11.7

Robert Boyle's experiment. In the mid 1600s, Robert Boyle experimentally determined the volume v. pressure relationship for gases. He used the apparatus shown here and increased the pressure on the gas trapped in the "J" tube by adding mercury to the open tube. The pressure exerted on the trapped gas is the sum of atmospheric pressure and the additional pressure exerted by the unbalanced mercury column. (A 1 cm column of mercury exerts a pressure of 1.33 kPa.)

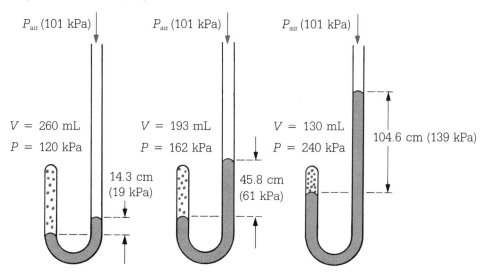

Table 11.3 Boyle's Experiment: Sample Pressure and Gas Volume Data

Pressure (kPa) (Independent Variable)	Volume (L) (Dependent Variable)	Controlled Variables
120	0.261	(a) temperature
145	0.218	(b) number of
162	0.193	moles of gas
180	0.171	
200	0.159	
216	0.145	
240	0.130	
258	0.120	

Data Analysis

Examining the data in Table 11.3, you will notice that the volume of the gas decreases as the pressure exerted on it increases. For example, doubling of the pressure from 120 kPa to 240 kPa causes the volume to shrink from 0.261 L to half its original volume, 0.130 L. (Note that all experimental data have some degree of error and 0.130 L may be considered as half of 0.261 L.) This type of relationship is described as an inverse variation. When these data are plotted on a volume vs. pressure graph, a curved line, rather than a straight line, is obtained. (See Figure 11.8.) The equation of a curved line cannot be obtained as readily as the equation of a straight line.

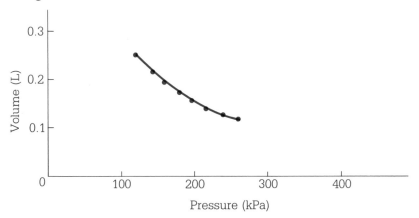

Figure 11.8
Gas volume v. pressure. A curved line graph such as this indicates that the dependent variable (volume) is inversely related to the independent variable (pressure).

Since the relationship is inverse (the volume decreases as the pressure increases), perhaps replotting the graph using the inverse values of pressure ($1/P$) as calculated in Table 11.4 would be helpful. This graph does produce a straight line. (See Figure 11.9.)

Table 11.4 Data Analysis – Inverse Pressure and Gas Volume (temperature and amount of gas held constant)

Pressure (kPa)	Inverse Pressure (1/P) (kPa^{-1})	Volume (L)	$P \times V$ (kPa·L)
120	8.33×10^{-3}	0.261	31.3
145	6.90×10^{-3}	0.218	31.6
162	6.17×10^{-3}	0.193	31.3
180	5.56×10^{-3}	0.171	30.8
200	5.00×10^{-3}	0.159	31.8
216	4.63×10^{-3}	0.145	31.8
240	4.17×10^{-3}	0.130	31.3
258	3.88×10^{-3}	0.120	31.0

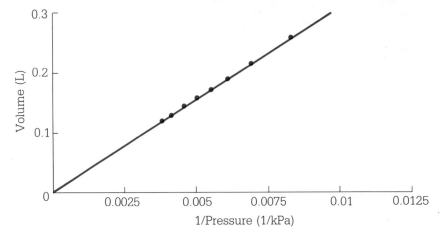

Figure 11.9
Gas volume v. inverse pressure. A straight line graph is obtained by plotting volume v. the inverse values of pressure (1/P).

The equation of the straight line graph in Figure 11.9 can be obtained as follows.

The general equation of a straight line is:

$$y = mx + b$$

Replace x and y in the equation with the variables being measured on these axes, inverse pressure $(1/P)$ and gas volume (V) respectively.

$$V = m(1/P) + b$$

Determine the slope (31.3 kPa·L) and y intercept (0) from the straight line graph in Figure 11·9. Substitute these into the last equation.

$$V = (31.3 \text{ kPa·L}) (1/P) + 0$$

or $$V = 31.3 \text{ kPa·L} \times \frac{1}{P}$$

Rearranging the equation, you have

$$PV = 31.3 \text{ kPa·L}$$

This equation indicates that the volume of a sample of gas multiplied by its pressure is a constant, provided the temperature stays the same. Multiplying the volume by the pressure for the data as listed in Table 11.4 does produce a constant (31.3 kPa·L) within experimental error. For a given sample of gas maintained at a constant pressure; $PV = m$, where m is a constant.

Exercise

5. An experiment carried out on a sample of oxygen gas at 25°C resulted in the following equation:

$$PV = 2.1 \times 10^3 \text{ kPa·L}$$

Calculate the volume of this sample of oxygen gas when the pressure is 500 kPa and the temperature is 25°C.

The Variation of Gas Volume with Pressure: Boyle's Law

The $PV = m$ relationship, rearranged as $V = m/P$, was first stated as a law by Robert Boyle in the mid 1600s. **Boyle's law** states that *the volume of a fixed mass of gas at constant temperature varies inversely as the applied pressure.*

The mathematical expression of Boyle's Law is

$$V \propto 1/P$$

Using Boyle's Law

Boyle's law can be used to solve problems involving gas volume and pressure. For example, if the pressure on 12 L of gas goes from 100 kPa to 200 kPa (pressure doubles), then the gas would be squeezed into half the original volume, 6 L, provided the temperature remains constant. Similarly, if the pressure triples, the gas volume would be only one third of its original value. Whatever the pressure factor increase, the volume decreases by the same factor. (See Table 11.5.)

Table 11.5 Gas Volume Change as Pressure Changes (at constant temperature)

Pressure (kPa)	Pressure Change Factor	Volume Change Factor	Volume (L)
100	–	–	12
200	200/100 = 2	1/2	6
300	300/100 = 3	1/3	4
600	600/100 = 6	1/6	2
50	50/100 = 1/2	2	24
25	25/100 = 1/4	4	48
171	171/100	100/171	?

What is the volume of the gas sample discussed in Table 11.5 if the pressure is changed to 171 kPa? The pressure change factor is 171/100. Since volume varies inversely as the pressure, the volume change factor must be the inverse of 171/100, that is, 100/171. In other words, the conversion factor for determining the new volume is 100/171. The original gas volume must be multiplied by this factor (12 L × 100/171) to obtain the new volume (7 L). This procedure can be summarized as follows.

new volume = original volume × conversion factor

$$V_2 = V_1 \times \frac{P_1}{P_2}$$

Note that the conversion factor for the volume is the inverse of the pressure change factor; if the pressure increases, the conversion factor is less than one and hence the gas volume decreases.

Sample Problem 11.3

What is the volume of 4.8 L of hydrogen gas if the pressure exerted on it increases from 55 kPa to 127 kPa? Assume that the temperature remains constant.

Solution

Step 1

List the given information and identify the unknown.

$$V_1 = 4.8 \text{ L} \qquad V_2 = ?$$
$$P_1 = 55 \text{ kPa} \qquad P_2 = 127 \text{ kPa}$$

Step 2

Identify the conversion factor, substitute the data, and calculate the answer.

$$V_2 = V_1 \qquad \times \qquad \frac{P_1}{P_2}$$

$$V_2 = 4.8 \text{ L} \times \frac{55 \text{ kPa}}{127 \text{ kPa}}$$

(Remember, if the pressure increases, the volume must decrease. The conversion factor is therefore less than one.)

$$V_2 = 2.1 \text{ L}$$

Therefore, the new volume is 2.1 L.

Exercise

6. The pressure on 15.0 L of gas is increased from 80.0 kPa to 320 kPa. If the temperature remains constant, calculate the new volume of the gas.

7. 1.00 L of helium gas in a cylinder under 5.68×10^4 kPa pressure fills a balloon at 163 kPa pressure at the same temperature. What is the volume of the balloon?

8. What volume of air at 100 kPa pressure would be required to fill a 35.0 L car tire to a pressure of 285 kPa if the temperature remains constant?

9. The barrel of a bicycle pump has a volume of 108 mL and contains air at 102 kPa pressure. What volume of the barrel would the gas occupy if its pressure increases to 630 kPa? Assume that the temperature is constant.

10. A 16.0 L fire extinguisher contains 15.0 L of water and 1.00 L of compressed air. When in use, the fire extinguisher must expel the last bit of water at a pressure of 110 kPa. What should the original compressed air pressure in the fire extinguisher be? Assume that the temperature is constant.

11.4 A Combination Gas Law

Since Boyle's law relates pressure and volume of a fixed mass of gas, and Charles' law relates its temperature and volume, it should be possible to combine the two laws to see how the volume of a fixed mass of gas changes when both temperature and pressure change at the same time. (See Figure 11.10.)

Figure 11.10
If the pressure exerted on a sample of gas is changed and then is followed by a change in temperature, the resulting gas volume is the same as when both the temperature and pressure change at the same time.

(a) sequential changes

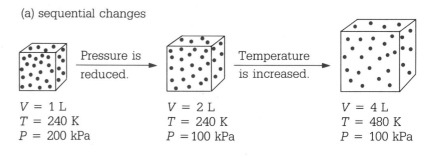

(b) combined change

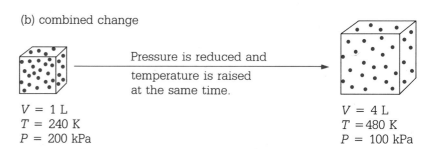

The combination of Boyle's and Charles' laws is called the **combined gas law**. It is used to solve problems involving the volume of a fixed mass of gas when both pressure and temperature change. Calculating the new volume of a sample of gas when both temperature and pressure change requires the use of two conversion factors.

$$V_2 = V_1 \times \left(\begin{array}{c} \text{conversion factor due} \\ \text{to temperature change} \end{array} \right) \times \left(\begin{array}{c} \text{conversion factor due} \\ \text{to pressure change} \end{array} \right)$$

Charles' Law
This factor must cause the volume to increase if the temperature increases (or the volume to decrease if the temperature decreases). In other words, this factor is a direct variation.

Boyle's Law
This factor must cause the volume to decrease if the pressure increases (or the volume to increase if the pressure decreases). In other words, this factor is an inverse variation.

$$V_2 = V_1 \times \frac{T_2}{T_1} \times \frac{P_1}{P_2}$$

Sample Problem 11.4

A firefighter's air tank contains 12.0 L of air compressed to 1.40×10^4 kPa at 22°C. What volume of air will this tank provide when it is used in a hot, smoke-filled building where the temperature is 42°C and the pressure 102 kPa?

Solution

Step 1

List the given information, identify the unknown, and convert temperatures to kelvins.

$$V_1 = 12.0 \text{ L} \qquad\qquad V_2 = ?$$

$$P_1 = 1.40 \times 10^4 \text{ kPa} \qquad P_2 = 102 \text{ kPa}$$

$$T_1 = 22°C \qquad\qquad T_2 = 42°C$$

$$= 295 \text{ K} \qquad\qquad = 315 \text{ K}$$

Step 2

Identify the conversion factors, substitute the data, and calculate the answer.

$$V_2 = V_1 \qquad\times\qquad \frac{T_2}{T_1} \qquad\times\qquad \frac{P_1}{P_2}$$

$$V_2 = 12.0 \text{ L} \times \frac{315 \text{ K}}{295 \text{ K}} \times \frac{1.40 \times 10^4 \text{ kPa}}{102 \text{ kPa}}$$

(T increases, ∴   (P decreases, ∴
V increases,   V increases,
thus factor > 1)   thus factor > 1)

(direct variation)   (inverse variation)

$$V_2 = 1.76 \quad \times 10^3 \text{ L}$$

Therefore, the firefighter's air tank will provide 1.76×10^3 L of air in the hot building.

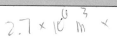
$2.7 \times 10^6 \text{ m}^3 \times$

Standard Temperature and Pressure

Since the volume of a fixed mass of gas depends on the temperature and pressure conditions under which it is kept, scientists have established **standard temperature and pressure (STP)** conditions when referring to gases. This allows the properties of different gases to be more easily compared. The average air pressure at sea level, 101.3 kPa, has been selected as **standard pressure**. An easily obtained temperature, the temperature of an ice-water mixture, 0°C (273 K), has been selected as **standard temperature**. The measured volume of a gas is often "converted" to its equivalent volume at standard temperature and pressure (STP).

Note that in combined gas law problems involving a change in the temperature and pressure conditions, the mass of gas must remain constant. (See Figure 11.11.)

Figure 11.11
A definite amount of gas under different conditions. The combined gas law can be used when a fixed mass of gas (fixed number of moles) is being used at different conditions of temperature and pressure.

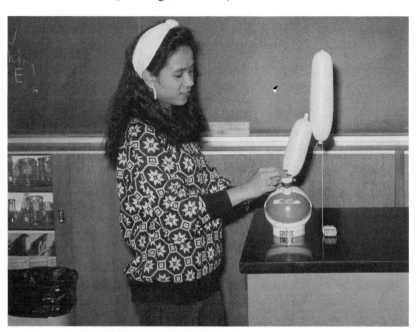

Exercise

11. A weather balloon contains 2.5 m³ of helium gas at 15°C and 98 kPa pressure. What volume would this gas occupy at STP?

12. A welder needs 5000 L of oxygen gas at 150 kPa pressure at a temperature of 21°C. To what pressure must a 50.0 L tank be filled at 13°C?

13. The helium tank shown in Figure 11.11 has a volume of 1.5 L, a pressure of 1.50×10^4 kPa, and is in a room kept at 23°C. The balloons being filled each have a volume of 2.55 L and maintain a temperature and pressure at 15°C and 115 kPa respectively. How many balloons is the tank capable of filling?

14. The pressure exerted on a diver by the water above increases by about 100 kPa for every 10 m of depth. A scuba diver uses air at the rate of 8.0 L/min at a depth of 10 m where the pressure is 200 kPa (100 kPa due to the atmosphere and 100 kPa due to water pressure) and the temperature 8°C. If the diver's 10 L air tank is filled to a pressure of 2.1×10^4 kPa at a dockside temperature of 32°C, how long can the diver remain safely submerged?

11.5 The Mole and Gases

Gases and Chemical Reaction

Many chemical reactions involve gases as reactants or products. Are there any regularities concerning gases when they react chemically? The results of numerous experiments have shown that there is a surprisingly simple regularity. Measured at the same temperature and pressure, gases react or are chemically produced in simple volume ratios. For example, the electrolysis of water always produces two volumes of hydrogen to each volume of oxygen produced, both gases measured at the same temperature and pressure. Many other chemical reactions involving gases produce similar results. For example, 1 L of hydrogen gas always reacts with exactly 1 L of chlorine gas to form 2 L of hydrogen chloride gas. See Table 11.6.

Laboratory Manual
How can you determine the mass of a standard volume of a gas? Find out in Experiment 11-3.

Table 11.6 Gas Volume Ratios in Chemical Reactions

Chemical Reaction	Volume Ratio
hydrogen + chlorine → hydrogen chloride 1 volume + 1 volume → 2 volumes	(H:Cl:HCl) 1:1:2
hydrogen + nitrogen → ammonia 3 volumes + 1 volume → 2 volumes	(H:N:NH₃) 3:1:2
water → hydrogen + oxygen (liquid) → 2 volumes + 1 volume	(H:O) 2:1

This volume relationship was recognized by the French scientist Joseph Louis Gay-Lussac in 1808. Gay-Lussac's **law of combining gas volumes** states that *whenever gases react or are produced, their volumes measured at the same temperature and pressure are in the ratio of small whole numbers.*

Scientists at first found the chemistry of these gaseous reactions puzzling. (See Figure 11.12.) What possible explanation involving

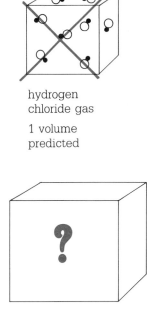

hydrogen chloride gas

1 volume predicted

2 volumes observed

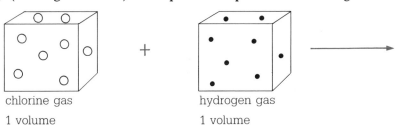

chlorine gas

1 volume

hydrogen gas

1 volume

Figure 11.12
The puzzle of reacting gases. A first hypothesis for the reaction of hydrogen and chlorine gases – that one hydrogen molecule reacts with one chlorine molecule – predicts that one volume of hydrogen chloride should be formed. Experiments however, show that two volumes of hydrogen chloride gas are actually produced.

gas particles or molecules could explain the formation of two volumes of hydrogen chloride gas from one volume of hydrogen gas and one volume of chlorine gas?

An Explanation – Avogadro's Hypotheses

The experimental results summarized in Gay-Lussac's Law of combining gas volumes were explained by the Italian scientist Amedeo Avogadro in 1811. Avogadro established a simple explanation by proposing two hypotheses as follows.

1. Equal volumes of gases measured at the same temperature and pressure contain equal numbers of molecules.
2. The particles making up gaseous elements such as hydrogen, oxygen, and chlorine are molecules containing an even number of atoms. (The simplest molecule, one containing two atoms per molecule, was assumed.)

Avogadro's hypotheses can be used to explain all of the gas volume results of chemical reactions. (See Figure 11.13.) None of the experimental evidence accumulated over the years since Avogadro proposed his hypotheses has required that his hypotheses be modified. Because of this, the first of these hypotheses is now known as **Avogadro's law.**

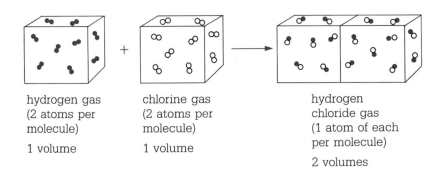

hydrogen gas
(2 atoms per
molecule)

1 volume

chlorine gas
(2 atoms per
molecule)

1 volume

hydrogen
chloride gas
(1 atom of each
per molecule)

2 volumes

Figure 11.13
Avogadro's explanation. If equal volumes of any gas at the same temperature and pressure contain equal numbers of molecules, and if these molecules have two atoms each (in the case of hydrogen and chlorine), then the experimental results are easily explained.

Exercise

15. Use Avogadro's ideas to explain how three volumes of hydrogen gas react with one volume of nitrogen gas to form two volumes of ammonia gas.

16. What is the chemical formula for ammonia?

THOUGHT EXPERIMENTS IN SCIENCE

Thought experiments in science have a very long and honoured tradition. In thinking about matter, Democritus, about 2500 years ago, carried out a thought experiment and generated the idea of an atom. He imagined a piece of material, a grain of salt, and asked himself what would happen if this grain of salt was divided in half, in half again, and again, etc. Could this process continue forever? Democritus concluded that this was not reasonable. An ultimately small, indivisible bit of salt must result. He called this smallest possible bit of matter an "atom". (*Atomos* in Greek means indivisible.)

Einstein also used thought experiments during the early development of his theory of relativity before much applicable experimental data had been obtained.

The great German physicist, Werner Heisenberg, called such experiments *gedanken* (imaginary) experiments. One of the most perplexing thought experiments concerning quantum theory and reality goes under the rather strange title of "Schrödinger's Cat". It is based on Heisenberg's uncertainty principle. You might be interested in finding out more about "Schrödinger's Cat" and its implications for that area of physics called quantum mechanics.

Avogadro's Law and Molar Mass

Avogadro's law states that equal volumes of different gases at the same temperature and pressure contain equal numbers of particles. Since the molecules of different gases have their own characteristic mass, we can predict that the mass of the same volume of different gases measured at the same temperature and pressure should be in the same ratio as their molar masses. Therefore, we can test Avogadro's Law by experiment. Avogadro's law can also be used to determine the molar mass of an unknown gas. This is done by comparing the mass of a volume of the unknown gas with the mass of an equal volume (at the same temperature and pressure) of a gas of known molar mass. Since the mass ratio and molar mass ratio are the same, the molar mass of the unknown gas can be calculated.

The operation of very high altitude balloons uses the principle of Avogadro's law. The balloon itself is made of thin plastic film, much like garbage bag plastic. The pressure and temperature are the same inside and outside the balloon. Inside the balloon, air particles (average molecular mass about 29 u) are replaced by lighter helium particles (molecular mass about 4 u). Since the balloon holds the same number of helium particles as air particles under the same temperature and pressure by Avogadro's Law, the density of the balloon is much less than that of the surrounding air. Therefore, the balloon rises and floats in the air.

Laboratory Manual
In Experiment 11-4, you will determine the molar mass of butane.

AIR – THE MOST IMPORTANT GAS

COMPOSITION

Our most vital and widely used gas is air. Air is a mixture of gases. The proportion of molecules making up this gaseous mixture is 78.08% nitrogen, 20.95% oxygen, 0.93% argon, and 0.03% carbon dioxide. Traces of neon, helium, krypton and hydrogen (along with variable amounts of water vapour and pollutant gases such as sulphur and nitrogen oxides) are also found in air.

USES

Air is used for a variety of purposes. Air is taken along by humans for breathing purposes when they venture into hostile environments ranging from the sea bottom to outer space. Compressed air is used to supply the energy for operating tools such as jack hammers and power wrenches. It is used as a "structural material" in car tires, rubber boats and camp mattresses. Air can support very large open structures such as domed stadiums and sports fields. And, air is the source of the most widely used chemical reactant, oxygen, for respiration and many combustion reactions.

Some Uses of Air

Use	Devices
Breathing	Scuba tanks; fireman rescue tanks
Energy source	Fire extinguishers; pneumatic "jack hammers"
Structural uses	Car tires; inflatable rubber boats; air mattress
Support medium	Aircraft; hot air balloons; hovercraft
Chemical reactant	Fire; combustion engines; sewage treatment

Mixed Gases – Dalton's Law of Partial Pressures

Consider a container held at a constant temperature and holding a fixed number of nitrogen molecules. The kinetic molecular theory states that the pressure on the walls of the container is caused by the nitrogen molecules striking and bouncing off. If twice the original number of nitrogen molecules is now injected into the container, the pressure in the container should theoretically double. This should occur because there are now twice as many particles striking the walls per second. We can continue this argument or "thought experiment" (see Special Topic 11.4) to generate the results shown in Table 11.7.

If the extra particles injected into the container of nitrogen gas were oxygen molecules, would the pressure be any different? Avogadro's law predicts that the total pressure would be the same as when the extra molecules were nitrogen molecules. It does not matter what the gas particles are – only how many there are.

Experiments carried out by John Dalton in 1800 showed that the total pressure exerted by a mixture of non-reacting gases was the sum of the individual pressures exerted by each of the gases in the mixture. The pressure caused by the particles of one of the gases in a mixture of gases is called the **partial pressure** of that gas. **Dalton's Law of partial pressures** states that *the total pressure of a mixture of non-reacting gases is the sum of the partial pressures of the component gases.* (See Figure 11.14 on page 302.)

Laboratory Manual
You will study the pressure exerted by a mixture of non-reacting gases in Experiment 11-5.

Table 11.7 Gas Pressure and Gas Molecules

Pure Gas		Total Pressure (units)	Mixed Gas					
No. of N$_2$ (molecules)	Sketch		No. of molecules		Sketch	Pressure (units)		
			N$_2$	O$_2$		N$_2$	O$_2$	
4		4	1	3		1	3	
8		8	2	6		2	6	
16		16	2	14		2	14	

Figure 11.14
Dalton's law of partial pressures. The total pressure of a mixture of non-reacting gases is the sum of the partial pressures of the mixture's component gases.

1 volume

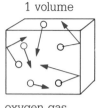

oxygen gas
pressure = 2 units

1 volume

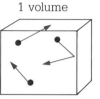

helium gas
pressure = 1 unit

1 volume

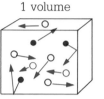

oxygen-helium gas mixture

total pressure = 3 units
partial pressure of oxygen = 2 units
partial pressure of helium = 1 unit

Exercise

17. What is the partial pressure of nitrogen in the atmosphere on a day when the pressure is 100 kPa? See Special Topic 11.5.

18. What is the partial pressure of oxygen on the same day?

Standard Molar Volume

Since many chemical reactions involve gases as reactants and products, the volume occupied by a mole of gas particles is of considerable interest. According to Avogadro's law, equal numbers of gas particles at the same temperature and pressure occupy the same volume. Therefore, 6.02×10^{23} gas particles (one mole of particles) at 0°C and 101.3 kPa (STP) should have a fixed volume regardless of the nature of the particles. Dividing molar mass by gas density gives the molar volume. (See Table 11.8.)

Table 11.8   The Molar Volume of Some Gases at STP

Gas	Density (g/L)	Molar Mass (g/mol)	Molar Volume (L/mol)
O_2	1.429	32.00	22.39
He	0.1785	4.003	22.43
N_2	1.251	28.01	22.39
NH_3	0.7621	17.03	22.35
H_2	0.08988	2.016	22.43
Ar	1.784	39.95	22.39

The accepted value for the **standard molar volume** (V_m), the volume occupied by one mole of a gas at STP, is 22.4 L. (See Figure 11.15.)

Figure 11.15

The standard molar volume. The volume occupied by 2.016 g of hydrogen gas (1 mol or 6.02×10^{23} molecules) at STP is 22.4 L. The volume occupied by one mole of molecules of any gas at the same temperature and pressure is 22.4 L.

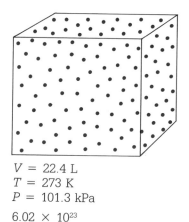

$V = 22.4$ L
$T = 273$ K
$P = 101.3$ kPa

6.02×10^{23}
hydrogen molecules

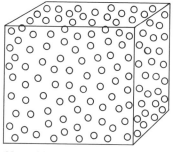

$V = 22.4$ L
$T = 273$ K
$P = 101.3$ kPa

6.02×10^{23}
oxygen molecules

At the same T and P conditions, the volume of a gas depends only on the *number* of molecules it contains, and not on what the particles are.

Sample Problem 11.5

A 2.60 L container was evacuated and its mass was found to be 655.53 g. The container was then filled with carbon dioxide gas at a pressure of 100 kPa and a temperature of 24°C. The container and gas were found to have a mass of 660.13 g. Calculate the standard molar volume of carbon dioxide.

Solution

Step 1

List the given information, identify the unknown, and determine the mass of CO_2.

$$m_1 = 655.53 \text{ g} \qquad V_m = ?$$

$$m_2 = 660.13 \text{ g} \qquad P = 100 \text{ kPa}$$

$$T = 24°C$$

$$= 297 \text{ K}$$

$$V = 2.60 \text{ L}$$

Mass of CO_2,

$$m = m_2 - m_1$$

$$= 660.13 \text{ g} - 655.53 \text{ g}$$

$$= 4.60 \text{ g}$$

Step 2
Convert the volume to STP.

$$V_1 = 2.60 \text{ L} \qquad V_2 = \text{?}$$

$$P_1 = 100 \text{ kPa} \qquad \left.\begin{array}{l} P_2 = 101.3 \text{ kPa} \end{array}\right\} \text{STP}$$

$$T_1 = 297 \text{ K} \qquad T_2 = 273 \text{ K}$$

$$V_2 = V_1 \quad \times \left(\begin{array}{c} \text{conversion factor} \\ \text{due to} \\ \text{pressure change} \end{array}\right) \times \left(\begin{array}{c} \text{conversion factor} \\ \text{due to} \\ \text{temperature change} \end{array}\right)$$

$$= 2.60 \text{ L} \times \quad \frac{100 \text{ kPa}}{101.3 \text{ kPa}} \qquad \times \qquad \frac{273 \text{ K}}{297 \text{ K}}$$

$$\qquad\qquad\qquad (\text{Boyle's law}) \qquad\qquad (\text{Charles' law})$$

$$= 2.36 \text{ L}$$

Step 3
Convert the volume, V_2, to the molar volume, V_m. To convert volume to molar volume, use the ratio of molar mass to mass as the conversion factor ($M_{CO_2} = 44.0$ g/mol).

$$V_m = V_2 \quad \times \frac{M}{m}$$

$$V_m = 2.36 \text{ L} \times \frac{44.0 \text{ g/mol}}{4.60 \text{ g}}$$

$$= 22.6 \text{ L/mol}$$

Thus, from this set of experimental data, the volume occupied by 1 mol of CO_2 gas at STP conditions is 22.6 L. (Remember that the accepted value for the molar volume is 22.4 L.)

Exercise

19. The empty container in Sample Problem 11.5 was filled with butane gas at 102 kPa and 20°C. It was then found to have a mass of 661.84 g. Determine the molar mass of butane.

Sample Problem 11.6

(a) Determine the volume that 2.5 mol of chlorine gas occupies at STP.
(b) How many individual chlorine molecules are there in the 2.5 mol?

Solution

(a) Since each mole of gas molecules occupies 22.4 L at STP, then 2.5 mol would occupy

$$22.4 \text{ L/mol} \times 2.5 \text{ mol} = 56 \text{ L}$$

(b) One mole of gas contains 6.02×10^{23} molecules. Thus, 2.5 mol would contain

$$2.5 \text{ mol} \times 6.02 \times 10^{23} \text{ molecules/mol} = 1.51 \times 10^{23} \text{ molecules}$$

Exercise

20. (a) How many moles of oxygen molecules are there in 50.0 L of oxygen gas at STP?
    (b) How many individual oxygen molecules are there in the 50.0 L of gas?

21. What volume does 0.80 mol of nitrogen gas occupy at STP?

22. Determine the volume that 1.00×10^{24} helium gas atoms occupy at STP.

Thus, as the previous examples and problems confirm, the physical behaviour of a gas concerning temperature, pressure, and volume depends only on the number of particles, and not on what the particles are. For example, the same number of particles of *any* gas injected into a container of fixed volume and at a constant temperature would always exert the same pressure.

11.6 Gas Law Limitations

As with all scientific theories and laws, the gas laws have limitations. These laws are based on experiment. Their equations were developed using extrapolations to 0 K and included the assumptions that gas molecules have zero volume and exert no attractive force between each other. This type of "ideal gas" does not exist. Real gases approach this "ideal" state when pressures are moderate (less than several hundred kilopascals) and temperatures are well above the condensation point of the gas. (See Figure 11.16 on page 307.)

As Figure 11.16 shows, the molecules of real gases under high pressure do occupy a significant proportion of the volume the gas is occupying. At lower pressures this molecular volume is insignificant compared to the total gas volume.

PUTTING CHEMISTRY TO WORK

MARGARET ALDEN / CONSTRUCTION SUPERVISOR

The handling of natural gas begins at its source, where it is pumped from the ground and transported to a refinery. But to be used in homes or industries, natural gas must be safely and continuously supplied to consumers. Supervising the construction of gas transport and utilization systems is the job of Margaret Alden, a civil engineer.

"Safety is my foremost concern," Margaret explains. "Natural gas is flammable and explosive — mistakes or miscalculations could be dangerous. Using the proper methods, however, the gas can be safely piped to wherever it is required. This is important because natural gas must go directly to the point at which it will be used, including homes."

The use of natural gas is growing dramatically as the price of alternative energy sources increases. "Our customers should be able to take their supply of natural gas for granted," Margaret says. "Everything from warmth in a home to industrial processes depends on it." Margaret makes certain, for example, that gas supplies will not be interrupted by other phases of the construction process.

From the moment Margaret leaves her office at 7:30 am until she is finished at whichever construction site is currently in progress, she needs a knowledge of people as well as technical expertise. "I'm in charge of everything from training people in safety procedures to scheduling work and supplies. These are as much a part of my job as designing the system. It's a good thing that I really enjoy working with people."

A while ago, on a job site, Margaret met a former classmate. "He was building a bridge; I was installing a gas main," she recalls. That meeting reminded Margaret of one of the most interesting aspects of civil engineering. "From my graduating class there are people designing and building landfill sites, airports, highrises, subways, water treatment plants — the variety is truly endless."

Getting dirty is just part of the job. "I am out in the field — where things are happening," says Margaret. "But I experience a real sense of accomplishment when I complete a construction project and know that I was able to overcome all of its problems, even those which seemed almost impossible to resolve."

Figure 11.6
Ideal and real gases

(a) An ideal gas

Ideal gas molecules have zero volume.

There is zero attraction between ideal gas molecules.

(b) A real gas (at low pressure and high temperature)

Real gas molecules have *some* volume. At low pressures, the volume of all the molecules is small compared to the volume of the container.

Real gas molecules do attract each other. At high temperatures (and low pressure), the kinetic energy of the molecules makes this attraction insignificant.

(c) A real gas (at high pressure and low temperature)

At high pressures, the volume of all real gas molecules *is* a significant fraction of the container's volume.

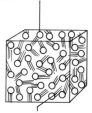

At low temperatures, the attraction among gas molecules becomes significant. High pressure forcing many molecules close together also increases the significance of this attraction.

Since gases eventually condense as temperature drops or as pressure increases, gas molecules must have some attraction for each other. At low pressures the molecules are generally very far apart and the attraction between the particles is insignificant. At temperatures well above the condensation point, the kinetic energy (speed) of gas molecules is so great that it overwhelms the attractive forces between them. Thus, the combined gas law gives good results only at relatively low pressures (up to about 200 kPa) and at temperatures well above the condensation point.

Looking Back – Looking Ahead

In this chapter you have used some of the techniques scientists employed to develop mathematical models or relationships to describe gases and their behaviour. In the next chapter you will use these gas laws to extend to gases your ability to calculate the relative amounts of substances used or produced in chemical reactions.

Vocabulary Checklist

You should now be able to define or explain the meaning of the following terms.

Avogadro's law	combined gas law	law of combining	standard molar volume
Boyle's law	Dalton's law of	gas volumes	standard temperature
Charles' law	partial pressures	partial pressure	and pressure (STP)

Chapter Objectives

You should now be able to do the following.

1. Appreciate the essential role of gases in every-day life and activity.

2. Identify the problems associated with measuring the quantity of a gas.

3. Appreciate some of the factors and difficulties involved in experimental design and in the control of variables in carrying out experiments.

4. Design experiments to gather information on the physical behaviour of gases concerning temperature, pressure, volume, and quantity (number of moles) of a gas.

5. Use experimental results to derive mathematical formulas to describe the behaviour of gases.

6. Use Boyle's and Charles' laws to solve both qualitative and quantitative gas law problems.

7. Use the combined gas law to solve quantitative problems involving gas temperature, pressure and volume.

8. Identify standard temperature and pressure (STP) conditions and the standard molar volume (22.4 L), and use the molar volume to calculate the number of moles of a gas at STP.

9. Identify the conditions under which the ideal gas assumptions are not valid for real gases and explain why these limitations arise.

Understanding Chapter 11

REVIEW

1. Identify some of the problems or difficulties associated with measuring a definite quantity of a gas.

2. What is meant by the term "variable"?

3. For each of the following laws or terms, write the appropriate equation, variation, or numerical value associated with it.
   (a) mole
   (b) STP
   (c) Boyle's law
   (d) molar volume
   (e) Dalton's law
   (f) Charles' law
   (g) Avogadro's law

4. (a) List both of the ideal gas assumptions.
   (b) Identify the conditions under which the gas laws derived using these assumptions do not yield reliable results.
   (c) Explain why the gas laws are not reliable under these conditions.

APPLICATIONS AND PROBLEMS

5. An experiment was designed to obtain information concerning the behaviour of gases. An amount of air was trapped in a bicycle pump and the volume of the air in the pump was systematically reduced. The resulting pressure was read for each volume reduction. Care was taken to ensure that the temperature of the air had reached 25°C before each reading was taken. In this experiment, identify each of the following. (Give reasons for your answers.)
   (a) the controlled variable
   (b) the independent variable
   (c) the dependent variable

6. The results for the experiment described in question 5 are to be graphed. Identify the variables that you would put on the horizontal (x) and vertical (y) axes, and explain your choices.

7. An old Fahrenheit thermometer was used in an experiment to determine the variation of gas volume with temperature.
   (a) Use the data recorded below to derive the equation for this variation and to determine absolute zero on the Fahrenheit scale.
   (b) Which gas law does this variation follow?

Temperature (°F)	Volume (cm³)
212	256
180	244
140	229
100	213
65	200
32	187
0	175

8. Copy and complete the following table in your notebook concerning possible investigations of gases using these variables: temperature (T), pressure (P), volume (V), and number of molecules (N). *Do not write in your textbook.*

Independent Variable	Dependent Variable	Controlled Variables	Law
		P and N	Charles'
			Boyle's
T	P		Gay-Lussac's (T–P)
N			Avogadro's
	P	T and V	Dalton's

9. Determine the number of moles in each of the following samples of gas.
   (a) 100 L of oxygen at STP
   (b) 50 mL of hydrogen at STP

10. A 200 mL aerosol can containing a petroleum based furniture polish carries this warning: "CONTENTS UNDER PRESSURE. Keep away from flame or sparks. Do not store near sources of heat or incinerate." Explain why this caution is necessary.

11. Complete the following statements in your notebook. *Do not write in your textbook.*
    (a) The Kelvin temperature of a 3 L sample of oxygen gas is tripled while the pressure is held constant. The new volume of this sample of oxygen is ____ L.
    (b) Air trapped in a syringe at a pressure of 100 kPa has its volume decreased to 1/8 its former volume. If the temperature is held constant, the pressure in the syringe would be ____ kPa.

$$\frac{P_1 V_1}{T_1} = \frac{P_2 V_2}{T_2}$$

Since volume is constant

$V_1 = V_2$.

(c) An "empty" aerosol can still contains some gas. If its temperature rises from 7°C to 107°C, the pressure in the can goes up by ____ %.

12. Use Boyle's law, Charles' law, or the combined gas law to determine the missing information from (a) to (h) in the following table. Record your answers from (a) to (h) in your notebook.

T_1(°C)	P_1(kPa)	V_1(L)	T_2(°C)	P_2(kPa)	V_2(L)
20	Constant	10	100	Constant	(a)
Constant	50	2.3	Constant	200	(b)
(c)	Constant	5.0	127	Constant	20
Constant	100	30	Constant	(d)	6.0
7.0	90	1.5	47	180	(e)
−20	(f)	3.5	22	100	250
0	110	0.50	(g)	250	0.80
−73	80	Constant	327	(h)	Constant

13. A car tire filled with compressed air has a pressure reading of 230 kPa when the temperature is 17°C. After an hour of high speed driving on a hot road, the temperature of the tire has gone up to 47°C. What would the tire pressure reading be now? Assume that the volume of the tire does not change.

14. A 10.5 L class A fire extinguisher (used only on paper or wood fires) contains 9 L of water and 1.5 L of air compressed to 900 kPa. When the fire extinguisher is operated, water is ejected through the hose by the air pressure. If there is no significant change in temperature, calculate the pressure remaining in the fire extinguisher when the last bit of water has just been expelled.

15. (a) Calculate the air pressure in the globe when the piston shown in the diagram below is pushed in to the end of the syringe body.

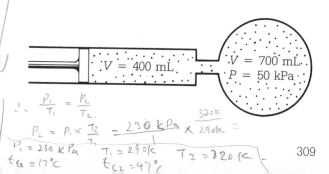

$\therefore \dfrac{P_1}{T_1} = \dfrac{P_2}{T_2}$

$P_2 = P_1 \times \dfrac{T_2}{T_1} = 230\,kPa \times \dfrac{320k}{290k} =$

$P_1 = 230\,kPa \qquad T_1 = 290k \qquad T_2 = 320k$

$t_{C_1} = 17°C \qquad t_{C_2} = 47°$

(b) Calculate the air pressure in the right hand globe shown below when the valve connecting the two globes is opened.

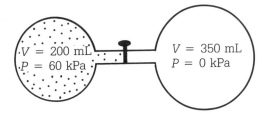

(c) For the apparatus shown below, calculate the pressure in the centre globe after:
(i) valve S_1 is opened and then closed after 2 min,
(ii) valve S_2 is then opened.

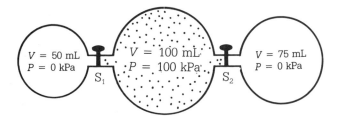

16. In passing through a turbine electrical generator, 1 L of steam at 300°C and 3500 kPa pressure expands to 55 L and has its temperature lowered to 110°C. What is the pressure of the steam leaving the turbine?

CHALLENGE

17. (a) The temperature and pressure of the gas mixture produced by the burning of excess hydrogen in oxygen (6 kg of hydrogen to 1 kg of oxygen) in the main engines of the space shuttle are 2000°C and 5.0×10^8 kPa respectively. Calculate the volume to which each cubic metre of this gas mixture expands when it has been ejected into the upper atmosphere where the pressure is 10 kPa and the temperature -30°C.

(b) Why might the actual volume of the gas produced be different from the answer you calculated in (a)?

(c) To be completely burned, 0.125 kg of hydrogen requires 1.0 kg of oxygen. Why do you think so much excess hydrogen

is used in the reaction that drives the shuttle engines?

18. Each of the four pistons in an automobile engine compresses a mixture of air and gasoline vapour from 700 cm³ to 100 cm³ during its compression stroke. (This is a 7 to 1 compression ratio.)

An automobile engine's compression and power strokes

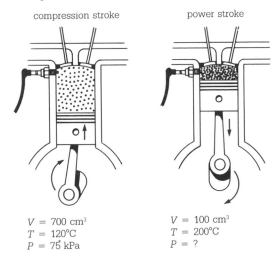

(a) If the gas mixture just before the compression stroke had a pressure of 75 kPa (normal for an engine) – why? – and a temperature of 120°C, calculate the pressure in the cylinder after the compression stroke. The heat produced by compression raises the gas temperature to 200°C.

(b) Calculate the number of moles of molecules in the cylinder.

(c) When the spark plug ignites the gas mixture, each mole of gasoline and air react to produce 1.31 mol of gaseous products, mostly carbon dioxide and steam. This "explosion" in the cylinder produces a peak gas temperature of 1200°C. Calculate the maximum pressure in the cylinder just as the piston starts to move back down the cylinder.

(d) What would the maximum pressure be if the car had a turbocharger which raises the initial pressure in the cylinder before the compression stroke to 100 kPa?

12

GASES AND CHEMICAL CALCULATIONS

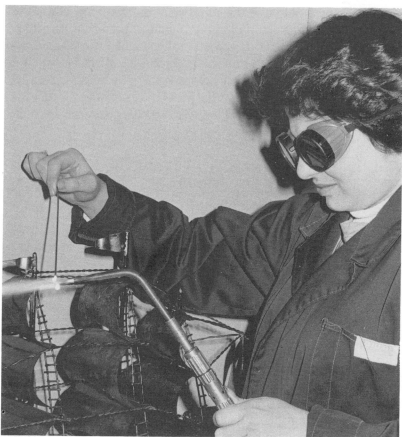

Many chemical reactions involve gases as reactants or products. The industrial production of ammonia, for example, involves the gaseous reactants hydrogen and nitrogen and a gaseous product ammonia. Furnaces and engines make use of combustion reactions involving natural gas, propane, or vaporized gasoline mixed with air. Chemists and chemical engineers must be able to determine the proper volumes of these gases needed to ensure efficient and complete reaction.

In Chapter 9, you learned to do calculations involving the relative masses of reactants needed and products formed in chemical reactions (stoichiometry). In this chapter you will see how the gas laws developed in Chapter 11 are used to develop an "ideal gas" law. This will give you a more direct method of calculating the number of moles of a substance in the gaseous state. You will then be able to include gases in stoichiometric calculations.

12.1 The Ideal Gas Law

Another Thought Experiment

We can use Avogadro's law to do a thought experiment to determine the relationship between the volume of a given gas and the number of particles it contains. If a gas has a volume of 12 L, what should happen to this volume if the number of gas particles is doubled while the temperature and pressure remain constant? What should happen to the volume if the number of gas molecules is decreased to half at the same temperature and pressure? Recall that Avogadro's law states that equal volumes of gases at the same temperature and pressure contain the same number of particles. Using this law, we can deduce the logical answers to these and other, similar questions. (See Table 12.1)

Table 12.1 Avogadro's Law and a Thought Experiment

Number of Gas Particles (Independent Variable)	Volume (Dependent Variable)	Controlled Variables
		(a) temperature (b) pressure
x	12 L	
2x	24 L	
1/2 x	6 L	
3x	36 L	
1/6 x	2 L	

What type of variation do the results in Table 12.1 indicate? Examining the data, you will observe that whatever happens to the number of particles, the same factor applies to the volume. For example, if the number of particles triples, the volume triples. This is a direct variation – the volume of a gas is directly proportional to the number of molecules it contains. If V represents the volume and N the number of molecules, we can write $V \propto N$ or $V \propto n$, the number of moles of molecules.

Putting it All Together

In the experiments for determining the various gas laws discussed earlier in Chapter 11, all but two variables had to be held constant. In many real situations involving gases, however, all of the variables – temperature, pressure, volume, and number of moles of molecules – can and do change. Is there any way to combine the set of gas laws derived earlier? Consider three of the relationships.

$V \propto T$ Charles' law (Volume varies directly as the absolute temperature.)

$V \propto 1/P$ Boyle's law (Volume varies inversely as the applied pressure.)

$V \propto n$ from Avogadro's law (Volume varies directly as the number of moles of gas particles.)

In Chapter 11, you combined Boyle's law and Charles' law and saw how the volume of a fixed mass of gas varies with changes in temperature and pressure. In a similar way, you can combine the three relationships listed above. When all three variables, temperature, pressure, and number of particles change, the volume responds as if each variable had changed separately. (See Figure 12.1.) You can write the combined relationship as

$$V \propto \frac{nT}{P}$$

You can express the relationship as an equation by replacing the "$\propto$" sign with an "$=$" sign and using constant "R" in the relationship:

$$V = \frac{RnT}{P}$$

Rearranging the variables, the equation becomes

$$PV = nRT$$

This equation, $PV = nRT$, is called the **ideal gas law** equation. It can be used to calculate the value of one of the variables (T, P, n, or V) when the other three variables are known or can be calculated by other means.

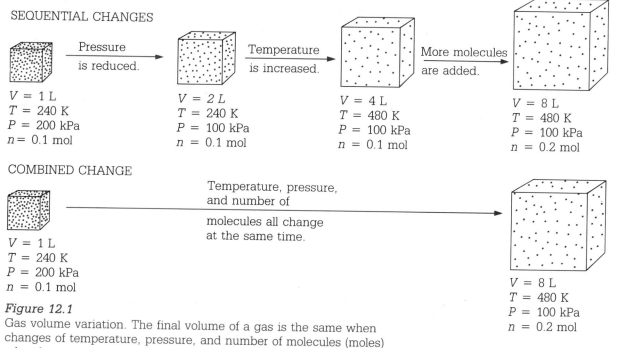

SEQUENTIAL CHANGES

Pressure is reduced. → Temperature is increased. → More molecules are added. →

$V = 1$ L
$T = 240$ K
$P = 200$ kPa
$n = 0.1$ mol

$V = 2$ L
$T = 240$ K
$P = 100$ kPa
$n = 0.1$ mol

$V = 4$ L
$T = 480$ K
$P = 100$ kPa
$n = 0.1$ mol

$V = 8$ L
$T = 480$ K
$P = 100$ kPa
$n = 0.2$ mol

COMBINED CHANGE

Temperature, pressure, and number of molecules all change at the same time.

$V = 1$ L
$T = 240$ K
$P = 200$ kPa
$n = 0.1$ mol

$V = 8$ L
$T = 480$ K
$P = 100$ kPa
$n = 0.2$ mol

Figure 12.1
Gas volume variation. The final volume of a gas is the same when changes of temperature, pressure, and number of molecules (moles) take place one after another or when they occur all at once.

You can calculate the value of the ideal gas law constant, R, by using standard molar volume information – 1 mol of gas molecules occupies 22.4 L at STP.

At STP conditions:

$$V = 22.4 \text{ L}$$
$$T = 273 \text{ K}$$
$$P = 101.3 \text{ kPa}$$
$$n = 1.00 \text{ mol}$$
$$R = ?$$

Rearrange the variables in the ideal gas law equation, $PV = nRT$.

$$R = \frac{PV}{nT}$$

Substitute the known values of the variables.

$$R = \frac{101.3 \text{ kPa} \times 22.4 \text{ L}}{1.00 \text{ mol} \times 273 \text{ K}}$$

$$R = 8.31 \frac{\text{kPa·L}}{\text{mol·K}}$$

This constant, R, is known as the **ideal gas constant**.

Sample Problem 12.1

Calculate the volume that 6.30 mol of carbon dioxide gas at 23°C and 550 kPa pressure occupy.

Solution

Step 1
List the given information, identify the unknown and determine the Kelvin temperature.

$$V = ?$$
$$P = 550 \text{ kPa}$$
$$n = 6.30 \text{ mol}$$
$$T = 23°C$$
$$= (23°C + 273) \text{ K}$$
$$= 296 \text{ K}$$

Step 2
Rearrange the ideal gas law equation.

$$PV = nRT$$

$$V = \frac{nRT}{P}$$

Step 3
Substitute and work out the answer, cancelling units wherever possible.

$$V = \frac{6.30 \; \cancel{mol} \times 8.31 \frac{kPa \cdot L}{\cancel{mol} \cdot \cancel{K}} \times 296 \; \cancel{K}}{550 \; \cancel{kPa}}$$

$$= 28.2 \; L$$

Therefore, 6.30 mol of carbon dioxide gas occupies a volume of 28.2 L at 23°C and 550 kPa pressure.

Note: In problems involving equations that combine many variables, solutions using the conversion factor method can become cumbersome. In problems such as these, the method involving formula rearrangement, substitution, and unit cancellation is more practical.

Exercise

1. A person can inhale a maximum of 0.115 mol of air per breath. Calculate the maximum volume of air a person can inhale in one breath if the atmospheric pressure is 100 kPa and the person's body temperature is 37°C.

2. A 5.00 L balloon contains 0.200 mol of air at 120 kPa pressure. What is the temperature of the air in the balloon?

12.2 Stoichiometry and Gases

In Chapter 9, you used a stoichiometric calculation procedure to determine the mass of reactants or products involved in chemical reactions. (See Figure 12.2.) In this section you will apply the ideal gas law in calculating the volume of gaseous elements or compounds used or produced in chemical reactions.

Laboratory Manual
In Experiment 12-1, you will study quantitative relationships in the reaction of hydrochloric acid with magnesium.

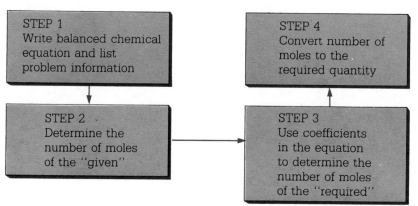

Figure 12.2
A stoichiometric calculation procedure

EXTENDING MODELS

Many mathematical models (equations) can be extended into regions where they could produce poor or inaccurate results. In some cases, scientists have experimentally determined correction factors which are added to those equations.

Many processes involving gases, for example, occur at high pressure or at low temperature where the ideal gas law becomes unreliable. Can this law be modified to handle these conditions? The molecules of each specific gas have a characteristic volume and a characteristic attractive force for each other. One of the scientists who attempted to account for these factors was the Dutch scientist, J.H. van der Waals. In 1873, he modified the ideal gas law to account for these factors in two ways:

1. By subtracting the effective volume of one mole of gas molecules from the total volume occupied by the gas. The volume becomes $V - b$ where b is an experimental value characteristic of each specific gas.

2. By adding an "internal pressure" caused by the increasing amount of attractive force between molecules as they are forced closer together as the pressure increases. Pressure becomes $P + a$ where a is another experimentally determined factor specific for one mole of each type of gas.

Thus, for one mole of gas, an improved gas equation which is more reliable over an extended range is

$$(P + a)(V - b) = RT$$

In order to use this equation the a and b values for the gas concerned must be looked up in a reference book such as *The Handbook of Chemistry and Physics*.

Some gases deviate from ideal gas behaviour more than others. In addition, the deviation becomes worse as the pressure increases and the temperature decreases. At normal atmospheric pressure and room temperature the deviations are less than 1% for most gases.

hydrogen gas

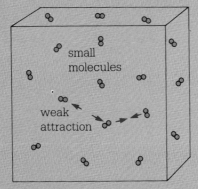

- very little attraction between molecules ($a = 0.244$)
- small molecules ($b = 0.0266$)

chlorine gas

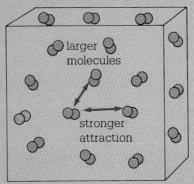

- much more attraction between molecules ($a = 6.493$)
- larger molecules ($b = 0.0562$)

Van der Waals correction factors for two different gases. The "internal pressure" correction factor, a, accounts for the attraction among gas molecules. A volume correction factor, b, is related to the volume of the gas molecules. In general, the larger the gas molecule, the larger the a and b correction factors are.

Sample Problem 12.2

Acetylene gas, C_2H_2, is normally prepared in the laboratory by reacting calcium carbide, CaC_2, with water.

$$CaC_2 + H_2O \rightarrow C_2H_2 + Ca(OH)_2$$

What mass of calcium carbide is required to produce 235 mL of acetylene gas at 0°C and 101.3 kPa pressure (STP)? (The molar mass of calcium carbide is 64.1 g/mol.)

Solution

Step 1

Write the balanced equation for the reaction and identify the data for both the "given" and "required" substances. Change the given data into the required units (volumes in litres and temperatures in kelvins). Organize the information about each substance under its location in the equation.

CaC_2	+	$2H_2O$	$\rightarrow$ C_2H_2	+	$Ca(OH)_2$
$m = ?$			$V = 235$ mL		
$M = 64.1$ g/mol			$= 0.235$ L		
			$T = 0°C$		
			$= 273$ K		
			$P = 101.3$ kPa		

Step 2

Calculate the number of moles of the "given" substance (acetylene in this case). Note that either the ideal gas law (solution (a)) *or* molar volume (solution (b)) could be used.

(a) Using the ideal gas law:

$$PV = nRT$$

$$n = \frac{PV}{RT}$$

$$n_{C_2H_2} = \frac{101.3 \text{ kPa} \times 0.235 \text{ L}}{8.31 \frac{\text{kPa·L}}{\text{mol·K}} \times 273 \text{ K}}$$

$$= 0.0150 \text{ mol}$$

(b) Since the volume of acetylene is given at STP, it can be easily converted to number of moles using the standard molar volume ($V_m = 23.4$ L/mol).

$$n_{C_2H_2} = V \times \frac{1}{V_m}$$

$$n = 0.235 \text{ L} \times \frac{1}{22.4 \text{ L/mol}}$$

$$= 0.0105 \text{ mol}$$

(Note: Gas volume must first be converted to STP conditions.)

Step 3
Use the coefficients from the balanced equation to determine the mole conversion factor and calculate the number of moles of the required substance (calcium carbide).

$$n_{CaC_2} = 0.0105 \text{ mol}_{C_2H_2} \times \frac{1 \text{ mol}_{CaC_2}}{1 \text{ mol}_{C_2H_2}}$$

$$= 0.0105 \text{ mol}_{CaC_2}$$

Step 4
Convert number of moles of calcium carbide to mass in g using the molar mass conversion factor (M).

$$m = n \times M$$

$$m_{CaC_2} = 0.0105 \text{ mol} \times 64.1 \frac{g}{mol}$$

$$= 0.673 \text{ g}$$

Therefore, 0.673 g of calcium carbide is needed to produce 235 mL of acetylene gas at 0°C and 101.3 kPa pressure (STP).

Exercise

3. Aluminum is produced by the electrolysis of a solution of aluminum oxide, Al_2O_3, dissolved in cryolite, Na_3AlF_6.

$$Al_2O_3 \rightarrow Al + O_2 \qquad \text{(unbalanced)}$$

What mass of aluminum metal is produced if, during the electrolysis, 150 L of oxygen gas at 350°C and 98 kPa pressure are released?

4. The equation below represents the process of photosynthesis.

$$6CO_2 + 6H_2O \rightarrow C_6H_{12}O_6 + 6O_2$$

Suppose a green plant produces 317 g of sugar.
(a) What volume of carbon dioxide gas at 23°C and 102 kPa is used?
(b) What volume of oxygen at the same conditions is produced?

Review

Let us review the steps involved in solving stoichiometric problems involving either mass or volume measurements of reactants and products. (Refer again to Figure 12.2 on page 315.)

Step 1

Write a balanced equation for the reaction involved, then identify the "given" and the "required" substances. List the data under these substances and, if necessary, change their units (to kelvins, grams, and litres). Note that the substances involved may be reactants or products.

Step 2

Change the quantity (mass or gas volume) of the "given" substance to number of moles of that substance.

Step 3

Use the coefficients in the *balanced* equation to determine a conversion factor and use the number of moles of the "given" substance to determine the number of moles of the "required" substance.

Step 4

Change the number of moles of the "required" substance to either the mass or gas volume as specified in the problem.

You have now learned how to extend the stoichiometric calculation procedure, first outlined in Chapter 9, to include gaseous reactants and products.

Sample Problem 12.3

When heated, baking soda (sodium hydrogen carbonate) decomposes according to the following equation.

$$NaHCO_3 \rightarrow Na_2CO_3 + CO_2 + H_2O \quad \text{(unbalanced)}$$

(Note: This reaction is especially useful in putting out grease fires in the kitchen. If baking soda is thrown onto a grease fire, the carbon dioxide released by heating will help to smother the flames. Using water would cause splattering and even spread the fire.)

What volume of carbon dioxide gas, measured at 65°C and 103.5 kPa, will be produced from the decomposition of 20.85 g of baking soda?

Solution

Step 1
(Balance chemical equation and list data)

$$2NaHCO_3 \quad \rightarrow \quad Na_2CO_3 \quad + CO_2 \quad + \quad H_2O$$

$m = 20.85$ g

$M = 84.0$ g/mol

$V = ?$

$P = 103.5$ kPa

$T = 65°C$

$ = 338$ K

Step 2
(Convert mass to number of moles of "given")

$$n = m \times \frac{1}{M}$$

$$n_{NaHCO_3} = 20.85 \text{ g} \times \frac{1}{84.0 \text{ g/mol}}$$

$$= 0.248 \text{ mol}$$

Step 3
(Determine moles of "required" substances using the coefficients in the balanced equation to obtain the conversion factor.)

$$n_{CO_2} = 0.248 \text{ mol}_{NaHCO_3} \times \frac{1 \text{ mol}_{CO_2}}{2 \text{ mol}_{NaHCO_3}}$$

$$= 0.124 \text{ mol}_{CO_2}$$

Step 4
(Convert number of moles to volume)
(a) Using the ideal gas law:

$$PV = nRTV$$

$$V = \frac{nRT}{P}$$

$$V_{CO_2} = \frac{0.124 \text{ mol} \times 8.31 \frac{\text{kPa·L}}{\text{mol·K}} \times 338 \text{ K}}{103.5 \text{ kPa}}$$

$$= 3.37 \text{ L}$$

(b) Using the combined gas law and molar volume:
  (i) Determine the gas volume at STP by using the moles to volume conversion factor.

$$V = n \times V_m$$

$$\text{at STP, } V_{CO_2} = 0.124 \text{ mol} \times 22.4 \frac{\text{L}}{\text{mol}}$$

$$= 2.78 \text{ L}$$

(ii) Convert the volume at STP to the required conditions.

$V_2 = V_1$	$\times$	conversion factor due to temperature change	$\times$	conversion factor due to pressure change
$= 2.78 \text{ L}$	$\times$	$\dfrac{338 \text{ K}}{273 \text{ K}}$	$\times$	$\dfrac{101.3 \text{ kPa}}{103.5 \text{ kPa}}$
		(factor > 1, T increases $\therefore V$ increases)		(factor < 1, P increases, $\therefore V$ decreases)
$= 3.37 \text{ L}$				

Thus, 20.85 g of sodium hydrogen carbonate will produce 3.37 L of carbon dioxide gas at 65°C and 103.5 kPa pressure.

Exercise

5. Hydrogen gas can be prepared conveniently in the lab by reacting zinc with hydrochloric acid. Zinc chloride, $ZnCl_2$, is also produced. What mass of zinc must be reacted to produce 850 mL of hydrogen gas collected at 95 kPa pressure and 17°C?

6. Lighter fluid is essentially butane, C_4H_{10}. Butane burns to produce carbon dioxide gas and water vapour. If 2.50 g of butane are burned, what volume of carbon dioxide at 47°C and 105 kPa will be produced?

12.3 Avogadro's Law and Stoichiometry

In stoichiometry problems, if both the "given" and "required" are gases measured at the same temperature and pressure conditions, the problems can easily be solved. Avogadro's law states that at the same temperature and pressure equal volumes of different gases contain the same number of particles (and therefore the same number of moles). Thus, the coefficients for gaseous substances in a balanced equation represent volume ratios as well as mole ratios (if temperature and pressure conditions are the same). (See Figure 12.3.)

Laboratory Manual
In Experiment 12-2, you will study the ratio of reacting gas volumes.

Figure 12.3
A chemical reaction involving only gases. The Haber process is applied industrially using hydrogen and nitrogen gases under high temperature and pressure to produce the commercially important gas ammonia.

Balanced Equation

Mole Ratio

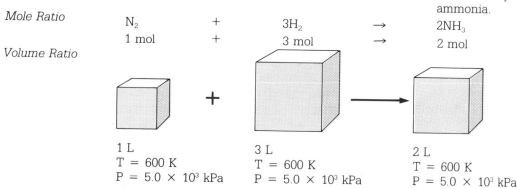

N_2	$3H_2$	$2NH_3$
1 mol	3 mol	2 mol

Volume Ratio

1 L
T = 600 K
P = 5.0 × 10³ kPa

3 L
T = 600 K
P = 5.0 × 10³ kPa

2 L
T = 600 K
P = 5.0 × 10³ kPa

Sample Problem 12.4

The Haber Process for the synthesis of ammonia, a key ingredient in the manufacture of fertilizers, is represented by the following equation.

$$N_2 + 3H_2 \rightarrow 2NH_3$$

What volume of nitrogen gas is required to react completely with 12.0 L of hydrogen to produce ammonia? All three substances are gases at the same conditions of 500°C and 5.0×10^4 kPa pressure in an industrial reaction chamber.

Solution

Since both the nitrogen and hydrogen are gases at the same temperature and pressure conditions, the coefficients of the balanced equation represent their volume ratios as well as mole ratios.

$$N_2 \quad + \quad 3H_2 \rightarrow \quad 2NH_3$$

| 1 volume of nitrogen | reacts with | 3 volumes of hydrogen |

The volume of nitrogen reacting with a given volume of hydrogen can be determined by using the volume ratio as the conversion factor.

$$\text{Volume of } N_2 = 12 \text{ L of } H_2 \times \frac{1 \text{ volume } N_2}{3 \text{ volumes } H_2}$$
$$= 4.0 \text{ L of } N_2$$

Therefore, 4.0 L of nitrogen are required to react with 12 L of hydrogen at the same temperature and pressure conditions.

Exercise

7. Hydrogen gas (H_2) and chlorine gas (Cl_2) react (slowly in diffused light and explosively in bright light) to produce hydrogen chloride gas (HCl).
   (a) What volume of hydrogen is needed to react with 7.5 L of chlorine gas at the same conditions of temperature and pressure?
   (b) What volume of hydrogen chloride gas will be produced under the same conditions?

8. Methane, CH_4, is the major constituent of natural gas. It burns in air to form carbon dioxide and water vapour. (All gas volumes are measured at STP.)
   (a) What volume of oxygen gas is required for the combustion of 12.5 L of methane gas?
   (b) What volume of air is required to burn completely 12.5 L of methane? (See Special Topic 11.5 on page 300.)

9. When propane, C_3H_8, is burned in a restricted supply of oxygen, one of the products is carbon monoxide. Since carbon monoxide is a deadly poison even in small quantities, it is

important that devices such as propane heaters operate in a plentiful supply of oxygen. If 4.25 L of propane gas measured at 25°C and 125 kPa is burned in limited oxygen (producing only carbon monoxide and water vapour) what volume of carbon monoxide gas, at the same temperature and pressure, will be produced?

SPECIAL TOPIC 12.2

CHEMICAL REACTIONS AND THE DESIGN OF USEFUL DEVICES

The operation of a device such as a car engine or a furnace depends on chemical reactions. These devices must be so designed as to provide for the intake of appropriate amounts of specific reactants and for the disposal of specific products. A gas furnace, for example, is designed to mix the correct volume of a combustible gas with at least enough air for complete and efficient combustion. A furnace designed to use natural gas will not burn propane gas properly because a litre of propane reacts with two and a half times as much oxygen (air) as does a litre of methane (natural gas). The volume ratios are indicated by the following balanced equations.

air intake
gasoline

$$CH_4 \quad + \quad 2O_2 \quad \rightarrow CO_2 + 2H_2O$$

1 volume	2 volumes
of methane	of oxygen

$$C_3H_8 \quad + \quad 5O_2 \quad \rightarrow 3CO_2 + 4H_2O$$

1 volume	5 volumes
of propane	of oxygen

Propane furnaces need to have larger air intakes for the same volume of gas burned. Periodic maintenance on gas burning devices such as furnaces and barbecues is necessary in order to prevent any obstruction of air or gas intakes by dirt or litter. A clogged air intake will cause the furnace or barbecue to use fuel inefficiently. As a result it will burn more fuel for the same amount of heat and it will produce incompletely burned products such as soot (carbon) or carbon monoxide.

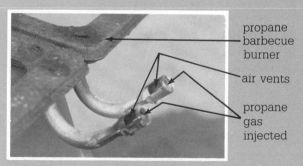
gas
air intakes

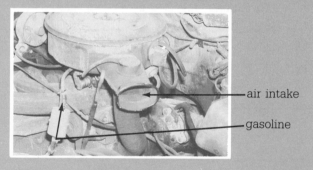
propane barbecue burner
air vents
propane gas injected

Devices using gaseous fuels are designed to supply the correct volumes of combustible gas and air (oxygen) for efficient combustion.

Looking Back – Looking Ahead

In this chapter, the gas laws developed in Chapter 11 have been used to develop the ideal gas law. This law enables you to calculate the amount of substances used or produced in chemical reactions involving gases. In Chapter 14, you will extend stoichiometry further to include chemical reactions in solutions.

Vocabulary Checklist

You should now be able to define or explain the meaning of the following terms.

ideal gas constant   ideal gas equation

Chapter Objectives

You should now be able to do the following.

1. Use Avogadro's law, Boyle's law, and Charles' law to derive the ideal gas law equation.

2. Use the ideal gas law equation to solve quantitative problems involving gas temperature, pressure, volume, and amount (moles) of gas.

3. Use the ideal gas law (or the molar volume and the combined gas law) to do stoichiometric problems involving gaseous reactants and products.

Understanding Chapter 12

REVIEW

1. For each of the following, write the appropriate equation, variation, or numerical value associated with it.
   (a) ideal gas volume
   (b) ideal gas law constant

APPLICATIONS AND PROBLEMS

2. Determine the number of moles in each of the following samples of gas.
   (a) 20.0 L of carbon dioxide at 17°C and 250 kPa pressure
   (b) a 12 m × 10 m × 3.0 m classroom of air at 100 kPa pressure and 24°C. (Remember: 1 L = 1 dm³)

3. Use the ideal gas law to answer the following questions.

(a) What is the volume of 1.5 mol of carbon dioxide gas at 68 kPa and −35°C?

(b) Calculate the number of moles of hydrogen gas in a 50 mL container at 120 kPa pressure and 12°C.

(c) If a 0.75 L cylinder holds 6.5 mol of nitrogen gas at 23°C, calculate the pressure in the cylinder.

(d) 0.50 mol of oxygen gas is introduced into an evacuated 10 L flask. The pressure of the gas in the flask is 105 kPa. Calculate the temperature of the flask.

4. Natural gas (methane) comes out of a gas meter at a pressure of 103 kPa corrected to a temperature of 16°C. How many moles of CH_4 molecules (methane) are there in 1.00 m³ of this gas?

5. Although the overall process is somewhat more complex, respiration can be considered as the oxidation of a carbohydrate. The following equation represents this simplified process.

$$C_6H_{12}O_6 + 6O_2 \rightarrow 6CO_2 + 6H_2O$$

(a) What volume of oxygen gas at body temperature, 37°C, and at a pressure of 98 kPa is required to completely oxidize 23.8 g of carbohydrate?

(b) What volume of carbon dioxide at the same temperature and pressure will be produced?

6. Sodium hydrogen carbonate (baking soda), $NaHCO_3$, reacts with vinegar, a dilute solution (about 5% by mass in water) of acetic acid, CH_3COOH.

$$NaHCO_3 + CH_3COOH \rightarrow$$
$$CH_3COONa + CO_2 + H_2O$$

What volume of carbon dioxide gas at 5°C and 97.5 kPa is produced by the reaction of 4.66 g of baking soda with an excess of acetic acid?

7. Nitric acid is a very important industrial chemical used in the manufacture of such products as explosives and fertilizers. The first step in the production of nitric acid is the oxidation of ammonia (the Ostwald process). The *unbalanced* equation for the reaction is as follows.

$$NH_3 + O_2 \rightarrow NO + H_2O$$

What volume of oxygen gas is required to react with 750 L of ammonia gas? All volumes are measured at 700°C and 100 kPa.

8. Stalagmites and stalactites form in limestone caves when a solution of calcium hydrogen carbonate evaporates leaving solid calcium carbonate, limestone, behind.

$$Ca(HCO_3)_2 \rightarrow CaCO_3 + CO_2 + H_2O$$

During the formation of a 35.0 kg stalactite what volume of carbon dioxide gas, measured at 10°C and 100 kPa, is produced?

9. Urea, $CO(NH_2)_2$, is used as a fertilizer. It is also involved in the manufacture of urea-formaldehyde resins — a type of plastic (bakelite) used for such things as electrical switch plates and electric kettle handles. Urea is made by the reaction of ammonia and carbon dioxide.

$$2NH_3 + CO_2 \rightarrow CO(NH_2)_2 + H_2O$$

(a) What volume of ammonia gas measured at 37°C and 105 kPa pressure is required to produce 600 g of urea by reaction with carbon dioxide?

(b) What volume of carbon dioxide gas, at the same temperature and pressure, is also required?

10. Match heads contain, among other substances, potassium chlorate. When the match is first ignited, the heat causes the potassium chlorate to decompose. This produces oxygen which in turn causes the match to burn more rapidly. This decomposition is represented by the following *unbalanced* equation.

$$KClO_3 \rightarrow KCl + O_2$$

What volume of oxygen gas, measured at 200°C and 104 kPa pressure, is produced when the 0.11 g of potassium chlorate in the head of one match decomposes?

11. The major cause of acid rain is the release of sulphur dioxide gas into the atmosphere from the combustion of sulphur-containing fossil fuels. One method of reducing sulphur dioxide output to the atmosphere is to pass smokestack gases over wet limestone. (Limestone, like marble and chalk, is a form of calcium carbonate.) The following reaction can occur.

$$CaCO_3 + SO_2 \rightarrow CaSO_3 + CO_2$$

Calculate the mass of calcium carbonate required to completely remove 3.0×10^6 L of sulphur dioxide stack gas measured at 85°C and 102 kPa pressure.

Challenge

12. Nitroglycerine, $C_3N_3H_5O_9$, is used both as a heart drug to relieve severe heart pains (called angina) and as a powerful explosive. The equation for the explosion of nitroglycerin is as follows.

$$4C_3N_3H_5O_9 \rightarrow 6N_2 + 12CO_2 + 10H_2O + O_2$$

All of the product are gases. In an explosion, these gases are produced at a high temperature in a very short time. The rapid expansion of these high pressure, high temperature gases constitute the explosion. Calculate the total volume of the gases at 250°C and 300 kPa when 1.00 kg (625 mL) of liquid nitroglycerine explodes.

VI SOLUTIONS

13 SOLUTIONS

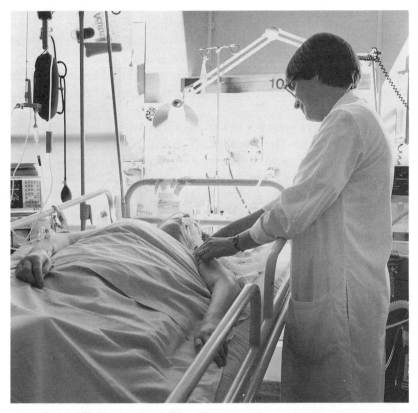

Like gases, liquid solutions are an important part of our daily lives. Foods such as apple juice and consommé are solutions. These and other types of food are used to produce solutions in our bodies which are essential to the functioning of our bodies, solutions such as blood plasma, digestive acids, spinal fluid, and urine. Some solutions are so necessary for life that they are used in hospitals to sustain the life of or heal seriously ill patients. For example, solutions containing dextrose (a form of sugar) are fed intravenously to patients who are not well enough to eat or drink on their own. Such patients obtain vital nutrients from dextrose solutions that are run slowly and directly into their veins.

Common household products such as window washing liquids, nail polish removers, and medicines such as cough syrups come in the form of solutions. Any medicine chest or cleaning product cupboard will contain many more examples of solutions.

Most of the liquids used in automobiles, such as gasoline, radiator antifreeze, and transmission fluids, are also solutions. You could probably readily list many more solutions that you use or see every day.

Although often useful, solutions can sometimes cause problems. Acid precipitation, for example, is caused by the dissolving of gaseous oxides of sulphur and nitrogen in rain or snow. A large percentage of these oxides comes from industrial processes and automobile exhausts. Water is able to dissolve most substances to some degree. Drinking water often contains small amounts of dissolved substances, some of which can be harmful. The water run-off from waste disposal dumps eventually makes its way into the rivers and lakes from which municipal water supplies are drawn. Thus, the safe disposal of ever-increasing amounts of municipal and industrial wastes, some of which are highly toxic, is an important issue facing society today.

In this chapter, you will become familiar with solutions by handling, preparing, and testing them. You will also come to appreciate some of the many useful applications of solutions, as well as some of the problems associated with them.

13.1 Investigating Solutions

Components of Solutions

Solutions consist of homogeneous mixtures of at least two substances. The proportions of the substances are not fixed – it is possible to make solutions of many different percentage compositions using the same substances. Sea water, for example, is a solution. A sample of Mediterranean sea water contains a different percentage of salt than does a sample of Atlantic sea water.

In any one solution, properties are the same throughout its volume. The solution is homogeneous and has one phase. For example, a 1 mL sample of a salt solution in a jar contains the same mass of salt as any other 1 mL sample taken from the same jar. (See Figure 13.1.) A particular salt solution is homogeneous and has one phase, but many different solutions containing different percentages of dissolved salt can be made.

Figure 13.1
Solutions have a variable composition – many solutions of different concentration can be made. However, any one solution is uniform throughout with different parts of the solution having identical concentrations.

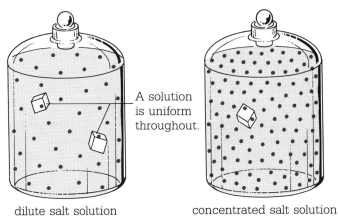

A solution is uniform throughout.

dilute salt solution   concentrated salt solution

Solutions are made up of at least two components. One is the **solvent**, the substance that does the dissolving. The second is the **solute**, the substance that dissolves in the solvent. For example, a salt solution is made up of salt as the solute and water as the solvent. (See Figure 13.2.) The **solution** refers to the combination of both the solvent and the dissolved solute.

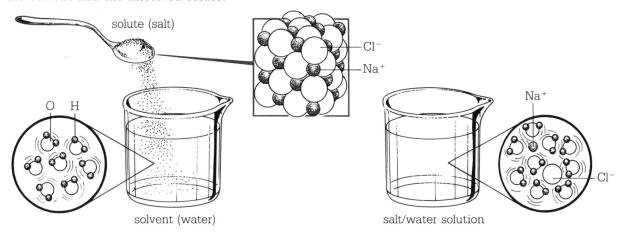

solute (salt)

Cl⁻

Na⁺

O   H

solvent (water)

Na⁺

Cl⁻

salt/water solution

Solutes are not necessarily solids; they can also be gases or liquids. A gas such as carbon dioxide dissolves quite readily in water under moderate pressure to form soda water. Flavouring and sweeteners are dissolved in this solution to produce different types of soda pop. Fish and other aquatic animals survive by absorbing oxygen that has dissolved in water. Methanol, a poisonous liquid alcohol, is dissolved in water for use as windshield washer antifreeze for automobiles. Another liquid, ethylene glycol (also poisonous), is dissolved in water to make an automobile radiator antifreeze.

Water has been called the universal solvent because it dissolves so many solutes. Almost every substance will dissolve, to some small degree at least, in water. Because of the abundance and availability of water and its ability to dissolve so many substances, water is the most important and widely used solvent.

Water may be the most familiar and essential solvent, but solvents other than water abound. For example, perchloroethylene is used as a dry cleaning fluid because it easily dissolves greasy or oily substances. Varsol, turpentine, and alcohol are other common household solvents used to dissolve oils, paints, and greases, substances that do not dissolve well in water.

Figure 13.2

A solute dissolves in a solvent to form a solution. For example, the ions of a salt crystal completely separate and mix with the molecules of water.

Exercise

1. Kool Aid drink is a solution. Identify the solvent and at least two solutes that are present in Kool Aid.

2. What is the solvent in "acid rain"? What is a possible solute?

Concentrated and Dilute Solutions

A solution that is composed of a high percentage of solute is said to be **concentrated**. In Figure 13.1, the solution that contains a higher percentage of dissolved ions has them squeezed closer together, that is, the ions are more concentrated within the solution. On the other hand, a **dilute** solution is made up of a low percentage of solute. The dilute solution in Figure 13.1 has comparatively fewer ions. The ions are on average, farther apart than in the concentrated solution. The concentrated salt solution can be diluted by adding more water, since this will allow the ions to spread out more.

Solubility

Laboratory Manual
In Experiment 13-1, you will determine the solubility of a substance over a range of temperatures.

The **solubility** of a substance in a solvent is the maximum amount of that substance (the solute) which will dissolve in a fixed quantity of the given solvent at a specified temperature. One way of expressing solubility is to state the mass of solute that dissolves in 1 kg of solvent at a given temperature.

The solubility of a solute changes with temperature. The graph in Figure 13.3 illustrates the relationship between temperature and the solubility of several different compounds in water.

You can see that the solubility of each compound is different at different temperatures. In addition, the solubility of each compound changes in a different way as the temperature changes. This characteristic change of solubility with temperature can sometimes be used as a basis for separating different substances from each other. Consider a solution that contains the maximum amount of potassium chlorate and sodium chloride that it can normally dissolve at 90°C. If the solution is cooled to 10°C, then most of the potassium chlorate will crystallize out of solution because of its greatly decreased solubility at 10°C. The solubility of sodium chloride, however, changes very little with temperature. Thus, most of it remains dissolved in solution at 10°C. By filtering or pouring off the solution, we attain a reasonably effective separation of the potassium chlorate (the crystals) from the sodium chloride (still dissolved).

Exercise

3. Use the graphs in Figure 13.3 to answer the following questions.
   (a) What is the solubility of potassium chlorate at each of the following temperatures?
      (i) 30°C   (ii) 70°C
   (b) At what temperature is the solubility of lead(II) nitrate equal to each of the following?
      (i) 960 g/kg of water   (ii) 440 g/kg of water

(c) You are given a jar of a white crystalline substance that is a mixture of potassium chloride and potassium nitrate. Explain how you could separate (most of these two compounds from each other.

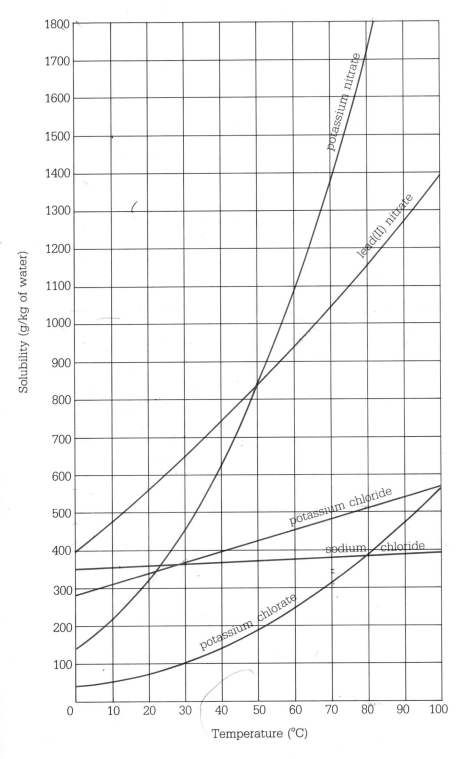

Figure 13.3
Solubility of some substances in water

Unsaturated, Saturated, and Supersaturated Solutions

Not all solutions contain the maximum amount of dissolved solute at a given temperature. Solutions, like the shopper shown in Figure 13.4, can "carry" different amounts of solute.

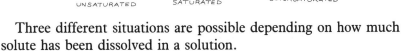

UNSATURATED   SATURATED   SUPERSATURATED

Figure 13.4
Unsaturated, saturated, and supersaturated

Three different situations are possible depending on how much solute has been dissolved in a solution.

1. An **unsaturated solution** is one in which still more solute can be dissolved while the temperature remains constant. An unsaturated solution can be compared to a shopper leaving a grocery store carrying only one bag of groceries – there is still the capacity to pick up and carry another bag. An unsaturated solution for a particular solute is indicated by any point in the region *under* the solute's solubility curve, as shown in Figure 13.5.

2. A **saturated solution** is one in which no more solute can be dissolved into the solution at a fixed temperature. This saturated solution can be compared to a shopper carrying two large bags of groceries – the shopper cannot pick up any more bags without assistance. A saturated solution for a particular solute is indicated by any point *on* the solubility curve of that solute, as shown in Figure 13.5.

3. A **supersaturated solution** is a solution containing more dissolved solute than it normally could dissolve at a particular temperature. A supersaturated solution is like a shopper who has been loaded up with four or five bags of groceries, one in each arm and two or three more stacked on top of these. As shown in Figure 13.5, any point in the region *above* the solubility curve represents a supersaturated solution.

The supersaturated condition shown in Figure 13.4 is an unstable situation. Chances are the shopper's stacked bags will tumble off the lower bags and crash to the ground. Similarly a supersaturated solution is unstable. The clear, supersaturated solution may suddenly form solid crystals of the solute. These crystals settle, leaving only a saturated solution behind. How can such an unstable situation occur? In the case of the shopper, another person had to help lift

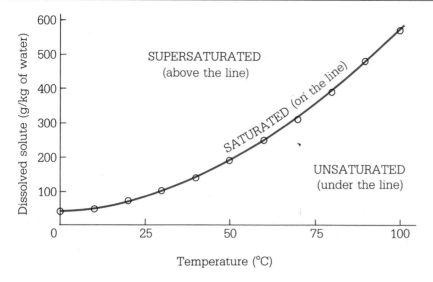

Figure 13.5
Unsaturated, saturated, and supersaturated solutions

up and stack the extra bags on top of the two bags the shopper could pick up alone. In the case of the supersaturated solution, a saturated solution has to be formed at a high temperature where the solubility of the solute is greater. If this solution is then cooled to a lower temperature (where solubility is less), it will contain more dissolved solute than it normally could at the lower temperature. This results in a supersaturated solution.

Precipitation

Not all samples of saturated solutions can be cooled down to form supersaturated solutions. In most cases the solute "comes out" of the solution as a solid. The sudden appearance of a solid in a liquid is called precipitation. The solid that forms is called the **precipitate**. Precipitation in a supersaturated solution can be initiated by adding a single **seed crystal** of the solute. The seed crystal provides a nucleus around which the surplus dissolved solute can start crystallizing. Sometimes an impurity such as dirt, or even a sharp tap on the container, is sufficient to initiate precipitation. Whether or not a precipitate appears after the addition of an extra crystal of solute is a test for a supersaturated solution.

Exercise

4. You are given a sample of a potassium chlorate solution. Describe what you would do to determine if the solution is unsaturated, saturated or supersaturated.

5. Suggest a hypothesis to explain the observation that tea, without sugar or milk, is clear when it is first poured, but is often cloudy after sitting for 15 or 20 minutes.

Sample Problem 13.1

Use the graph in Figure 13.3 to determine whether a solution containing 200 g of potassium chlorate dissolved in 1 kg of water at 70°C is unsaturated, saturated, or supersaturated.

Solution

The 70°C, 200 g/kg point is located under the solubility curve for potassium chlorate. The region under a solubility curve represents an unsaturated condition, since more solute could still be dissolved in the solution at that temperature. Therefore the given potassium chlorate solution is unsaturated.

Exercise

6. Use the graph in Figure 13.3 to determine whether each of the following solutions is saturated, unsaturated, or supersaturated.
   (a) 700 g of lead(II) nitrate are dissolved in 1 kg of water at 30°C.
   (b) 300 g of sodium chloride are dissolved in 1 kg of water at 12°C.
   (c) 480 g of potassium chlorate are dissolved in 1 kg of water at 90°C.

13.2 Predicting Solubilities

Chemists usually express solubility and solution concentration as moles of solute dissolved per litre of solution formed (mol/L) because it is much more convenient to use for chemical equation calculations. As you have seen, different compounds have different solubilities in water. It is generally agreed that if a solute has a solubility of greater than 0.1 mol/L of solution in a particular solvent, it is classified as soluble in that solvent. Below a solubility of 0.1 mol/L, the solute is classified as insoluble in the solvent. Using this information, a set of qualitative "solubility rules" can be determined experimentally.

Organizing data from experiments and looking for trends and regularities have led to some general solubility rules as summarized in Table 13.1.

Laboratory Manual
In Experiment 13-2, you will verify some of the solubility rules shown in Table 13.1

Table 13.1 Qualitative Solubility Rules

Rule No.	Rule Statement	Summary
1	Hydrogen, ammonium, and all Group I ions form soluble compounds with all negative ions.	
	H^+, NH_4^+, Li^+, Na^+, K^+, Rb^+, Cs^+, Fr^+	All compounds are soluble
2	Acetate and nitrate ions form soluble compounds with all positive ions.	
	CH_3COO^-, NO_3^-	All compounds are soluble
3	Sulphide ion forms soluble compounds only with the ions listed in rule 1 and with Group II positive ions.	
	S^{2-} with Mg^{2+}, Ca^{2+}, Sr^{2+}, Ba^{2+}, Ra^{2+}, and ions in rule 1	Soluble compounds
	S^{2-} with all other positive ions	Forms precipitates
4	Chloride, bromide, and iodide ions form compounds that have low solubility with silver, lead(II), mercury(I), copper(I) and thallium positive ions only.	
	Cl^-, Br^-, and I^- with Ag^+, Pb^{2+}, Hg_2^{2+}, Cu^+, Tl^+	Form precipitates
5	Sulphate ion forms compounds that have low solubility with calcium, strontium, barium, radium, and lead(II) positive ions only.	
	SO_4^{2-} with Ca^{2+}, Sr^{2+}, Ba^{2+}, Ra^{2+}, Pb^{2+}	Forms precipitates
6	Hydroxide ion forms compounds that are soluble only with the positive ions listed in rule 1 and with strontium, barium, radium, and thallium positive ions.	
	OH^- with Sr^{2+}, Ba^{2+}, Ra^{2+}, Tl^+, and ions in rule 1	Soluble compounds
	OH^- with all other positive ions	Forms precipitates
7	Phosphate, carbonate, and sulphite ions form compounds that have low solubility with all positive ions except those listed in rule 1.	
	PO_4^{3-}, CO_3^{2-}, SO_3^{2-} with positive ions in rule 1	Soluble compounds
	PO_4^{3-}, CO_3^{2-}, SO_3^{2-} with all other positive ions	Form precipitates

Formation of Precipitates

A precipitate can result from the formation of a new, insoluble compound when a chemical reaction takes place in solution. Notice that the solubility rules listed in Table 13.1 can be used to predict whether or not a precipitate will form. A precipitate will form when positive ions and negative ions that make up a compound of low solubility are added together from two different sources. (See Figure 13.6.) (Note, however, that reliable predictions concerning precipitate formation can only be made if the ion concentrations after mixing are above 0.1 mol/L.)

Figure 13.6
The formation of a precipitate. If ions that form an insoluble substance are mixed from two different sources, they combine to form a solid precipitate.

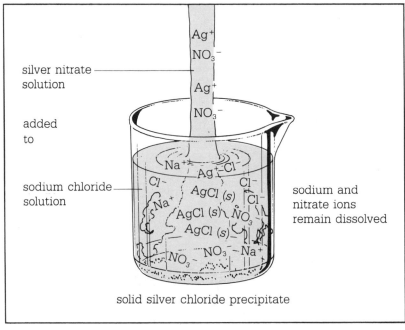

silver nitrate solution

added to

sodium chloride solution

sodium and nitrate ions remain dissolved

solid silver chloride precipitate

This concept gives us another experimental method of verifying the general solubility rules stated in Table 13.1. For example, if a solution of silver nitrate (a soluble compound according to rule 2) is added to a solution of sodium chloride (also a soluble compound according to rule 1), four ions are present in the mixed solution. The sodium and nitrate ions always form soluble compounds (rules 1 and 2). But what about the silver and chloride ions? In this case, a precipitate of silver chloride forms. This is in agreement with rule 4, which predicts that the positive silver ion will form a compound of low solubility with the negative chloride ion. If silver and chloride ions are both in the same solution, they join to produce silver chloride as a solid precipitate. The other solubility rules may be verified by experimenting in a similar way using the appropriate ions.

The solubility rules listed in Table 13.1 can be used to determine methods of identifying specific ions. They can also be used to develop methods of separating solutions containing several different ions.

Laboratory Manual
You will observe the effect of electric current on solutions of ions in Experiment 13-3.

Sample Problem 13.2

A solution might contain one, or all, of the following ions: Ag^+, Ba^{2+}, and/or Mg^{2+}.

(a) Devise a test to identify which of these ions are present in the solution.

(b) If all of these ions are present, propose a laboratory procedure that could be used to separate each of them from the others.

Solution

(a) (i) Add chloride ions (from a sodium chloride solution, for example) to a test sample of the unknown solution. If a precipitate forms, you may conclude that silver ions are present. If a precipitate does not form, then silver ions are not present (rule 4). Barium and magnesium ions do not form a precipitate with chloride ions. Sodium ions from the sodium chloride solution, like all group I ions, form soluble compounds in all cases (rule 1). They do not interfere with other precipitation reactions.

(ii) To test for the presence of barium ions, add a solution containing sulphate ions to the test sample, for example, a potassium sulphate solution. If a precipitate forms, barium ions are present (rule 5). Neither silver nor magnesium ions form a precipitate with sulphate ions.

(iii) Add excess chloride and sulphate ions to the test sample to precipitate virtually all the silver ions and barium ions. Then add hydroxide ions by adding, for example, a dilute solution of sodium hydroxide. Magnesium ions are present if a new precipitate forms when hydroxide ions are added (rule 6).

(b) In order to separate the three metal ions, follow the steps in (a). Add excess precipitating ion to the solution one at a time and filter the resulting mixture after each step. The residue on the filter paper from filtration after adding chloride ions will contain silver chloride only; the second residue after adding sulphate ions will contain barium sulphate only; and the third will contain mostly magnesium hydroxide.

In working out a separation procedure, you are attempting to obtain only one precipitate at a time. Filtration will then yield a compound containing only one of the ions to be separated. In the case of Sample Problem 13.2, either chloride or sulphate ion could be added first since each would produce only one precipitate. If hydroxide ion had been added first, a mixed precipitate of silver, barium, and magnesium hydroxides would have been produced. Separation would not have been achieved. (See Figure 13.7.)

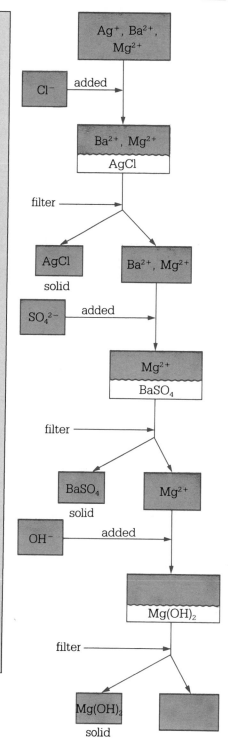

Figure 13.7
An ion separation flow chart. Adding solutions containing ions that selectively cause one precipitate at a time to form can be used to separate a mixture of ions.

Exercise

7. Drinking water often contains dissolved chloride and/or calcium ions. Devise a procedure to test which of these ions is present in a sample of tap water.

8. A solution contains at least 0.8 mol/L of each of the following ions: Pb^{2+}, Ag^+, Mg^{2+}, Fe^{2+}. Use a flow chart to illustrate a procedure that could be used to separate these ions from each other.

13.3 Water – The Universal Solvent

As mentioned earlier, water is a most important and common solvent. Water is found almost everywhere on earth, and life has evolved in and around water. All biological systems, many chemical reactions, and many everyday activities involve water solutions. What are the special characteristics of water that make it so important?

The Water Molecule

The water molecule itself has a very important characteristic. It has a permanent electrostatic charge on each end. One end or pole of the molecule is always slightly positive, while the other end or pole is slightly negative. (See Figure 13.8.) As discussed in sections 5.5 and 5.6 in Chapter 5, molecules like water, which have two oppositely charged ends, are said to be dipolar or polar molecules.

As shown in Figure 13.8, the water molecule is "V" shaped. The oxygen atom attracts the electron pair being shared with hydrogen more than the hydrogen atom does. (Oxygen has a higher electronegativity than does hydrogen.) This electron shift results in the oxygen end of the water molecule being slightly negative and the hydrogen end being slightly positive.

Water as a Solvent

The molecular polarity of water helps to explain its ability to dissolve so many ionic compounds. Figure 13.9 shows that the slightly negative ends of water molecules are attracted to positive ions in an ionic crystal. This attraction tends to pull the positive ions away from the crystal, thereby "dissolving" it. The slightly positive ends of water molecules behave similarly with negative ions. When ions are surrounded by water molecules in solution, they are referred to as **hydrated** or **aqueous ions**. They are designated in writing by (*aq*). For example, $Na^+(aq)$ designates the aqueous sodium ion. Similarly, the aqueous chloride ion is written as $Cl^-(aq)$. (See Figure 13.10.)

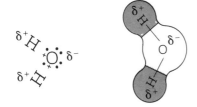

Figure 13.8
The water molecule. The shared pairs of electrons are, on average, closer to the more electronegative oxygen atom than to the less electronegative hydrogen atoms. This produces a permanent slight negative charge on the oxygen atom and a slight positive charge on the hydrogen atoms.

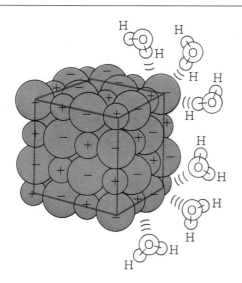

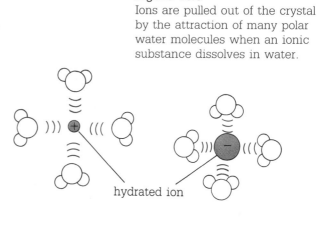

Figure 13.9
Ions are pulled out of the crystal by the attraction of many polar water molecules when an ionic substance dissolves in water.

hydrated ion

The solubility of an ionic compound depends, among other things, on the strength of the bonds holding the ions together in the crystal. If this bonding is stronger than the attractive forces exerted by the polar water molecules, then the substance would be expected to dissolve only slightly. On the other hand, if the attractive force of the polar water molecules is greater than the crystal bond strength, the compound will be quite soluble.

When some highly polar substances such as hydrogen chloride gas are dissolved in water, their molecules are pulled apart by polar water molecules tugging at both ends. The molecules are said to *dissociate*, forming aqueous ions in water. (See Figure 13.11.) Thus, hydrogen chloride is said to ionize in water. Polar molecules, whether they ionize or not, tend to be soluble in polar solvents such as water, alcohol, or liquid ammonia.

Figure 13.10
Hydrated or aqueous sodium and chloride ions

aqueous sodium ion

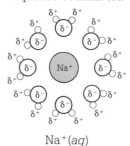

$Na^+(aq)$

aqueous chloride ion

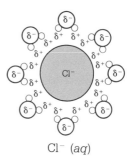

Cl^- (aq)

◄ *Figure 13.11*
Hydrogen chloride molecules are ionized by water. Polar water molecules exert attractions on both ends of a polar molecule and in some cases can pull the molecule apart (dissociate it).

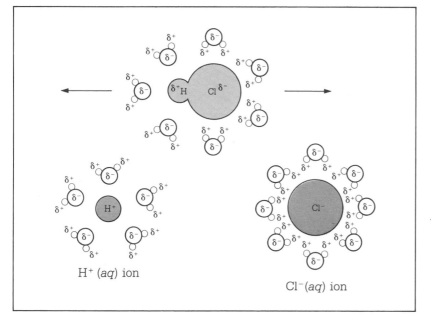

H^+ (aq) ion

Cl^-(aq) ion

Evidence for Aqueous Ions

The tendency for water molecules to hydrate ions results in some interesting changes in the properties of pure water. For example, when salt (sodium chloride) is spread on ice, the ice "melts" even when the temperature of the ice is below 0°C. The attraction between the polar water molecules in the ice and the ions in the salt pulls the water molecules out of the solid ice crystal. A solution is formed. Hence, the ice "melts".

Changes in temperature are often observed when solutions are prepared. For example, the process of diluting concentrated sulphuric acid can be very hazardous since a large amount of heat is produced in the process. So much heat is produced that if water is added to the acid, the temperature of the water being added can in fact reach its boiling point. The water boils into steam, which can splatter the concentrated acid solution. Serious acid burns can result. However, concentrated sulphuric acid has a much higher boiling point than water. Thus, concentrated sulphuric acid can be safely added to water without this splattering effect. This is why you should always add acid to water, never the reverse. (See Fig. 13.12.)

What causes the heat effects when a solution is prepared? Two factors are at work. Energy is absorbed when the ions in an ionic crystal (or the atoms in a polar molecule) are pulled apart to form individual ions. At the same time, however, energy is being released as the polar water molecules move in to hydrate the ions. (See Figure 13.13.) For any given solute, these two energies are usually different. In the dilution of sulphuric acid, for example, the energy produced when the water molecules hydrate the hydrogen and sulphate ions is much greater than the energy used to dissociate the acid into hydrogen and sulphate ions. This extra potential energy is liberated as kinetic energy which increases the temperature of the solution.

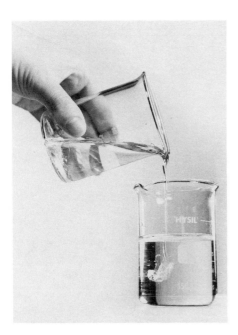

Figure 13.12
Safe dilution of a concentrated acid. Always add the acid to the water – not the other way around!

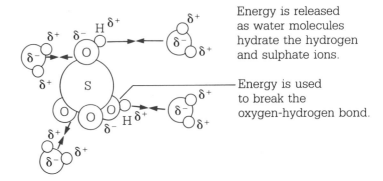

Energy is released as water molecules hydrate the hydrogen and sulphate ions.

Energy is used to break the oxygen-hydrogen bond.

Figure 13.13
Heat of solution. When sulphuric acid dissolves, energy is used to pull the hydrogen atoms away from its molecules. At the same time, energy is released as hydrated ions form. For sulphuric acid, the energy of the hydration is greater than the dissociation energy, and the solution becomes hot.

Sample Problem 13.3

Sodium chloride can melt ice at temperatures as low as $-18°C$. It has no effect on ice below $-18°C$. However, calcium chloride ($CaCl_2$) can melt ice at temperatures considerably lower than $-18°C$. Explain why this is so.

Solution

The sizes of the calcium and sodium ions are about the same. The charge on the calcium ion, however, is twice as much as that on the sodium ion. Therefore, the attraction between a polar water molecule and a calcium ion is greater than the attraction between a water molecule and a sodium ion. More energy is released when a calcium ion is hydrated. The calcium ions are able to pull the water molecules out of ice crystals (which therefore "melt") at a lower temperature than sodium ions are able to. (See Figure 13.14.)

Figure 13.14
Different salts can melt ice below 0°C

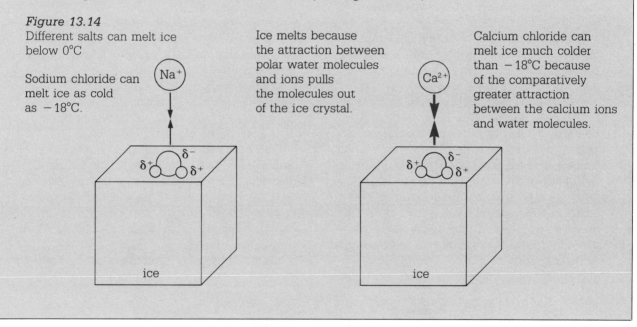

Sodium chloride can melt ice as cold as $-18°C$.

Ice melts because the attraction between polar water molecules and ions pulls the molecules out of the ice crystal.

Calcium chloride can melt ice much colder than $-18°C$ because of the comparatively greater attraction between the calcium ions and water molecules.

Exercise

9. Considering the effect that ions have on water molecules in an aqueous solution, explain why a solution of salt water has a higher boiling point than that of pure water alone.

10. (a) Explain why the solution temperature can drop to a few degrees below 0°C when a concentrated aqueous solution of ammonium chloride is prepared from 20°C water.
    (b) Why does the solution not freeze to a solid?

PUTTING CHEMISTRY TO WORK

ERIC KIISEL / DEVELOPMENT TECHNOLOGIST

Did you realize that you have probably eaten or worn something that has been cut with water? Eric Kiisel is involved in the development of a very unusual tool — the water jet cutter.

"Water at high pressure had been used for cleaning and cutting for many years," Eric recalls. "What makes the water jet cutter different is the extremely high pressures (up to about 400 MPa) we use." (The pressure in a household tap averages about 400 kPa.)

Water under such high pressure can be forced out in an extremely thin and powerful stream called a jet. If a fine abrasive such as crushed garnet is added to the water, the jet is capable of cutting through substances as varied as ice cream, rubber, or steel. "The truly unique thing about using water as a cutting tool," Eric notes, "is that absolutely no heat is added to the material being cut. This is very important in industries such as aircraft manufacturing, because the properties of a metal can be changed by heat, which might lower its quality. The water jet cutter eliminates that problem completely."

Eric is responsible for running and main-taining the sophisticated equipment used in water jet cutting for Atomic Energy of Canada. The nuclear industry began developing this tool for its own use, since it needed a means of cutting metal which did not add potentially damaging heat. "The cutter proved to be so useful, we decided to offer the technology to other industries. Part of my job now is to demonstrate the cutter and to help customers find ways to use it in their processes."

Eric came to his job after obtaining a diploma in mechanical engineering technology from community college. "I had always been interested in mechanical things and how they worked," he remembers. "I majored in drafting and machine shop in high school. I never could have predicted that I would be working with water!"

Eric plans to continue working on the water jet cutter. "I am taking part-time courses in mechanical engineering. This will help me in designing and testing industrial robots to control the cutter. A robot can cut very intricate shapes such as automobile dashboards very easily and very quickly. The future is wide open."

13.4 Electrolytes

Solutes that form solutions which are able to conduct electricity are called **electrolytes**. All types of chemical batteries contain electrolytes. For example, the electrolyte used in automobile batteries is sulphuric acid. Great care must be exercised when handling these batteries to avoid causing serious acid burns. Examples of industrial electrolytes include chromium and nickel compounds that are used to make the electroplating "baths" used in chrome plating. Another example is the electrolyte aluminum oxide dissolved (at high temperatures) in molten cryolite (Na_3AlF_6). Electrolysis of this solution produces aluminum metal. The chemical reaction that occurs when electricity passes through a molten ionic compound or through an electrolyte solution is called **electrolysis**.

For a solution to conduct electricity, it must contain ions that are free to move (mobile ions). Ions are able to donate or accept electrons, thus allowing electricity to "flow" through the solution. (See Figure 13.15.) The ions donate or accept electrons on the surface of conducting plates or rods called **electrodes**.

The negative electrode is called the **cathode**, and the positive electrode is called the **anode**. Positive ions, called **cations**, are attracted to the cathode, while negative ions, called **anions**, are attracted to the anode. The positive ions gain electrons from the cathode and the negative ions give up electrons to the anode.

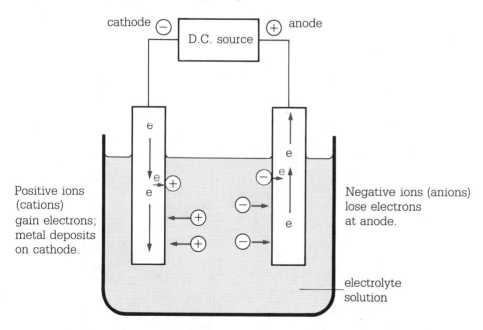

Figure 13.15
An electrolyte solution conducts electricity. Positive ions gain electrons at the cathode, and negative ions lose electrons at the anode. This transfer of electrons has the same effect as a flow of electrons, and the solution is said to conduct electricity.

ELECTROPLATING

Electroplating is the coating of an object with a thin film of metal deposited from an electrolytic solution by electrolysis. Electroplating is often used to produce an attractive and durable finish on articles such as knives and forks (silver plate) and car bumpers (chrome plate). It is also used to protect corrosion-prone metals. Relatively unreactive cadmium metal, for example, is often plated onto steel parts to inhibit rusting.

In the electroplating process, the article to be plated is used as the cathode and the metal being plated onto the article is used as the anode. The electrolytic solution or "bath" contains a salt of the metal being plated. A low voltage electrical current causes metal ions from the bath to gain electrons at the cathode and to deposit as a metal coating on the cathode (the article). At the same time it also causes metal atoms in the anode to lose electrons and go into the bath as ions. As the plating proceeds, the anode gradually disappears and maintains the metal ion's concentration in the bath.

Electroplating operations involve the use of many toxic solutions. Articles to be plated are thoroughly cleaned of all grease and dirt, using concentrated acidic or basic solutions. The cleaning solutions eventually become ineffective and must be disposed of. The electroplating baths themselves contain a variety of toxic substances such as the heavy metals cadmium, chromium, nickel and copper. Deadly poisons such as cyanides are also used. These electrolytic solutions become contaminated over time and must be replaced.

Treatment of these waste solutions to remove or inactivate the toxic substances they contain has not always been carried out as carefully as it should be. In the summer of 1986, for example, an electroplating company in Toronto, Ontario was found to have repeatedly released untreated toxic wastes into the city sewer system. The company was heavily fined and its president sentenced to jail (the jail sentence was overturned after an appeal) for many violations of provincial water pollution regulations.

Electroplating is a very valuable industrial process but its use requires costly and consistently effective treatment of the wastes it produces.

Many consumer and industrial products are electroplated

Strong and Weak Electrolytes

Electrolytes can be either strong or weak. A **strong electrolyte** is a soluble solute composed totally of ions, or composed of molecules that break up almost entirely to form ions in water. Sodium chloride, for example, is a soluble ionic compound that dissolves to form large numbers of aqueous ions. It is therefore a strong electrolyte. Almost all hydrogen chloride molecules when dissolved in water dissociate to form aqueous hydrogen ions and chloride ions. Thus, hydrogen chloride is also a strong electrolyte. On the other hand, most acetic acid molecules remain as molecules when they dissolve in water. Only a small percentage of these molecules will dissociate to form aqueous hydrogen ions and aqueous acetate ions. Acetic acid is therefore a **weak electrolyte**. (See Figure 13.16.)

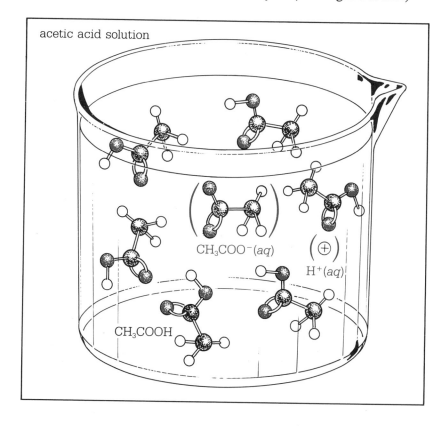

acetic acid solution

$CH_3COO^-(aq)$

$H^+(aq)$

CH_3COOH

Figure 13.16
Acetic acid – a weak electrolyte. Most acetic acid molecules remain intact in a water solution. Only a few of the molecules dissociate, forming aqueous ions.

Hazardous Electrolytes

Since water is such a good solvent for so many electrolytes, the dissolved substances in ordinary tap water considerably increase its electrical conductivity. This property of water makes it a highly dangerous substance around electrical equipment. Using electrical equipment in a wet environment could cause electrocution. Since 80% by mass of the human body is aqueous solution, electricity can pass easily through the human body to the ground.

ELECTROLYTES IN THE BODY

Solutions in the living body such as blood plasma, cell protoplasm, and the fluid between cells all contain a variety of electrolytes. The concentrations of these electrolytes are different inside and outside the cells.

The difference between the electrolyte concentrations inside and outside of body cells is important to the functioning of the body. This difference is a necessary condition for the transfer of signals between cells, the regulation of osmotic pressure in cells and organs, and the maintenance of body structures such as bones.

Even moderate changes in the normal concentration of body electrolytes can affect major functions of the body. For example sodium ion concentration can affect such things as the strength of the heartbeat, transmission of nerve signals, the functioning of the brain, and the production of secretions by various glands.

CALCIUM AND OSTEOPOROSIS

Calcium, an important element for normal development of bones and teeth, is transported through the body as an electrolyte in fluids.

Muscles, nerves, blood, and cell membranes all require calcium to function properly.

The skeleton is the body's calcium storehouse. If there is insufficient dietary calcium or inefficient calcium absorption by the body, calcium is taken from bones, weakening the skeleton. If this persists and calcium "robbing" continues, the bones become progressively more porous, brittle, and fragile. These weakened bones fracture easily – in serious cases, even rolling over in bed can break ribs.

This deteriorating bone condition is called osteoporosis and it especially affects women in later years. Dietary calcium, vitamin D, and specific hormone levels need to be adequate to prevent the development of osteoporosis. The recommended daily intake of calcium is 800 mg (about 600 mL of milk) per day. Vitamin D assists in the absorption of calcium from the digestive tract.

The hormones, estrogen in women and testosterone in men, stimulate the production of calcium regulating substances in the thyroid gland. Women after menopause are often put on an estrogens replacement program, sometimes specifically to prevent osteoporosis.

Electrolyte Concentrations Inside and Outside Human Cells

Electrolyte Concentration (mmol/L) in Fluid Outside Cell	Ions Present	Electrolyte Concentration (mmol/L) in Fluid Inside Cell
144	Sodium	10
4	Potassium	160
2.5	Calcium	1
1	Magnesium	13
108	Chloride	3
0.5	Sulphate	10
29	Bicarbonate	10
0.7	Phosphate	33

13.5 Acids and Bases

Acids are substances which, when dissolved in water, produce the aqueous hydrogen ion, $H^+(aq)$. Acids are therefore electrolytes. They are generally sour tasting and corrosive substances. They can react with many substances such as active metals (for example, zinc) and carbonate-based rock such as limestone and marble. See Figure 13.17(a). Acids affect the colour of certain natural and synthetic dyes and can chemically break down or "burn" some organic materials such as skin and paper. These properties of acids are all related to the presence of the aqueous hydrogen ion.

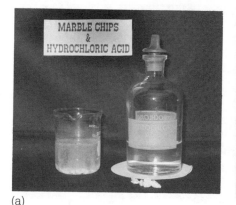

(a)

(b)

Bases are substances which, when dissolved in water, produce the aqueous hydroxide ion, $OH^-(aq)$. Bases are also electrolytes. Basic solutions tend to have a slippery feel, taste bitter, and also affect the colour of many dyes. Strong bases like sodium hydroxide (lye, a constituent of many drain cleaners) are very corrosive and chemically break down organic substances such as fat, hair, and vegetable matter. See Figure 13.17(b). Bases react with $H^+(aq)$ ions present in acids to form water.

An **indicator** is a dye that shows a different colour in an acidic solution than it does in a basic solution. Indicators are used to determine how acidic or basic a solution is. For example, the synthetic indicator phenolphthalein is a dye that is colourless in acidic solutions but bright red or purple in basic solutions. (See Table 13.2.) Litmus, a natural purple dye extracted from lichens, is blue in basic solutions and red in acidic solutions.

Figure 13.17

(a) Acids can dissolve zinc and marble

(b) A strong base such as sodium hydroxide can dissolve a variety of organic matter such as the hair, fat, and vegetable matter that often constitutes the "drain gunk" that plugs sinks

Table 13.2 Acid-Base Indicators

Indicator	Colour in	
	Acid	**Base**
Phenolphthalein	Colourless	Red
Litmus	Red	Blue
Bromothymol Blue	Yellow	Blue

ACIDS, BASES, AND THE pH SCALE

The pH or "power of hydrogen ion" scale was introduced by S.P. Sorenson in 1909. It is the power (to the base ten) of the hydrogen ion concentration written as a fraction. For example, the pH of a solution that has a hydrogen ion concentration of

$\dfrac{1}{1000}$ mol/L $\left(\dfrac{1}{10^3}\right.$ or 1×10^{-3} mol/L) is 3. The power of 10 in the denominator is the pH (i.e., the negative of the power when it is written in standard form).

NEUTRAL SOLUTIONS

A neutral solution is one in which the concentration of acidic ions ($H^+(aq)$) is equal to the concentration of basic ions ($OH^-(aq)$). In pure water, a small proportion of water molecules (less than one in 500 million) is dissociated, producing both hydrogen and hydroxide ions.

$$H_2O(l) \xrightarrow[\text{in } 5.6 \times 10^8]{\text{1 water molecule}} H^+(aq) + OH^-(aq)$$

The concentration of hydrogen ions (and hydroxide ions) in pure, neutral water has been measured as 1×10^{-7} mol/L. Pure water and all neutral solutions have a pH of 7.

ACIDIC SOLUTIONS

Any solution with a pH of *less* than 7 is acidic – it has a *higher* hydrogen ion concentration than a neutral solution. For example, a solution of pH 3 has a hydrogen ion concentration of 1×10^{-3} mol/L – ten thousand times higher than the neutral concentration of 1×10^{-7} mol/L.

A difference of one in pH number represents a ten-fold difference in hydrogen ion concentration. For example, a solution with a pH of 2 has an aqueous hydrogen ion concentration of 1×10^{-2} mol/L, ten times the hydrogen ion concentration of a solution with a pH of 3, which is 1×10^{-3} mol/L.

BASIC SOLUTIONS

Basic solutions have lower hydrogen ion concentration than neutral solutions (and correspondingly higher concentration of hydroxide ions). They have pH values greater than 7. A pH of 9, for example, indicates that the solution has a very small concentration of hydrogen ions, 1×10^{-9} mol/L. A solution that has a pH of 9 is ten times more basic than a solution that has a pH of 8.

Measuring Acid-Base Strength

The strength of an acidic or basic solution is usually measured on the **pH scale**. (See Figure 13.18.) On the pH scale, a neutral solution has a pH of 7. Numbers lower than 7 indicate an acidic solution. The stronger the acid the lower the pH number. For example, a solution of pH 3 is a stronger acid than a solution of pH 4. Since the pH scale is based on a power of ten, a solution with a pH of 3 would be ten times more acidic than a solution with a pH of 4.

Figure 13.18
The pH scale is used to measure the strengths of acids and bases.

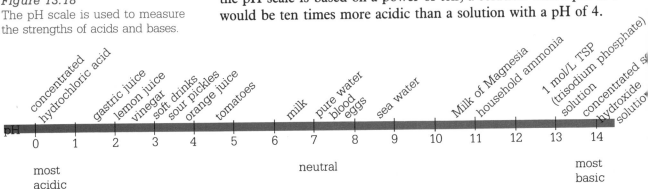

ACIDS AND BASES IN COOKING

Acids and bases play a significant role in the technology of food packaging, preparation, and cooking. Vinegar, wine, cream of tartar, and lactic acid are some of the acidic compounds common in food preparation. Baking soda is the only common base found in the kitchen. Egg white is the only common food that is naturally basic.

EGGS
Egg white, which is mostly made up of a protein called albumen, gradually changes from a nearly neutral pH of 7.7 to a definite basic pH of 9.2 or more as it ages. This pH change occurs as a result of the egg slowly losing carbon dioxide gas through its shell as it ages. (Why should the loss of carbon dioxide gas result in a more basic solution?) It is commercial practice to coat egg shells with a thin film of mineral oil to slow down this loss of carbon dioxide and hence slow down aging.

The albumen protein molecules in egg white have negative charges which tend to keep the individual molecules separated. The positive hydrogen ions in more acidic fresh eggs neutralize these negative charges and the albumen molecules tend to stick together and coagulate more easily during cooking. In older eggs, there is a lower hydrogen ion concentration. The negative charges on the albumen molecules are not neutralized and older egg white is runnier, clearer, and takes longer to coagulate when cooking. Cooks sometimes add an acid to the cooking liquid when cooking eggs to promote coagulation. Cream of tartar, vinegar, lemon juice, or the juice of other fruits or vegetables all lower the pH of the egg. Recipes for poached eggs, for example, often call for the addition of vinegar to the poaching liquid.

When egg whites are beaten to make meringue or souffles, a small amount of acid, usually cream of tartar, is added to stabilize the egg foam.

MILK
As in egg protein, the coagulation of milk protein, mostly casein, is dependent on the pH of the milk. The production of cheese, yogurt, and sour cream all involve the growth of bacteria which produce lactic acid. The lactic acid lowers the pH of the milk and the extra hydrogen ions from the lactic acid neutralize the negative charges on the casein molecules which then combine to make cheese curds or the thickened liquids, yogurt and sour cream.

Poaching an egg in water containing some vinegar causes the egg white to coagulate faster, preventing it from spreading too much.

PICKLES

Pickling is the preservation of foods by soaking them in acid which discourages the growth of most bacteria. Pickling can be accomplished in two ways. The most direct method is to add an acid, usually vinegar, to a raw or partially cooked food. The acid permeates the food, inhibiting the growth of almost all bacteria. Moulds and yeasts, however, can grow at low pH and could contaminate the surface.

Another method of pickling is by fermenting. A vegetable, such as shredded cabbage, is put in a brine (salt) solution. This solution is concentrated enough to prevent growth of undesirable bacteria, but dilute enough for several lactic acid-producing varieties to grow. The increasing lactic acid pickles the vegetable. An example is sauerkraut which is shredded cabbage fermented in brine.

BASES AND COOKING

Bases are used less frequently than acids in cooking but a basic (high pH) cooking environment does have some useful effects. The hemicellulose in the cell walls of foods such as beans is more soluble in a basic environment than in an acidic one. Thus, kidney beans used to make chili should be completely cooked before adding acidic chili sauce or the beans will remain hard no matter how long the chili is cooked afterward. However, baking soda (sodium hydrogen carbonate, a basic substance), should not be added to the water used to cook beans since many vital nutrients are also soluble in a basic solution. These nutrients will be leached out during cooking if baking soda is used.

Laboratory Manual
In Experiment 13-4, you will examine acids and bases and test the pH of several substances.

Any pH value greater than 7 indicates a basic solution. In this case, the higher the number, the stronger the base. Again, because of the mathematical basis for the scale, an increase of one in pH number, for example, from pH 8 to pH 9, indicates a base that is ten times stronger.

Sample Problem 13.4

(a) Which of the substances listed below are electrolytes?
(b) Which substance(s) will form acid in aqueous solution?
(c) Which substance(s) will form base in aqueous solution?

$Fe(OH)_2$, Na_2SO_4, $Ba(OH)_2$, PbS, $CuCl$, HNO_3, $C_6H_{12}O_6$ (sugar)

Solution

(a) Na_2SO_4, $Ba(OH)_2$, and HNO_3 are electrolytes because they dissolve in water and produce ions which conduct electricity. (See solubility rules 1, 2 and 6 on page 335.) With the exception of sugar, the other substances in the list are not soluble in water (solubility rules 3, 4 and 6) even though they consist of ions. Sugar dissolves in water but does not produce ions (refer to Chapter 5).

(b) Only HNO_3 will dissolve and ionize in water to produce the acidic aqueous hydrogen ion, $H^+(aq)$, forming an acid.

(c) Only $Ba(OH)_2$ will dissolve in water to produce the basic aqueous hydroxide ion, $OH^-(aq)$, forming a base.

Exercise

11. Use diagrams to explain why an alcohol-water solution does not conduct electricity, whereas a sulphuric acid solution does.

12. Identify each of the following solutions as acidic, basic, or neutral.
    (a) pH = 5.4
    (b) pH = 12.5
    (c) pH = 9.0
    (d) pH = 7.0
    (e) pH = 1.9
    (f) pH = 7.4 (the approximate pH of blood)

    Which of the above solutions is the strongest acid? Which is the strongest base? Give reasons for your answers.

13.6 Chemical Equations for Aqueous Reactions

Dissolving

Ionic equations are chemical equations that specify the individual aqueous ions existing in a solution. Often, reactions in water can be best described using ionic equations. Ionic equations are used to describe the dissociation of a solute into ions in water. For example, the ionic equation for the dissolving of solid potassium hydroxide in water is

$$KOH(s) \xrightarrow[\text{in H}_2\text{O}]{\text{dissolving}} K^+(aq) + OH^-(aq)$$

solid potassium hydroxide   dissolved potassium hydroxide

Exercise

13. Use chemical equations to illustrate the dissolving of the following solutes in water:
    (a) $K_2SO_4(s)$ (b) $Ba(NO_3)_2(s)$ (c) $(NH_4)_2SO_4(s)$.

Neutralization

The ionic equation for the reaction between a potassium hydroxide solution and a hydrochloric acid solution is

$$K^+(aq) + OH^-(aq) + H^+(aq) + Cl^-(aq) \rightarrow H_2O(l) + K^+(aq) + Cl^-(aq)$$

(KOH solution)   (HCl solution)   (KCl solution)
a base   an acid   a salt

Note that in the ionic equation on page 351, two of the ions appear identically on both sides of the equation. These are called **spectator ions**, because they have not been involved in the reaction. Spectator ions have no bearing on the actual reaction taking place, and thus can be left out of the ionic equation, resulting in a **net ionic equation**. A net ionic equation records only those substances or ions that have actually reacted or have been changed in some way in the reaction. Thus, the ionic equation

$$\cancel{K^+(aq)} + OH^-(aq) + H^+(aq) + \cancel{Cl^-(aq)} \rightarrow H_2O(l) + \cancel{K^+(aq)} + \cancel{Cl^-(aq)}$$

becomes the net ionic equation

$$OH^-(aq) + H^+(aq) \rightarrow H_2O(l)$$

Laboratory Manual
You will study the neutralization of two acids in Experiment 13-5.

The above net ionic equation represents an acid-base **neutralization reaction**. This specific neutralization reaction ocurs whenever any acid (a source of $H^+(aq)$) and any base (a source of OH^- (aq)) are mixed together. In general, a neutralization reaction is any reaction that destroys the acidic properties of a substance and the basic properties of another substance.

Sodium carbonate (soda ash) is a base because it will produce hydroxide ions when it is dissolved in water. It is often used in its solid form to neutralize acid spills. The ionic equation for the neutralization of sulphuric acid solution by solid sodium carbonate is

$$\underset{\substack{\text{sodium} \\ \text{carbonate}}}{Na_2CO_3(s)} + \underset{\substack{\text{sulphuric acid} \\ \text{solution}}}{2H^+(aq) + SO_4^{2-}(aq)} \rightarrow 2Na^+(aq) + CO_2(g) + H_2O(l) + SO_4^{2-}(aq)$$

Exercise

14. Write the net ionic equation for the neutralization of an acid using sodium carbonate.

15. For each of the following situations, write the ionic, and then the net ionic equations for the chemical reaction that occurs.
    (a) A sulphuric acid solution is mixed with a sodium hydroxide solution.
    (b) A person suffering from "heartburn" takes an antacid preparation containing a suspension of solid magnesium hydroxide. (Assume "stomach acid" is mostly hydrochloric acid.)

Precipitation

The net ionic equation for the production of a given precipitate also summarizes many possible specific reactions. For example, the reaction between silver ions and chloride ions, regardless of their source, is always the same:

$$Ag^+(aq) + Cl^-(aq) \rightarrow AgCl(s)$$
(a precipitation reaction)

Thus, for example, if a silver nitrate solution (containing $Ag^+(aq)$ and $NO_3^-(aq)$ ions) is mixed with a sodium chloride solution (containing $Na^+(aq)$ and $Cl^-(aq)$ ions), a precipitate of silver chloride results. (See Figure 13.19.) The sodium and nitrate ions are spectator ions and do not show up in the net ionic equation.

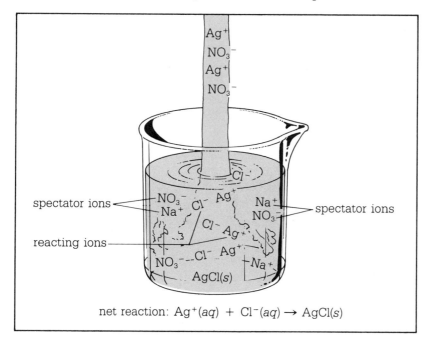

net reaction: $Ag^+(aq) + Cl^-(aq) \rightarrow AgCl(s)$

Figure 13.19
Spectator ions in the precipitation of silver chloride. Only silver and chloride ions undergo a change as they join to form a precipitate when they are mixed. Sodium and nitrate ions remain exactly the same – they are spectator ions.

Applications of Precipitation Reactions

Precipitation reactions are used extensively in industry, both to produce and to purify products. For example, sodium hydrogen carbonate (baking soda) is produced as a precipitate in the final step of a series of reactions known as the Solvay process.

Exercise

16. General Chemical Corporation has a large plant in Amherstburg, Ontario which manufactures sodium carbonate using the Solvay process.
    (a) Using reference books, look up the set of reactions that make up the Solvay process.
    (b) Where appropriate, write the ionic and net ionic equations for each reaction.
    (c) Identify the reaction type (refer to sections 8.3–8.6, Chapter 8) in each of the steps in the Solvay process.

(d) Identify the two main products that are made using this process.

(e) List several uses for each of these products.

Water Softening and Precipitation Reactions

A precipitation reaction can also be used to soften hard water. Hard water contains aqueous ions such as magnesium, calcium, and iron(II) ions which cause soap to form a scum rather than a lather. In many communities the only water which is economically available is hard water obtained from underground sources where the water percolates through rock. The water dissolves some of the rock's minerals which causes "hardness". One method used in home water softeners to remove these ions is to precipitate them from the water. Sodium carbonate solution is mixed with the incoming hard water and the ions which cause hardness are precipitated out by reaction with carbonate ions.

$$2Na^+(aq) + CO_3{}^{2-}(aq) \ + \ Ca^{2+}(aq) \ \rightarrow \ CaCO_3(s) \ + \ 2Na^+(aq)$$

sodium carbonate solution	ion causing hardness	a solid precipitate	this ion does not cause hardness

The resulting precipitate can be filtered off and disposed of. In effect, sodium ion, which does not interfere with the action of soap, has been exchanged for calcium ion.

Exercise

17. Write the ionic and the net ionic equation for the reactions that will occur if the following solutions are mixed together.

(a) A calcium nitrate solution is mixed with an ammonium sulphate solution.

(b) An iron(II) nitrate solution is mixed with a sodium carbonate solution.

Looking Back – Looking Ahead

As we have seen in this chapter, solutions have many useful properties. Many commercial and household products come in the form of solutions and many chemical processes operate in solutions. The study of solutions forms a major part of the study of chemistry. In the next chapter, you will learn how to prepare and work with solutions of known concentration.

Vocabulary Checklist

You should now be able to define or explain the meaning of the following terms.

acid	dilute	neutralization reaction	solution
anion	electrode	pH scale	solvent
anode	electrolysis	precipitate	spectator ion
aqueous ion	electrolyte	saturated solution	strong electrolyte
base	hydrated ion	seed crystal	supersaturated solution
cathode	indicator	solubility	unsaturated solution
cation	ionic equation	solute	weak electrolyte
concentrated	net ionic equation		

Chapter Objectives

You should now be able to do the following.

1. Identify several uses for solutions in everyday life.

2. Identify several problems or hazards associated with solutions.

3. Describe and carry out a test to distinguish among saturated, unsaturated, and supersaturated solutions.

4. Use a set of solubility rules to design and carry out experiments to identify an unknown ion, and to separate a given ion from a solution of several ions.

5. Appreciate that water is an extremely important solvent that is essential for life and that drinking water can be polluted by a variety of soluble contaminants.

6. Understand the importance of water as a solvent and describe how water molecules and ions form hydrated or aqueous ions.

7. Explain reactions of solutions in terms of ionic and net ionic equations, spectator ions, precipitation of insoluble ionic compounds, and acid-base reactions.

Understanding Chapter 13

REVIEW

1. List two examples of solutions for each of the following categories.
   (a) foods
   (b) beverages
   (c) cleaning products
   (d) cosmetics
   (e) home health products
   (f) automotive uses
   (g) hospital uses
   (h) industrial uses
   (i) agricultural and gardening products

2. (a) Identify three harmful or unwanted solutions and explain why they are considered undesirable.
   (b) Identify where these harmful solutions might have originated and how their production might be reduced or eliminated.

3. What properties of solutions distinguish them from the other categories of matter? (If necessary, refer to Chapter 1.)

4. What are the minimum number of components of a solution? Give several examples of these components. (Use all three states of matter in your examples.)

5. Classify the following list of properties as acidic, basic, both acidic and basic, or neutral.
   (a) a pH of 9.5
   (b) a sour taste
   (c) a pH of 7.0
   (d) affects the colour of some dyes
   (e) a pH of 2.5
   (f) corrosive
   (g) reacts with and dissolves magnesium metal
   (h) can be a non-electrolyte
   (i) reacts with vegetable matter
   (j) feels slippery
   (k) a pH of 11
   (l) tastes bitter
   (m) reacts with limestone
   (n) is an electrolyte
   (o) a pH of 5.0

APPLICATIONS AND PROBLEMS

6. Two popular Canadian foods are maple syrup and maple sugar. Both of these are produced from the sap of maple trees.
   (a) Explain in terms of solutions how a thick concentrated syrup is made from this dilute watery sap.
   (b) How is solid maple sugar obtained from the syrup?

7. The wine, brewing, and distilling industries all produce solutions.
   (a) What could be considered the solute in all of these solutions?
   (b) Which product (wine, beer, or distilled spirits) is the most concentrated? Explain your answer.
   (c) Which is the most dilute? Explain.

8. The cork in a wine bottle allows oxygen from the air to slowly diffuse into the wine. Explain why a sediment sometimes slowly precipitates out as a corked bottle of wine ages.

9. Good nutrition depends on maintaining a reasonable concentration of both water soluble and fat or oil soluble vitamins in the body. Under the same conditions, the body's concentration of vitamins B and C (which are water soluble) decreases much more quickly than the concentrations of vitamins A, D, and E (which are fat soluble).
   (a) Suggest one reason why this occurs.
   (b) Consuming large amounts of fat soluble vitamin A over an extended period of time can lead to a harmful buildup of its concentration. If similar large amounts of vitamin C are consumed, no such harmful buildup will occur. Explain.

10. Fresh honey is a syrupy liquid. A sealed jar of fresh honey at a fixed temperature will slowly form solid sugar crystals. What type of solution is fresh honey? Explain your answer.

11. Determine the solubilities of the following compounds in 1 kg of water at the temperatures given. Use the graph in Figure 13.3.
    (a) potassium chlorate at 25°C.
    (b) lead(II) nitrate at 73°C
    (c) potassium chloride at 15°C
    (d) potassium nitrate at 5°C
    (e) sodium chloride at 88°C

12. At what temperature does each of the following compounds have a solubility of 450 g/kg of water? Use the graph in Figure 13.3 to determine your answers.
    (a) lead(II) nitrate
    (b) potassium chlorate
    (c) potassium nitrate
    (d) potassium chloride
    (e) sodium chloride

13. If 1 kg of water is saturated with potassium chloride at 90°C, calculate the mass of precipitate that will form if the solution is cooled to 10°C. Use the graph in Figure 13.3.

14. Explain why bubbles of air often form on the inside surface of a glass containing cold tap water as the water slowly warms up to room temperature.

15. During the hot summer months, people who fish often try to locate the deeper areas of a river or small lake. The fish tend to prefer these deeper, cooler waters rather than the warmer regions. Propose an explanation for this. (Remember that fish, like people, need oxygen to live.)

16. Lead is a poisonous substance. Design a test to determine if a sample of water contains any soluble lead(II) ion. Make use of the solubility rules given in Table 13.1.

17. Describe in terms of ions and water molecules what happens when the ionic substance lithium bromide dissolves. (Diagrams may be helpful in your explanation.)

18. The polar covalent compound, hydrogen bromide gas, is very soluble in water and is an electrolyte. Describe in terms of the molecules what happens as the hydrogen bromide dissolves. (Diagrams may be helpful in your explanation.)

19. When sodium hydroxide, a major constituent of drain cleaning products, is dissolved in water, the solution's temperature rises dramatically. (This is why you should never use hot water when making a concentrated NaOH solution – the boiling point of the solution can easily be reached and the solution may splatter.) Explain this temperature increase in terms of water molecules and the nature of sodium hydroxide solid.

20. Some home ice cream makers freeze the ice cream mix by using an ice and salt mixture packed around the ice cream container. The temperature of the ice may start out near 0°C, but can reach as low as − 18°C as the salt is added and starts to "melt" the ice. Explain how this can happen.

21. Chemical cold packs are often used by trainers of sports teams to treat injuries immediately. These cold packs can be stored at room temperature, yet they are capable of producing cold temperatures when they are needed. Explain how such a product might be constructed and identify the types of substances that must be present inside the packs.

22. A moisture meter may be used to test whether plants need watering. Such a meter consists of a battery, an electrical current flow meter (ammeter) and two metal prongs that are inserted into the soil to be tested. Explain how the meter can determine if the plant's soil needs water.

23. Classify the following as either electrolytes or non-electrolytes. Give reasons for your answers.
    (a) CH_4
    (b) $MgCl_2$
    (c) CCl_4
    (d) HI
    (e) CH_3OH (methyl alcohol)

24. Explain why fire extinguishers that use water should not be used to extinguish electrical fires.

25. Write both the ionic and the net ionic equations for each of the following reactions.
    (a) A calcium hydroxide solution is added to a sulphuric acid solution.
    (b) A copper(II) sulphate solution is added to metallic zinc. Copper metal is one of the products, and the other product is soluble.
    (c) Solutions of lead(II) nitrate and potassium iodide are mixed. (Refer to the solubility rules in Table 13.1.)

CHALLENGE

26. In dry regions of the world, farmland must be consistently irrigated with fresh water. One problem associated with this constant irrigation is that the salinity of the soil increases. (That is, the amount of salt in the soil increases.) This greater salinity lowers the land's productivity.
    (a) Explain why the salinity can keep rising as the land is irrigated.
    (b) Suggest one possible answer to this problem. How can the land continue to be irrigated without changes in salinity which affects its productivity?

27. 1 kg of water has been saturated with both potassium chlorate and lead(II) nitrate at 85°C. What is the total mass of the precipitate that forms if the solution is cooled to 15°C? Use the graph in Figure 13.3.

28. A compound known to consist of two of the ions referred to in the solubility rules in Table 13.1 has been dissolved in water forming a solution.
    (a) Use the results of the following tests to identify the compound's possible composition.
       (i) A precipitate forms if samples of the solution are added to solutions containing hydroxide, carbonate, or lead(II) ions.
       (ii) No precipitate forms when solutions containing chloride or sulphide ions are added.
    (b) Suggest a further precipitation test (or tests) you could use to identify the compound.

29. The waste water dumped into a river from Plant A contains dissolved magnesium sulphide. A second plant, Plant B's waste water contains strontium hydroxide in solution. A third plant, Plant C, has some silver sulphate dissolved in the water it discards into the river. When

these waste waters eventually mix, various precipitates form.

(a) Identify all of the precipitates which form and are deposited as sludge on the river bottom.

(b) How could Plant C treat its waste water to recover a very valuable product?

30. The "run off" water from heavily fertilized farm fields may contain phosphate ions. These ions also fertilize river and lake water and cause large, smelly and unsightly blooms of algal growth. What treatment could be used to remove these phosphate ions? (Try to consider the availability and cost of the substances needed for your proposed treatment.)

31. Sodium chloride is ineffective when spread over road ice if the temperature is −18°C or lower. Use your knowledge of the kinetic molecular theory and the effect of temperature on particles to propose an explanation for the fact that sodium chloride will not "melt" ice that is below −18°c. (Remember to consider what happens as aqueous ions form.)

32. The substances known as PCBs, polychlorinated biphenyls, are extremely stable and toxic compounds. PCBs were previously used as a solute in an oil solvent for use in electrical equipment. These toxic solutions have excellent cooling and electrical insulating properties and some large electrical transformers and capacitors still contain them.

(a) Are PCBs electrolytes? Justify your answer.

(b) Why are PCBs no longer used as a solute for cooling and insulating oil solutions in new electrical equipment?

(c) Use a library or other sources to find out how PCB solutions have to be treated for safe disposal.

14 SOLUTIONS – CALCULATIONS AND REACTIONS

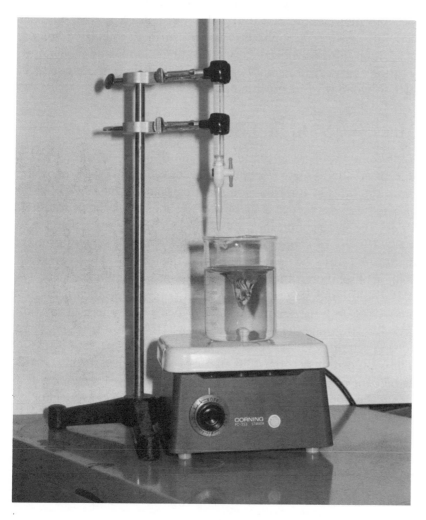

You have examined solutions, their involvement in chemical reactions, and their importance in everyday life. Chemists are also interested in the quantitative aspects of solutions, such as the amount of solute contained in a certain volume of solution. Since many chemical reactions take place in solution, stoichiometric calculations involving solutions are important in designing equipment such as analytical apparatus and industrial chemical reactors. In this chapter, you will learn how to express solution concentration and how to calculate the concentration of a solution. You will also develop analytical techniques such as titration, and extend your ability to carry out stoichiometric calculations to include solutions.

14.1 Solution Concentration

As you have seen earlier, solutions can be qualitatively described as dilute or concentrated. Quantitatively, concentration is measured in moles per cubic metre. Since solutions in the laboratory are made in relatively small quantities, chemists often state the concentration of solutions as moles per litre of solution (mol/L). In this textbook concentrations will be stated in mol/L. A required number of moles of solute can be easily obtained by measuring a calculated volume of solution. Since chemical reactants combine in mole-to-mole ratios, solutions expressed as mol/L are, as you will find in section 14.2, easy to work with in stoichiometric calculations.

Mass Fraction and Solution Percentages

Expressing the mass of solute in kg that dissolves in a kilogram of solvent is called a mass fraction. It is a convenient way of comparing the solubilities of different solutes. For example, at 20°C, 0.35 kg of NaCl will dissolve in 1 kg of water to make 1.35 kg of solution. Under the same conditions, only 0.07 kg of $KClO_3$ will dissolve to make 1.07 kg of solution. Although this is useful solubility information, it cannot be used to determine the mass of solute in a given volume of solution without further data such as the solution's density.

If a solution is described in terms of percent solute by mass, the amount of both the solute and solvent in any given mass of solution is easily determined. The solute components of most household and garden products sold as solutions are expressed as percent by mass. The composition of some solid mixtures such as alloys are also stated as percent by mass.

The composition of liquid-liquid solutions such as alcohol or acetic acid in water are often expressed as percent solute by volume. Vinegar, for example, is a 5% by volume solution of acetic acid. (See Figure 14.1.)

Figure 14.1
Ways of expressing solution composition

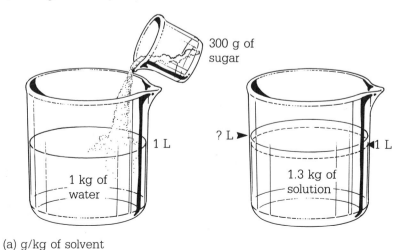

300 g of sugar

1 kg of water

1.3 kg of solution

1 L

? L

1 L

(a) g/kg of solvent

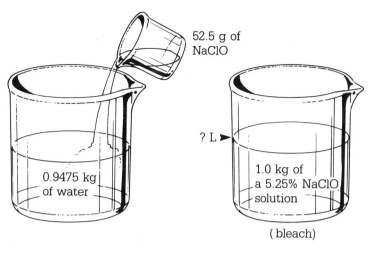

(b) % by mass

52.5 g of
NaClO

0.9475 kg
of water

? L ▶

1.0 kg of
a 5.25% NaClO
solution

(bleach)

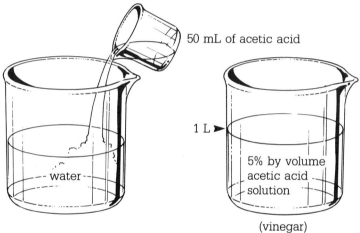

(c) % by volume

50 mL of acetic acid

water

1 L ▶

5% by volume
acetic acid
solution

(vinegar)

Exercise

1. Examine the labels of at least three household, pool, or garden products sold as solutions. Identify the solutes they contain and report their composition.

Sample Problem 14.1

Describe how you would make 650 g of a 5% by mass sugar solution in water.

Solution

The required solution must contain 5 g of sugar for every 100 g of solution. Therefore you need 5 g of sugar and 95 g of water to make 100 g of solution.

The masses of sugar and water required to make 650 g of the solution are determined as follows.

$$\text{Mass of sugar} = 650 \text{ g solution} \times \frac{5 \text{ g sugar}}{100 \text{ g solution}}$$

$$= 32.5 \text{ g sugar}$$

$$\text{Mass of water} = 650 \text{ g solution} \times \frac{95 \text{ g water}}{100 \text{ g solution}}$$

$$= 617.5 \text{ g water}$$

(Note that the % of the component is the conversion factor in each case.)

Therefore, 32.5 g of sugar must be added to 617.5 g of water to make the required 650 g of 5% sugar solution.

Exercise

2. If 325 g of a 12.8% by mass solution of salt water is evaporated to dryness, calculate the mass of salt that would be left behind.

3. 50 mL (50 g) of pure water is added to 250 g of a 25% by mass potassium nitrate solution. What is the percent by mass of potassium nitrate in the new, diluted solution?

4. A concentrated hydrochloric acid solution is 36.0% by mass HCl. Its density is 1.18 g/mL. What mass of HCl is dissolved in 1.00 L of this solution?

5. 400 g of ammonium chloride dissolves in 1.00 L (1 kg) of water at 30°C. Express this as percent solute by mass.

6. What volume of vinegar (5.00% acetic acid by volume in water) can be prepared from 600 mL of pure acetic acid?

Very Dilute Solutions

Toxic substances can be harmful even in very dilute solutions. The quantity of these substances present in our air, water, and food is of great concern to us. These substances are usually measured as micrograms (1 μg $= 10^{-6}$ g) per gram of solution (or per gram of such substances as food, blood, or flesh). For example, the amount of chlorine used as a disinfectant in drinking or swimming pool water is measured in micrograms per gram of water (μg/g). This is often expressed as parts per million (ppm). Safe swimming pool water must contain about 1 μg/g (1 ppm) of chlorine in water.

TOXIC SUBSTANCES IN THE ENVIRONMENT

Many toxic substances occur naturally in the environment. Heavy metals such as cobalt, cadmium, and copper leach out of their ores and show up in water in creeks and waterways. The gaseous products from forests and forest fires produce many substances such as dioxins that are considered pollutants.

Using increasingly more sensitive instruments, chemists are able to measure toxic substances and pollutants down to such low levels that their results may indicate the natural or background level of some of those substances. For example, is the 50 fg/g of dioxin found in fresh fruits and vegetables a naturally occurring level or is it caused by human activities?

Much environmental research remains to be done to sort out the background or natural level of toxic substances from any extra amounts of these substances being added to the environment by human activities. A career in analytical or environmental chemistry could involve you in this branch of science.

Extremely small quantities of contaminants are often expressed in terms of nanograms (1 ng $= 10^{-9}$ g), picograms (1 pg $= 10^{-12}$ g), or even femtograms (1 fg $= 10^{-15}$ g) per gram of solution or mixture. For example, the occurrence of the toxic substances, dioxins, has been found to be about 50 fg/g in some foods. For very toxic substances, even extremely small quantities in food, air, or water could be harmful.

The occurrence of contaminants can also be expressed in mass per volume or volume per volume units. For example, government regulation has set the maximum allowable amount of ammonia gas in air at 50 μL/L. Solid pollutants dispersed as dust are measured in milligrams per cubic metre (mg/m^3) of air. Cobalt metal, used to produce very hard tool steels, causes an illness called "hard metal disease". Industrial safety regulations limit the amount of cobalt dust in factory air to 0.1 mg/m^3.

Exercise

7. Titanium dioxide is a bright white solid compound used as a paint pigment (a colouring agent). Industrial safety regulations limit titanium dioxide dust to 15 mg/m^3 of air. What is the maximum mass of titanium dioxide dust allowed in the air in a paint factory which has a volume of 2000 m^3?

8. A government regulation sets the maximum allowable amount of chlorine gas in air at 1.0 μL/L. What volume of chlorine gas at STP must escape into a room measuring 7.5 m wide by 10.0 m long and 3.5 m high to just exceed regulated limits? (1 dm^3 = 1 L)

9. What mass of dioxins is there in a 300 g apple if dioxin is present at 50 fg/g in the apple?

14.2 Mole-Based Solution Calculations

A solution with a concentration of 1.0 mol/L contains 1.0 mol of solute in 1.0 L of solution. If 0.50 mol of solute is added to enough water to make 1.0 L of solution, a 0.50 mol/L solution is formed. (See Figure 14.2.) If, however, the 0.50 mol of solute is added to enough water to make only 0.50 L of solution, then a 1.0 mol/L solution is formed.

Consider the information given in Table 14.1 How many moles of solute are needed to make 250 mL of a 0.25 mol/L solution? If you study the first four solutions in the table, you will detect a regularity. The solution concentration, c, in mol/L, multiplied by the solution volume, V, in L, gives the number of moles, n, needed to make the required solution.

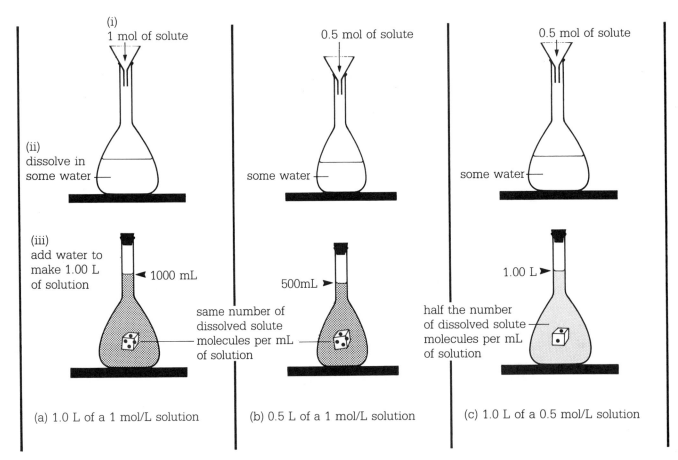

Figure 14.2
In making solutions of known concentration (i), the correct number of moles of solute are added to a volumetric flask partially filled with water (ii). The solute is dissolved and (iii) enough water is added to make the required volume solution. Different volumes of solutions with the same concentration are made by changing the number of moles added.

Table 14.1 Moles of Solute Necessary to Make a Solution of Specified Volume and Concentration

Concentration (c)	Volume of Solution (V)	Moles of Solute Needed (n)
1.00 mol/L	1.00 L	1.00 mole
1.00 mol/L	0.50 L (500 mL)	0.50 mole
0.50 mol/L	1.00 L	0.50 mole
0.50 mol/L	0.50 L (500 mL)	0.25 mole
0.25 mol/L	0.25 L (250 mL)	? mole

For any specified volume and concentration of solution, the number of moles is calculated by multiplying the volume, V, by the conversion factor, which is the concentration, c.

$$n = V \times c$$

Therefore, to find the missing value in Table 14.1:

$$n = 0.25 \text{ L} \times 0.25 \frac{\text{mol}}{\text{L}}$$

$$= 0.0625 \text{ mol}$$

Sample Problem 14.2

(a) Calculate the number of moles of sodium chloride needed to prepare 250 mL of a 0.850 mol/L salt solution.
(b) What mass of NaCl does this represent?
(c) Describe how you would prepare the solution.

Solution

First organize the information.

$$m_{\text{NaCl}} = ?$$

$$V = 0.250 \text{ L}$$

$$c = 0.850 \text{ mol/L}$$

$$M_{\text{NaCl}} = 58.5 \text{ g/mol}$$

(a) Convert volume of solution to moles of solute.

$$n = V \times c$$

$$n = 0.250 \text{ L} \times 0.850 \frac{\text{mol}}{\text{L}}$$

$$= 0.212 \text{ mol}$$

Therefore 0.212 mol of NaCl is needed to make the required solution.

Figure 14.3
Preparing solutions of given concentrations

(a) Measure the required mass (number of moles) of solute.

(b) Use a powder funnel to add the solute to the volumetric flask partially filled with distilled water.

(c) Use a squeeze bottle to flush all traces of the solute into the flask.

(d) Stopper the flask and mix well until the solute has completely dissolved.

(e) Add water to within a few millilitres of the etched line, mix well, and allow the solution to stand until it reaches room temperature.

(f) Use a squeeze bottle to add water until the bottom of the meniscus is on the etched line. Mix well and properly label the container holding the solution.

(b) Convert moles to mass

$$m = n \times M$$

$$m = 0.212 \text{ mol} \times 58.5 \; \frac{\text{g}}{\text{mol}}$$

$$= 12.4 \text{ g}$$

Therefore 12.4 g of NaCl are needed to make the required solution.

(c) (i) Measure 12.4 g of sodium chloride and transfer it to a 250 mL **volumetric flask** which is partially filled with water. A volumeteric flask has a large bottom and a narrow neck. See Figure 14.3(a), (b), (c). A single etched line on the narrow neck indicates precisely a volume of 250 mL.

(ii) Swirl or invert the stoppered flask several times until all of the solute has been dissolved (d).

(iii) Add water to within a few mL below the etched line in the flask (use a squeeze bottle for the last few millilitres) and invert several times to make the solution uniform (e).

(iv) Allow the flask to stand until it reaches room temperature (this may take several hours in some cases), then use a squeeze bottle to fill it with water up to the etched mark. This final step compensates for any changes (due to thermal expansion or contraction) in the solution volume caused by the heat evolved or absorbed when the solute dissolves (f).

Exercise

10. Calculate the mass of solute needed to make each of the following solutions.
    (a) 250 mL of a 1.25 mol/L lithium bromide, LiBr, solution
    (b) 500 mL of a 0.450 mol/L magnesium nitrate, $Mg(NO_3)_2$, solution
    (c) 2.00 L of a 0.0500 mol/L potassium carbonate, K_2CO_3, solution
    (d) 100 mL of a 0.850 mol/L sodium phosphate (TSP, trisodium phosphate), Na_3PO_4, solution (a TSP solution is used to wash walls before repainting)
    (e) 50.0 mL of a 2.30 mol/L aluminum chloride, $AlCl_3$, solution

14.3 Stock Solutions

Not all solutes are solids whose mass can be conveniently measured. For example, the hydrochloric acid solutions you used earlier in this course were made by diluting concentrated hydrochloric acid solutions with water. Hydrochloric acid, a solution of gaseous hydrogen chloride in water, is usually purchased as a 12 mol/L concentrated solution. This commercially available, concentrated solution is referred to as a **stock solution**.

Another common stock solution, sulphuric acid, is an 18 mol/L solution of sulphur trioxide gas in water. A stock solution can be any solution made up with a specific concentration for laboratory use.

Dilution of Stock Solutions

As discussed in the previous chapter, many concentrated solutions are hazardous. For many laboratory or industrial applications, dilute solutions are preferred or required. How, then, can we determine the volume of concentrated solution that must be added to water to make the necessary volume of a specified dilute solution?

To make a dilute solution from a concentrated stock solution, you have to determine the volume of concentrated solution, V_c, that must be added to water to make the necessary volume of the dilute solution, V_d.

If a given volume of a solution of known concentration (c_c) is poured into a beaker, the number of moles of solute in the beaker can be easily calculated. For example, if 150 mL of 12 mol/L hydrochloric acid solution is poured into a beaker, the beaker contains 1.8 moles of HCl solute.

Laboratory Manual
In Experiment 14-1, you will prepare solutions of known concentrations.

$$n = ?$$

$$V_c = 150 \text{ mL} \quad \text{(the volume of the concentrated solution)}$$

$$= 0.150 \text{ L}$$

$$c_c = 12 \text{ mol/L} \quad \text{(the concentration of the concentrated solution}$$

$$n = V_c \times c_c \quad \text{(convert volume to moles)}$$

$$= 0.150 \text{ L} \times 12 \frac{\text{mol}}{\text{L}}$$

Figure 14.4
Diluting a solution. The number of moles of solute in a small volume of a concentrated solution is equal to the number of moles in a larger volume after it has been diluted.

Thus, $\quad n = 1.80 \text{ mol}$

If this amount of hydrochloric acid is added to water the total volume of solution increases but the number of moles of dissolved solute remains the same. (See Figure 14.4.)

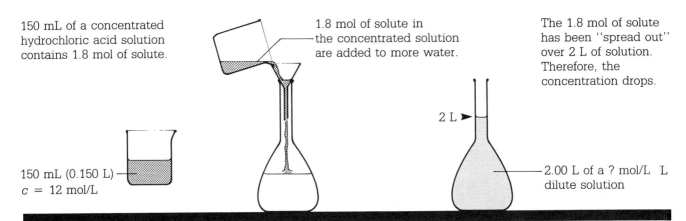

150 mL of a concentrated hydrochloric acid solution contains 1.8 mol of solute.

1.8 mol of solute in the concentrated solution are added to more water.

The 1.8 mol of solute has been "spread out" over 2 L of solution. Therefore, the concentration drops.

2 L ▶

150 mL (0.150 L)
$c = 12$ mol/L

2.00 L of a ? mol/L L dilute solution

What, then, is the final (diluted) concentration (c_d) if the acid is added to some water in a 2 L (2000 mL) volumetric flask and enough water is added to fill the flask to the 2 L mark?

$$n = 1.80 \text{ mol} \quad \text{(number of moles of solute added)}$$

$$V_d = 2.00 \text{ L} \quad \text{(final volume of the diluted solution)}$$

$$c_d = ? \quad \text{(concentration of diluted solution)}$$

$$n = V_d \times c_d \quad \text{(volume to moles conversion equation)}$$

$$c_d = \frac{n}{V_d} \quad \text{(rearranging } n = V \times c)$$

$$= \frac{1.80 \text{ mol}}{2.0 \text{ L}}$$

$$c_d = 0.90 \text{ mol/L}$$

Thus, the concentration of the diluted solution is 0.90 mol/L. This procedure may be summarized as follows:

number of moles of solute obtained from the concentrated solution	=	number of moles of solute that will be present in the dilute solution

$$n_{concentrated} = n_{dilute}$$

or

$$n_c = n_d$$

From this relationship, we can derive a simple procedure to calculate the volume of concentrated solution needed to make any given volume of a specified dilute solution.

$$n_c = n_d$$

Since

$$n_c = V_c \times c_c$$

and

$$n_d = V_d \times c_d$$

Then

$$V_c \times c_c = V_d \times c_d$$

We can rearrange this equation to obtain the volume of concentrated solution required:

$$V_c = V_d \times \frac{c_d}{c_c}$$

(conversion factor)

Sample Problem 14.3

What volume of 16.0 mol/L stock nitric acid solution is needed to prepare 500 mL of a 2.50 mol/L nitric acid solution?

Solution

First organize the information.

$$V_c = ?$$ $$V_d = 500 \text{ mL}$$
$$= 0.500 \text{ L}$$

$$c_c = 16.0 \text{ mol/L}$$ $$c_d = 2.50 \text{ mol/L}$$

Determine the volume of concentrated solution.

$$V_c = V_d \times \frac{c_d}{c_c}$$

$$V_c = 0.500 \text{ L} \times \frac{2.50 \text{ mol/L}}{16.0 \text{ mol/L}}$$

$$= 0.0781 \text{ L}$$

$$= 78.1 \text{ mL}$$

Thus, 78.1 mL of the stock 16.0 mol/L nitric acid solution is needed to prepare 500 mL of 2.50 mol/L nitric acid solution.

Exercise

11. Calculate the volume of stock solution that must be used to make each of the following solutions.
    (a) 500 mL of a 0.750 mol/L solution of sulphuric acid (sulphuric acid stock solution = 18.0 mol/L)
    (b) 100 mL of a 3.40 mol/L solution of ammonium hydroxide (ammonium hydroxide stock = 15.0 mol/L)
    (c) 250 mL of a 0.120 mol/L solution of acetic acid (acetic acid stock solution = 18.0 mol/L)

Calculating the Concentration of Purchased Stock Solutions

The concentration of stock solutions can be calculated from the percent of solute by mass and specific gravity (density) information listed on the bottle labels. (See Figure 14.5.) The density (or specific gravity) is used to calculate the mass of 1 L of the solution. The percentage of solute by mass in the solution is then used to determine the mass of solute in 1 L of the solution. Finally, this mass is converted to number of moles, which gives the concentration of the solution in mol/L.

Figure 14.5
Information supplied with commercially available sulphuric acid

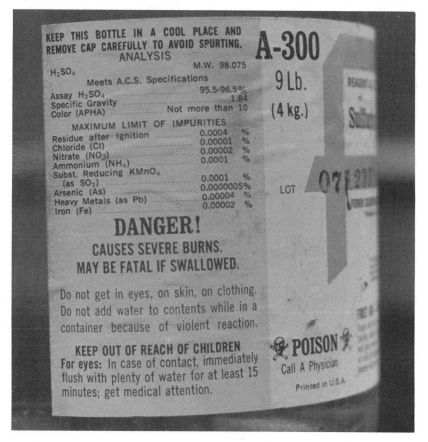

Sample Problem 14.4

Calculate the concentration, in mol/L, of the sulphuric acid solution in the stock bottle shown in Figure 14.5.

Solution

Step 1
Organize the information.

$$c = ?$$

$$\rho = 1.84 \text{ g/mL} \quad \text{(specific gravity or relative density from the label)}$$

$$\text{Percent } H_2SO_4 \text{ by mass} = 96.0\% \quad \text{(from the label)}$$

$$M_{H_2SO_4} = 98.0 \text{ g/mol}$$

Step 2
Calculate the mass of 1 L of solution.

$$m = V \times \rho$$

Use the density (volume to mass conversion) equation and assume a volume of 1 L (1000 mL)

$$m_{solution} = 1000 \text{ mL} \times 1.84 \text{ } \frac{g}{mL}$$

$$= 1840 \text{ g}$$

Step 3
Calculate the mass of solute per litre of solution.

$$m_{H_2SO_4} = m_{solution} \times \%_{H_2SO_4}$$

$$= 1840 \text{ g} \times \frac{96.00 \text{ g}}{100 \text{ g}}$$

$$= 1766 \text{ g}$$

Step 4
Calculate the number of moles of solute per litre of solution.

$$n = m \times \frac{1}{M}$$

$$n_{H_2SO_4} = 1766 \text{ g} \times \frac{1}{98.0 \text{ g/mol}}$$

$$= 18.0 \text{ mol}$$

Thus, the concentration of the stock sulphuric acid is 18.0 mol/L.

Exercise

12. Calculate the concentration in mol/L for a stock hydrochloric acid solution that has the following information on its label:
Specific Gravity: 1.18
% HCl by mass: 36.0%

14.4 Stoichiometry and Solutions

Stoichiometry, the calculation of the quantity of material used and/ or produced in a chemical reaction, is one of the most important types of chemical calculations. In Chapters 9 and 12, stoichiometric calculations were carried out using the mass of solids or pure liquids, and the volume of gases under stated conditions. We can also carry out stoichiometric calculations using solutions of known concentration. An important part of any stoichiometric calculation is the determination of the number of moles of substance being used or produced. We now have four different ways of determining the number of moles of substance. These are summarized in Table 14.2.

Figure 14.6
A stoichiometric calculation

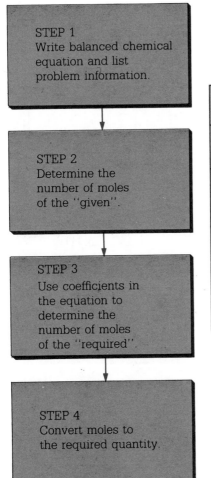

Table 14.2 Determining Number of Moles of a Substance

Measurement or Given Quantity	Determining Moles
Mass	$n = m \times \dfrac{1}{M}$
Volume of a gas under given temperature and pressure conditions	$n = \dfrac{P \times V}{R \times T}$
Number of atoms or molecules	$n = \dfrac{N}{N_A}$
Volume of a solution whose concentration is expressed as mol/L	$n = V \times c$

Depending on the information given, any of the four relationships listed in Table 14.2 can be used to determine the number of moles of the substance being used or produced in a chemical reaction. The balanced chemical equation is used to determine the number of moles of the required substance. Finally, one of the relationships in Table 14.2 is used to calculate the required quantity (mass, volume of a gas, or concentration or volume of a solution) of reactant or product. (See Figure 14.6.)

Sample Problem 14.5

400 mL of a 0.800 mol/L hydrochloric acid solution is poured over excess zinc metal. Calculate the following.
(a) mass of zinc chloride produced
(b) volume of hydrogen gas produced at 100 kPa pressure and 27°C

Solution

Step 1 Write a balanced equation and list the information.

$$Zn(s) + 2HCl(aq) \rightarrow \qquad ZnCl_2(aq) \qquad + \qquad H_2(g)$$

(a)		(b)
$V = 400$ mL	$m = ?$	$V = ?$
$\quad = 0.40$ L	$M = 137$ g/mol	$P = 100$ kPa
$c = 0.80$ mol/L		$T = 27°C$
		$\quad = 300$ K
		$R = 8.31 \dfrac{\text{kPa} \cdot \text{L}}{\text{mol} \cdot \text{K}}$

Step 2 Determine the number of moles of given.

$$n = V \times C$$
$$n_{HCl} = 0.40 \text{ L} \times 0.80 \text{ mol/L}$$
$$\quad = 0.32 \text{ mol}$$

Step 3 Use the balanced equation to convert n_{HCl} to number of moles required.

$$n_{ZnCl_2} = 0.32 \text{ mol}_{HCl} \times \frac{1 \text{ mol}_{ZnCl_2}}{2 \text{ mol}_{HCl}}$$
$$\quad = 0.16 \text{ mol}$$

Similarly, $n_{H_2} = 0.16$ mol

Step 4 Convert number of moles to the required quantity.

(a) $m = n \times M$
$m_{ZnCl_2} = 0.16 \text{ mol} \times 137 \text{ g/mol}$
$\quad = 22$ g

(b) $V = \dfrac{nRT}{P}$

$$V_{H_2} = \frac{0.16 \text{ mol} \times 8.31 \dfrac{\text{kPa} \cdot \text{L}}{\text{mol} \cdot \text{K}} \times 300 \text{ K}}{100 \text{ kPa}}$$
$$\quad = 4.0 \text{ L}$$

Therefore, 400 mL of 0.80 mol/L hydrochloric acid poured over excess zinc metal will produce 22 g of zinc chloride and 4.0 L of hydrogen gas at 27°C and 100 kPa.

Exercise

13. 15.0 g of sodium oxide was added to water. If the final volume of solution was 350 mL, calculate the sodium hydroxide concentration in mol/L.

$$Na_2O(s) + H_2O(l) \rightarrow 2NaOH(aq)$$

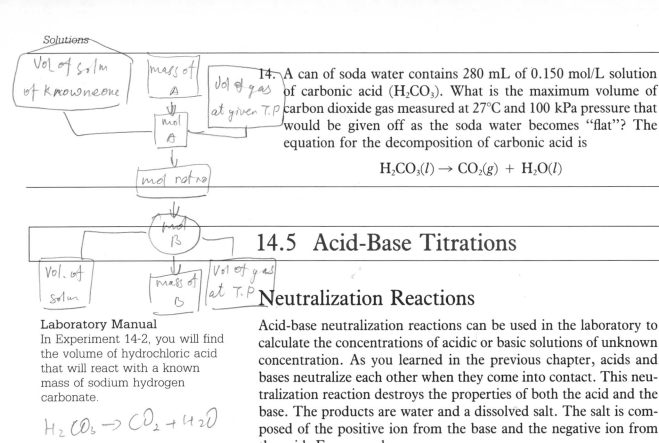

14. A can of soda water contains 280 mL of 0.150 mol/L solution of carbonic acid (H_2CO_3). What is the maximum volume of carbon dioxide gas measured at 27°C and 100 kPa pressure that would be given off as the soda water becomes "flat"? The equation for the decomposition of carbonic acid is

$$H_2CO_3(l) \rightarrow CO_2(g) + H_2O(l)$$

14.5 Acid-Base Titrations

Neutralization Reactions

Laboratory Manual
In Experiment 14-2, you will find the volume of hydrochloric acid that will react with a known mass of sodium hydrogen carbonate.

Acid-base neutralization reactions can be used in the laboratory to calculate the concentrations of acidic or basic solutions of unknown concentration. As you learned in the previous chapter, acids and bases neutralize each other when they come into contact. This neutralization reaction destroys the properties of both the acid and the base. The products are water and a dissolved salt. The salt is composed of the positive ion from the base and the negative ion from the acid. For example:

$$K^+(aq) + OH^-(aq) + H^+(aq) + Cl^-(aq) \rightarrow H_2O(l) + K^+(aq) + Cl^-(aq)$$

| KOH solution | HCl solution | | potassium chloride |
| (a base) | (an acid) | (water) | (a dissolved salt) |

The neutralization reaction can be summarized as a net ionic equation:

$$OH^-(aq) + H^+(aq) \rightarrow H_2O(l)$$

In other words, 1 mol of hydroxide ion from the base reacts with 1 mol of hydrogen ion from the acid.

Titration

A laboratory procedure called **titration** can be used to quantitatively carry out acid-base neutralizations. Titration is the progressive addition of a solution (the **titrant**) from a graduated tube called a **burette** to a known volume or mass of a second solution or substance until the **end-point**, indicated by a colour change of an added indicator or by some other detectable change, is reached. (See Figure 14.7.)

Indicators such as phenolphthalein or bromothymol blue change colour when a solution is close to being neutral. An acid-base titration involves the addition of a basic solution to a flask containing a measured volume of an acidic solution (or vice-versa) to which a few drops of an indicator has been added. The first appearance of

Figure 14.7

Equipment used to prepare and measure the volume of solutions.
Volumetric flasks (a) are used to accurately prepare solutions of known
concentrations. Pipettes (b) are used to deliver set volumes of solutions
with very high accuracy. Burettes (c) are used to deliver variable
volumes of solutions with good accuracy.

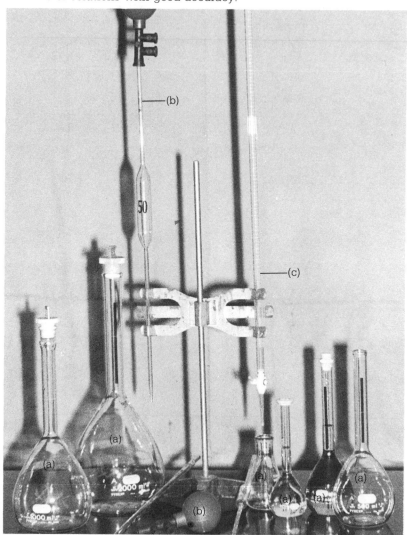

of a permanent colour change in the mixed solution indicates the
end-point of the titration. For indicators such as phenolphthalein
or bromothymol blue, this end-point is also the **neutralization point**
since the colour change occurs when the solution is close to being
neutral. The neutralization point is the point at which the number
of moles of hydroxide ion added is the same as the number of moles
of hydrogen ion originally present in the flask.

$$n_{\substack{\text{hydroxide} \\ \text{ion added}}} = n_{\substack{\text{hydrogen} \\ \text{ion present}}}$$

It follows that if the concentration of one of the solutions is known,
the concentration of the second solution can then be calculated.

CHEMICAL TECHNIQUES – TITRATION

EQUIPMENT

The performance of a laboratory titration involves the use of equipment specially designed to measure the volume of solutions to a high degree of accuracy. Such equipment is called *volumetric glassware*. Volumetric flasks are used to prepare solutions of specified concentrations. *Pipettes* are glass tubes, some with thicker bulbs in the middle, that are calibrated to deliver specified known volumes of solutions accurately. A *burette* is a graduated glass tube with a valve or stopcock on the bottom. The burette is used to add a variable amount of solution to a flask. The amount added can be accurately read from the graduations on the burette.

PROCEDURE

A titration could be used to determine, for example, the concentration of a sodium hydroxide solution. A solution of an acid of known concentration (such as 0.100 mol/L hydrochloric acid), an acid-base indicator, and volumetric glassware are all required.

Step 1

A pipette (or a burette) is used to transfer a definite, highly accurate volume (for example, 10.00 mL) of an acid solution of known concentration to an Erlenmeyer flask. A few drops of an indicator solution, such as phenolphthalein, are then added.

Using a pipette

(a) A suction bulb (or other suction device) is placed on the top end of the pipette. Never draw up a solution by mouth.

(b) Air is squeezed out of the bulb's top valve to produce a vacuum. The solution is drawn into the pipette up to the etched line by controlling the lower valve on the bulb. The valves are opened by squeezing on the valve grips.

(c) The solution in the pipette is delivered to the flask by operating the bulb's side-arm valve.

(d) The last drops of solution are ▶ delivered by touching the tip of the pipette to the flask wall.

Step 2

The sodium hydroxide solution of unknown concentration is used to rinse out and then fill a burette. The initial reading is recorded. Holding a white card with a black mark behind the burette so that the black mark is just below the bottom of the meniscus makes the meniscus more visible. Sodium hydroxide solution is slowly run from the burette to the acid solution in the Erlenmeyer flask. The contents of the flask are swirled constantly and gently to mix the solutions. The addition of sodium hydroxide is stopped when the first persistent pink colour (the end-point) appears in the flask. Placing the flask in front of a white background will enable you to detect the appearance of the colour change more easily. The new burette

reading is taken and the volume of sodium hydroxide added is then calculated. At the end-point, the number of moles of hydroxide ion added from the burette (that is, the titrant) is equal to the number of moles of hydrogen ion originally pipetted into the flask.

In order to detect any errors that may have crept into the procedure, the titration should be repeated until at least three very close (within 0.2 or 0.3 mL) burette readings have been obtained. The experimental readings are used to calculate the unknown concentration of the sodium hydroxide solution.

Titration procedure

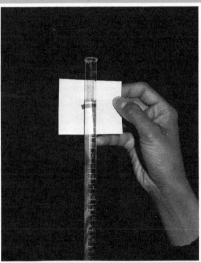

(a) Rinse a clean burette with a ▶ few millilitres of solution and discard. Fill the burette with solution, remove any drops from the tip, and read the initial volume. A white card with a black mark on it makes the meniscus more visible.

(b) Be sure to add a drop or two of the indicator before you start to titrate.

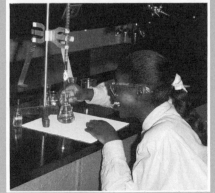

◀ (c) Add solution from the burette until colour change just starts to appear. Add more titrant slowly and constantly swirl the flask to mix the solutions.

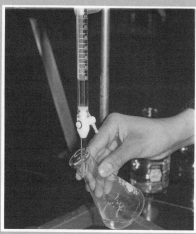

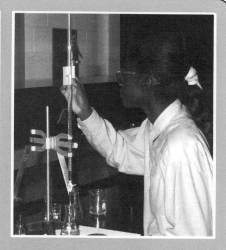

(d) Near the end-point (permanent appearance of a colour change), single drops of titrant can be "grown" on the burette tip and transferred to the flask by

touching the flask wall. A good titration produces a permanent colour change with the final addition of a single drop of titrant.

(e) Read the final volume of solution in the burette. The volume of titrant added will be this reading minus the initial one.

Sample Problem 14.6

If 20 mL of hydrochloric acid of unknown concentration requires 35 mL of a 0.25 mol/L sodium hydroxide solution to reach the end-point in a titration, calculate the concentration of the hydrochloric acid solution.

Solution

Write a balanced chemical equation and check the stoichiometry.

$$NaOH(aq) + HCl(aq) \rightarrow NaCl(s) + H_2O(l)$$
$$\text{1 mol} \qquad \text{1 mol}$$

1 mol of base provides 1 mol of OH^- to neutralize 1 mol of H^+ provided by 1 mol of acid.

Organize the information.

$$V_{NaOH} = 35 \text{ mL} \qquad\qquad V_{HCl} = 20 \text{ mL}$$
$$= 0.035 \text{ L} \qquad\qquad\qquad = 0.020 \text{ L}$$
$$c_{NaOH} = 0.25 \text{ mol/L} \qquad c_{HCl} = ?$$

Use the point of neutralization relationship.

$$n_{OH^-} = n_{H^+}$$

Since 1 mol of base provides 1 mol of OH^- and 1 mol of acid provides 1 mol of H^+, and since $n = V \times c$,

$$V_{NaOH} \times c_{NaOH} = V_{HCl} \times c_{HCl}$$

It follows that

$$c_{HCl} = c_{NaOH} \times \frac{V_{NaOH}}{V_{HCl}}$$

$$c_{HCl} = 0.25 \frac{mol}{L} \times \frac{0.035 \cancel{L}}{0.020 \cancel{L}}$$

$$= 0.44 \ mol/L$$

Therefore, the concentration of the hydrochloric acid solution is 0.44 mol/L.

Exercise

15. A titration was performed to determine the concentration of acetic acid in vinegar. It took 43.2 mL of a 0.350 mol/L sodium hydroxide solution added to 20.00 mL of vinegar for the phenolphthalein indicator to change colour. What was the concentration of acetic acid in the vinegar?

    $$CH_3COOH(aq) + NaOH(aq) \rightarrow CH_3COONa(aq) + H_2O(l)$$

    acetic acid $\qquad\qquad\qquad$ sodium acetate

16. A solid acid, benzoic acid, produces 1 mol of hydrogen ion for each mole of acid dissolved. A solution was made by mixing 1.00 g of this solid acid with 50 mL of distilled water in a flask. Several drops of phenolphthalein indicator were added. The solution was titrated using a 0.250 mol/L sodium hydroxide solution. The first permanent pink colour appeared after 32.8 mL of titrant had been added. Use this titration information to calculate the molar mass of benzoic acid.

17. Sulphuric acid produces 2 mol of hydrogen ion for each mole of sulphuric acid. A student performed titrations to determine the concentration of sulphuric acid in two different car batteries, one brand new and fully charged, and the other old and completely dead. The concentration of the NaOH solution she used was 1.20 mol/L. It took 40.3 mL of the NaOH solution to neutralize 5.00 mL of the sulphuric acid from the new battery, and 28.0 mL of NaOH solution to neutralize 10.00 mL of the acid from the old battery. What is the concentration, in mol/L, of sulphuric acid in each battery?

Laboratory Manual
You will practise analysing substances by titration in Experiment 14-3.

Looking Back – Looking Ahead

In this chapter you have developed the skills needed to know how to prepare solutions of known concentrations and to use these solutions in acid-base titrations. You are now able to perform stoichiometric calculations on chemical reactions involving solutions. You may be able to apply these skills and abilities when you study the applications of chemistry in industry and their impact on society in Chapters 15 and 16.

Vocabulary Checklist

You should now be able to define or explain the meaning of the following terms.

burette	neutralization point	titrant	volumetric flask
end-point	stock solution	titration	

Chapter Objectives

You should now be able to do the following.

1. Solve numerical problems involving solutions using different methods of expressing amount of solute in a solution.

2. Appreciate that toxic substances in the environment are dangerous even at very low concentrations.

3. Recognize the responsiblity of government and industry to establish and regulate levels of toxic substances in the workplace and in the environment.

4. Prepare, using volumetric equipment, specified volumes of solutions of specified concentration.

5. Perform acid-base titrations and calculate the concentration of an unknown acid or base.

6. Solve stoichiometric problems involving solutions.

Understanding Chapter 14

REVIEW

1. Explain how you would make each of the following sodium chloride (NaCl) and water solutions. Clearly indicate how you calculate the amount of solute (and solvent if applicable) involved.
   (a) 2.4 kg of solution containing 200 g of sodium chloride per kg of water
   (b) 1.0 kg of a 20% by mass solution
   (c) 10 L (10 kg) of 65 μg/g solution
   (d) 1.0 L of a 3.5 mol/L solution

2. Copy the following chart into your notebook and fill in the missing values. *Do not write in your textbook.*

Solute	Mass (m) in Grams	Number of Moles (n)	Volume (V)	Concentration (c) in mol/L
$AgNO_3$		0.350	500 mL	
NaCl			1.50 L	0.250
NH_4Cl		0.850		3.40
KI	120		2.00 L	
NaOH	9.20			2.30
$CaSO_4$			250 mL	0.0500

APPLICATIONS AND PROBLEMS

3. Enough water is added to 4.0 L of pure ethylene glycol to make up 16 L of solution needed to fill a car's cooling system. What is the percent by volume of ethylene glycol in this solution?

4. Municipal water supplies must have 0.50 μg of chlorine added to each gram of water in order to disinfect drinking water properly. What mass of chlorine must be used to disinfect 2500 m^3 of water used by a small city every day? (1 m^3 of water has a mass of 1×10^6 g.)

5. A 20% by mass solution of calcium chloride (used as an antifreeze solution in some kinds of heat pumps) has a density of 1250 kg/m^3. What mass of calcium chloride is needed to prepare 0.80 m^3 of solution needed to fill such a heat pump system?

6. A pollutant found downstream from Sarnia in Lake St. Clair, OCS (octachlorostyrene), was reported to be present in a water sample at 1.5 mg/g. What mass of OCS would there be in 1.0 m^3 of Lake St. Clair water? (1 m^3 of water has a mass of 1×10^6 g.)

7. A "normal saline solution" used in hospitals for bathing wounds and as an intravenous fluid is 0.850% sodium chloride by mass. How much sodium chloride is needed to make up 1.00 L of this normal saline solution? (Density, ρ, is 1.0043 g/mL)

8. 12.0 L of hydrogen chloride gas measured at 17°C and 110 kPa pressure is dissolved in enough water to produce 500 mL of solution. What was the concentration of the hydrochloric acid solution?

9. What volume of stock solution would be needed to prepare each of the following solutions?
   (a) 250 mL of a 1.00 mol/L hydrochloric acid solution; stock HCl(aq) is 12.0 mol/L
   (b) 500 mL of 4.20 mol/L sulphuric acid solution; stock sulphuric acid is 18.0 mol/L
   (c) 1.0 L of a 0.50 mol/L ammonium hydroxide (aqueous ammonia) solution; stock NH_4OH is 15 mol/L

10. Calculate the concentration in mol/L of a stock acetic acid solution that has the following information on its label:

    % CH_3COOH by mass: 99.5%
    Specific Gravity: 1.05

11. What mass of zinc will completely react with 400 mL of a 1.00 mol/L solution of hydrochloric acid?

12. Calculate the volume of oxygen gas that could be produced from 50 mL of a 9.8 mol/L hydrogen peroxide solution (30% H_2O_2 by mass). Hydrogen peroxide decomposes according to the following equation.

    $$2H_2O_2(l) \rightarrow O_2(g) + 2H_2O(l)$$

13. A 75.0 mL sample of nitric acid, HNO_3, was titrated using a 0.100 mol/L potassium hydroxide, KOH, solution. Determine the concentration of the nitric acid solution if it took 31.5 mL of the KOH solution to reach the end point.

14. In a titration, it took 42.5 mL of a 0.100 mol/L sodium hydroxide solution to neutralize 15.0 mL of a solution of sulphuric acid of unknown concentration. Calculate the concentration of the sulphuric acid. (Remember that each mole of sulphuric acid produces 2 mol of hydrogen ions.)

15. Calculate the mass of aluminum sulphate, $Al_2(SO_4)_3$, required to prepare 300 mL of 0.220 mol/L aluminum sulphate solution.

16. What volume of 0.0300 mol/L sodium sulphate, Na_2So_4, solution can be prepared from 145 g of solid Na_2SO_4?

17. What mass of potassium bromide (KBr) is left if 200 mL of a 1.20 mol/L KBr solution is left to evaporate?

18. Describe how to prepare 100 mL of a 0.0250 mol/L zinc chloride, $ZnCl_2$, solution using a 2.50 mol/L stock solution of $ZnCl_2$.

CHALLENGE

19. What is the concentration of a dilute solution of copper(II) sulphate made by adding enough water to 20.0 mL of 1.65 mol/L copper(II) sulphate solution to make 500 mL of the dilute solution?

20. Five litres (5.00 L) of a solution containing 0.300 g of substance X was found to have a concentration of 1.80×10^{-4} mol/L. Calculate the molar mass of substance X.

21. 200 mL of a 0.500 mol/L sodium hydroxide solution is added to 300 mL of a 0.300 mol/L hydrochloric acid solution. What is the concentration of the sodium chloride solution that results? (Hints: Determine which of the reactants is in excess, and the volume after the two solutions have been mixed.)

22. If 300 mL of a 0.02 mol/L silver nitrate solution were added to 500 mL of a 2.0 mol/L sodium chloride solution, calculate the mass of any precipitate that would form. Refer to the solubility rules in Table 13.1 on page 335 and assume that any insoluble product totally precipitates.

23. You wish to determine the mass of pure acetylsalicylic acid (A.S.A.) in a headache tablet. If the tablet has a mass of 0.33 g and it takes 26.8 mL of 0.0500 mol/L sodium hydroxide solution to reach an end point, calculate the mass of the tablet that is pure A.S.A. What percentage of the tablet is pure A.S.A.? (The molar mass of A.S.A. is 180 g/mol and each mole of A.S.A. produces 1 mol of hydrogen ions.)

24. One possible way of combatting the effects of acid rain on some small lakes is to spread limestone (calcium carbonate) in the lakes to neutralize the acid. The following chemical reaction illustrates the neutralization reaction that occurs.

$$CaCO_3(s) \ + \ 2H^+(aq) \rightarrow Ca^{2+}(aq) \ + \ CO_2(g) \ + \ H_2O(l)$$
calcium   acid
carbonate

The cottagers on a lake which averages 2.0 km long by 1.0 km wide by 2.0 m deep wish to neutralize the acid in their lake. One of the cottagers, an enterprising high school student, brings a sample of lake water to school and performs a titration to determine the concentration of hydrogen ions in the water. The student found that it took 40.0 mL of a 1.00×10^{-3} mol/L sodium hydroxide solution to neutralize 1.00 L of lake water. What is the minimum number of tonnes of calcium carbonate the cottagers need to neutralize the acid present in the lake? (1 t = 10^6 g)

25. Calculate the resultant concentration of sodium chloride in a solution made by mixing 300 mL of a 0.400 mol/L NaOH solution with 200 mL of a 0.500 mol/L HCl solution.

26. (a) 2.1 g of magnesium is added to 250 mL of a 0.50 mol/L phosphoric acid (H_3PO_4) solution. What volume of hydrogen gas collected at 17°C and 105 kPa pressure will be produced from the reaction?

    (b) What is the concentration of phosphoric acid left in solution after the reaction is over?

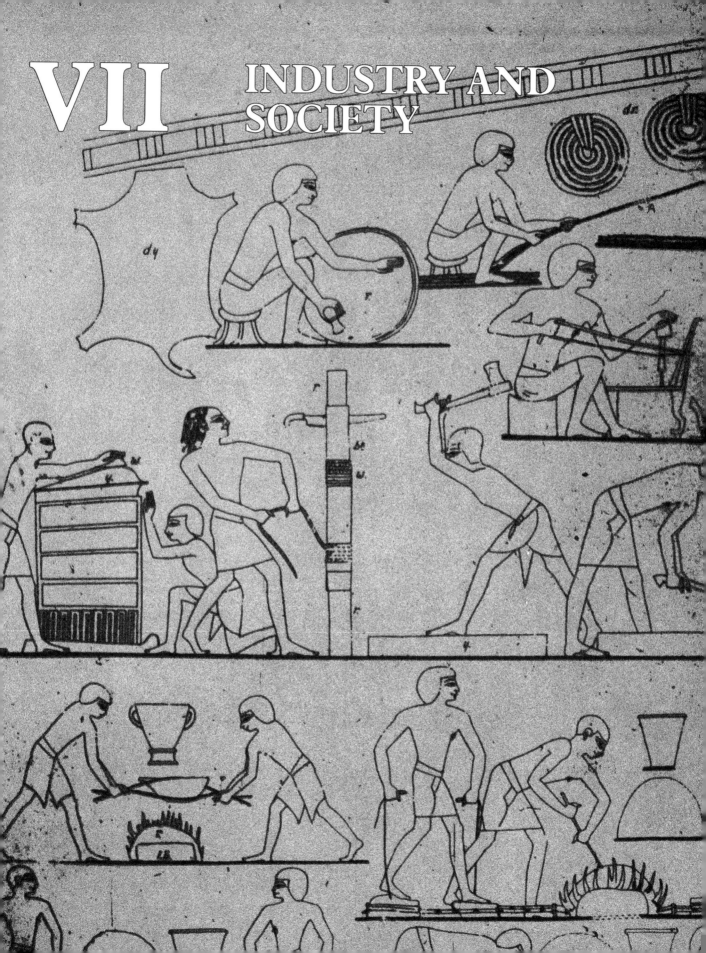

15   SCIENCE, INDUSTRY, AND SOCIETY

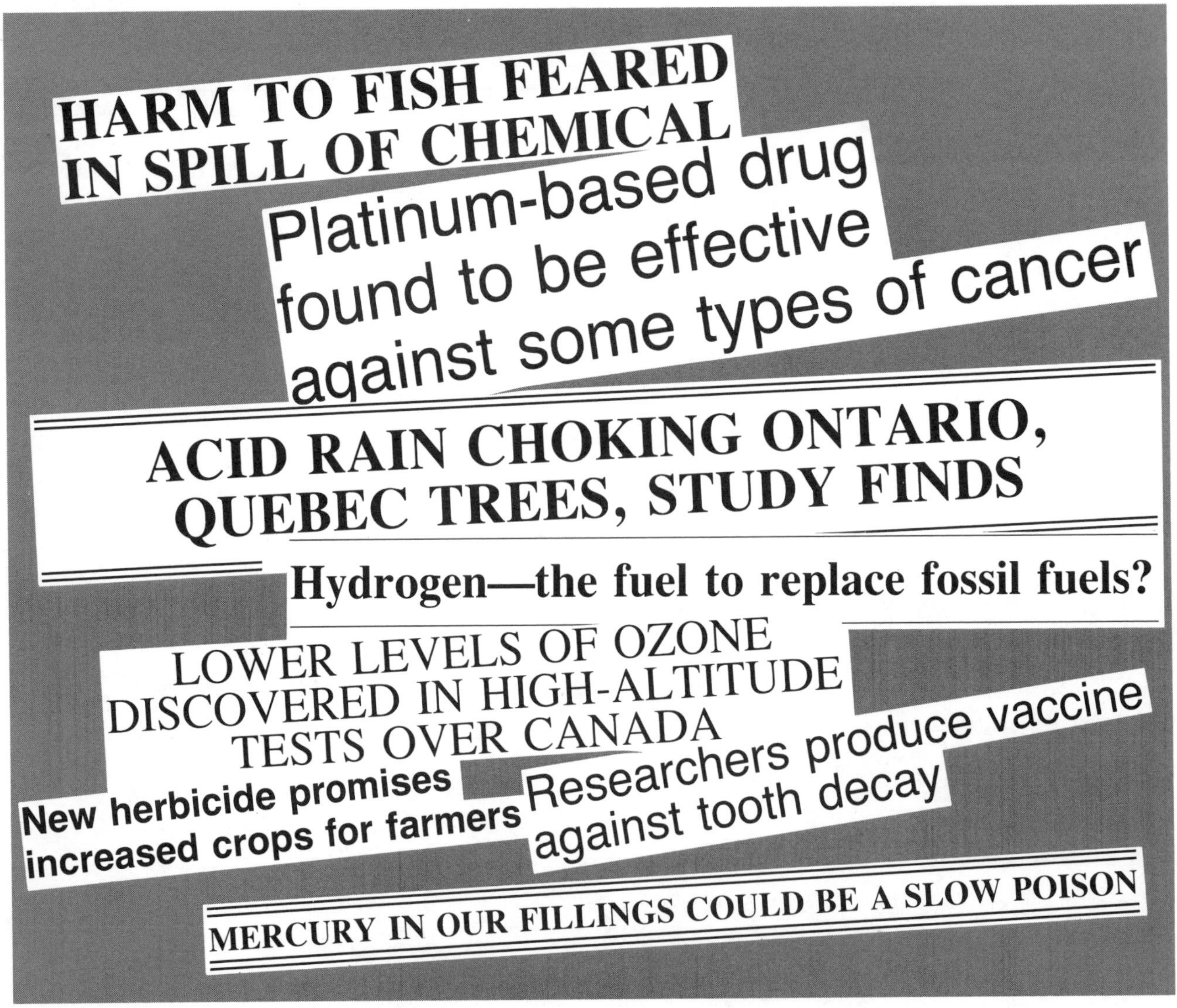

HARM TO FISH FEARED IN SPILL OF CHEMICAL

Platinum-based drug found to be effective against some types of cancer

ACID RAIN CHOKING ONTARIO, QUEBEC TREES, STUDY FINDS

Hydrogen—the fuel to replace fossil fuels?

LOWER LEVELS OF OZONE DISCOVERED IN HIGH-ALTITUDE TESTS OVER CANADA

Researchers produce vaccine against tooth decay

New herbicide promises increased crops for farmers

MERCURY IN OUR FILLINGS COULD BE A SLOW POISON

Try to imagine a world without the products of the chemical industry. Certainly, headlines such as those shown here would disappear from our newspapers. Toxic chemicals such as dioxins, PCBs, and methyl isocyanate would no longer be cause for concern. Environmental issues such as acid rain, mercury poisoning, and the transportation of hazardous materials would not exist. Gone, too, would be such familiar products as paints, plastics, fertilizers, drugs, and pesticides.

Is the kind of world described above possible? Is it desirable? Any serious attempt to answer these questions involves an analysis of the complex relationships among science, industry, and society. This chapter looks at the Canadian chemical industry – its products, natural and human resources, and its benefits and hazards. You will not find many answers in this chapter. But you *will* gain a new appreciation for the need to ask those questions.

15.1 Canada's Chemical Industry

The chemical industry in Canada is composed of two distinct groups: chemical manufacturers and chemical formulators. Chemical manufacturers are involved in the chemical transformation of raw materials into new chemicals. For example, a company that produces ammonia from the raw materials nitrogen and hydrogen is considered to be a chemical manufacturer. Most of the chemicals produced by this group are sold to chemical formulators.

Chemical formulators carry out additional chemical changes on materials supplied by manufacturers. Formulators are responsible for products such as paints, drugs, adhesives, and personal care preparations. The two groups together make up what is referred to as the Chemical and Chemical Products Industries. The types of chemicals produced by these industries are shown in Figure 15.1. Table 15.1 on page 386 lists the annual production of the most important industrial chemicals in Canada.

Figure 15.1
The chemical industry produces a wide variety of products.

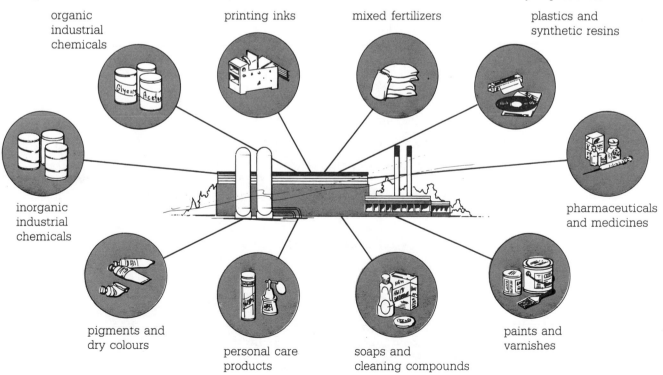

organic industrial chemicals

printing inks

mixed fertilizers

plastics and synthetic resins

inorganic industrial chemicals

pharmaceuticals and medicines

pigments and dry colours

personal care products

soaps and cleaning compounds

paints and varnishes

Table 15.1 Annual Production (in tonnes) of Selected Chemicals for 1986

Chemical	Production (t)
Ammonia	3 540 071
Sulphuric acid	3 536 062
Urea	2 329 720
Ethylene	1 908 731
Sodium hydroxide	1 696 489
Chlorine	1 536 325
Ammonium phosphate	1 149 400
Polyethylene	1 128 961
Nitric acid	1 023 921
Ammonium nitrate	947 808
Benzene	635 378
Propylene	627 721
Phosphoric acid	547 443
Toluene	393 307

Source: *Industrial Chemicals and Synthetic Resins*. Catalogue 46-002 Monthly, December 1986, Statistics Canada.

People and Canada's Chemical Industry

The chemical industry in Canada employs thousands of people in a variety of occupations. (See Table 15.2.) Chemists, chemical engineers, and technicians work in production and in research and development. People with a background in business administration manage the business affairs of the companies. To inform consumers

Table 15.2 Selected Data on the Chemical and Chemical Products Industries for 1984

Number of establishments:	1263
Number of employees:	
(a) in production and related areas	45 568
(b) in administration and other non-manufacturing areas	42 184
Salaries and wages:	
(a) for production	$1 204 029 000
(b) for administration	$1 410 334 000
Cost of fuel and electricity	$1 212 618 000
Cost of materials and supplies used	$9 262 612 000
Value of goods manufactured	$17 174 939 000

Source: *Manufacturing Industries of Canada*. Catalogue 31-203 Annual, 1984, Statistics Canada.

about its products, the industry employs many people in sales and marketing. With a background in the sciences, you may find a career in one of these areas.

Exercise

1. The chemicals listed in Table 15.1 are not normally used directly by the consumer. Instead they are used by chemical formulators to manufacture consumer products. For the first five chemicals on the list, find at least one consumer product manufactured using each chemical.

2. Identify the five chemicals in Table 15.1 that contain the element nitrogen. What important consumer product do you think is manufactured from these nitrogen-containing chemicals?

3. Use Table 15.2 to calculate the following.
   (a) the average salary per production employee
   (b) the average salary per non-manufacturing employee
   (c) the average value of goods manufactured per establishment
   (d) the percentage of the total work force involved in
       (i) production and   (ii) administration

Factors Affecting Chemical Industries

Although there is a great diversity within the chemical industry, the basic objective is the same as that of any other business – to operate at a profit. The money obtained from the sale of the product(s) must exceed the total costs involved. Some of the costs are related to the following aspects:

1. purchasing the land and building the plant
2. obtaining and processing the raw materials
3. transportation of raw materials and product
4. energy use
5. research and development
6. pollution control and waste disposal
7. wages and salaries
8. sales and marketing

The decision to manufacture a chemical on a large scale is not always an easy one. There are two conditions that must be met. First, there must be a demand for the product, either actual or potential. (It is highly unlikely, for example, that there would be a demand for a garlic-flavoured toothpaste, even if it could be shown to absolutely prevent cavities when used once a month!) Second, the manufacturer must be able to show that the enterprise has a good chance of being profitable within a reasonable period of time.

PUTTING CHEMISTRY TO WORK

BONNIE COLDHAM / HUMAN RESOURCES PROFESSIONAL

There are careers associated with chemistry which do not result from a study of chemistry in school. Bonnie Coldham, for instance, is a human resources professional with Atomic Energy of Canada. It is her job to consult with the scientifically trained research staff at Atomic Energy of Canada.

"I'm the people expert," she says with a smile. "We have the brains and expertise here in every aspect of the nuclear industry, but the bottom line is that they are people." Bonnie's work includes matching professional scientists and engineers to the projects underway at AEC. She also helps supervisors trained in science manage their own staff. "I have to look into the future, too," she notes. "I try to place people into jobs that will help them grow and learn so that five years later, for example, they themselves are ready to meet the changing needs of the company."

Working with people — influencing their careers, and in some way guiding their futures — is a great responsibility as well as a rewarding one. "The nuclear industry faces hard economic times the same way any other business does. Sometimes I have to tell people that their jobs are over even though they have been great workers. This is harder than releasing someone who really hasn't performed their job well." But there has been a bright side. "Fortunately, people have become more aware of the need to update their skills, to be curious about new advances in chemistry and engineering. In this way they are ready to adapt when technology changes."

Bonnie sees a great potential in her work. "The business of science requires human resources professionals to provide guidance in managing staff. People are the most important resource in any scientific endeavor." A person interested in becoming a human resources professional needs college or university training in management and personnel. "To work in a science area, such as I do, it is definitely helpful to have some science training as well. I need all of my high school chemistry — and a great deal of 'on-the-job' training — to communicate with the science staff." She has valuable advice to offer: "Learn everything you can, and always seek new challenges."

15.2 Science and Technology

The chemical industry combines the principles of both chemistry and technology. It is important to understand the distinction between these two terms. Chemistry is a science; it is a pursuit of knowledge for its own sake. The "product" of science is new information or knowledge. **Technology**, on the other hand, uses established knowledge to produce something which society considers useful. Technology is an applied science.

An example may help to reinforce the distinction between chemistry and technology. When two different metals are placed in a solution of an electrolyte and connected by a wire, electrons flow from one metal to the other through the wire. The research chemist asks questions such as: Why do electrons flow? What determines the direction of flow? Why do different combinations of metals produce different voltages? Having asked the questions, the chemist then designs experiments to answer them. The technologist is only concerned with the fact that electrons do flow, in a specific direction, with a specific energy that can be measured in volts. The technologist uses the information to produce a device that can, for example, power calculators or flashlights. In this case, technology uses the results of science to produce a dry cell or battery.

In most cases, the science comes first and technology puts the science to use. It is reasonable to think of technology as the link between science and society. (See Figure 15.2.)

Note that the relationships shown in Figure 15.2 are two-way relationships. Advances in the field of technology may in fact allow discoveries in the area of science. As an example of this, you may recall from Chapter 2 that the discovery of the electron and other sub-atomic particles (new scientific knowledge) depended on the availability of equipment to provide high voltages and low pressures (devices of technology).

Figure 15.2
Science and society are related through technology.

Technology and Society

Society usually, but not always, benefits from the results of technology. Most people would regard the development of television, computers, plastics, and polyester as beneficial to society. Thus, society "rewards" the creators of these products by purchasing them. But when the results or by-products of technology are harmful, society (through governments usually) can persuade technologists to devise more acceptable alternatives. For example, western society (particularly in North America) uses huge amounts of energy for heating, lighting, and transportation. The production of this energy results in widespread pollution, which society finds increasingly unacceptable. Technology must then respond by developing methods of reducing this pollution. The electrostatic precipitator was developed as one method of reducing the pollution from smokestacks.

It is easy to blame science and/or technology for many of the problems of modern society, such as the threat of nuclear weapons, air and water pollution, birth defects, etc. But is this fair? Very often, the same scientific discovery can lead to totally different outcomes. Consider the discovery that the fission of a uranium-235 atom releases large amounts of energy. This has led to the technological development of atomic bombs and thus, the potential for nuclear warfare. However, this discovery has also led to the development of nuclear reactors, which now produce a significant percentage of the world's energy needs. (See Figure 15.3.) Who is responsible for deciding which outcome is more important? Should either the bomb or the reactor have been developed? Do the benefits outweigh the risks?

Figure 15.3
The same scientific discovery can lead to two entirely different outcome.

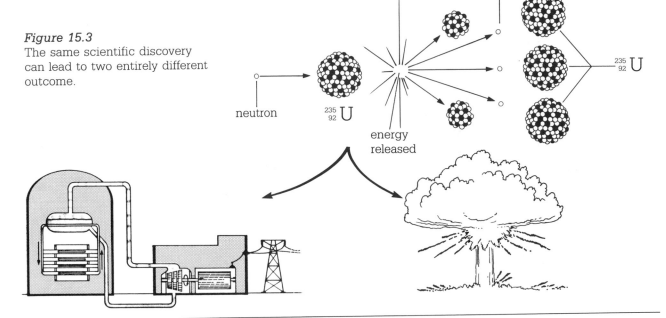

Exercise

4. What is meant by the term nuclear fission?

5. How is nuclear fission different from a chemical decomposition reaction?

6. How is the reaction in a nuclear reactor controlled?

7. In what way(s) does the operation of a nuclear reactor differ from that of an atomic bomb?

8. What dangers are associated with a nuclear reactor?

9. What advantages are associated with nuclear reactors?

10. In your opinion, should the scientists who discovered that a nuclear reaction released huge amounts of energy have kept their discovery quiet? Explain your reasoning.

11. Who do you think should bear responsiblity for the development of the atomic bomb?

12. Do you think that nuclear reactors for peaceful purposes should have been developed? Justify your answer.

15.3 The Design of a Chemical Plant

The starting point in the development of a chemical plant is usually the result of research at a university or at an existing industrial chemistry research laboratory. This research usually involves small amounts of materials, probably less than 50 g in many cases. If the industrial chemist or chemical engineer thinks that the chemical reaction has some potential market value, further investigation will take place. This involves the design of a pilot plant, which is a scaled down version of the final plant.

The Pilot Plant

The design of a pilot plant is one of the most important steps in the development of a chemical plant. The **pilot plant** includes all of the processes that will be carried out in the main plant. It allows chemists and chemical engineers to investigate experimentally each process in the overall reaction, but on a much smaller scale. The pilot plant is not a large laboratory consisting of test tubes, flasks, beakers, and Bunsen burners. Rather, it consists of the equipment, on a smaller scale, that would be found in a full size chemical plant. (See Figure 15.4.) The pilot plant allows the process to be studied economically and safely before a decision on full scale construction is made.

Figure 15.4
A pilot plant is *not* a large laboratory!

During the pilot plant stage, chemists and engineers can experiment with changing certain variables and see the effect(s) on the percentage yield of the product and on the rate of its production. The variables usually include temperature, pressure, the mole ratio of the reactants, the rate at which the reactants are introduced into the reaction chamber, agitation or stirring, and the form of the catalyst if one is used in the reaction. An increase in pressure, for example, may increase the yield of product. It is important to know how much the yield can be increased in this way, because higher pressure requires more expensive equipment. The benefit of increased yield has to be balanced against the increased equipment costs. The major aims of the pilot plant study are as follows:

1. To show that the chemical reaction carried out in the laboratory is feasible on a large scale.
2. To provide data that can aid in the design of the chemical plant.

When the chemists and engineers are satisfied that the reaction can be carried out on a commercial scale and that this can be done profitably, the design and construction of the full scale plant can proceed.

15.4 The Location of Industrial Plants

Approximately 85% of all chemical industry plants in Canada are located in Ontario and Quebec. Why is this the case? In deciding where to locate a plant, a manufacturer must consider many factors, listed as follows.

1. Raw materials: Is there ready access to the raw materials?
2. Transportation: Is the location close to major transportation routes?
3. Human resources: Is there a suitable workforce in the area, for both plant construction and operation?
4. Energy needs: Are sufficient energy resources readily available?
5. Environmental concerns: Can the chemical plant carry out its operations without unacceptable damage to the environment?
6. Water supplies: Is there an adequate water supply in the area?

We will examine each of these factors in greater detail.

Raw Materials

Raw materials are the starting materials in the manufacturing process from which the final product is obtained. For example, aluminum is obtained from the raw material bauxite, gasoline is obtained from petroleum, or crude oil, and chlorine is obtained from sodium chloride. (See Figure 15.5.) There is obviously a relationship between the cost of the raw materials and the cost of the final product.

Figure 15.5
All chemical plants require raw materials.

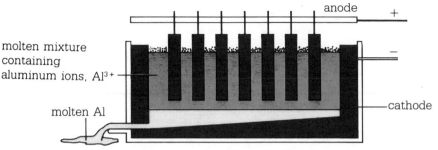

(a) The raw material for making aluminum is aluminum oxide, Al_2O_3, which is mixed with other compounds and electrolyzed at about 1000°C. Aluminum ions are attracted to the negatively charged carbon tub where they gain electrons to become aluminum atoms. Molten aluminum (melting point, 660°C) is run off into moulds. Oxide ions move to the positive electrode where they give up electrons to form oxygen atoms. (The oxygen also reacts with the carbon electrode, and is lost as carbon dioxide and carbon monoxide.)

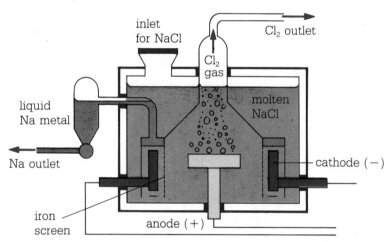

(b) The raw material for making sodium is molten sodium chloride, which is electrolyzed. The cathode is a circular ring that surrounds the anode. The electrodes are separated from each other by an iron screen. During operation of the electrolytic cell, molten sodium collects at the top of the cathode compartment, from which it is periodically drained. Chlorine gas bubbles out of the anode compartment and is also collected.

All raw materials are obtained from the earth, from the atmosphere, or from the oceans. Concentrated deposits of various materials are found in the earth. Besides bauxite, petroleum, and sodium chloride mentioned above, other examples of raw materials from the earth include sulphur, metal ores, and coal. The atmosphere contains nitrogen and oxygen, both of which are used in the manufacture of other chemicals. The oceans are a source of important materials such as sodium chloride and bromine. (See Special Topic 15.1 on page 394.)

CHEMICALS FROM THE OCEANS

As we exhaust our natural resources in the earth, the oceans are receiving more attention as an alternative source of raw materials.

Significant amounts of some resources are already being obtained from the oceans – either from the water or from the bottom of the ocean. For example, about 20% of the world production of petroleum is obtained by drilling into the ocean bed. The ocean floor yields about 10% of the world production of sand and gravel. The sediment on the ocean floor contains aluminum, iron, copper, nickel, cobalt, and titanium compounds. It is not yet economically feasible to mine these sources. However, as the conventional sources become depleted, ocean mining becomes more attractive.

Among the more interesting deposits on the ocean floor are manganese nodules. These are roughly spherical particles, usually a few centimetres in diameter. They are quite rich in manganese and iron, with much smaller amounts of nickel, cobalt, and copper. It is for these last three elements that manganese nodules are mined.

Sodium chloride is one of the most important minerals obtained from sea water. In addition, bromine, magnesium, and magnesium salts are also extracted. In some parts of the world, the oceans are a source of fresh water. In desalinization plants, the water is separated from the salt by distillation. To avoid the use of expensive fossil fuels to heat the water, solar energy is usually used.

Some scientists have suggested using the waste heat from nuclear reactors to distill the water. At the present time, water produced from sea water is more expensive than water from a fresh water treatment plant.

The ocean floor is covered with quantities of tiny, manganese-rich nodules. These nodules are only a few centimetres long.

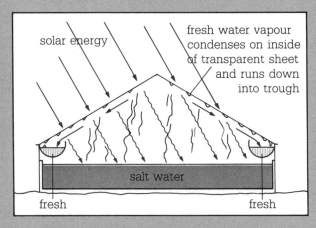

One design for a solar still for producing fresh water from salt. Large salt water ponds are covered by tents of plastic or glass, which keep the solar heat inside. The water evaporates, condenses on the inside of the tent, and drops into the fresh water troughs along the sides. The hot fresh water exits through pipes running through the incoming salt water, helping to heat it in advance.

It is important to recognize that our planet contains limited quantities of the raw materials used by the chemical industry. Canada is particularly blessed with an abundance of **natural resources** – substances that we use to produce energy or useful materials. Resources include minerals, fresh water, fossil fuels, forests, and land suitable for agriculture. As these natural resources are depleted, they are replaced either very slowly or not at all. When all of the earth's supply of bauxite is used up, for example, no more aluminum can be produced from that source. It has been estimated that if we continue to use petroleum at our present rate, we will exhaust the known supplies by the middle of the 21st century. Estimates have been made of the amount of natural resources that are available for our use in the future. These available amounts are known as **reserves**. Although the chemical industry uses large amounts of our natural resources, they do so to meet the needs of society. It is the responsibility of society, not only the chemical industry, to ensure that we manage our resources wisely. If society is prepared to reduce its demand for energy, then the reserves of petroleum, natural gas, and coal will last longer. We can use **renewable resources**, such as water or forests, instead of **non-renewable resources**, such as fossil fuels. The recycling of some materials, aluminum and iron for example, can greatly extend the estimated "lifetime" of our resources. As a society, we must adopt an attitude of conservation toward our valuable natural resources.

Exercise

13. Many chemical plants are located close to the source of the required raw materials. On the other hand, it may be necessary to locate a plant hundreds or even thousands of kilometres away from the source of raw materials. Suggest at least one reason why this may be done.

Transportation

In deciding where to locate a chemical plant, a manufacturer must consider proximity to transportation routes and the location of both the source of raw materials and the market for the product. The raw materials have to be transported to the plant, and the final product must be transported to the customer or distributor. The four main forms of transportation are railway, truck, water, and pipeline. (See Figure 15.6 on page 396.) Transportation by water is relatively cheap. Because of this, many chemical plants are located close to major shipping channels. Pipelines are used to transport oil and gas in large quantities, usually over land, but sometimes under water.

Petrochemical plants, many of which use gaseous raw materials such as ethene or propene, are usually located close to the source of the raw materials. It is more difficult to transport gases than it is to transport liquids or solids. Thus, it is more economical to locate the plant close to the source of raw materials. For example, the raw material for the manufacture of the plastic polyethylene is the gas ethene. Ethene is obtained from petroleum, so polyethylene plants are usually found near petroleum refineries.

Figure 15.6
What are the advantages and
disadvantages of each of these
forms of transportation?

It is sometimes more economical for a manufacturer to locate
close to the market for the product rather than near the raw materials. For example, sulphuric acid plants are normally located near
the market; it costs more to transport sulphuric acid than the main
raw material, sulphur. Sulphuric acid requires elaborate, corrosion-
proof containers. Sulphur can be more easily transported by truck
or railway car.

Exercise

14. Sodium carbonate (soda ash) is manufactured from sodium
    chloride and calcium carbonate (limestone).

$$2NaCl + CaCO_3 \rightarrow Na_2CO_3 + CaCl_2$$

(a) What is the total mass of raw materials required to produce 100 kg of soda ash?

(b) In view of your answer to (a), where would you expect a soda ash plant to be located: close to the market or close to the source of the raw materials? Explain your answer.

Human Resources

Earlier in this chapter, we mentioned that the chemical industry employs chemists, chemical engineers, technicians, administrators, and people in sales and marketing. This represents only a partial list of careers in the chemical industry. We could add other professionals such as biologists, physicists, electrical and mechanical engineers, and accountants. In some of the larger chemical plants, people trained in robotics and computer automation may be required. In addition to these professional positions, a large number of workers are required for the daily operation of the plant. A work force of this diversity is therefore likely to be found only near major centres of population.

Energy Needs

Large amounts of energy are required for the production of many industrial chemicals. In some cases, proximity to a plentiful supply of energy may be the determining factor in deciding where a chemical plant should be located. For example, aluminum is produced from bauxite by passing electricity through a molten mixture of purified bauxite and cryolite. This process requires such a large amount of energy that aluminum producing plants in Canada are located where electrical energy is available in great quantities, and fairly inexpensively. Arvida, Quebec, and Kitimat, British Columbia both have huge hydroelectric generating stations; aluminum smelters are found nearby. (See Figure 15.7 on page 398.)

Environmental Concerns

The chemical reactions involved in changing raw materials into a final product usually result in the formation of other substances in addition to the desired chemical. These by-products may also be useful, in which case they can be separated from the product mixture and sold. For example, during the roasting of nickel ore, sulphur dioxide is produced. This can be separated and sold to sulphuric acid manufacturers. In many cases, however, the by-products of a reaction are of little use and must be disposed of by the manufacturer. In the past, it was common to place chemical wastes in barrels. These barrels were then stored in huge quantities, either underground or on the surface. In many cases, the barrels eventually

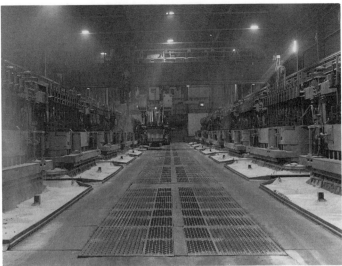

(b) Interior of the aluminum
smelting operation at
Kitimat, British Columbia

Figure 15.7

(a) The power line from Kemano
to Kitimat, in British
Columbia. Without the
electricity carried to Kitimat
by this line, the aluminum
smelter there could not
operate.

corroded, the contents leaked out, and the environment was contaminated. One of the worst examples of this kind of environmental damage took place in Niagara Falls, New York. Hundreds of barrels of chemical wastes were buried and forgotten on the site of a disbanded waterway called the Love Canal. Homes were later built on the site. In the 1970s it became obvious that there were serious health problems in the area. The number of birth defects was much greater than in other areas. Eventually, those homes had to be demolished and the families were moved to a new neighbourhood.

Many chemical plants have, in the past, disposed of liquid wastes simply by allowing the wastes to flow into rivers or lakes. This was a quick and inexpensive way of getting rid of waste products, but obviously added a great deal of pollution to the water. To discourage this illegal method of waste disposal, legislation has been passed to fine companies that dump waste in this manner. Executives of companies found guilty of polluting the environment have even been sentenced to prison terms for continued abuse of the law and of the environment.

In some chemical plants, gaseous products escape into the atmosphere through smokestacks. For example, sulphur dioxide is produced in a nickel smelter. As you know, when this gas combines with water in the atmosphere, acid rain is produced. Enormous amounts of money are now being spent to develop ways to remove the sulphur dioxide before it escapes into the air. (See Figure 15.8.)

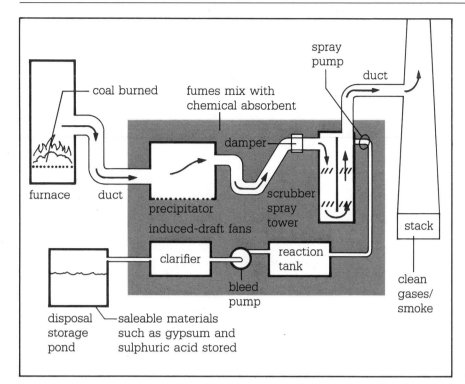

Figure 15.8
One method for preventing acid rain

Many industrial chemical reactions produce heat; they are exothermic reactions. Since the reactions are carried out on a large scale, great amounts of heat are released into the surroundings. Most chemical companies use cold water from a river or lake to cool the products or smokestack gases. The water becomes warm and when it is returned to the river or lake, it raises the temperature of the river or lake by a significant amount. This is called **thermal pollution**. As the temperature of the water increases, the amount of dissolved oxygen in the water decreases. This can affect plant and animal life in the water. Some chemical companies now cool the water before returning it to its source.

Water Supplies

As described in the previous paragraph, many chemical plants use water for cooling purposes. Plants that require large amounts of cooling water are usually located close to large bodies of water. The quality of the water is also important in determining overall operating costs. For example, because of its greater corrosive ability, sea water requires more expensive, corrosion-resistant tubing than fresh water. If the water used by the plant is hard water (containing dissolved salts of calcium, magnesium, or iron), it probably has to be treated before use. Otherwise, the insides of pipes and tubes carrying the heated water gradually become coated with a hard scale. (See Figure 15.9.) The removal of this scale can be expensive and may require a partial or complete shutdown of the plant.

Figure 15.9
This water pipe is almost completely blocked due to the formation of calcium carbonate.

PUTTING CHEMISTRY TO WORK

DR. JIM PIMENTA /
WASTE MANAGEMENT SPECIALIST

To properly manage wastes – and prevent pollution – people must learn and study the complex interactions among chemicals and the environment. Unfortunately, poorly managed industrial wastes, particularly toxic chemical wastes, are the current major source of pollution in the environment. Jim Pimenta is a government specialist in the field of industrial waste management.

"New regulations monitor an industry's waste 'from cradle to grave'," Jim says seriously, "that is, from the industry itself (the waste generator) to the final disposal site of the waste. This is essential if we are to prevent any pollution from industrial chemical waste." Jim's work consists of analyzing samples of industrial waste, sewage, and water from observation wells at landfill sites.

One vital aspect of Jim's job is to ensure that hazardous waste is disposed of in a way that is safe for the environment. "For example, at landfill sites I analyze leachate caused by rain water." Leachate is any water soluble chemical carried away from its original location by moving water. The presence of dangerous chemicals in the leachate is a warning that improperly disposed of waste might be contaminating the ground water – and drinking supplies.

Sewage sludge, the indigestible remains from wastewater treatment plants, is sometimes used as fertilizer since it is high in usable nutrients for plants. Jim cautions: "Sludge must first be checked for contamination by metals such as mercury or chromium. Otherwise these toxic elements could get into the food web through crops." If an industry is found to be willfully polluting the environment, Jim gives expert testimony in court. "Penalties for polluting used to be so trivial that they were ignored, but

that's no longer the case. Now the top executives of a guilty company face large fines and even jail terms."

Proper waste management may be the ultimate solution to human pollution of the environment. "As people become more aware of the hazards of industrial waste materials, the need for practical research such as I perform will expand tremendously," Jim notes. "Both industry and government will require trained specialists to provide real, workable solutions to industrial waste management problems. After all, we owe future generations a safe environment in which to live."

Other Factors

Other questions that must be answered before deciding where to locate a chemical plant include the following. Is the climate of the area suitable? Is land available near the proposed site at a reasonable cost? Is there any incentive offered by governments to locate in a particular region? Is there a recent history of labour unrest in the area? Are the general living conditions in the area suitable for the people involved who will be relocating to the area? If the plant is to be located in another country, what is the rate of import and/or export duty that must be paid? How stable is the government of that country? What is the exchange rate of the country's currency?

The final decision on where to locate a chemical plant involves a careful analysis of all of the factors discussed. Plant planners cannot single out any factor as being the most important, although they often give more weight to the location of the raw materials and the markets for the product, along with transportation costs. Sometimes the factors oppose each other. For example, it may be considered important to locate a plant close to a large population centre, because of the labour force requirements, and the need to be close to transportation routes. The source of raw materials, however, may be a great distance away, and this would obviously increase the costs for the transportation of the raw materials. The final decision really involves a compromise among all of the factors discussed.

Exercise

15. Select a chemical plant situated in or near your community or a plant with which you are familiar. Briefly outline how you think the factors discussed in section 15.4 played a part in determining the location of the plant.

15.5 Research and Development

The chemical industry is a very competitive industry. A particular product will sell if the price and quality are attractive to the buyer. Research departments try to find methods to decrease production costs and enhance the quality of the company's products. Researchers also devise new production processes for existing products. They design production processes for new products, discover new uses for existing products, investigate methods to reduce wastes and pollution, modify existing products to satisfy the needs of customers, and support sales staff by offering technical assistance to customers. The role of an industrial research chemist is varied and can be extremely challenging.

Research is critical for the continued success of any chemical company. It is only by being committed to producing a better and/or cheaper product that a company can remain competitive or even survive. Without research, the chemical industry in Canada would stagnate.

Factors Affecting Research

Size

Most chemical companies in Canada are small. In 1984, according to Statistics Canada, more than 50% of the 1263 companies in the Chemical and Chemical Products Industries employed fewer than 20 people. Only 15% of the companies had more than 100 employees, and slightly more than 2% had more than 500 employees. In small companies, research departments, if they exist at all, tend to be small, with limited funding. The research carried out may be restricted to only one or two small areas.

Foreign Ownership

Many of the larger chemical companies are subsidiaries of foreign owned corporations. These corporations are often unwilling to duplicate in Canada the full research and development facilities of the parent company. The Canadian production plants are expected to make use of the research and development carried out by the parent company.

Government Funding

Until recently, government in Canada did not seem to regard research and development as a high priority item. For example, the United States spends about 2.8% of its Gross Domestic Product (GDP) on research and development. Canada spent only 1.35% of GDP, according to 1986 Statistics Canada figures. This places Canada 19th among 24 western industrialized nations, based on percentage of GDP spent on research and development. Since 1975, the federal government's share of total funding for research and development has dropped from 43% to 34%.

Because of the three factors mentioned above – namely, the large number of small companies, foreign ownership, and limited government funding – the amount of money per capita committed to research and development in Canada has been less than one half of that spent in the United States.

15.6 Making Decisions

The benefits obtained from some chemicals must often be balanced against the dangers associated with the production, transportation, and use of those chemicals. The dangers of explosions in chemical plants, spills during transportation, and waste disposal are all very real concerns for society. (See Figure 15.10.)

Decisions on whether to allow the production and transportation of dangerous chemicals are not easy. In 1985, a pesticide producing plant owned by Union Carbide in Bhopal, India accidentally released the deadly compound methyl isocyanate into the atmosphere. More than 2000 people were killed and about 60 000 more were injured. (See Figure 15.11.) Even if catastrophic accidents like this could be prevented, toxic substances would still find their way into the environment through carelessness and unavoidable leaks.

The pesticide industry is an excellent example of an industry facing a dilemma. Hundreds of people are adversely affected every year by exposure to pesticides and related chemicals. Health authorities suspect that the list of injuries and deaths is even longer, but it is difficult to prove a link between a particular illness and exposure to pesticides. In a world already short of food, however, we do need these products to control insect populations. Otherwise, crops would be destroyed, and millions of people would die of starvation.

Figure 15.11
The Union Carbide pesticide plant at Bhopal, India. The failure of a valve on an underground storage tank resulted in the release of deadly methyl isocyanate into the air. The cloud of poisonous gas drifted through the residential area that had been established just outside the plant.

Figure 15.10
In November 1979, a freight train pulling tank cars of propane and chlorine derailed in the heart of Mississauga, Ontario. The propane caught fire and the escape of an entire rail car of extremely poisonous chlorine was feared. The city was evacuated until the threat was over. Do you think the transportation of dangerous chemicals should be permitted through heavily populated areas?

403

PUTTING CHEMISTRY TO WORK

TOM KOBOLAK /
CANADA CUSTOMS TARIFF ADMINISTRATOR

All materials and goods which cross the border into Canada must first pass through Canada Customs. This is due to trade and tariff (tax) regulations, and also because government restrictions on certain products differ from country to country. Tom Kobolak is a tariff administrator in the Chemicals and Plastics Unit in Ottawa.

"What I do is to classify and value imported goods," Tom explains. "This may sound simple enough, but those goods can be anything — including, for example, plastics, pharmaceuticals, oils, adhesives, or any sort of chemical." A person or business firm wishing to import a previously unclassified plastic or chemical must contact Canada Customs.

Tom began his career with Canada Customs as an inspector. "One of my most vivid memories of being an inspector is of the time I jumped briskly into someone's van to inspect it, only to awaken a very large Great Dane. Although the dog turned out to be friendly, I realized then I should perhaps work in a slightly different area of customs."

As a tariff administrator, Tom uses all of his university and high school education. "I have to be able to read highly technical material and to understand it thoroughly," Tom says. "What I particularly enjoy about this job is the research I have to do when contacted about some new substance or product." Most tariff officials have a university education, and, depending upon the area of Customs worked in, their university degrees vary from business or public administration to science, such as in Tom's case. I have to use my chemistry knowledge everyday."

The chemical industry is fast growing and complex. "I have to keep up-to-date with chemical products and that can be a real challenge when even the manufacturing processes themselves are changing rapidly. I'm quite literally still "in school" all the time. Because of this, I would say the most important skill a student could bring to any aspect of chemistry would be the ability to manage her or his time. If you learn to be well organized and disciplined, you will be able to keep ahead in this vast and interesting field."

If some dangerous chemicals are so essential that they must be produced, perhaps the best decision would be for society to insist on stringent safety regulations and procedures for their production, transportation, and use. This means that it would be expensive to build, maintain, and operate chemical plants and transportation facilities, resulting in a more expensive product. Is society willing to pay for the increased costs? It is tempting to consider manufacturing dangerous chemicals in some third world countries where there are less stringent regulations and chemical plants would be less expensive to operate. Should we give in to this temptation?

Society must make decisions concerning the production, transportation, and use of chemicals based on knowledge and understanding. It is the responsibility of every citizen in our increasingly technological society to obtain the scientific background necessary to make such decisions concerning the chemicals that we require.

Exercise

16. Carry out the following investigation in groups of four or five students.

    Chlorine is an extremely poisonous substance. Imagine that a proposal has been made to abandon its production and use. Identify the costs and the benefits of such a ban. Make a decision to either support or oppose the proposal. Your decision must be backed by facts, not emotion.

Looking Back – Looking Ahead

In this chapter, you have seen how science, technology, and society are interrelated. The importance of the chemical industry to the Canadian economy has been demonstrated in terms of both people and dollars.

You should now know something about the steps involved in developing a chemical manufacturing plant. In addition, you should now have a greater appreciation of the need to be informed so that decisions regarding the production, transportation, and use of dangerous but essential chemicals can be made.

In the next chapter, you will have the opportunity to prepare a report on a specific chemical industry. You will have to consider many of the issues that have been discussed in this chapter, and find out how the industry you have chosen has dealt with them.

Vocabulary Checklist

You should now be able to define or explain the meaning of the following terms.

natural resources	pilot plant	reserves
non-renewable resources	raw materials	technology
	renewable resources	thermal pollution

Chapter Objectives

You should now be able to do the following.

1. Discuss the importance of the chemical industry to the Canadian economy.

2. Explain how science, technology, and society are interrelated.

3. Demonstrate your knowledge of the variety of careers available in the chemical industry.

4. List several factors that influence the decision as to where to locate a chemical plant, and explain how each factor plays a part.

5. Discuss the importance of research and development in industry.

6. Outline the steps involved in the development of a chemical plant.

7. Distinguish between natural resources and reserves, and explain how the wealth of a nation depends on its natural resources.

8. Discuss the need for the wise management of our natural resources.

9. Discuss some of the environmental concerns associated with the chemical industry.

10. Think critically about the continued production and use of some chemicals considered to be dangerous.

11. Discuss the need for citizens to be knowledgeable about science.

Understanding Chapter 15

REVIEW

1. Explain the difference between a chemical manufacturer and a chemical formulator.

2. List five of the most important industrial chemicals in Canada, and give one example of its use for each.

3. A chemical company has only one source of income – the sale of its product. There are many costs involved in manufacturing a chemical. List six of these costs.

4. Explain the difference between science and technology.

5. State two of the major functions of a pilot plant.

6. What are the major sources of raw materials used by the chemical industry?

7. (a) What is meant by thermal pollution?
   (b) What causes this type of pollution?
   (c) What are some of the consequences of thermal pollution?

8. Research and development departments within chemical companies are responsible for a number of functions. List three of these functions.

9. (a) What is the difference between renewable resources and non-renewable resources?
   (b) List three examples of each type of resource.

10. Name four factors affecting chemical reactions that can be investigated during the pilot plant stage of a chemical production process.

11. Some chemical plants are located close to the source of the raw materials needed for the chemical reactions. Other plants are located close to the market for the final product. What influences the decision whether to locate the plant closer to the raw materials or closer to the market?

APPLICATIONS AND PROBLEMS

12. Approximately 85% of all chemical plants in Canada are located in Ontario and Quebec. Suggest a number of reasons for this fact.

13. List three examples of technological discoveries that are considered to be (a) beneficial to society, and (b) harmful to society.

14. In what way is a pilot plant in the chemical industry different from (a) a chemistry laboratory, and (b) a production plant?

15. Why is it important that we use our natural resources wisely?

16. The industrial production of almost every chemical involves some degree of risk to the environment.
    (a) Give three ways in which the environment may be affected.
    (b) Why does society usually accept the risks associated with the production of chemicals?

17. (a) What is meant by "hard" water?
    (b) Why is hard water considered to be a problem in chemical production plants?
    (c) Find out one way of treating hard water to make it soft.

18. What do you think is meant by the following statement: The annual production figures for sulphuric acid are often considered to be an economic indicator for the state of an industrial nation.

19. Ammonia, NH_3, is manufactured from the raw materials nitrogen and hydrogen according to the following equation:

    $N_2 + 3H_2 \rightarrow 2NH_3$

The reaction is carried out at high temperature and pressure and in the presence of a catalyst.
(a) What is the maximum mass of ammonia that can be produced from the reaction of 127.5 kg of ntirogen with excess hydrogen?
(b) If 45.50 kg of ammonia is actually produced, calculate the percentage yield.

CHALLENGE

20. Why do you think it is important that all citizens should be knowledgeable about chemistry, at least to some degree?

21. Gold is found in sea water in extremely low concentrations. At the present time, it is not economically feasible to extract the gold. Imagine that you have just discovered a method of isolating gold from sea water using chemical reactions. Your method is relatively inexpensive. Assuming that your method does work on a large scale, outline some of the considerations that are important in deciding where to locate the plant.

22. Imagine that you are the president of a company that wishes to build a chemical production plant in a mainly non-industrial town. You are aware that there will be some opposition from the local government and from the town residents. Outline a report that you would make to the local council, pointing out the advantages of your project for the area.

23. Suppose that you are the spokesperson for an environmental group for the town referred to in question 22. Your group is opposed to the construction of the chemical plant. Outline a report that you would make to the local council, pointing out the disadvantages of the project for the area.

24. Find out what the raw materials and uses are for the following:
    (a) sulphuric acid
    (b) sodium hydroxide
    (c) urea
    (d) nitric acid
    (e) ammonium nitrate

16 UNDERSTANDING CHEMICAL INDUSTRIES

Throughout this textbook, you have considered many examples of the interplay of science, technology and society. You yourself are an important and necessary part of this interrelationship. As a student of science you have learned about some of the methods scientists use to obtain and organize information, and about the ways this information is used to generate new scientific knowledge and theories. You have also learned a great deal about matter, its composition, and its physical and chemical properties.

As a human being in an age of rapid technological development, your knowledge of science and of the methods of science will have an impact on the development of your careers. You live in an environment surrounded by materials both beneficial and harmful. Should some chemicals and processes be banned? What benefits would be lost if they were banned? As a member of society, the decisions you make and the methods you use to promote the acceptance of these decisions can influence your own health, safety, and quality of life. Your knowledge of the relationship between science and technology, and the implications it holds for society, can make you a more informed and concerned citizen better able to work for the protection and improvement of the quality of life for all.

This chapter gives you an opportunity to examine further the relationship among science, technology, and society. It is designed to help you prepare a report on a chemical-based industry. Unlike the other chapters in this textbook, this one does not contain exercises or questions at the end of the chapter. Instead, in section 16.1, you will first be given a brief overview of industries that rely on chemistry. This will allow you to consider an industry for your research report.

One particularly important industry is missing from section 16.1 – the petroleum and natural gas industry. Since materials based on petroleum products are extremely common, we have chosen to discuss this industry in more detail in sections 16.2, 16.3, and 16.4. This more detailed discussion is not exhaustive, nor is it intended to be. Its purpose, rather, is to provide background information for the *Possible Topics for Further Investigation* you will find in these three sections. You may wish to choose one or several related topics for a report on the industry. Section 16.5 contains useful guidelines and hints for producing an informative and successful report.

16.1 Chemical-Based Industries

Chemistry is concerned with matter – its composition, its properties, and the changes it undergoes. Therefore, we can say that almost every industry that handles or processes matter in most forms is involved in chemistry. Table 16.1 lists some of these industries and shows some of the chemical materials and processes that they handle.

Table 16.1 Selected Industries and their Involvement in Chemistry

Industry	Product, Raw Material, or Process	Possible Areas of Investigation

Cleaning Products	Soaps and detergents
	Fabric softeners
	Bleaches
	Dry cleaning fluids
	Spot removers
	Metal cleaners
	Dirt preventers:
	  Scotch Guard
	  tarnish remover
	Waxes and polishes
	Glass cleaners
		Chemical composition, action, and pollution considerations
		
		$\longrightarrow$

Construction	Lumber and wood products
	Sheeting:
	  styrofoam
	  sheetrock
	  siding
	Pipes, wire, paint, cement, roofing etc.
	Glass, fibreglass insulation
	
	Cement setting
	Plaster hardening
	Paint "drying"
	Glues and adhesives setting
		Properties related to composition

Cosmetics	Perfumes	Sources of raw materials Chemical structure of these compounds

	Colours:
	  lipsticks
	  blushes
	  hair dyes
		How they work

	Hygiene:
	  soaps and shampoos
	  deodorants
	  toothpastes
	  mouthwash
		Compounds involved and their action

	Skin & hair treatment:
	  skin cream
	  bath oil
	  hair conditioner
	  acne ointments
		Compounds involved and their action

Electronics	Microchips
	Semiconductors
	Components:
	  transistors
	  picture tubes
	  speakers
		Production
		
		Elements involved
		
		How they work

Food production	Fertilizers	Types, essential elements, pollution

	Animal feed and additives	Types, effect on animal growth, residual effects
	Pesticides: herbicides insecticides	Types, chemical composition, effects, pollution
Food processing and packaging	Preservatives Enhancers: taste and flavours texture appearance sweeteners	Types, chemicals involved, action, effects on humans
	Packaging	Materials and methods (freezing, canning, radiation, sterilization etc.)
	Cooking fats and oils	Production and properties
	Prepared foods: cheese ketchup Synthetic foods	Production
Heavy chemicals	Sulphuric acid Sodium carbonate Sodium hydroxide Chlorine Salts	Production and use
Manufacturing Industries	Automobiles Appliances Furniture Electronics	Materials and processes used
Mining and refining	Mining Concentrators Refineries Explosives	Processes and chemicals used
Pharmaceuticals	Drug types Sources of raw materials Production (a specific product)	Chemicals involved, their production or purification; How they work
	Testing the product	Analysis procedures
	Research and development	Steps in bringing a new drug to market Patent protection →

Pulp and paper	Pulp Paper: bleaching binding	Processes Chemicals used and properties needed
Steel industry	Blast-furnace Steel-making	Raw materials and the chemistry involved
	Alloys	Composition and properties
	Coatings: electroplating electropainting	Chemicals and processes used to clean and coat steel
Textiles	Natural fibres	Processing, preparation, and properties
	Synthetic fibres	Production and raw materials
	Blended cloth Natural/synthetic fibre combinations	Properties

16.2 The Petroleum and Natural Gas Industry

Laboratory Manual
Experiment 16-1 will give you an opportunity to perform chemical reactions that are used in a particular chemical industry. You will also prepare a report on the steps needed to develop and manufacture products related to these chemical reactions.

The petrochemical industries, as the name suggests, are the chemical industries based on petroleum. Oil refineries and the petrochemical plants that grow up around them are the largest scale physical and chemical processing plants.

Petroleum, natural gas, and petrochemicals are immensely important to Canada and the rest of the world. In 1985, Canada produced more than 7.2×10^7 m^3 of crude oil and 5×10^{10} m^3 of natural gas. The prosperity of oil-producing regions such as Alberta depends on the price of oil and gas. On the world scene, the oil market can make or break the economies of countries like Mexico and many smaller, third world producers such as Libya or Nigeria.

Most of the petroleum and natural gas produced is burned in the engines and furnaces that keep our society functioning. Tractors, cars, planes, factories, businesses, and homes use petroleum products and gas as their sources of energy. But not all oil and gas goes up the chimney or out of the exhaust pipe. In Canada, more than 5% of all oil and gas produced is used to manufacture petrochemicals. Thousands of products such as plastics, paints, solvents, and even foodstuffs are derived from crude oil and natural gas. (See Figure 16.1.)

Because of its importance, the petroleum and natural gas industry will be considered in much greater detail here than the other industries listed in Table 16.1. The following discussion is intended to be a starting point for individual or group investigations into many of the petroleum based sub-industries. The industry itself can be divided into three segments. The natural gas industry deals with gaseous materials. The petroleum or oil industry processes mainly liquids. The petrochemical industry deals with both gases and liquids as raw materials for the manufacture of other chemicals.

Petroleum

It is generally believed that slow chemical change deep down in the earth has produced much of the treasure we know as crude oil and natural gas. Sediments containing the remains of plants and animals were subjected to high pressure and temperature for millions of years. Gradually, the complex molecules of the organisms were converted to much simpler molecules containing only carbon and hydrogen. These molecules, called hydrocarbons, are found in the porous rock deposits that underlie oilfields. (See Figure 16.2.)

Figure 16.1
Uses of oil and natural gas

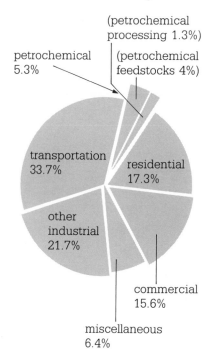

(petrochemical processing 1.3%)
petrochemical 5.3%
(petrochemical feedstocks 4%)
transportation 33.7%
residential 17.3%
other industrial 21.7%
commercial 15.6%
miscellaneous 6.4%

Figure 16.2
Oil and gas deposits consist of oil- or gas-soaked porous rock formations. The core samples shown here illustrate the different types of rock that could contain oil or gas. The crude oil contained in these deposits can range from very heavy, tarry substances to very light, thin liquids.

Peat and coal also were formed from matter that was once living. Oil, gas, peat, and coal are all therefore known as fossil fuels. Unlike peat and coal, oil and gas can percolate upward through gravel and porous rock. An oil or gas field is formed if a layer of solid, impermeable rock stops this upward flow. In some cases the oil flows all the way up to the surface and can soak into surface sands.

Crude oil deposits are tapped by drilling wells down from the surface. Sometimes the oil flows upward in the drill hole due to underground pressure, producing what is known as a "gusher". More often the oil must be pumped out. (See Figure 16.3.) Natural gas always flows upward on its own.

Figure 16.3
A number of large pumps are needed to raise heavy oil to the surface.

In some hydrocarbon deposits the oil is locked up in thin films between rock layers. These deposits are called oil shales. Other deposits consist of a mixture of sand, water, and oil called tar sands. In tar sands, water surrounds each individual sand grain. A film of oil in turn surrounds the wet sand particle. The world's largest deposits of tar sands lie in northern Alberta. To get the sand from the ground, expensive equipment such as that shown in Figure 16.4 is required. Further resources (both monetary and human) are needed to separate the oil from the sand. Huge volumes of undesirable waste sand (called tailings) are produced along with the oil. The "synthetic crude oil" produced in this way is therefore expensive. Investors will not commit money to build such plants if the price of oil is low.

Many physical and chemical processes are used in producing and cleaning up natural gas and crude oil. Some natural gas wells produce gas that is contaminated with undesirable sulphur compounds. These must be removed before the gas can be shipped to market. Older, depleted oil wells produce very little oil by conventional pumping. However, additional amounts of crude oil can be obtained

Figure 16.4
The Syncrude Plant at Fort McMurray, Alberta. Very costly plants and
earth-moving equipment are needed to produce "synthetic" crude oil
from tar sands.

from them by using chemical or physical recovery techniques. Steam
heating to promote faster oil flow (see Figure 16.5 on page 417) is
just one of many methods currently used or being investigated.
Recovery of very heavy crude oil depends on complex and expensive
techniques right from the start.

Much of Alberta's tar sands lie under thick layers of soil. These
tar sands are considered resources rather than reserves because it
costs more to produce crude oil from them than the oil would be
worth at present market values. Undeveloped oil fields such as the
Hibernia field under the ocean off the coast of Newfoundland are
also resources that become reserves when enough investment has
been committed to bring them into production.

MINING OIL

Petroleum engineer Leo Ranney, who pioneered the mining for oil process some 50 years ago

Some fifty years ago, Leo Ranney, a petroleum engineer living in Petrolia, Ontario, pioneered and patented an oil drilling process involving horizontal drill holes rather than the more conventional vertical wells. The relatively inexpensive and easily obtainable oil supplies available since that time did not encourage the development of Ranney's ideas until the more recent oil price increases. A small oil company, Devran Petroleum, supported both by a major oil company (Shell) and government grants and incentives, is developing Ranney's method by combining the technology of both the mining and the oil industries.

Devran hopes to revive the depleted oil fields near the location of Canada's first commercial oil well (in 1857) in the Petrolia–Oil Springs, Ontario region. The procedure being developed involves three major steps.

1. A mine shaft is sunk to the oil formation.

2. A production station is excavated at the point where the shaft passes through or reaches the oil containing rock formations.

3. The production station becomes the site from which holes are drilled horizontally in a "wagon wheel" pattern into the rocks holding the oil.

Each production station with its set of wells drains several hundred acres of oil fields. Oil flows through the horizontal holes and is collected underground. It is then pumped to the surface for treatment, storage, and transport to refineries.

This "mining of oil" technology has important implications. Depleted oil fields can be brought into production again and the life of current fields extended. Heavy oil areas and other reservoirs where oil does not flow easily or cannot be pumped from the surface may also become productive using this combination of mining and oil production technologies.

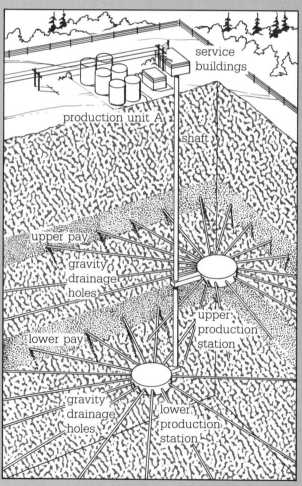

An "oil mine"

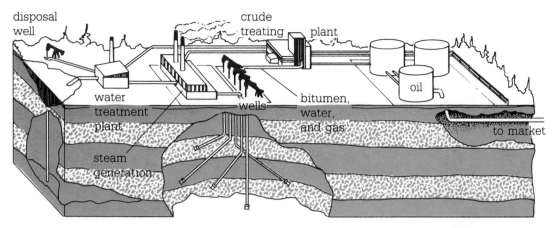

Figure 16.5
Steam injected into heavy oil bearing formations heats up the oil, enabling it to flow more easily. The oil moves toward the producing wells and is pumped to the surface.

Transportation

Fluid hydrocarbons such as natural gas and crude oil are transported most efficiently by pipeline. Even the oil produced from the sea bed is pumped ashore via underwater pipelines. Canada has a large and complex network of pipelines connecting often remote production facilities to markets thousands of kilometres away. (See Figure 16.6.)

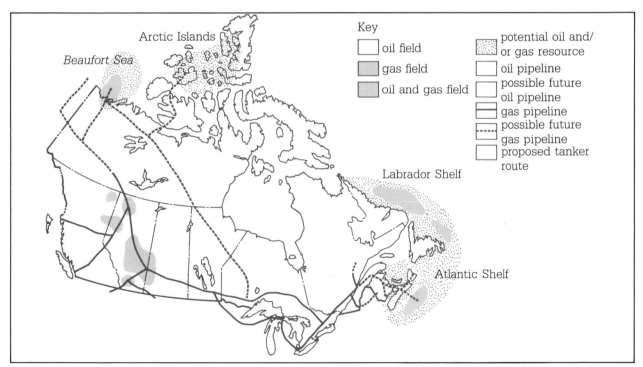

Figure 16.6
Canada's oil and gas pipeline networks. In Canada, the major oil and gas production occurs in Alberta. Central Canada is the major market for these products. Vast pipeline networks are needed to transport millions of cubic metres of oil and gas per day, thousands of kilometres across the continent.

Smaller quantities of liquid products such as gasoline, liquefied propane, and furnace oil are delivered to the consumer by truck or rail tanker. Ocean tankers and supertankers are used to transport large quantities of crude oil for export to countries half a world away.

Transporting flammable fluids is a risky business. Accidental spills, fires, and explosions are an almost daily occurrence. Oil slicks damage oceans and beaches. During the transportation and storage of these liquids and gases, explosions and spectacular and often deadly fires break out on highways, rail lines, and in towns and cities. (See Figure 16.7.)

Figure 16.7
Accidents and spills during the transportation of petroleum and natural gas can cause problems ranging from the fouling of beaches to catastrophic events like the natural gas explosion which levelled much of the downtown core of Essex, Ontario on Febuary 14, 1980.

Good equipment and training can reduce the frequency of accidents and minimize their severity. Perfection, however, is never possible. Society must continue to balance the risks of transporting and storing petroleum and natural gas against the benefits of their uses.

Possible Topics for Further Investigation

1. Production of synthetic crude oil from either tar sand or oil shale deposits

2. Continued production of oil from depleted oil wells

3. Production and upgrading of heavy crude oil to a product which refineries can more easily handle

4. "Sweetening" of sour (sulphur containing) natural gas

5. Transportation of petroleum and its products

6. The time it takes for the formation of hydrocarbons:
   (a) Depletion of natural (non-renewable) sources
   (b) Possible economic synthesis of hydrocarbons from plant materials (renewable sources)

PUTTING CHEMISTRY TO WORK

PAUL DRUET / PRODUCTION ENGINEER

Mention oil and gas fields to people and many will think of Alberta or the Arctic, but there are also large reserves of both oil and gas resources in Ontario. Paul Druet manages the development and operation of all the on-land fields owned by Consumers Gas Company. His job is to ensure that as much of the oil or gas as possible is removed from the ground.

"As oil becomes more scarce and so much more valuable," Paul says, "we have to look to engineering to expand our supplies." Paul is responsible for making sure that his company gets the maximum return from its investment in the exploration and development of a field. This involves new ways to use reserves such as tar sands which were once thought too difficult to obtain. It also includes going back to fields which were considered exhausted and closed. "In the past," Paul explains, "a typical oil reservoir recovered only about 20% to 40% of the oil, leaving from 60% to 80% in the ground. Today it makes economic sense to develop ways of getting out that remaining supply."

Paul entered university without a clear goal. "Like a lot of my fellow students," Paul remembers, "I really did not know what I wanted to do. I did know I enjoyed my science-oriented classes. Fortunately, I discovered that engineering included all of the aspects of science that I loved, combined with the challenges of applying science to real situations." Paul chose to become a geological engineer.

One of the pleasures for Paul in his job is that he uses different aspects of science all the time. "Oil, gas, and/or water can be found within rock (called a reservoir) either separately or together. I need chemistry to understand the way their different phases interact," Paul comments. "In addition, I need physics to predict the flow of these fluids through rock as well as through any pipes that are installed. And mathematics is vital when I make predictions based upon models of each situation." Paul also draws upon his English courses. "About 80% of my work is writing reports and communicating directions to personnel."

Paul's work provides him with the satisfaction of using his skills to fulfill an important role for his company and its customers. "The nice thing about oil field engineering is that, since most fields last from 20 to 30 years, I can make a prediction about a field's production and, although I go on to other projects, I can keep track of how good my prediction was."

16.3 Petroleum Refining

Fractional Distillation

The boiling points of similar hydrocarbons depend on their molecular masses. In general, the greater the molecular mass, the higher the boiling point. Hydrocarbons have specific names depending on the number of carbon atoms they contain and how they are arranged in their molecules. The lightest hydrocarbons, methane and ethane, have boiling points of $-161.5°C$ and $-88.6°C$ respectively. These compounds are gases at room temperature and pressure and are often found in permeable rock above crude oil deposits. They are collected directly from gas wells. Propane and butane, with boiling points of $-42.1°C$ and $-0.5°C$ respectively, may also come off with the natural gas. More often, however, propane and butane are dissolved under pressure in crude oil. These substances tend to boil out of petroleum as it is pumped to the surface. For many years they were considered to be waste products and were simply burned off at the oil well!

The hydrocarbons in crude oil are separated in refineries. Oil refineries are, essentially, large fractional distillation units. (See Figure 1.13 in Chapter 1.) Crude oil contains hundreds of different hydrocarbon compounds. A distillation unit separates the compounds into groups according to their boiling points. These groups are called fractions. A fraction is a solution of different hydrocarbons with similar but slightly different boiling points. (See Table 16.2.)

The molecules making up any one fraction can be quite diverse. For example, some of the molecules making up naphtha gas (5 or 6 carbon atoms) are the following:

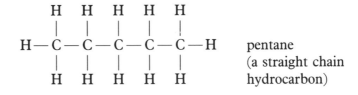

pentane
(a straight chain
hydrocarbon)

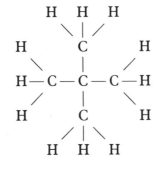

dimethylpropane
(a branched
hydrocarbon)

Table 16.2 Petroleum Refining and Its Products

No. of Carbon Atoms In The Molecules of the Fraction	Boling Range (°C)	Product Name	Uses
1–2 (Methane and Ethane)	−162 to −88 (separate in the ground)	Natural gas	Fuel for furnaces, cars, and trucks; raw material for petrochemicals
3–4 (Propane and Butane)	−42.1 to −0.5 (mostly separated at the oil well)	LPG (liquefied petroleum gases)	Portable fuel for heaters in campers, barbecues, and torches; a motor fuel and a petrochemical raw material
5–6	30 to 65	Naphtha gas or petroleum ether	Liquid fueled camp stoves; solvent
4–12	60 to 150	Gasoline	Motor fuel
10–16	170 to 230	Kerosene	Fuel for lamps and heaters; jet fuel, light diesel fuel
14–18	250 to 400	Light oils	Diesel fuel, furnace oil
18–21	over 400	Heavy oils	Industrial furnace oil, lubricating oils and greases
22–40	Residual (melts above 50°C)	Paraffin wax	Candles, canning wax, fireplace logs
40+	Residual solids	Asphalt	Road surfaces and roofing compounds

2,3-dimethylbutane

As well as physically separating the hydrocarbons by boiling point, refineries also process some of these hydrocarbons chemically. Molecules are rearranged to obtain greater yields of the more desirable fractions such as gasoline.

Gasoline

Gasoline is one of the petroleum fractions most in demand. Direct fractional distillation of petroleum produces only a portion of the gasoline sold. (See Figure 16.8.) Some of the molecules of other petroleum fractions are transformed chemically into gasoline.

Figure 16.8
The components found in a typical crude oil do not meet the market demand for these products. More gasoline, diesel fuel, and furnace oil, and less kerosene and heavy oils are required than are found naturally in crude oil.

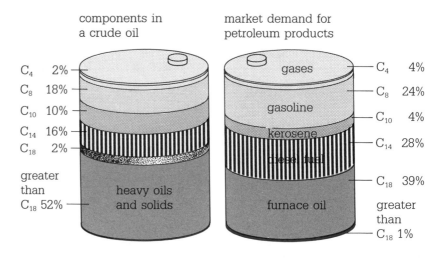

components in a crude oil

market demand for petroleum products

If required, the smaller molecules in lower-boiling fractions can be joined to form larger molecules. This is a useful technique for turning some less-valuable light fractions into gasoline, a process called alkylation. Alkylation is just one of many reactions involving hydrocarbons. (Hydrocarbon reactions, in turn, are just part of the whole range of carbon chemistry that is called organic chemistry. A brief introduction to organic chemistry is provided in Chapter 17.) A typical alkylation reaction is shown below.

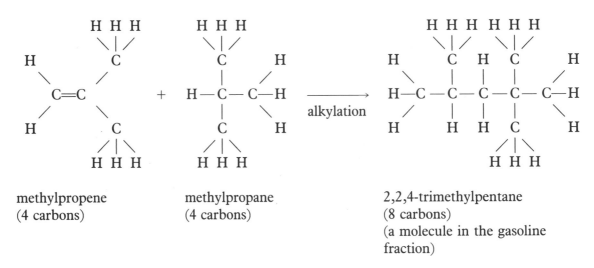

methylpropene
(4 carbons)

methylpropane
(4 carbons)

2,2,4-trimethylpentane
(8 carbons)
(a molecule in the gasoline fraction)

A simpler way of writing this chemical equation is

$$C_4H_8 \quad + \quad C_4H_{10} \quad \longrightarrow \quad C_8H_{18}$$

Even though alkylation reactions can be used to obtain gasoline, this useful substance is more often produced from molecules in the higher fractions. They are broken down or "cracked" into smaller molecules using high temperatures and appropriate catalysts. In some cracking processes, hydrogen may also be added as a reactant.

$$C_{14}H_{30} \ + \ H_2 \ \xrightarrow[\text{heat}]{\text{catalyst}} \ 2C_7H_{16}$$

tetradecane   heptane
(furnace oil fraction)   (gasoline hydrocarbon)

The composition of gasoline is also very important. Gasoline in car engines must burn evenly and ignite only when the engine's spark plugs fire. Straight chain hydrocarbons tend to burn too quickly. They also tend to ignite before the pistons of the car's engine have compressed the air-gasoline vapour mixture sufficiently. When this happens, the engine makes a "pinging" noise described as knocking. Knocking wastes energy and can cause serious damage to a car's engine. Branched chain gasoline molecules, on the other hand, tend to burn more evenly and ignite only when the spark plug fires. The air-gasoline mixture burns more smoothly to produce a smooth running and efficient engine.

Most of the gasoline fraction obtained from crude oil consists of straight chain molecules. The anti-knocking qualities of this fraction must be improved before it can be sold. One way of doing this is to increase the proportion of branched chain molecules in the gasoline mixture. Straight chain molecules passed over a hot platinum catalyst in the presence of hydrogen (to prevent unwanted side reactions) will cause the molecules to "reform" or rearrange into branched chain varieties. An example is shown below.

octane
(straight chain molecule that causes
"knocking")

2,2,4-trimethylpentane
(a branched chain "anti-knock"
molecule)

This is a fairly expensive operation and accounts for some of the extra cost associated with unleaded gasolines.

LEAD IN THE ENVIRONMENT

Lead poisoning has been a public health hazard for thousands of years. Historians speculate that lead contamination of food, wine, and cosmetics caused a serious decline in the fertility of Rome's women and was a factor in the decline and fall of the Roman Empire. The Romans had found that wine prepared in lead vessels tasted sweeter and lasted longer than wine prepared in other types of containers. They found that sour wine (which contained acetic acid as vinegar) reacted with lead to produce lead(II) acetate (called sugar of lead because of its sweet taste). Lead(II) acetate was used by the Romans to sweeten wine and some foods. It was also used for the same purposes as late as the middle ages in Europe.

With the industrial revolution, lead found its way into manufacturing processes and products. The important chemical, sulphuric acid, was produced in lead chambers. Lead-containing products such as lead pipes, lead shot (bullets), pewter metal used for wine goblets and plates, solder, and paints were made. Lead poisoning was recognized, however, only when its symptoms had become painfully obvious. These symptoms included severe stomach pain (known as colic) and a body-wasting disease called cachexia.

Today lead is widely distributed in the environment, and is inhaled and ingested from a variety of sources.

Potential Sources of Lead In the Environment

Ingested	Inhaled
Lead-based paint (now banned)	Burning of leaded objects
Interior: walls, window sills, floors, furniture	Automobile batteries
Exterior: door frames, fences, porches, siding	Newspaper logs and coloured paper
	Candles containing a lead core in the wick
Plaster, caulking	Sanding and scraping of lead-based painted surfaces
Unglazed pottery	Automobile exhaust
Coloured newsprint	Cigarette smoke
Painted food wrappers	Leaded gasoline fumes
Cigarette butts and ashes	Dust in poorly cleaned urban housing
Foods or liquids from cans soldered with lead	Contaminated clothing and skin of workers in smelting factories
Household dust	
Soil, especially along heavily travelled roadways	
Food grown in contaminated soil	

A cheaper method of improving the anti-knock properties of gasoline is to add a small amount of tetraethyl lead, $Pb(C_2H_5)_4$ to the gasoline. There is, however, a hidden cost to society of this more economical method. Burning leaded gasolines distributes lead, a toxic material, into the environment. Even low levels of lead in children causes irritability, short attention spans, and has been associated with learning difficulties in school. Following public pressure, governments, mostly in developed countries, are gradually lowering the permissible levels of tetraethyl lead in gasolines.

The main source of lead pollution today is the use of leaded gasolines and lead in the batteries of automobiles.

As lead pollution in the environment climbs steadily, we are beginning to recognize the danger. Even at very low levels in the body, lead can cause problems. Children are especially at risk. As they explore their environment, toddlers like to taste everything. Lead poisoning has sometimes been traced to the paint on an old crib. Children's play areas are sometimes close to travelled roads, where the dust and dirt contain significant amounts of lead.

Although the effects of lead poisoning are serious for any body system, the most serious and irreversible effects are on the nervous system. The first indications of lead toxicity are behavioural changes. Hyperactivity, aggressiveness, impulsiveness, lethargy, irritability, short attention span, distractibility, and learning difficulties are all symptoms of lead in the body. In extreme cases, the result may be convulsions, mental retardation, paralysis, blindness, coma, and even death.

Lead compounds from automobiles using leaded gasoline enter the environment as gases and as solid particles.

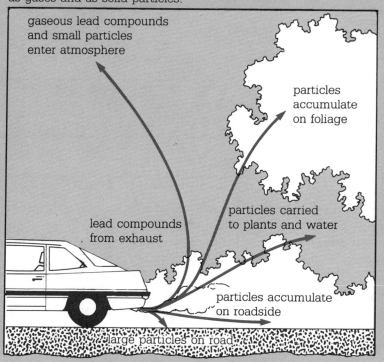

gaseous lead compounds and small particles enter atmosphere

particles accumulate on foliage

lead compounds from exhaust

particles carried to plants and water

particles accumulate on roadside

large particles on road

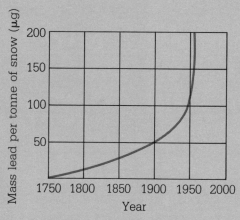

The snowfall in some arctic regions remains packed in layers for centuries. These snow layers can be dated. The lead in these layers has increased since the industrial revolution and markedly since the introduction of automobiles.

Refinery Location

The location of an oil refinery depends on many considerations. (See section 15.4 on page 392.) A source of crude oil, transportation facilities, and a ready market are all crucial factors. It is usually far easier to transport one major raw material, in this case crude oil, over long distances to a refinery and then to deliver the many different products of the refinery to nearby customers than the other way around.

Sometimes transportation facilities make it more attractive to build refineries close to oil fields. World scale refineries such as the Aruba plant off the coast of oil-producing Venezuela in South America are located near excellent deep sea ports. These refineries ship their products to markets around the globe. A decision to locate a refinery involves, once again, a balance of costs, risks, and benefits. Refineries operating with the most recent technology can be clean, safe and virtually odour-free during normal operation. But accidents and production problems can produce objectionable odours and spills.

Possible Topics for Further Investigation

1. The operation of an oil refinery

2. Transforming hydrocarbons – the chemical conditions necessary to reform, crack, and synthesize molecules used in gasoline

3. Lead in gasoline – the true costs

4. Balancing costs and benefits – the location of an oil refinery

16.4 Raw Materials From Petroleum and Natural Gas

Many chemical industries depend on natural gas and petroleum products for their raw material, or feedstocks. Some of the products of an oil refinery are piped directly to nearby plants for further processing. Figure 16.9 illustrates the flow of feedstocks among the major petrochemical plants near Sarnia, Ontario. Some of these plants are truly enormous. For example, Polysar's Sarnia works at Corunna, Ontario, can supply about 40% of Canada's petrochemical needs.

There is a very large number of petrochemical feedstocks. Two of these will be considered in more detail.

Natural Gas

Natural gas is a major primary feedstock for the chemical industry. Sulphur, one of the by-products from sweetening sour natural gas, is used to make sulphuric acid. Sulphuric acid in turn is one of the most widely used substances produced by the chemical industry. It is used in the manufacturing of fertilizers, explosives, dyestuffs, and other acids. It is also used as an electrolyte in car batteries and as a cleaner for metals prior to applying surface coatings or films.

Figure 16.9
Natural gas and crude oil pipelines from western Canada bring in the primary feedstocks fuelling the massive, multi-company petrochemical operation near Sarnia, Ontario. These plants both supply and feed each other different petrochemicals through a complex web of sub-pipelines.

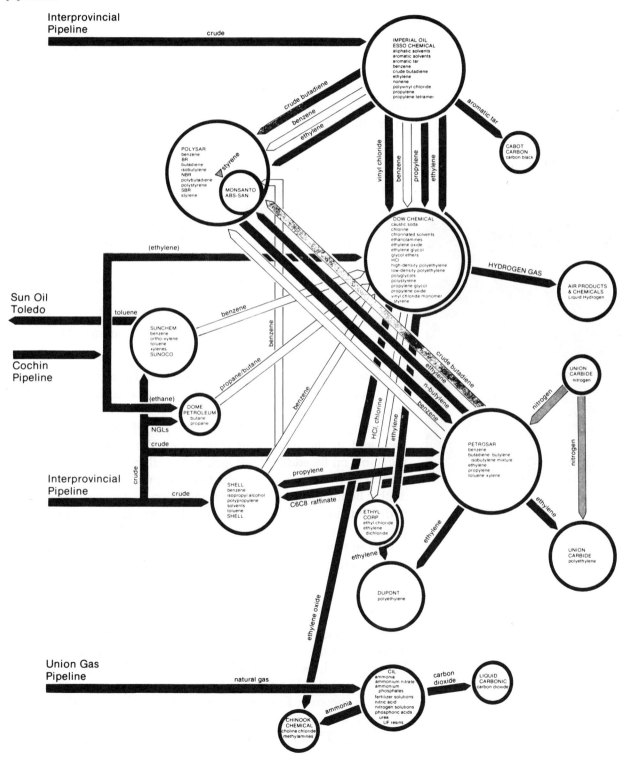

Methane, the major component of natural gas, is reacted with steam to produce a mixture of carbon monoxide and hydrogen, called synthesis gas.

$$CH_4 + H_2O \xrightarrow[\text{catalyst}]{\text{high temperature}} CO + 3H_2$$

$$\underbrace{}_{\text{methane}} \qquad \underbrace{}_{\text{synthesis gas}}$$

The synthesis gas mixture, as well as hydrogen and carbon monoxide separately, are feedstocks for a large number of chemical products. For example, carbon monoxide and hydrogen are used to produce methanol.

$$CO + 2H_2 \xrightarrow{\text{catalyst}} CH_3OH \text{ (methanol)}$$

Methanol is used as a solvent, a fuel, and a feedstock for other products. It is used to produce chemicals such as acetic acid, CH_3COOH, and formaldehyde, CH_2O. These compounds in turn are used to synthesize a wide variety of resins, plastics, and synthetic fibres.

Hydrogen is one of the raw materials used in the production of ammonia.

$$3H_2 + N_2 \xrightarrow[\text{high temperature}]{\text{catalyst}} 2NH_3 \text{ (ammonia)}$$

Ammonia is used directly as a fertilizer and as a refrigerant fluid. It is used as a feedstock in the production of solid fertilizers, nitric acid, explosives, and rocket ruel.

Hydrogen gas is also used in some oil refining processes, as a fuel, and in the making of margarine. Margarine is made from vegetable oil by adding hydrogen to the oil molecules at high pressure. The hydrogen atoms add to the oil molecules at the sites of double bonds, turning them into single bonds. This process is called hydrogenation. (See Figure 16.10.) The resulting substance (margarine) is a solid with a higher melting point than the oil.

Oils and fats that contain double bonds are said to be unsaturated. The molecules of saturated fats contain only single bonds.

Figure 16.10
Hydrogen, chemically added to the double bonds between some of the carbon atoms in edible oil, converts it into a higher melting point fat, margarine.

Benzene

Benzene is just one of many primary petrochemicals that is itself a feedstock for the production of other chemicals. (See Figure 16.11.) Benzene, C_6H_6, is a six-membered ring of carbon atoms, with each carbon atom also bonded to a hydrogen atom. Its structure can be drawn as shown in Figure 16.12 with three double and three single bonds between the carbon atoms. This picture is useful, but a little misleading. The carbon-carbon bonds in benzene are known to be

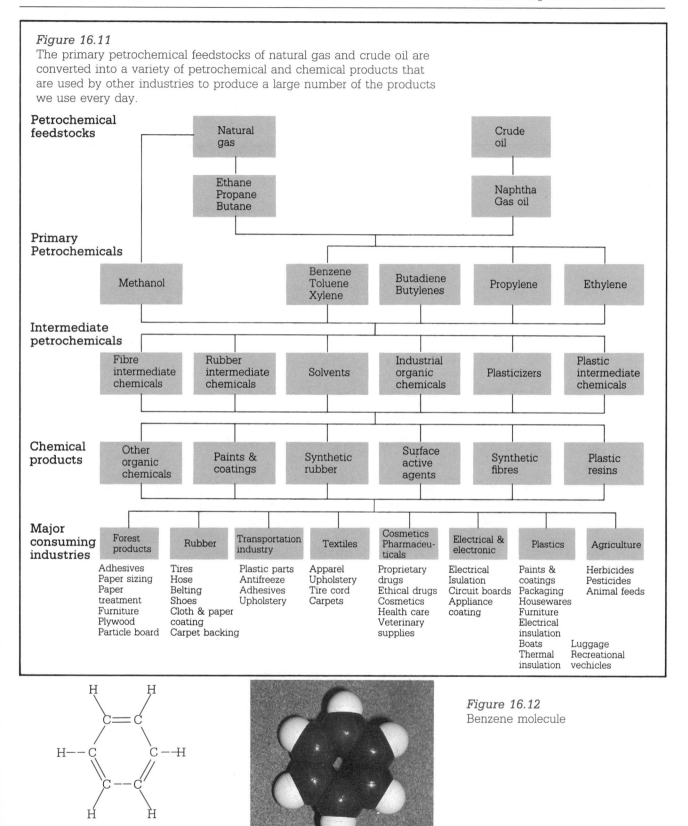

Figure 16.11

The primary petrochemical feedstocks of natural gas and crude oil are converted into a variety of petrochemical and chemical products that are used by other industries to produce a large number of the products we use every day.

Petrochemical feedstocks	Natural gas — Crude oil
	Ethane Propane Butane — Naphtha Gas oil
Primary Petrochemicals	Methanol — Benzene Toluene Xylene — Butadiene Butylenes — Propylene — Ethylene
Intermediate petrochemicals	Fibre intermediate chemicals — Rubber intermediate chemicals — Solvents — Industrial organic chemicals — Plasticizers — Plastic intermediate chemicals
Chemical products	Other organic chemicals — Paints & coatings — Synthetic rubber — Surface active agents — Synthetic fibres — Plastic resins

Major consuming industries

Forest products	Rubber	Transportation industry	Textiles	Cosmetics Pharmaceuticals	Electrical & electronic	Plastics	Agriculture
Adhesives	Tires	Plastic parts	Apparel	Proprietary drugs	Electrical Isulation	Paints & coatings	Herbicides
Paper sizing	Hose	Antifreeze	Upholstery	Ethical drugs	Circuit boards	Packaging	Pesticides
Paper treatment	Belting	Adhesives	Tire cord	Cosmetics	Appliance coating	Housewares	Animal feeds
Furniture	Shoes	Upholstery	Carpets	Health care		Furniture	
Plywood	Cloth & paper coating			Veterinary supplies		Electrical insulation	
Particle board	Carpet backing					Boats	Luggage
						Thermal insulation	Recreational vechicles

Figure 16.12
Benzene molecule

the structural formula of benzene

different from both single and double bonds. (We will not go into the details here, but you can learn more about benzene's structure in Chapter 17.)

One of the results of benzene's bonding arrangement is its relative chemical stability. The benzene ring is not easily broken during chemical reactions. A benzene-like ring is a component in hundreds of useful compounds.

Benzene occurs naturally in crude oil and more is obtained by the chemical reforming of cyclohexane:

$$C_6H_{12} \xrightarrow[\text{catalyst}]{\text{heat}} C_6H_6 + 3H_2$$

cyclohexane   benzene

Benzene and its derivatives form the basis of a host of products. Resins and glues, drugs, dyes, food additives, insecticides, and herbicides are only a few of the useful types of compounds which have the stable benzene ring structure. (See Table 16.3.)

Table 16.3 Benzene and Its Derivatives

Substance Name and Structural Formula		Uses
Phenol		Used in the production of plastics, resins and glues for plywood and furniture
Aspartame		Artificial, low calorie sweetener

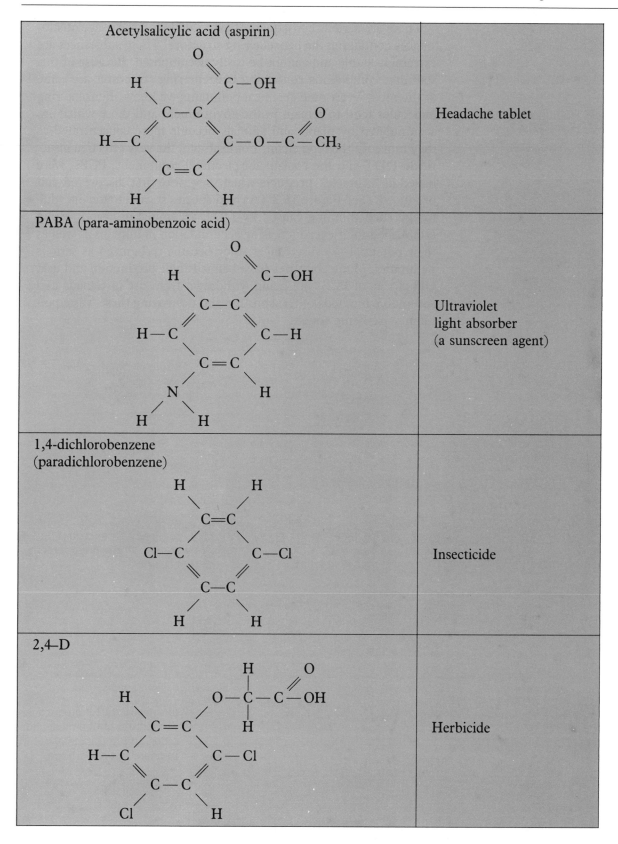

Acetylsalicylic acid (aspirin)	Headache tablet
PABA (para-aminobenzoic acid)	Ultraviolet light absorber (a sunscreen agent)
1,4-dichlorobenzene (paradichlorobenzene)	Insecticide
2,4–D	Herbicide

There are serious consequences associated with the usage of substances containing the benzene ring structure. These substances are extremely stable and cannot be easily decomposed. Because of this stability, compounds containing benzene ring structures are some of the most worrisome chemical pollutants we know. Benzene ring molecules tend to remain in the environment and those which are accumulated by plants and animals become more concentrated as they move up the food chain. For example, the very effective insecticide DDT and the coolant for electrical equipment, PCBs, were hailed as "miracle" products when they were first discovered and produced. (See Figure 16.3.) As bird species started to die out (due to egg shell thinning caused by DDT accumulation), and testing established their ever-increasing accumulation in humans as well as their potential to cause cancer, they became recognized as serious pollutants. Many countries have banned their production and use. Old stocks of PCBs are being withdrawn from use in Canada and acceptable methods of transporting and destroying these waste pollutants are being sought.

Figure 16.3
Benzene derivative pollutants

DDT

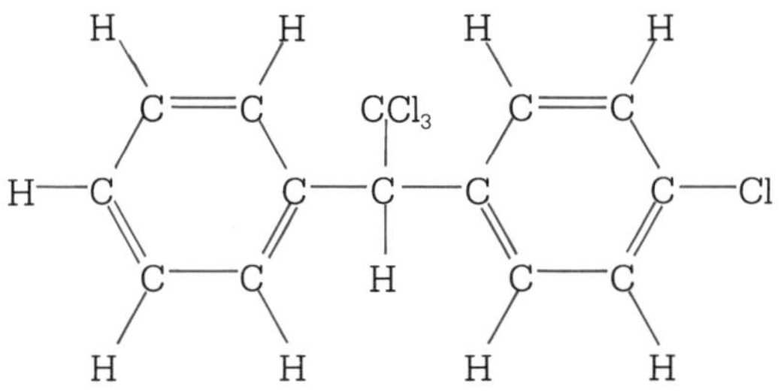

originally an excellent insecticide

a PCB

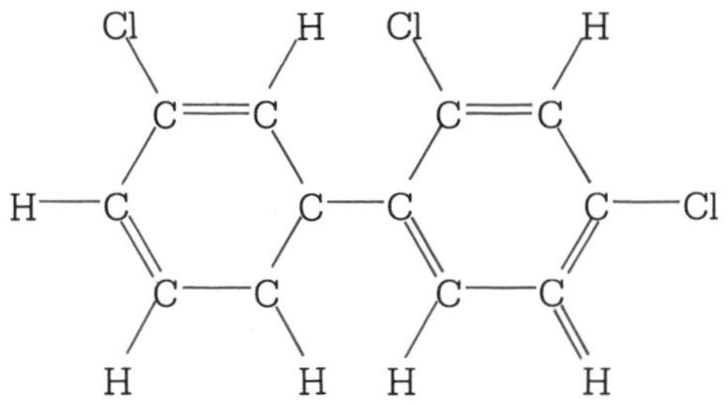

originally an excellent and stable electrical equipment coolant

a dioxin

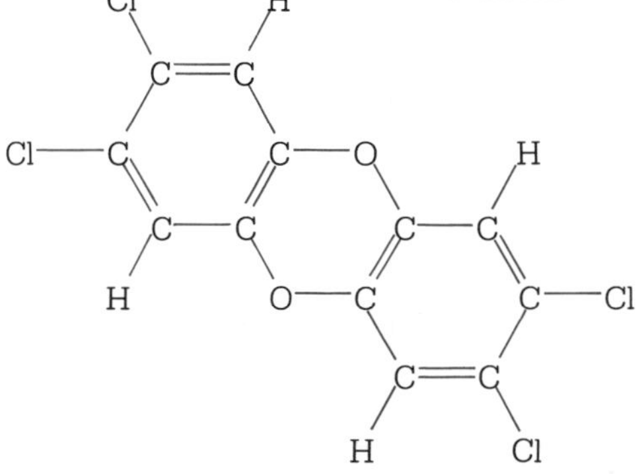

an unwanted by-product in the production of chemicals (insecticides and herbicides) and in the incineration of garbage

Possible Topics for Further Investigation

1. DDT – a case study

2. Benzene ring derivatives – structure and uses

3. PCBs and society's dilemma – how to transport and destroy them safely

4. The use of sulphur in the chemical industry

5. Feedstocks – one company's waste, another's gold

16.5 Preparing a Report on a Chemical Industry

Gathering Information

Once you have selected an industry for study, your first step is to gather as much relevant information as possible. Raw materials, transportation needs, production methods, wastes, product and product uses are all important factors for considering any industry. The industry's research focus and methods should be investigated as well. How profitable is the business? What is its impact on communities and the nation? Are samples of raw materials and products available to you?

Finding Existing Information

Table 16.4 on page 434 suggests several ways of gathering information. Not all of these approaches may be suitable for the industry you choose. Even if you use only two or three of these information gathering procedures, much information can be accumulated.

Generating Your Own Information

You can obtain your own information by performing experiments. Remember that experiments need the approval and the supervision of your teacher. Review the procedure for designing experiments discussed in Special Topic 11.1 on page 280. Your experiment can illustrate a production process or test the properties of a product. For example, you can test the effectiveness of a sunscreen lotion, or you can demonstrate the production of aspirin or a nail polish remover. When your teacher has approved your procedure, carry out your experiments carefully. (See Figure 16.14.) Save all your products, and record all the relevant data. Remember that controlled variables, as well as test variables, must be noted.

Figure 16.14
The person shown here is testing the strength of a fixed size of gypsum wallboard. Does she appear to be doing so safely? What safety precautions do you think she should be following?

Table 16.4 Techniques for Gathering Information

Resources	Activities
Telephone and the yellow pages	Phone the public relations department of local companies in the industry requesting information and samples. If the industry isn't local, contact a dealer or sales representative.
Telephone and the blue pages	Phone local government agencies and regulatory bodies (such as the Ministry of Food and Agriculture) for such items as industry wide statistics, and pollution and product standards and regulations.
Human resources	Talk to your teacher or librarian for leads such as addresses or special information packages. Phone, or interview, a knowledgeable person in the industry (for example, a plant manager or sales person). Your parents or family friends may also be knowledgeable about the industry you choose.
Library resource centre	Look up specific product or process information in standard references such as encyclopedias and chemical reference books. Use the card catalogue/computer data base to locate books and articles dealing with the industry in question.
Pen and paper	Write to specific companies in the industry, industry organizations (such as the Canadian Chemical Producers Association), and government agencies requesting information or samples.
Personal visit	Arrange for a tour of a plant or factory in the industry being studied.
Computer and modem	Conduct (or arrange for) a computer search of appropriate data bases. (Note: This could be expensive! Investigate the potential costs first.)

Organization

The checklist in Table 16.5 may be used to ensure the completeness of your report. Depending on the industry being studied, not all of the components listed will apply.

Full value can only be obtained from data if they are properly organized and presented. Organization is a key element in the production of your report. A table of contents is essential. The readers of your report will need it later, and you need it now as you try to

Table 16.5 Report Checklist

Does your report consider or answer the following questions concerning the industry you chose?
Raw Materials • What raw materials are needed? • What are the sources of the raw materials? • How are the raw materials produced? • Is there a concern about the possible depletion of the resource? • Are the raw materials dangerous? • Are there pollution problems associated with their production or transportation?
Plant and Process • What technology is currently being used? • What physical processes are carried out? • What chemical reactions take place? • What wastes or by-products are produced? • Are there any special factors that determined the location of the plant? • What safety and pollution control equipment was installed? • What additional or advanced technology is being considered to upgrade the plant or process to make them safer or more competitive?
Product • What is the product? What is its composition? • What is the product used for? • Is the product or its use dangerous? • Does the product cause pollution either during its use or when it is discarded? • How can it be safely disposed of after use?
Research and Development • What research does this industry engage in? What new products or procedures has it developed recently? • How is quality control maintained? How are the raw materials and products tested? How are the manufacturing processes monitored? • Have you included the results of your experiments?
Economic Impact • Is the industry profitable? • What economic impact does the industry have on its local community? on Canada? on the world?
Societal Implications • Does the product represent a net benefit for society? • Should society allow its continued production? • If you were in charge of research in this industry, what areas or projects would you favour?

organize your data. Be sure to review Chapter 15 for some of the factors that affect industrial activities. Briefly jot down some of the points you wish to discuss under the appropriate headings. Identify any graphs, charts, tables, flowcharts, diagrams, or pictures you will be using.

Producing Your Report

The production of a report is a big undertaking, but it can be very satisfying. Almost any career involves effective communication; for some careers, communication is crucial. As society moves further into the "information age", more new careers and a larger part of the more traditional careers will involve information processing.

The actual production of the report calls upon your writing, grammar, and spelling skills. Being able to communicate information in chart or graph form is another valuble skill.

Much of the above activity can be carried out using a computer. Word processing programs allow you to change and rearrange your report at will; some will check spelling, punctuation, and even grammar. Programs that construct and print out graphs, and bar and pie charts are also available. Of course, you do not need a computer for your report. Pens, pencils, typewriters, graph paper, and rulers are more than adequate tools.

Later, in your chosen career, you may find that having the right answer is no guarantee of success. You must be able to convince others that you are right, and that your plan of action is the best one. Good, crisp writing and attractive layout in your reports are often as important as the contents.

VIII ORGANIC CHEMISTRY

17

ORGANIC CHEMISTRY

In Chapter 7, you were introduced briefly to the difference between organic and inorganic compounds. **Organic compounds** are the compounds of the element carbon, excluding carbon monoxide, carbon dioxide, cyanides, and carbonates. Inorganic compounds include all those compounds not classified as organic. At one time, scientists believed that nature was the only source of organic compounds. They thought that some "vital force", found only in living things, was necessary to form organic compounds. In 1828, Friedrich Wöhler, a German chemist, was able to produce the organic compound, urea, by heating ammonium thiocyanate, NH_4CNO, an inorganic compound. From this date on, scientists gradually began to abandon the "vital force" theory.

Today chemists have been able to identify over six million organic compounds. Thousands of new organic compounds are discovered each year. Although many organic compounds are found in plants or animals, the majority are manufactured in laboratories. The science of organic chemistry is devoted to a study of these compounds.

In this chapter, you will learn how the study of organic chemistry is made easier by classifying organic compounds into groups or families based on similar chemical reactions and molecular structure. You will also see that organic chemistry has a greater impact on our lives than any other field of chemistry. In fact, the chemistry of life itself is part of organic chemistry. This is because all known life forms are based on the element carbon.

438

17.1 Carbon – The Unique Element

Carbon is a unique element. Its atoms can easily bond to one another to form rings, long chains, and complex branched chain molecules. Since carbon atoms can form four covalent bonds, they can be part of a chain and still bond strongly to other non-metal atoms such as hydrogen, oxygen, nitrogen, and the halogens. (See Figure 17.1.) No other element has these properties. These properties enable carbon to form the "backbone" of an almost limitless number of compounds. Figure 17.2 shows the formulas of a few organic compounds you may have heard of.

Figure 17.1
Carbon atoms are able to form chains by bonding to other carbon atoms, and at the same time bonding to non-metal atoms.

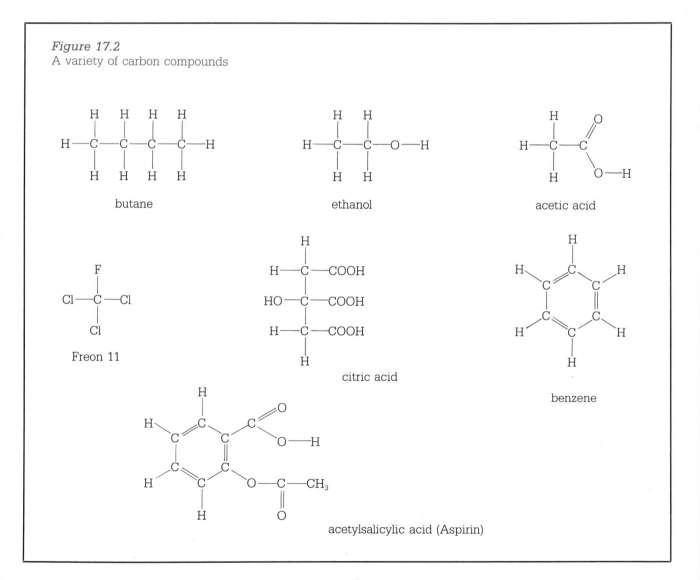

Figure 17.2
A variety of carbon compounds

17.2 Hydrocarbons

We mentioned in the introduction to this chapter that organic compounds are classified into groups according to molecular structures. One group of organic compounds is the **hydrocarbons**. These are compounds consisting only of the elements carbon and hydrogen. Hydrocarbons can be further subdivided into two major categories, **aliphatic hydrocarbons** and **aromatic hydrocarbons**. Aliphatic hydrocarbons consist of chains or rings of carbon atoms. These carbon atoms are bonded to one another by ordinary single, double, or triple covalent bonds. Hydrogen atoms are attached to the carbon atoms. Aromatic hydrocarbons consist of carbon atoms arranged in six-membered rings with no more than one hydrogen atom attached to each carbon. The compound benzene, C_6H_6, consists of one such six-membered ring. All aromatic compounds are based on benzene. The bonding in aromatic hydrocarbons cannot be adequately described using our present bonding model. The bonding in aromatics will be discussed in section 17.6.

Aliphatic hydrocarbons can be subdivided into alkanes, alkenes, and alkynes. (See Figure 17.3.)

Figure 17.3
A classification scheme for hydrocarbons

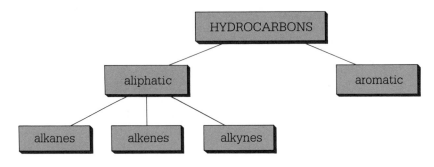

Alkanes consist of chains or rings of carbon atoms joined by single covalent bonds. Hydrogen atoms are attached to the carbon atoms where necessary. Some examples of alkanes are shown in Figure 17.4.

Figure 17.4
Examples of alkanes

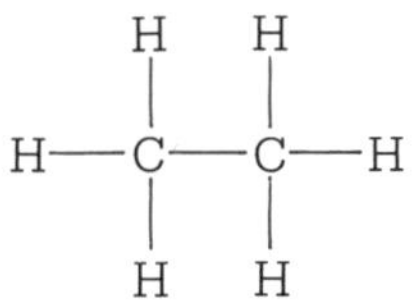

ethane 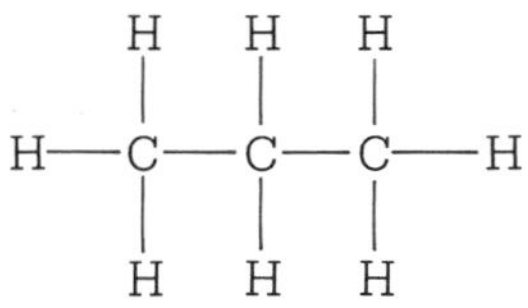   propane

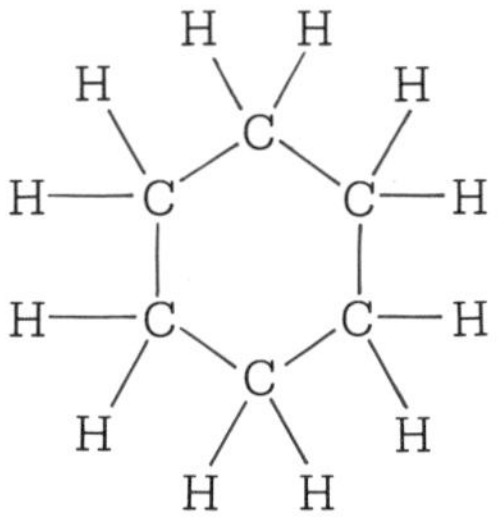

cyclohexane

Alkenes are similar to alkanes, except that two of the carbon atoms in the chain or ring are attached by a double bond. (See Figure 17.5.)

Figure 17.5
Examples of alkenes

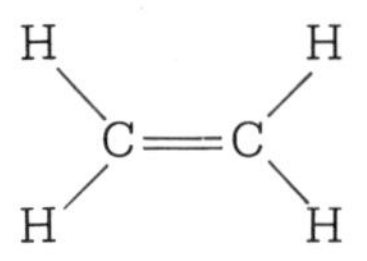

ethene

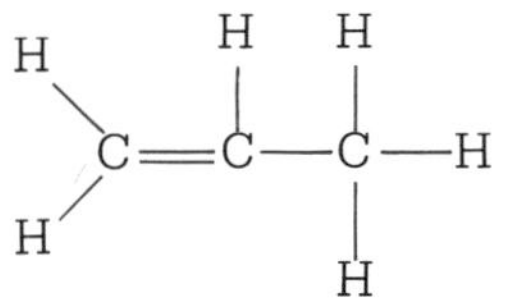

propene

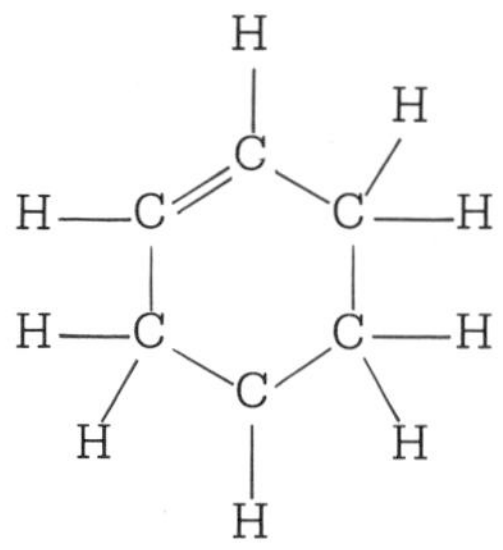

cyclohexene

In **alkyne** molecules, two of the carbon atoms are bonded by a triple bond. (See Figure 17.6.)

Figure 17.6
Examples of alkynes

ethyne

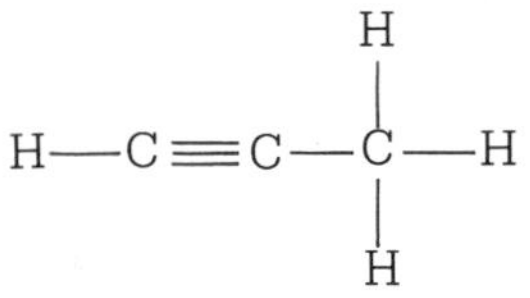

propyne

Hydrocarbons with only single bonds in the molecule (the alkanes) are said to be **saturated**. It means that the molecule has the maximum number of hydrogen atoms bonded to it. The alkenes and alkynes, on the other hand, have double or triple bonds, and have room for more hydrogen atoms to be attached. (This would happen only if the double or triple bond were changed to a single bond.) These compounds are therefore described as **unsaturated**. In the following four sections, you will find out more about the different kinds of hydrocarbons.

17.3 Alkanes

Alkanes are relatively unreactive compounds. They are important as the raw materials for the manufacture of other organic compounds and as fuels. Alkanes burn in air. Propane, for example, burns in a plentiful supply of oxygen according to the following equation.

$$C_3H_8(g) + 5O_2(g) \rightarrow 3CO_2(g) + 4H_2O(g)$$

The other alkanes react in a similar way. In each case, the chemical products of combustion are neither useful nor valuable compounds; however, the energy released during combustion is vital to our energy-consuming society. Methane, the simplest alkane, is the major component of natural gas, which is used in home furnaces and stoves. Propane is used for gas barbecues, and also as a motor vehicle fuel. Butane is used in lighters and as fuel for camping stoves. Longer chain alkanes are found in kerosene, diesel fuel, various oils, and in waxes such as candle wax. Table 17.1 lists the first ten straight chain alkanes together with their formulas and boiling points.

Table 17.1   Names, Formulas, and Boiling Points of the First Ten Alkanes

Formula	Name	Boiling Point (°C)
CH_4	Methane	− 161
C_2H_6	Ethane	− 89
C_3H_8	Propane	− 44
C_4H_{10}	Butane	− 0.5
C_5H_{12}	Pentane	36
C_6H_{14}	Hexane	68
C_7H_{16}	Heptane	98
C_8H_{18}	Octane	125
C_9H_{20}	Nonane	151
$C_{10}H_{22}$	Decane	174

Notice that all of the names end in the suffix "-ane". The prefix for each compound represents the number of carbon atoms in the chain. The prefix "eth-" means two carbon atoms, "prop-" means three carbon atoms, "but-" means four carbon atoms, and so on. These same prefixes are used in other organic compounds, so it is important that you learn them well now.

If you examine the formulas of the alkanes, you should see that there is a constant difference between adjacent compounds. Each compound has one more carbon atom and two more hydrogen atoms than the preceding compound. The formulas differ by a CH_2 unit. In general, the molecular formulas of all of the alkanes can be represented by the formula C_nH_{2n+2}, where n is the number of carbon atoms in the formula. A series of compounds with the same general formula, and with a CH_2 – unit difference between members is called a **homologous series**.

Exercise

1. Write the formulas of the alkanes containing the following number of carbon atoms.
   (a) 13 carbon atoms per molecule
   (b) 18 carbon atoms per molecule
   (c) 24 carbon atoms per molecule

Isomers

When the number of carbon atoms in an alkane is four or more, an interesting phenomenon arises. We find that it is possible to combine the carbon atoms and hydrogen atoms in more than one way. For example, there are two possible structures for butane, C_4H_{10}.

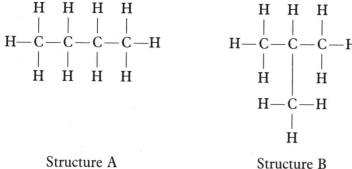

Structure A   Structure B

Although the compounds have the same molecular formula, C_4H_{10}, the structural arrangements of the atoms are different. Such compounds are called **isomers**. Isomers are actually different compounds with slightly different properties. (See Table 17.2.)

Table 17.2   Melting Points and Boiling Points of Isomers of Butane

	Isomer A	**Isomer B**
Melting point	− 138°C	− 159°C
Boiling point	− 0.5°C	− 12°C

Exercise

2. Draw the structural formulas of the three isomers of pentane, C_5H_{12}.

Structure and Bonding of the Alkanes

Since the alkanes consist only of the non-metals carbon and hydrogen, the bonds between the atoms are covalent. In addition, since carbon and hydrogen have approximately the same electronegativity, the bonds in alkanes have little polarity. As you will see later in this section, the physical properties of the alkanes are consistent with this type of bonding.

Examine the structural formula diagrams for methane and ethane shown below.

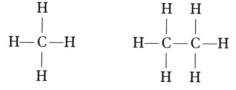

These diagrams might imply that alkane molecules are flat or two-dimensional. This, however, is not the case. Structural formula diagrams drawn on a flat page are useful, because they show us how the atoms are bonded to each other. But they do not give an

Figure 17.7
Ball-and-stick models and space filling models for methane and ethane.

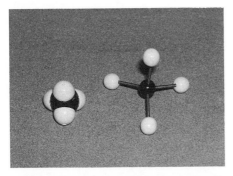

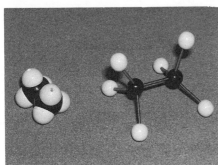

accurate picture of the shape of the molecule. Three-dimensional molecular models, on the other hand, indicate the shape of the structure. The two main types of molecular models are ball-and-stick models and space-filling models. Figure 17.7 shows both types of models for methane and ethane.

Molecular models are much more difficult to draw, but they give more information about the relative positions of the atoms in the molecule. For our purposes, structural formula diagrams are usually adequate. However, you must remember that these diagrams show only which atoms are joined together and do not indicate the shape of the molecule.

Exercise

3. Refer back to Exercise 2. Construct either ball-and-stick or space-filling models for the three isomers of pentane.

Physical Properties of the Alkanes

The bonds in alkane molecules are essentially nonpolar bonds. This means that the molecules have a net dipole of close to zero. The molecules are essentially nonpolar. As a result, the forces of attraction between alkane molecules are relatively weak. The lighter alkanes – methane, ethane, propane, and butane – are gases at room temperature. As the number of carbon atoms in the molecule increases, the intermolecular forces of attraction increase. Consequently, the boiling points of the alkanes increase as the molar mass increases. (Refer again to Table 17.1.) The straight chain alkanes containing from 5 to 16 carbon atoms are liquids. Alkanes with more than 16 carbon atoms are soft, waxy solids with fairly low melting points.

The low melting points of solid alkanes and their ability to burn are put to use in candles. Candle wax consists of alkanes with about 24 carbon atoms per molecule. The wax melts at about 50°C. As the candle burns, the liquid wax is transported up the candle wick. At the tip of the wick, the wax combines with the oxygen in air at high temperature, and gives off heat and light.

Because they are nonpolar, alkanes are insoluble in water and other polar solvents. Long-distance swimmmers make use of this property of alkanes. They coat their bodies with grease which is a mixture of alkanes. Since the grease does not dissolve in water, it remains on the swimmers' bodies and acts as an insulating material, reducing heat loss.

Gasoline and most lubricating oils consist mainly of alkanes. Have you ever tried to remove grease or oil from your hands or clothing using only water? If so, you would have observed the inability of water to dissolve alkanes. Wax paper is coated with a thin layer of solid alkanes. Because the wax does not dissolve in water, water does not pass through the paper. Alkanes may be used for other waterproofing purposes. People who make jam at home usually seal the product with a layer of wax.

Since there are no ions or free electrons in alkanes, they are non-conductors of electricity.

Chemical Properties of the Alkanes

Combustion Reactions

You learned earlier that the most important chemical reaction of the alkanes is combustion. For example, methane burns in a plentiful supply of oxygen to form carbon dioxide and water vapour.

$$CH_4(g) + 2O_2(g) \rightarrow CO_2(g) + 2H_2O(g)$$

When a hydrocarbon burns in excess oxygen, the reaction is referred to as **complete combustion**. If the supply of oxygen is restricted, the products of the combustion are different. Instead of carbon dioxide, either carbon monoxide or carbon is produced.

$$CH_4(g) + \frac{3}{2}O_2(g) \rightarrow CO(g) + 2H_2O(g)$$

or

$$CH_4(g) + O_2(g) \rightarrow C(s) + 2H_2O(g)$$

The amount of oxygen available determines the reaction that occurs.

When a hydrocarbon burns in a limited supply of oxygen, the reaction is called **incomplete combustion**. This type of combustion is inefficient and potentially dangerous, since carbon monoxide (one product of incomplete combustion) is a poisonous gas. If the combustion occurs in an enclosed area, anyone in that area is in danger of gas poisoning. It is extremely important that home heating furnaces and gas stoves be properly maintained and ventilated. If the supply of oxygen to the furnace or stove is restricted, the consequences could be disastrous. The same caution applies to the combustion of any hydrocarbon. Whether it is propane in a gas barbecue or gasoline in a car engine, combustion in a limited supply of oxygen will result in the formation of significant amounts of carbon monoxide.

Substitution Reactions

Alkanes also react with the halogens. For example, in the presence of light or ultraviolet radiation, alkanes react with chlorine and bromine. One or more hydrogens in the alkane are replaced by halogen atoms. A typical reaction is shown on page 446.

$$H-\underset{\underset{\displaystyle H}{|}}{\overset{\overset{\displaystyle H}{|}}{C}}-H \;+\; Cl_2(g) \;\rightarrow\; H-\underset{\underset{\displaystyle H}{|}}{\overset{\overset{\displaystyle H}{|}}{C}}-Cl \;+\; HCl(g)$$

or $CH_4(g) \;+\; Cl_2(g) \rightarrow CH_3Cl(g) \;+\; HCl(g)$

The product CH_3Cl is called chloromethane. This type of reaction, in which a hydrogen atom in a hydrocarbon is replaced by another atom, is called a **substitution reaction**.

In the above reaction, the substitution of chlorine for hydrogen can continue, forming a variety of products. If two hydrogen atoms are substituted, the product is CH_2Cl_2, dichloromethane. The substitution of three hydrogen atoms produces $CHCl_3$, trichloromethane. This compound is often called by its old name, chloroform. If all four hydrogen atoms are replaced by chlorine atoms, the product is CCl_4, tetrachloromethane (or carbon tetrachloride).

Compounds produced by substituting halogen atoms for hydrogen atoms in alkanes have a wide variety of uses. They are used as solvents, anesthetics, and refrigerants. (See Special Topic 17.1.)

Cracking

Crude oil contains many different hydrocarbons. (See Table 16.2, page 421.) The demand for specific hydrocarbons does not always match the composition of the available crude oil. Oil as it comes from the ground contains too many "heavy" molecules but not enough of the lighter molecules of gasoline. Chemists have learned to reorganize hydrocarbon molecules, thereby increasing the yield of the more valuable products. One of the earliest reorganizing techniques used by the oil refiners is called "cracking". To improve the yield of gasoline, the refiners "crack", that is break up, some of the longer chain hydrocarbons into smaller chains. For example:

$$C_{16}H_{34}(l) \rightarrow C_8H_{18}(l) \;+\; C_8H_{16}(l)$$

The reaction is carried out by heating the long chain hydrocarbons in the presence of a catalyst such as alumina, Al_2O_3, or silica, SiO_2. The number of carbon atoms in the product molecules is in the correct range for gasoline ($C_5 - C_{12}$). (See Figure 17.8.)

Figure 17.8
Cracking of long chain hydrocarbons produces shorter molecules. Compounds composed of shorter molecules flow more easily.

17.4 Alkenes

The alkenes are a family of hydrocarbons that contain two fewer hydrogen atoms per molecule than the corresponding alkanes. The general formula for the alkenes is C_nH_{2n}. The characteristic feature of the alkenes is the presence of a double bond between two carbon

SPECIAL TOPIC 17.1

FREONS AND THE OZONE LAYER

Freons are compounds composed of carbon, chlorine, and fluorine. They are derived from the alkanes by substituting chlorine and fluorine atoms for the hydrogen atoms. Freons are chlorofluorocarbons, or CFCs. Two examples of freons are shown below.

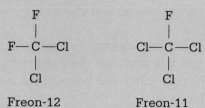

Freon-12   Freon-11

Freons are widely used as refrigerants, both in refrigerators and in air conditioners. Until the late 1970s, they were also used as the propellant material in aerosol cans. The properties of freons made them ideal for these applications. They are volatile (easily evaporated), chemically unreactive, odourless, tasteless, and non-poisonous.

Ozone is a gas found in fairly large concentrations in the stratosphere, about 15 km to 25 km above the earth. It is a form of oxygen, with a formula O_3. Ozone is formed at these high altitudes when oxygen absorbs ultraviolet radiation from the sun.

$$3O_2 \xrightarrow[\text{radiation}]{\text{ultraviolet}} 2O_3$$

Ozone absorbs harmful ultraviolet radiation, preventing it from reaching earth. (Over-exposure to ultraviolet radiation can cause skin cancer.)

What is the connection between CFCs and ozone? During the 1970s, a number of scientists, working with models of the atmosphere, predicted that CFCs were reacting with ozone in the stratosphere. As a result of this reaction, the ozone layer was being depleted, and more ultraviolet radiation was reaching earth. There were pessimistic forecasts that this would result in an increase in the incidence of skin cancer, and might even affect the climate on earth. Scientists believed that the major culprit in this ozone depletion was chlorine atoms, produced when the CFCs absorbed ultraviolet radiation. They believed that the chlorine atoms reacted with the ozone, producing oxygen.

Because freons take several years, perhaps as long as 100 years, to reach the stratosphere, it is difficult to prove or disprove the effect of these compounds on the ozone layer. Although not all scientists agreed that CFCs were depleting the ozone layer, there was enough concern that the use of freons as propellants in aerosol cans has been banned in North America. Also, in the summer of 1987, about 30 countries signed an international agreement to limit the production of CFCs. No satisfactory substitute has yet been found for freons used in refrigerators and air conditioners. But the freons in these applications are in sealed systems, and therefore they pose a minimal risk to the environment.

atoms in the molecule. Alkenes are said to be unsaturated compounds. The relationship between ethane and ethene is shown in the structural formula diagrams below.

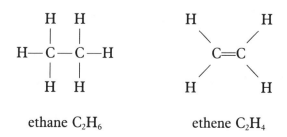

ethane C_2H_6 ethene C_2H_4

Exercise

4. Build molecular models for ethene, C_2H_4, and propene, C_3H_6.

Naming the Alkenes

The names of the straight chain alkenes are based on the names of the corresponding alkanes. The same prefix is used to indicate the number of carbon atoms in the chain, and the ending "-ene" is used in place of "-ane".

If you draw a structural formula diagram or build a molecular model for butene, C_4H_8, you will find that there are two possible straight chain butenes. The only difference between the molecules is the position of the double bond.

$$
\begin{array}{cccc}
H & H & H & H \\
| & | & | & | \\
C=C-C-C-H \\
| & & | & | \\
H & & H & H
\end{array}
\qquad
\begin{array}{cccc}
H & H & H & H \\
| & | & | & | \\
H-C-C=C-C-H \\
| & & & | \\
H & & & H
\end{array}
$$

These compounds are isomers of butene. To distinguish between them, they are named differently. The compound on the left is called 1-butene, since the double bond is at the first position, counting from the left. The compound on the right is 2-butene, since the double bond is at the first position, counting from the left. It may be tempting to draw a third isomer of butene, with the bonds arranged as C—C—C=C. But this is just the 1-butene molecule viewed from the other side. It does not represent a different molecule.

Physical Properties of Alkenes

Like alkane molecules, alkene molecules are essentially nonpolar. The physical properties of alkenes are quite similar to those of the alkanes.

The lighter alkenes are gases at room temperature. The heavier alkenes are liquids or solids with fairly low melting points.

Alkenes are insoluble in water and other polar solvents, but dissolve in other liquid hydrocarbons.

Chemical Properties of Alkenes

Combustion Reactions

Like the alkanes, the alkenes burn. In a plentiful supply of oxygen, ethene, C_2H_4, burns according to the equation shown below.

$$C_2H_4(g) + 3O_2(g) \rightarrow 2CO_2(g) + 2H_2O(g)$$

Laboratory Manual
In Experiment 17-1, you will compare the reactivity of an alkane with the reactivity of an alkene.

Addition Reactions

The presence of a double bond in alkene molecules gives the alkenes a greater reactivity than the alkanes. Alkenes undergo **addition reactions** in which a molecule adds on to the alkene molecule. A portion of the molecule being added attaches to each carbon of the double bond. Some examples of addition reactions to ethene are shown below.

1. Addition of hydrogen

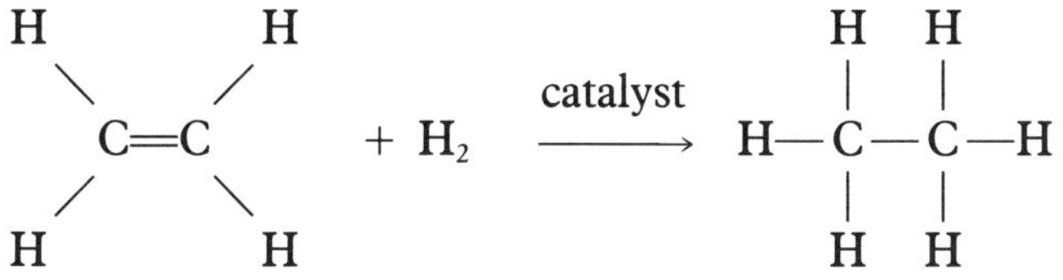

ethane

2. Addition of a halogen

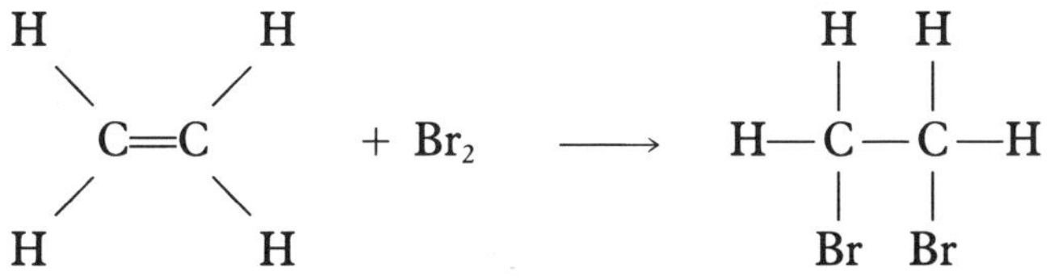

1,2-dibromoethane

3. Addition of HCl

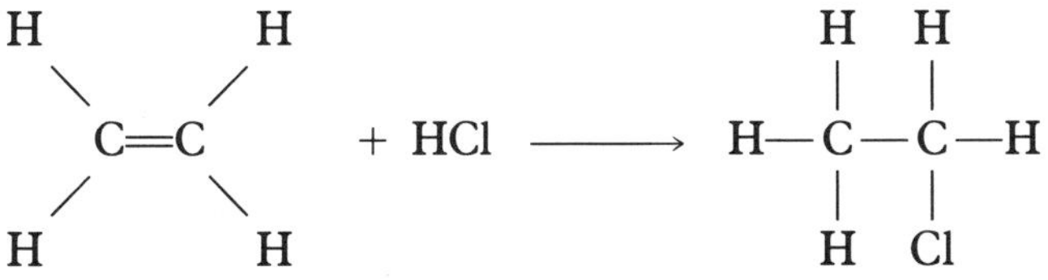

chloroethane

4. Addition of water

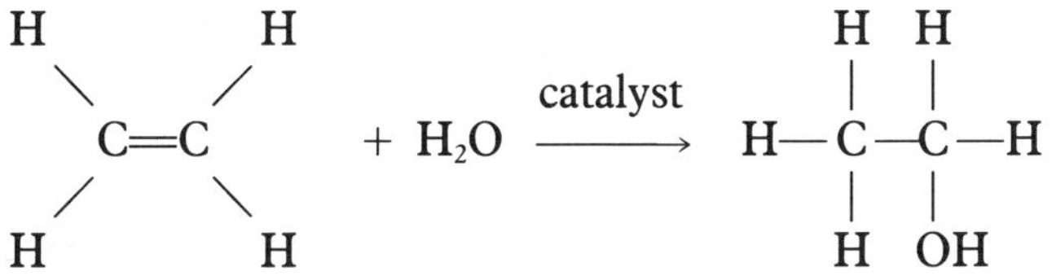

ethanol

Because alkenes are chemically reactive compounds, they are widely used as starting materials for the production of many useful organic compounds.

Polymerization

One of the most interesting, and perhaps the most useful, properties of the alkenes is their ability to polymerize. This refers to the ability of alkene molecules to join together to form very long chains. These long chain molecules are called **polymers**. The simple molecules from which they are formed are called **monomers**.

Under certain conditions, ethene will polymerize to form polyethylene. An old name for ethene is ethylene, and this name has been retained in the polymer. The equation for the reaction is shown as follows.

$$n\ H_2C{=}CH_2\ \rightarrow\ (-H_2C-CH_2-)_n$$
$$\text{ethene} \qquad\qquad \text{polyethylene}$$

n is a large number, usually much greater than 500.

Polyethylene is an extremely useful material. It is a flexible plastic used in the manufacture of plastic sheets, bottles, tubes, ice cube trays, and many other items. Its resistance to attack by most chemicals makes it ideal for these purposes.

Propene also undergoes polymerization to form polypropylene.

$$\begin{array}{ccc} CH_3 & & CH_3 \\ | & & | \\ n\ H_2C{=}CH & \rightarrow & (-H_2C-CH-)_n \\ \text{propene} & & \text{polypropylene} \end{array}$$

Polypropylene is used in the manufacture of indoor-outdoor carpeting, piping, and food containers.

17.5 The Alkynes

The alkynes differ from the alkanes and alkenes in that each alkyne molecule has a triple bond between two carbon atoms. The alkynes are unsaturated compounds. Each alkyne contains two less hydrogen atoms than the corresponding alkene. The general formula for the alkynes is C_nH_{2n-2}. The simplest alkyne is ethyne, C_2H_2, commonly known as acetylene. The structural formula for ethyne is

$$H-C{\equiv}C-H$$

The names of the straight chain alkynes are based on the names of the alkanes with the same number of carbon atoms. (See Table 17.3.)

Table 17.3 Names and Formulas of Some Alkanes, Alkenes, and Alkynes

Alkane	Alkene	Alkyne
Methane CH_4	——	——
Ethane C_2H_6	Ethene C_2H_4	Ethyne C_2H_2
Propane C_3H_8	Propene C_3H_6	Propyne C_3H_4
Butane C_4H_{10}	Butene C_4H_8	Butyne C_4H_6

Exercise

5. Why is there no alkene or alkyne corresponding to methane?

6. Build molecular models for propyne and butyne (straight chain molecules only). How many straight chain isomers of butyne are there?

Physical and Chemical Properties of the Alkynes

The physical properties of the alkynes are similar to those of the alkanes and alkenes. Alkynes are gases, liquids, or low melting solids at room temperature, and do not dissolve in water.

The chemical properties of the alkynes are similar to those of the alkenes. Remember that the alkynes, like the alkenes, are unsaturated compounds. The alkynes burn in air or oxygen. (See Figure 17.9.) They also undergo addition reactions with hydrogen, the halogens, and with hydrogen halides such as hydrogen chloride.

Figure 17.9
Ethyne, more commonly called acetylene, burns in oxygen with a very hot flame. One practical application of this is welding.

Laboratory Manual
In Experiment 17-2, you will prepare acetylene and examine its combustion.

Exercise

7. Write balanced equations for the following.
   (a) the complete combustion of ethyne
   (b) the incomplete combustion of ethyne to form carbon monoxide.

17.6 Aromatic Hydrocarbons

As you can probably guess, the term aromatic refers to the odour of a substance. In the early days of chemistry, chemists often classified substances according to their physical properties, such as colour or odour. Substances with an odour were referred to as aromatic compounds. Today, an aromatic compound is one which is based on the compound benzene, a hydrocarbon with the molecular formula C_6H_6.

The Structure of Benzene

How the structure of the benzene molecule was determined is a fascinating tale. Although benzene was discovered in 1825, and its molecular formula was determined in 1834, it was not until 1865 that a satisfactory structure was proposed. The proposal resulted not from some brilliant experiment but from a dream!

Prior to 1865, the molecular structures that had been proposed for benzene included double or triple bonds. In every suggested structure, the carbon atoms were in a chain. (See Figure 17.10.)

Figure 17.10
Some suggested structures for the benzene molecule prior to 1865

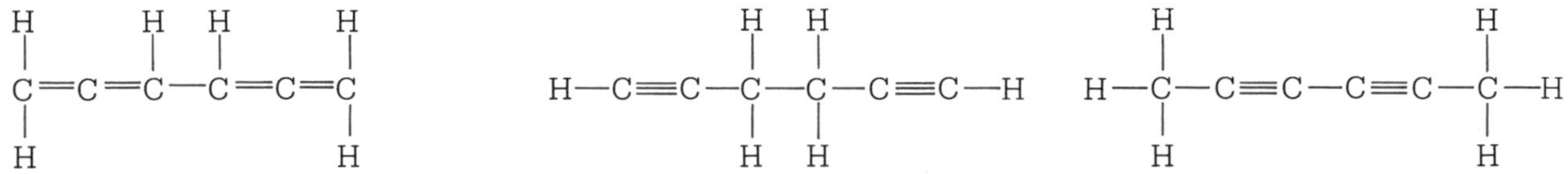

Such molecules are unsaturated, and would be expected to have the properties typical of unsaturated compounds. Benzene, however, does not behave like an unsaturated compound. It does not react by addition. Instead, it behaves as though it is a saturated compound, reacting by substitution. How can this be?

The solution to this problem was provided by Friedrich August Kekulé, a German chemist and former architecture student. Kekulé had been trying unsuccessfully for some time to establish a molecular structure for benzene which was consistent with its properties. One day, while dozing in front of a fire, Kekulé dreamed that he saw atoms and molecules moving around. He saw chains of atoms twisting around like snakes. Suddenly, according to Kekulé's own account of his dream, one of the snakes grabbed its own tail, forming a ring-like structure. Kekulé woke at once. With the vision of a ring of carbon atoms in his head, he developed a proposed structure for the benzene molecule.

Kekulé's proposed structure had six carbon atoms in a ring, with one hydrogen atom bonded to each carbon. (See Figure 17.11.) To satisfy the bonding requirements of carbon, Kekulé suggested that there were alternating single and double bonds in the ring.

According to this proposal, benzene should behave like a compound with three double bonds. Unfortunately for Kekulé, the fact is that benzene undergoes substitution reactions, typical of the alkanes. If Kekulé's double bond theory had been correct, benzene should undergo addition reactions, like the alkenes. Despite this shortcoming, his ring structure idea was basically sound. Other researchers were able to modify it to make it fit the facts. To make sense, the ring structure has to be drawn in two different ways, as shown in Figure 17.12.

Modern theory says that neither of these structures is correct by itself. Studies have shown that in the benzene molecule, the distances between adjoining carbon atoms are all the same. The actual structure of the benzene molecule is thought to be somewhere between the two extremes of the Kekulé structures. This "in between" structure is called a **resonance hybrid**.

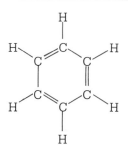

Figure 17.11
Kekulé's proposed structure for the benzene molecule

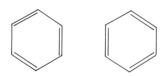

Figure 17.12
Two equivalent Kekulé structures for benzene

Representations of Benzene

Benzene is often represented by the structure shown in Figure 17.13.

Each vertex consists of a carbon atom with a hydrogen atom attached. The circle within the hexagon indicates that all of the carbon-carbon bonds are identical. Note, however, that they are neither single nor double bonds, but something in between.

Figure 17.13
A modern representation of the benzene molecule

Physical Properties of Benzene

Because of its shape, and the similar electronegativities of carbon and hydrogen, the benzene molecule has a zero net dipole. The physical properties of benzene are typical of a nonpolar compound. It is a liquid at room temperature, does not conduct electricity, and is insoluble in water. Benzene is soluble in nonpolar solvents.

Chemical Properties of Benzene

Benzene reacts by substitution reactions. One or more of the hydrogen atoms attached to the ring of carbon atoms is replaced by some other atom or group of atoms. The ring of carbon atoms is very stable and remains intact during most reactions of benzene. Table 17.4 on page 454 shows the formulas and names of a number of compounds derived from benzene.

Table 17.4 Some Compounds Derived from Benzene
(The common name or use is shown in brackets.)

Formula	Name
Cl — benzene ring	chlorobenzene
NO_2 — benzene ring	nitrobenzene
CH_3 — benzene ring	methylbenzene (toluene – used to make TNT)
OH — benzene ring	hydroxybenzene (phenol – used as a disinfectant)
NH_2 — benzene ring	aminobenzene (aniline – used to make dyes)
H—C=O benzene ring	benzaldehyde (flavour of almonds)
Cl — benzene ring — Cl	paradichlorobenzene (mothballs and air freshener)
O_2N, CH_3, NO_2, NO_2 benzene ring	2,4,6,-trinitrotoluene (TNT – an explosive)
O=C—O—H benzene ring, O=C—CH_3	acetylsalicylic acid (aspirin)

| O═C─OCH₃ OH | methyl salicylate (oil of wintergreen) |

(structure: benzene ring with OH group and C bonded to $=O$ and OCH_3)

17.7 Alcohols

If one of the hydrogen atoms in a hydrocarbon is replaced by a hydroxyl group, –OH, an **alcohol** results. Here are two examples.

$$H-\underset{\underset{H}{|}}{\overset{\overset{H}{|}}{C}}-H \rightarrow H-\underset{\underset{H}{|}}{\overset{\overset{H}{|}}{C}}-OH$$

methane → methanol (or methyl alcohol)

$$H-\underset{\underset{H}{|}}{\overset{\overset{H}{|}}{C}}-\underset{\underset{H}{|}}{\overset{\overset{H}{|}}{C}}-H \rightarrow H-\underset{\underset{H}{|}}{\overset{\overset{H}{|}}{C}}-\underset{\underset{H}{|}}{\overset{\overset{H}{|}}{C}}-OH$$

ethane → ethanol (or ethyl alcohol)

The alcohols derived from the alkanes form a homologous series with the general formula $C_nH_{2n+1}OH$. The hydroxyl group, –OH, is called a **functional group**. This is an atom or group of atoms which determines the properties of the compounds in which it is found. Different families of organic compounds have different functional groups. The study of organic chemistry can be thought of as the study of functional groups. This means that more than six million organic compounds can be classified into a relatively small number of families, based on the presence of certain functional groups. Members of a particular family have similar chemical properties, which can be attributed to a common functional group.

Naming the Alcohols

The names of the alcohols derived from the alkanes are obtained by removing the "–e" ending from the name of the alkane, and substituting "–ol" in its place. Older names for the alcohols are still sometimes used. These are obtained by removing the "–e" ending from the alkane, substituting "–yl" in its place, then adding the word alcohol to the name. The alcohol derived from methane is called either methanol or methyl alcohol.

Since the hydroxyl group can be placed on any of the carbon atoms in the chain, it is possible for a single alkane with three or more carbon atoms to give rise to two or more alcohols. Consider propane for example. It is possible to substitute the hydroxyl group either on one of the end carbon atoms or on the centre carbon atom.

| propane | 1-propanol | 2-propanol |

The two alcohols are isomers. The first isomer has the hydroxyl group on the first carbon atom (counting from the right), and is called 1-propanol. The hydroxyl group on the second isomer is on the second carbon atom (counting from either end) and this isomer is called 2-propanol.

Exercise

8. Build molecular models for the straight chain isomers of butanol, C_4H_9OH, and name the isomers.

Physical Properties of the Alcohols

The physical properties of the alcohols are very different from those of the alkanes from which they are derived. The differences are caused by the presence of the hydroxyl group in the molecule. Consider the boiling points of the four simplest alcohols, as shown in Table 17.5.

Table 17.5 Boiling points of First Four Alkanes and Corresponding Alcohols

Alkane	B.P.	Alcohol	B.P.
Methane	$-164°C$	Methanol	$65°C$
Ethane	$-89°C$	Ethanol	$78.5°C$
Propane	$-42°C$	1-Propanol	$97.4°C$
Butane	$-0.5°C$	1-Butanol	$117°C$

Back in Table 17.1, you saw that for related compounds, an increase in molecular mass results in an increase in boiling point. The molecular mass of each alcohol is greater than that of the corresponding alkane. For this reason alone, you would expect the boiling points of the alcohols to be higher than those of the related alkanes. The differences in molecular mass are not large enough, however, to explain the very large differences in boiling point shown in Table 17.5. Notice that even though a propane molecule is heavier than a methanol molecule (44 u compared to 32 u), the boiling point of methanol is still much higher than that of propane.

The high boiling points of the alcohols are explained in terms of increased forces of attraction between molecules, due to hydrogen bonding. This is the same reason that the boiling point of water is much higher than expected. (Refer to Chapter 5.) This similarity between alcohols and water is perhaps not too surprising, since both contain the hydroxyl group.

In methanol, for example, the O–H bond is quite polar due to the fact that oxygen has a much higher electronegativity than hydrogen. As a result, the hydrogen atom carries a slight positive charge, and the oxygen atom a slight negative charge. Because of their opposite charges, there is an electrostatic attraction between the oxygen atom of one molecule and the hydrogen atom of another molecule. (See Figure 17.14.) This extra attraction between molecules makes it relatively difficult to separate an alcohol molecule from its neighbours. Thus, methanol molecules cling to each other more tightly than do methane molecules. As a result, the boiling point of methanol is much higher than that of methane.

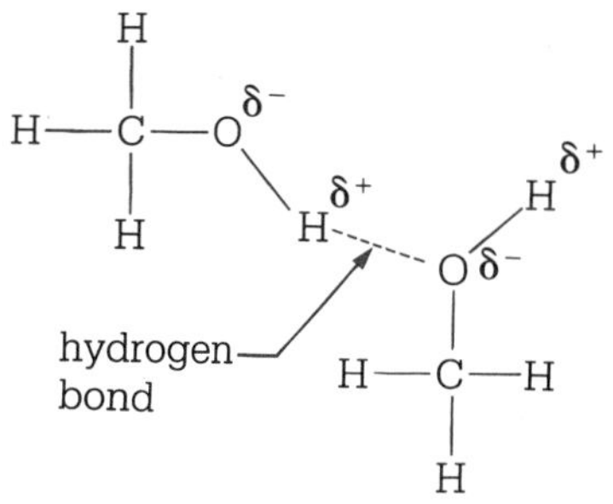

The lighter alcohols are extremely soluble in water. This property too is due to hydrogen bonding. Hydrogen bonds can exist between water molecules and methanol molecules as shown in Figure 17.15. This attraction accounts for the high solubility of methanol and water in each other.

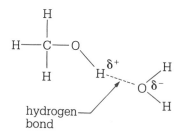

Figure 17.14
The attraction between methanol molecules

Figure 17.15
Hydrogen bonding between methanol and water molecules

Uses of Alcohols

Methanol, CH_3OH, is sometimes referred to as wood alcohol because it was originally obtained by heating wood to a high temperature in the absence of air. Methanol is used as a solvent for varnishes, as gas line antifreeze in automobiles, and as an engine fuel, either by itself or added to gasoline. It is a deadly poisonous liquid. In small quantities, methanol attacks the optic nerve, causing blindness. In larger quantities, methanol can cause death.

Ethanol, C_2H_5OH, is the "active" ingredient in alcoholic beverages. Compared to methanol, ethanol is relatively safe to drink, but prolonged use of ethanol can lead to permanent liver damage and eventually to death. Ethanol for beverages is produced by the fermentation of grains or other starch-containing materials. (See Special Topic 9.1, on page 240.)

Ethanol used in alcoholic drinks is subject to a high federal excise tax in Canada. Ethanol for industrial use is exempt from this tax. To make sure than industrial alcohol is unfit to drink, a small amount of a poisonous substance is added to denature or poison the ethanol. Methanol is one of the substances commonly used for this purpose. (See Figure 17.16.) In addition to its uses in alcoholic beverages, ethanol is used as an industrial solvent, as a solvent for perfumes and other personal care products, and as a raw material in the manufacture of other chemicals.

Figure 17.16
Methanol is sometimes used to denature ethanol.

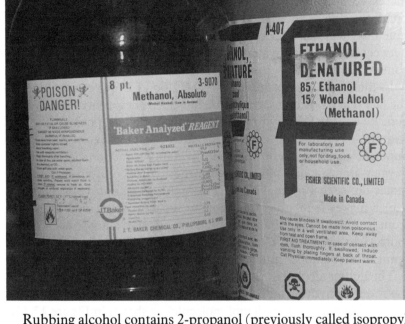

Rubbing alcohol contains 2-propanol (previously called isopropyl alcohol), an isomer of propanol.

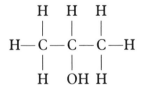

Like methanol, this is an extremely poisonous substance if taken internally. It can be used safely on the skin for such purposes as preventing bed sores and reducing body temperature.

Some more complex alcohols contain more than one hydroxyl group per molecule. Two examples are ethylene glycol, the major constituent of car radiator antifreeze, and glycerol, used in moisturizing lotions. (See Figure 17.17.)

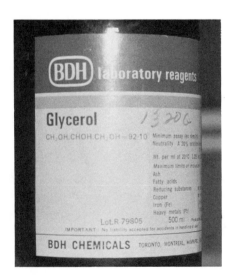

Figure 17.17
Glycerol is an example of an alcohol.

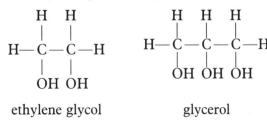

ethylene glycol   glycerol

17.8 Carboxylic Acids

Carboxylic acids contain the functional group

$$
\begin{array}{c}
\text{O} \\
\parallel \\
-\text{C}-\text{O}-\text{H}
\end{array}
$$

This is called the carboxyl group, named from carbonyl, $>$C=O, and hydroxyl, –OH. The carboxyl group is sometimes written as –COOH or –CO$_2$H. The general formula for **carboxylic acids** derived from the alkanes is $C_nH_{2n+1}COOH$, where n can take values 0,1,2,3,4, and so on.

The simplest carboxylic acid has a hydrogen atom attached to the carboxyl group.

$$
\begin{array}{c}
\text{O} \\
\parallel \\
\text{H}-\text{C}-\text{O}-\text{H}
\end{array}
\qquad \text{or HCOOH}
$$

This acid is derived from methane (one carbon atom). Its systematic name is obtained by substituting the ending "-oic" for the ending "-e", and adding acid. For example,

methan<u>e</u> → methan<u>oic acid</u>

A more common name for this acid is formic acid, so named because it is the acid responsible for the irritation caused by ant bites. (The Latin name for ant is *formica*.) Table 17.6 shows the names and formulas of other carboxylic acids derived from the alkanes. The common name for the acid is often derived from some property of the acid.

Table 17.6 Names and Formulas of Some Carboxylic Acids

Systematic Name	Formula	Common Name
Ethanoic acid	CH$_3$COOH	Acetic acid
Propanoic acid	C$_2$H$_5$COOH	Propionic acid
Butanoic acid	C$_3$H$_7$COOH	Butyric acid
Hexadecanoic acid	C$_{15}$H$_{31}$COOH	Palmitic acid

These carboxylic acids, or their salts, are found in some common substances. (See Figure 17.18.)

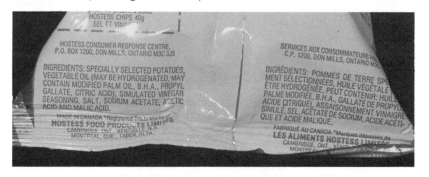

Figure 17.18

(a) Vinegar is a dilute solution, about 5% by volume, of ethanoic acid (acetic acid) in water.

(b) The calcium salt of propanoic acid, calcium propanoate, is used to inhibit the growth of mould on bread.

(c) Hexadecanoic acid (palmitic acid) is used in the manufacture of soap.

(d) Sodium ethanoate (sodium acetate) is added to some brands of potato chips to give the flavour of vinegar.

Physical Properties of Carboxylic Acids

The melting and boiling points of the saturated straight chain carboxylic acids are higher than those of the corresponding alcohols. (See Table 17.7.)

Table 17.7 Boiling Points of Alcohols and Carboxylic Acids Containing the Same Number of Carbon Atoms

Alcohol	B.P.	Carboxylic Acid	B.P.
Methanol	65°C	Methanoic acid	101°C
Ethanol	78.5°C	Ethanoic acid	118°C
1-Propanol	97.4°C	Propanoic acid	141°C
1-Butanol	117°C	Butanoic acid	164°C

Figure 17.19
The attraction between carboxylic acid molecules

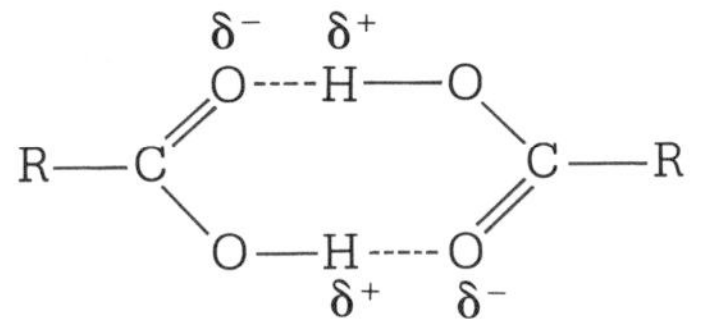

hydrogen bonding between molecules

The higher boiling points of the acids cannot be caused solely by increased molecular mass. For example, compare the boiling points of ethanol and methanoic acid. Both compounds have the same molecular mass, 46 u, but the boiling point of methanoic acid is higher. The acids have higher boiling points than the alcohols for the same reason that the alcohols have higher boiling points than the alkanes – because of hydrogen bonding. In the case of the alcohols, the hydrogen bonding is between one positive site in one molecule – the hydrogen atom – and one negative site in another molecule – the oxygen atom. (Refer back to Figure 17.14.) In the acid molecules, there are two polar bonds, the C=O bond and the O–H bond. (See Figure 17.19.) The positively charged hydrogen atom in one molecule attracts the negatively charged oxygen atom in another molecule. At the same time, the positively charged hydrogen atom in the second molecule attracts the oxygen in the first molecule. These extra forces of attraction between molecules result in higher melting and boiling points.

The lighter carboxylic acids are soluble in water. This solubility is also due to the hydrogen bonding between the acid molecules and water molecules. As the length of the carbon chain increases, the influence of the polar part of the molecule, the –COOH, decreases. The non-polar part of the molecule, the carbon chain, increases in length; the molecules become more like hydrocarbons, and the solubility decreases. Pentanoic acid, C_4H_9COOH, is only slightly soluble in water.

Chemical Properties of Carboxylic Acids

When dissolved in water, carboxylic acids ionize only to a slight degree. For example, if ethanoic acid (acetic acid), CH_3COOH, is added to water, about 0.5% of the molecules in the sample ionize. The majority of the molecules remain unaffected.

$$CH_3COOH(l) \Leftrightarrow H^+(aq) + CH_3COO^-(aq)$$

The double arrows, ⇔, show that the reaction is reversible. In a solution of ethanoic acid in water, all three species listed in the equation are present.

Because aqueous solutions of carboxylic acids contain hydrogen ions, $H^+(aq)$, they display the characteristic properties of acids. Their water solutions conduct electricity and they cause acid-base indicators to change colour to the colours typical of acids. (See section 13.5, page 347.) Since the concentration of the hydrogen ions is small, the reactions of the carboxylic acids are less vigorous than those of acids such as hydrochloric acid.

Carboxylic acids react with active metals such as magnesium. For example,

$$2CH_3COOH(aq) + Mg(s) \rightarrow (CH_3COO)_2Mg(aq) + H_2(g)$$

magnesium ethanoate
(acetate)

They also react with carbonates to produce carbon dioxide.

$$2CH_3COOH(aq) + Na_2CO_3(s) \rightarrow 2CH_3COONa(aq) + H_2O(l) + CO_2(g)$$

They also neutralize bases such as sodium hydroxide.

$$CH_3COOH(aq) + NaOH(aq) \rightarrow CH_3COONa(aq) + H_2O(l)$$

Why should carboxylic acids behave as acids when alcohols do not? The only structural difference is the doubly bonded oxygen atom in a spot occupied by two hydrogen atoms in alcohols. The answer is based on our understanding of the electronegativity scale. Oxygen is a highly electronegative element. In the carboxylic acids, the lone oxygen atom draws electrons toward it. This causes the O–H bond to become more polar, and the hydrogen itself more positive. A positive hydrogen atom in such an exposed position is easily attracted by passing water molecules. In some cases, the hydrogen is carried away by the water molecules leaving its electron behind. The presence of the H^+ ions in the solution causes it to behave as an acid.

17.9 Esters

Esters are the compounds formed when a carboxylic acid reacts with an alcohol. For example, when ethanoic acid (acetic acid) is heated with ethanol in the presence of some sulphuric acid, the following reaction occurs.

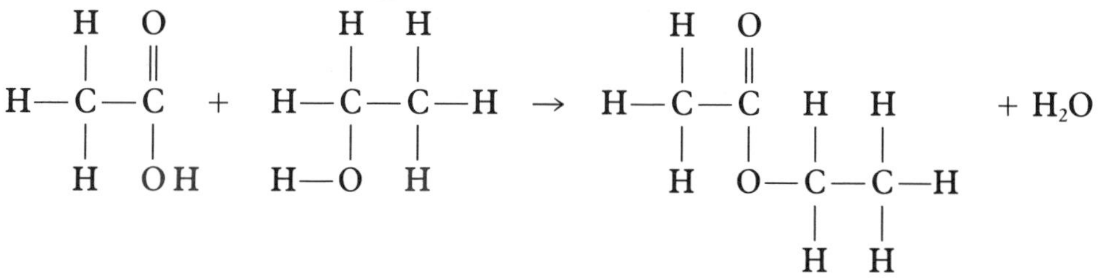

Another way of showing this reaction is

$$CH_3-\overset{\overset{\displaystyle O}{\|}}{C}-O-H + H-O-C_2H_5 \rightarrow CH_3-\overset{\overset{\displaystyle O}{\|}}{C}-O-C_2H_5 + H_2O$$

The ester produced by this reaction is called ethyl ethanoate (or ethyl acetate).

The general formula for an ester is

$$R-\overset{\overset{\displaystyle O}{\|}}{C}-O-R' \quad \text{or} \quad RCOOR'$$

The R and R' represent alkyl groups such as methyl, ethyl, propyl, and so on. In any one ester, the two alkyl groups may be the same or different. The R is the alkyl group associated with the acid, and the R' is associated with the alcohol.

It is interesting to note that water molecules produced in the formation of esters are assembled from a hydrogen atom from the alcohol and a hydroxyl group from the acid. This is the opposite of the process in inorganic neutralization reactions. In neutralizations, the hydrogen comes from the acid and the hydroxide from the base.

$$NaOH + HCl \rightarrow NaCl + H_2O$$

This difference serves as a reminder that the –OH groups in organic molecules are quite different from the OH^- ions in inorganic bases like NaOH.

Naming Esters

The first name of an ester is derived from the alcohol used in its preparation. The second name is taken from the name of the acid. For example, the ester $C_2H_5COOCH_3$ is called methyl propanoate. Its structural formula is

$$H-\overset{\overset{\displaystyle H}{|}}{\underset{\underset{\displaystyle H}{|}}{C}}-\overset{\overset{\displaystyle H}{|}}{\underset{\underset{\displaystyle H}{|}}{C}}-\overset{\overset{\displaystyle O}{\|}}{C}-O-\overset{\overset{\displaystyle H}{|}}{\underset{\underset{\displaystyle H}{|}}{C}}-H$$

Since the carbonyl group, $>C=O$, always comes from the acid, you can see from the structural formula that the acid used must have been C_2H_5COOH, or propanoic acid. The alcohol used must therefore have been CH_3OH, or methanol. This gives the name of the ester as methyl propanoate.

Exercise

9. Name the following esters.
   (a) $CH_3COOC_2H_5$
   (b) $C_3H_7COOCH_3$
   (c) $C_2H_5COOC_3H_7$

10. Write formulas for the following esters.
    (a) ethyl butanoate
    (b) propyl ethanoate
    (c) butyl propanoate

Properties of Esters

The most obvious property of the lighter esters is that they have extremely pleasant aromas. For example, pentyl butanoate smells like apricots, ethyl butanoate smells like pineapples, and octyl acetate smells like oranges. The fragrances of many fruits and flowers are often caused by esters.

An important chemical property of esters is that they undergo **hydrolysis**. This is the splitting apart of a molecule using water. When an ester undergoes hydrolysis, it splits apart into the acid and alcohol from which it was derived. For example,

$$C_2H_5COOC_3H_7 + H_2O \rightarrow C_2H_5COOH + C_3H_7OH$$
propyl propanoate   propanoic acid   propanol

An acid or base is usually used as a catalyst for the hydrolysis of an ester.

The hydrolysis of esters is important in the digestion of fats and oils in our diet. Fats and oils are esters of the alcohol glycerol (see section 17.6) and a number of different long chain carboxylic acids. They react with water in our digestive system, in the presence of enzymes, to form glycerol and a mixture of carboxylic acids.

Laboratory Manual
You will compare three different esters in Experiment 17-3.

17.10 Ethers

In section 17.7, we mentioned the similarity between water and the alcohols. Both contain the hydroxyl group, $-OH$. In fact, alcohols can be regarded as being derived from water by substituting an alkyl group for one of the hydrogen atoms in water.

$$H—O—H \qquad\qquad CH_3—O—H$$
water   methanol

If both of the hydrogen atoms in water are replaced by alkyl groups, a family of compounds called **ethers** results. The two alkyl groups may be the same or different. The simplest ether consists of two methyl groups joined by an oxygen atom.

$$CH_3—O—CH_3$$

Ethers are named by stating the two alkyl groups, in alphabetical order, followed by ether. If the alkyl groups are identical, the prefix "di–" is used. (See Table 17.8 on page 464.)

Table 17.8 Names and Formulas of Some Ethers

Formula	Name
CH_3—O—CH_3	Dimethyl ether
C_2H_5—O—CH_3	Ethylmethyl ether
C_2H_5—O—C_2H_5	Diethyl ether
C_2H_5—O—C_3H_7	Ethylpropyl ether

Diethyl ether was at one time commonly used as a general anesthetic. It has since been replaced by other safer substances. It is extremely flammable and must be handled with caution.

Properties of Ethers

Ether molecules contain electronegative oxygen atoms. Unlike water and alcohol molecules, however, the oxygen in ethers is not bonded to a hydrogen atom. As a result, hydrogen bonds cannot form between ether molecules. The boiling points of ethers are therefore relatively low.

Ether molecules do form hydrogen bonds with water molecules. The positively charged hydrogen atoms in water attract the negative oxygen atoms in the ether. Consequently, ethers dissolve in water to about the same extent as the corresponding alcohols.

Chemically, ethers are similar to the alkanes in that they undergo few chemical reactions other than combustion.

17.11 Aldehydes and Ketones

These two families of compounds will be considered together because they share certain structural features and chemical properties. **Aldehydes** and **ketones** both contain the carbonyl group, $>C=O$, that you first saw in the carboxylic acids. There is one major difference between aldehydes and ketones. With one exception, the carbonyl group in aldehydes is attached to an alkyl group and a hydrogen atom. The exception is the simplest aldehyde, methanal, where two hydrogen atoms are attached to the carbonyl group. In ketones, the carbonyl group is attached to two alkyl groups.

$$\begin{array}{cc} \overset{\displaystyle O}{\underset{\displaystyle \|}{}} & \overset{\displaystyle O}{\underset{\displaystyle \|}{}} \\ R-C-H & R-C-R' \\ \text{an aldehyde} & \text{a ketone} \end{array}$$

In the case of ketones, the two alkyl groups may be the same or different. Tables 17.9 and 17.10 list the names and formulas of some simple aldehydes and ketones.

Table 17.9 Names and Formulas of Some Aldehydes

Formula	Name
$\overset{\displaystyle O}{\overset{\displaystyle \|}{H-C-H}}$ or HCHO	methanal (formaldehyde)
$\overset{\displaystyle O}{\overset{\displaystyle \|}{CH_3-C-H}}$ or CH_3CHO	ethanal (acetaldehyde)
$\overset{\displaystyle O}{\overset{\displaystyle \|}{C_2H_5-C-H}}$ or C_2H_5CHO	propanal (propionaldehyde)
$\overset{\displaystyle O}{\overset{\displaystyle \|}{C_3H_7-C-H}}$ or C_3H_7CHO	butanal (butyraldehyde)

Table 17.10 Names and Formulas of the Two Simplest Ketones

Formula	Name
$\overset{\displaystyle O}{\overset{\displaystyle \|}{CH_3-C-CH_3}}$ or CH_3COCH_3	propanone (acetone)
$\overset{\displaystyle O}{\overset{\displaystyle \|}{CH_3-C-C_2H_5}}$ or $CH_3COC_2H_5$	butanone (methyl ethyl ketone)

The names shown in parentheses in Tables 17.9 and 17.10 are the common names for the aldehydes and ketones. The systematic names for the aldehydes use the prefix indicating the number of carbon atoms, and the suffix "–al". The systematic names for the ketones use the prefix indicating the number of carbon atoms, and the suffix "–one".

Methanal, or formaldehyde, is commonly used as an aqueous solution to preserve biological specimens. (See also Special Topic 17.2.) Propanone, or acetone, is used as nail polish remover.

UREA FORMALDEHYDE FOAM INSULATION

Under certain conditions, formaldehyde (methanal) reacts with urea, another organic compound, to form a polymer called urea formaldehyde resin. When this resin is mixed in the right proportions with a foaming agent, a catalyst, and compressed air, a foam with a consistency somewhat like shaving cream is produced. This urea formaldehyde foam gained some notoriety during the late 1970s as a home insulating material.

In 1977 the federal government approved urea formaldehyde foam insulation, or UFFI for short, as a home insulating material. The owners of poorly insulated older homes were able to obtain a grant from the government to subsidize the cost of insulating their homes. UFFI was a popular material for home insulation, because it could be sprayed directly into wall cavities, completely filling the space. It is estimated that about 50 000 homes insulated under the government plan were insulated with UFFI. In addition, there may have been as many as another 30 000 homes that had UFFI applied without government assistance.

Even while this program was in existence, studies carried out in the United States showed that there were serious problems associated with UFFI. It appeared that the UFFI, when exposed to moisture and high temperatures, decomposed in the wall cavities, releasing for-maldehyde into the home. Formaldehyde is known to cause the following symptoms: respiratory problems, ear, nose, and throat irritation, headaches, nausea, dizziness, skin rashes, allergies, and asthmatic attacks. These symptoms, in various combinations, were found in residents of homes where UFFI had been blown into the walls. In addition, there is a possibility that formaldehyde causes genetic defects, and may be carcinogenic.

By 1980, it was obvious that many Canadian homeowners who had insulated their homes with UFFI were suffering from overexposure to formaldehyde fumes. Faced with an increasing number of complaints about health problems, the federal government, in December of 1980, announced a ban on the use of urea formaldehyde foam as a home insulating material. The government subsequently announced a program to pay $5000 per home toward the cost of removing the insulation. It has been estimated that the cost of removing this insulation is at least $10 000 per home. In terms of health, however, it may be impossible to measure the final cost to homeowners.

Physical Properties of Aldehydes and Ketones

There are no O–H bonds in either aldehydes or ketones. There is, therefore, no hydrogen bonding between aldehyde molecules or between ketone molecules. Consequently, aldehydes and ketones have boiling points lower than alcohols with the same number of carbon atoms.

Water molecules can form hydrogen bonds with both aldehyde and ketone molecules. (See Figure 17.20.) As a result, aldehydes and ketones have approximately the same solubilities as alcohols with the same number of carbon atoms.

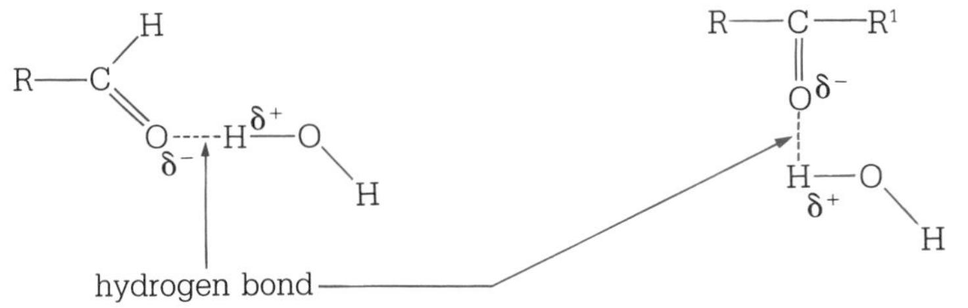

Figure 17.20
Hydrogen bonds can form between water molecules and aldehyde molecules, and between water molecules and ketone molecules.

17.12 A Summary of Functional Groups

The properties of an organic compound are determined by the presence of a functional group in the molecule. Table 17.11 summarizes the functional groups discussed in this chapter.

Table 17.11 Some Organic Functional Groups

Family	Functional Group	Example	Formula
Alkanes	None	Ethane	C_2H_6
Alkenes	$>C{=}C<$	Ethene	$CH_2{=}CH_2$
Alkynes	$-C{\equiv}C-$	Ethyne	$CH{\equiv}CH$
Alcohols	$-OH$	Ethanol	C_2H_5OH
Carboxylic acids	$-\overset{\overset{O}{\|\|}}{C}-OH$	Ethanoic acid (acetic acid)	$CH_3\overset{\overset{O}{\|\|}}{C}-OH$
Esters	$-C-\overset{\overset{O}{\|\|}}{C}-O-C-$	Methyl ethanoate (methyl acetate)	$CH_3-\overset{\overset{O}{\|\|}}{C}-O-CH_3$
Ethers	$-C-O-C-$	Dimethyl ether	CH_3-O-CH_3
Aldehydes	$-\overset{\overset{O}{\|\|}}{C}-H$	Ethanal (acetaldehyde)	$CH_3-\overset{\overset{O}{\|\|}}{C}-H$
Ketones	$-C-\overset{\overset{O}{\|\|}}{C}-C-$	Propanone (acetone)	$CH_3-\overset{\overset{O}{\|\|}}{C}-CH_3$

PUTTING CHEMISTRY TO WORK

DR. EDWARD ADAMEK / ORGANIC CHEMIST

Sophisticated chemical tests have found pollution from toxic substances in even remote areas of the environment. Many of the most dangerous of these substances are organic compounds such as polychlorinated biphenyls (PCBs), pesticides, herbicides, and chlorinated aliphatic and aromatic hydrocarbons (organic solvents used in industry). These compounds are dangerous both because they are toxic to living things and because some can accumulate in the bodies of organisms, entering the food web.

Dr. Adamek is in charge of several government laboratories in which samples of soil, vegetation, sediments, and living organisms are being analyzed for pollution by organic compounds. "My work includes advising government agencies, industry, universities, and the general public about my findings," he explains. Dr. Adamek also becomes involved in disputes over accidental spills of chemicals or deliberate dumping by industry. "I testify in court as an expert witness when necessary."

Despite a heavy workload, Dr. Adamek finds his job stimulating and interesting. "I am continuously learning and studying about many chemical-environmental problems. I must also stay aware of new technology in the field of organic chemistry."

Dr. Adamek became interested in chemistry while in secondary school in his native Austria. His thirst for knowledge led him to universities in Vienna and Innsbruck and eventually to Canada, where he studied organic chemistry at McGill University under Dr Purves, a world-renowned professor. During this time, and for several years afterward, Dr. Adamek worked for Dupont, the world's largest chemical company, as a research chemist. "Over the years I have produced many new chemicals and chemical processes," Dr. Adamek recalls. "These are patented and are used not only in Canada but in many other industrialized countries as well."

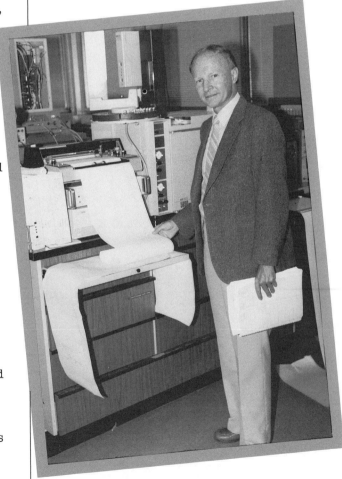

Looking Back – Looking Ahead

In this introduction to organic chemistry, you have found that the study of millions of compounds can be simplified by classifying them according to the presence of functional groups in their molecules. It is, however, impossible to discuss more than a very small percentage of organic compounds and reactions in a single introductory chapter. You may investigate organic chemistry in greater depth in later studies.

The next chapter also serves as an introduction to a subject of great interest and concern – not only to chemists and other scientists such as physicists, ecologists, and astronomers, but also to technologists and society as a whole. This subject, nuclear chemistry, will help to prepare you for similar, later studies. It will also serve to bring into focus once again the importance of the science-technology-society interrelationship you have been considering throughout this textbook.

Vocabulary Checklist

You should now be able to define or explain the meaning of the following terms.

addition reaction	aromatic hydrocarbon	hydrocarbon	polymer
alcohol	carboxylic acid	hydrolysis	resonance hybrid
aldehyde	complete combustion	incomplete combustion	saturated
aliphatic hydrocarbon	ester	isomer	substitution reaction
alkane	ether	ketone	unsaturated
alkene	functional group	monomer	
alkyne	homologous series	organic compound	

Chapter Objectives

You should now be able to do the followig.

1. Explain the difference between organic and inorganic compounds.

2. Account for the existence of an extremely large number of organic compounds.

3. Recognize the impact of organic chemicals on our lives.

4. Appreciate that a study of organic chemistry is made easier by grouping compounds according to the presence of functional groups.

5. Distinguish between aliphatic and aromatic hydrocarbons; and between saturated and unsaturated hydrocarbons.

6. Name and draw structural formulas for the first ten members of the alkanes, alkenes, and alkynes.

7. Recognize and draw structural formulas for isomers of some simple hydrocarbons.

8. Describe some of the physical and chemical properties of the alkanes, alkenes, and alkynes.

9. Relate the physical and chemical properties of the alkanes, alkenes, and alkynes to the molecular structure of the compounds.

10. State the uses for several hydrocarbons.

11. Identify and state the functional group in each of the following: alcohols, carboxylic acids, esters, ethers, aldehydes, and ketones.

12. Describe and explain some of the physical and chemical properties of the groups listed in objective 11.

13. Identify some common examples of the groups listed in objective 11.

Understanding Chapter 17

REVIEW

1. What unique characteristics of carbon are responsible for the large number of organic compounds?

2. (a) What is meant by the term hydrocarbon?
   (b) Give the names and formulas of three hydrocarbons, each containing three carbon atoms per molecule.

3. (a) What are isomers?
   (b) Draw structural formulas for three isomers of heptane, C_7H_{16}.

4. What is the major difference between alkanes and alkenes? Use examples to illustrate your answer.

5. (a) What chemical reaction is common to alkanes, alkenes, and alkynes?
   (b) What is the major use of alkanes?

6. (a) Write the names and formulas of the first three members of the alcohol family derived from the straight chain alkanes.
   (b) The three compounds you named are liquid at room temperature. The alkanes from which they are derived are gases at room temperature. Explain this difference.

7. Write equations for the following reactions.
   (a) lithium + ethanoic acid (acetic acid)
   (b) calcium hydroxide + methanoic acid (formic acid)

8. Write the names and formulas of the first three members of the carboxylic acid family.

9. In what main way do the carboxylic acids differ from inorganic acids such as hydrochloric acid?

10. Explain briefly the meaning of each of the following terms.
    (a) substitution reaction
    (b) addition reaction
    (c) saturated hydrocarbon
    (d) functional group
    (e) polymer
    (f) aromatic hydrocarbon

11. Why is a molecular structure consisting of alternating single and double bonds unsatisfactory for benzene?

12. (a) Aldehydes and ketones share a common molecular feature. Identify that feature.
    (b) In what way do molecules of aldehydes and ketones differ?
    (c) Draw structural formulas and give names for the aldehyde and ketone containing three carbon atoms per molecule.

APPLICATIONS AND PROBLEMS

13. Write the formula for each of the following.
    (a) the alkane containing 22 carbon atoms
    (b) the alkene containing 17 carbon atoms
    (c) the alkyne containing 13 carbon atoms

14. Write balanced equations for the following reactions.
    (a) the complete combustion of pentane, C_5H_{12}
    (b) the incomplete combustion of propene, C_3H_6

15. Ethane and ethene both react with bromine. Show, by means of equations, the reactions that occur. What is the essential difference between the two reactions?

16. Alkanes, alkenes, and alkynes are all insoluble in water. Explain this observation in terms of the structure of the molecules.

17. There is only one straight chain alkane containing four carbon atoms, but two straight chain alkenes containing four carbon atoms. Explain with the aid of structural formula diagrams why this is so.

18. Are alkynes more like alkenes or alkanes in their reactions? Explain.

19. Explain why methanol is more soluble in water than methane is.

20. You may remember the violent reaction that occurs when sodium is added to water. Hydrogen is produced in the reaction. Sodium and ethanol react much less vigorously when mixed together, but the reaction is similar. One way of getting rid of pieces of sodium is to add them to ethanol. Write an equation for the reaction between sodium and ethanol.

21. Ethanoic acid, CH_3COOH, is much more soluble in water than is hexanoic acid, $C_5H_{11}COOH$. Explain this observation.

22. The molecular formula, C_3H_8O, can represent either an alcohol or an ether. Draw structural formulas for each compound, and name each one.

23. The two compounds mentioned in question 22 have different boiling points. The alcohol has a higher boiling point than the ether. Explain.

24. Write a balanced equation to show the formation of the ester pentyl propanoate, $C_2H_5COOC_5H_{11}$.

CHALLENGE

25. Carboxylic acids have higher boiling points than alkanes of about the same molar mass. Their boiling points are also higher than alcohols with the same number of carbon atoms. Explain these observations.

26. The boiling points of aldehydes and ketones are lower than the boiling points of alcohols with the same number of carbon atoms. Explain this observation.

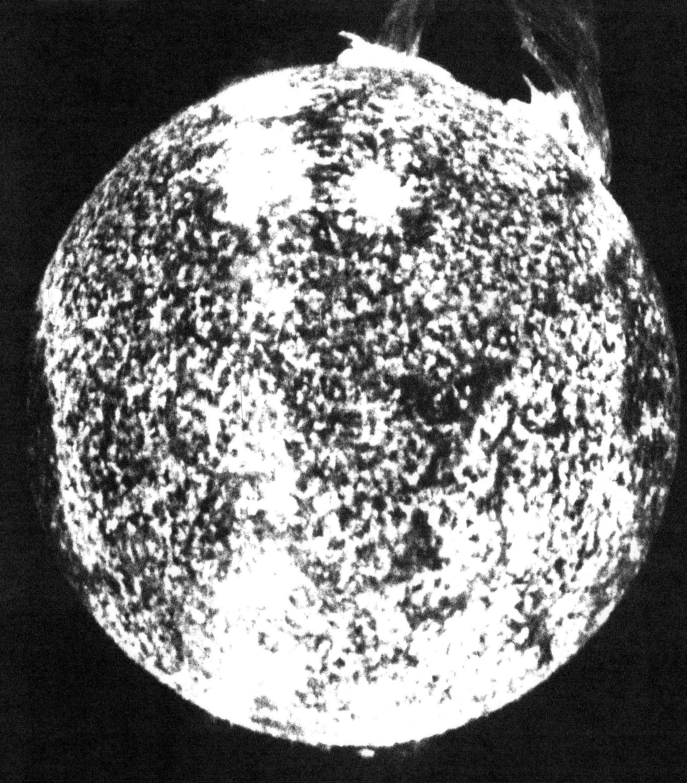

IX NUCLEAR CHEMISTRY

18 NUCLEAR CHEMISTRY

A large percentage of the energy used by our industrial society is produced by burning non-renewable fossil fuels – gas, oil, and coal. As the supplies of these resources dwindle, scientists and technologists have investigated and developed alternative energy resources such as solar energy, energy from wind and water, heat from the earth's crust (geothermal energy), energy from decaying vegetation and manure (biomass), and nuclear energy. Of these alternatives, the most controversial one is undoubtedly nuclear energy. Numerous questions surround this controversial issue. Can we justify more nuclear power stations? Are these nuclear power stations safe? Are we exposing ourselves to the danger of another disaster like the one at Chernobyl in the Soviet Union? Even if the nuclear power plants operate smoothly, can their radioactive by-products be reprocessed or stored safely for the forseeable future?

Nuclear science and technology have been applied to the development of a wide range of useful techniques and devices such as medical procedures, food preservation, fire detectors, and industrial control systems. Technology has also enabled the production of some of the most lethal weapons ever made. In this chapter, you will study some of the theories and terminology associated with nuclear energy and radioactivity. With a better understanding of nuclear science, you should be able to make your own judgements about benefits and risks. No form of technology is perfectly safe. Risk is a part of life. But how much risk is acceptable? It is up to you to decide.

PUTTING CHEMISTRY TO WORK

STEPHEN STRAUSS / SCIENCE REPORTER

It is a fact that people are affected by science and technology in everyday life. Chemistry in particular is often "hot news" — from the concerns about pollution or food additives to the benefits of newly developed processes or chemicals. Stephen Strauss is a science writer for the Globe and Mail newspaper. It is his job to report on the sciences in Canada and in the world.

"I tell people what is going on," Stephen says simply. "If I don't do too bad a job of it, the public knows more about things which will change their lives. If I do it poorly, I could create some strange and incorrect notions in people's minds."

Many people gain their knowledge about current science by reading newspaper accounts such as those prepared by Stephen. "I must take complex ideas *to* people," he explains. "Most readers are not scientists, but they are interested in science. So I often need non-science images to explain or illustrate an idea." One example of this imagery involved a new technique for producing penicillin. "The scientists were engineering molecules which could be made into a specific type of penicillin, formerly produced only by subjecting the original penicillin compound to a number of chemical steps," Stephen recalls. Stephen's first simile likened the process to cotton that wove itself into cloth while on a plant.

Stephen shakes his head. "It seemed to me to be a wonderful image. The scientists had trouble with it, though — wasn't chemical enough." Stephen went home and in the middle of the night had a vision. "Eureka! Pickles!" By comparing the new process to cucumbers turning to pickles while still on the vine, Stephen clarified the technology for his readers. "And the scientists thought it was great!"

Scientists do not always enjoy Stephen's articles. "I must approach what I do, and who I report on, very skeptically," he explains. "I am not a cheerleader for science — I am a reporter of facts and information. Sometimes the truth hurts."

Stephen relishes the freedom he has to pick and choose what he reports on. "My idea was to find an area in journalism I could do for many years without becoming bored or disillusioned. Science writing allows me to hop from topic to topic while at the same time continuing to learn and grow." As to the qualifications for a science reporter, Stephen notes: "The chemistry and physics courses I've taken have been very handy, but the most important qualifications for any writer are clear thinking and clear English."

18.1 Nuclear Forces

What holds the nucleus of an atom together? Think about an oxygen atom, for example. According to the laws of static electricity, the eight positive protons in the oxygen nucleus should exert a tremendous repulsion on each other. Why does the nucleus not explode? As discussed briefly in Chapter 4, there is a force even stronger than the electrostatic force. This stronger force is called the strong nuclear force. This force only becomes effective at very short distances within the nucleus. At these distances, the strong nuclear force between protons and neutrons can be up to one hundred times as strong as the electrostatic force.

The strong nuclear force and the electrostatic force operating in a nucleus can be compared to the forces that operate when the north poles of two magnets covered by velcro are brought close together. (See Figure 18.1.) Like two positive protons, the magnets shown in Figure 18.1 repel each other. However, if they are forced close enough together to engage the "velcro force", they will stick together firmly, even though the magnets are still repelling each other. In a stable nucleus, it is the strong nuclear force that causes protons and neutrons to cling together, despite the repulsive electrostatic force.

(a)

(b)

Figure 18.1
Nuclear forces – a comparison. The electrostatic and strong nuclear forces operating in a nucleus can be compared to the repulsion between the north poles of two velcro covered magnets. Just like two protons, the north poles of the magnets repel each other (a). However, if enough energy is expended in forcing them very close together, the very strong, short-range "velcro force" (strong nuclear force) arises and the magnets (protons) are held together (b).

18.2 The Nucleus

Early in this century, nuclear pioneers led by Rutherford, Chadwick, and Thomson showed that the atom is not the indivisible particle imagined by Democritus and Dalton. Scientists began to think about a new set of "indivisible" particles called protons, neutrons, and electrons. But the protons and neutrons can no longer be described as indivisible.

Half a century after Rutherford's discovery, specific structure was discovered in the **nucleons** which make up the nucleus of an atom. (Protons and neutrons are known collectively as nucleons.) Experiments with high energy proton beams have shown that nucleons themselves have three separate centres of electrical charge within them. (No such structure has so far been found in electrons.)

In the late 1960s, the American physicist Murray Gell-Mann developed a theory to explain observations such as those discussed above. He proposed the existence of a new kind of subnuclear particle which he called a **quark**. According to this theory, quarks would be influenced by the strong nuclear force, and combine in different ways to form protons, neutrons, and other kinds of subatomic particles.

High-energy collision experiments with various subatomic particles have provided evidence for six kinds of quarks as listed in Table 18.1. Note from the table that quarks have a *fractional* electric charge, such as $-\frac{1}{3}$ or $+\frac{2}{3}$ of a normal electron or proton charge. Protons and neutrons consist of only "up" and "down" quarks. (See Figure 18.2.) The fractional electric charges of quarks combine to produce an overall nucleon charge of either $1+$ for the proton or 0 for the neutron.

Figure 18.2
A neutron consists of one u quark and two d quarks. A proton consists of one d quark and two u quarks.

(a)

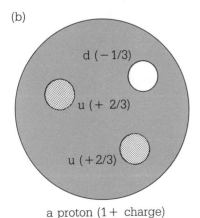

a neutron (0 charge)

(b)

a proton (1+ charge)

Table 18.1 Types of Quarks

Quark Name	Symbol	Electrical Charge
Up	u	$+\frac{2}{3}$
Down	d	$-\frac{1}{3}$
Strange (or sideways)	s	$-\frac{1}{3}$
Charmed	c	$+\frac{2}{3}$
Beauty (or bottom)	b	$-\frac{1}{3}$
Truth (or top)	t	$+\frac{2}{3}$

Note: The whimsical names given to the six types of quarks indicate that even "serious" scientists have poetry lurking in their souls.

Because the distance between the quarks in a nucleon is even less than the distance between nucleons, we would expect the force between quarks to be very strong indeed. Experiments show that the quark–quark strong force is about ten times as strong as the nuclear strong force between two nucleons.

There is a rough similarity between the way a nucleus is held together and the way an ice crystal is held together. *Within each* water molecule there are relatively large electrostatic forces (co-valent bonds) holding the atoms to each other. *Between* the polar water molecules there are lesser electrostatic forces (dipole-dipole attractions, about $1/10$ as strong) that bind the molecules to each other in the ice crystal. Similarly, *within each* nucleon there are very large "strong nuclear" forces holding the quarks to each other. *Between* the nucleons there are lesser "strong nuclear" forces that bind the nucleons together in the nucleus. (See Figure 18.3.)

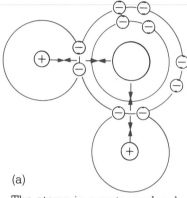

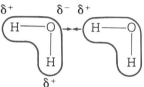

Figure 18.3
Electrostatic forces and molecules compared with strong nuclear forces and nucleons

(a)
The atoms in a water molecule are held together by relatively strong electrostatic attractions between electrons and atomic nuclei.

The water molecules in an ice crystal are held together by the much weaker attractions between the polar ends of water molecules.

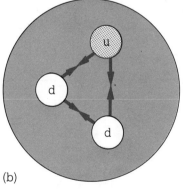

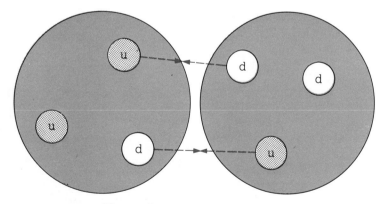

(b)
The quarks in a proton or neutron are held together by the very strong forces between quarks.

The nucleons in a nucleus are held together by strong nuclear forces operating over a longer distance, which weakens them

Nuclear Stability

Despite the effect of the strong nuclear force in holding the nucleus together, the structure can be overwhelmed by electrostatic repulsions if too many of the nucleons are protons. The positive electrical charge on protons causes them to repel each other. A nucleus made

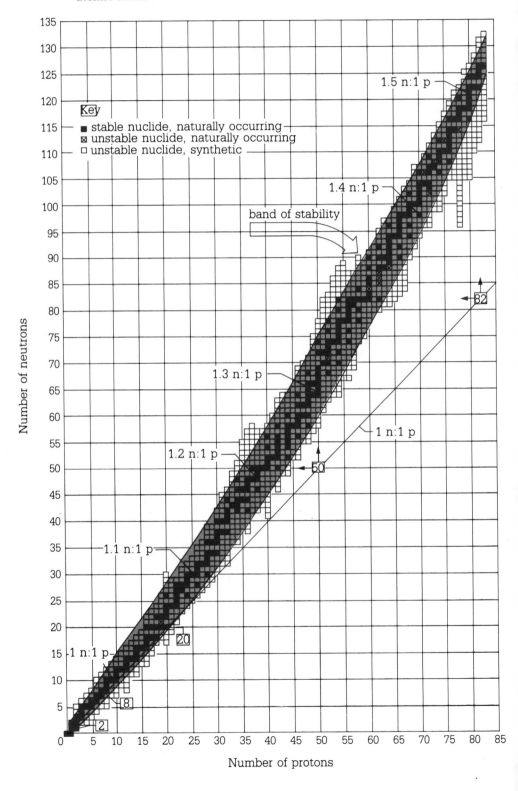

Figure 18.4
The region of stability for nuclei. A nucleus is stable only if it has a correct neuton-to-proton ratio. This ratio increases with increasing atomic mass.

of nothing but protons would have a very short life. Neutrons, on the other hand, have a stabilizing effect. With an electrical charge of zero, they contribute to the strong nuclear attractions without adding extra electrostatic repulsion. Thus, in virtually all stable nuclei, we find that at least half of the nucleons are neutrons. The graph in Figure 18.4 shows the number of neutrons versus the number of protons for all known stable and unstable isotopes (nuclides). You can see that as the atomic number increases, the ratio of neutrons to protons in a stable nucleus increases also. Surprisingly, nuclei with too high a neutron-to-proton ratio are also unstable. All of the stable nuclei as well as many unstable ones fall within a range called the **band of stability**. If the neutron-to-proton ratio is above or below the band of stability shown in Figure 18.4, however, the nucleus is always unstable. These unstable nuclei are **radioactive**. They will sooner or later change into more stable structures by "kicking out" subatomic particles (alpha or beta particles) or by losing excess energy by gamma radiation. This radioactive decay continues until the nucleus reaches a neutron-to-proton ratio corresponding to a stable nucleus.

Exercise

1. Use the graph in Figure 18.4 to determine which of the following are stable isotopes.

   (a) $^{92}_{40}Zr$ (b) $^{80}_{35}Br$ (c) $^{152}_{67}Ho$ (d) $^{200}_{80}Hg$

2. Why would $^{100}_{50}Sn$ be extremely unstable?

3. Why would $^{58}_{25}Mn$ be unstable?

18.3 Radioactivity

Unstable nuclei are called **radionuclides** or **radioisotopes**. These nuclei have excess energy which is distributed among the nucleons in a random and changing pattern. Every now and again this energy becomes concentrated in such a way that it can be eliminated from the nucleus by the ejection of either particles or gamma rays.

Scientists have found that they cannot predict when a particular radioactive nucleus will decay. What they have found, however, is that a fixed percentage of the nuclei in a sample will decay in a certain time interval. Thus, the greater the number of radioactive nuclei present, the greater the number that will disintegrate per second. The percentage that disintegrates per second is the same, regardless of the original number of nuclei present. This means that as time passes, the number of nuclei that decay per second in a given sample decreases. (See Figure 18.5 on page 480.)

Figure 18.5
Radioactive substances have a characteristic decay time. A definite, fixed proportion of the atoms in a given sample of a radioisotope undergo radioactive decay every second. A larger sample with more atoms has a larger number of decays per second than does a smaller sample. However, the percentage of nuclei that decay per second remains constant.

(a)

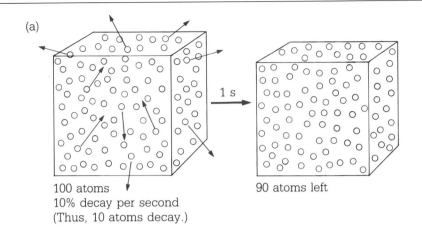

100 atoms
10% decay per second
(Thus, 10 atoms decay.)

90 atoms left

(b)

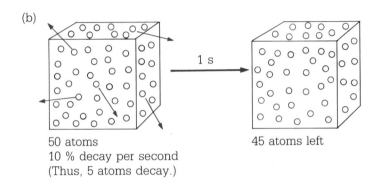

50 atoms
10 % decay per second
(Thus, 5 atoms decay.)

45 atoms left

(c)

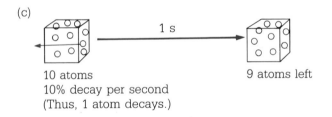

10 atoms
10% decay per second
(Thus, 1 atom decays.)

9 atoms left

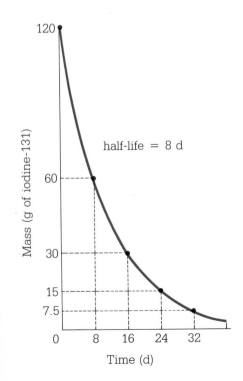

half-life = 8 d

Mass (g of iodine-131)

Time (d)

Each kind of radionuclide decays at a definite predictable rate. This rate is expressed as the half-life of the isotope. **Half-life** is defined as the time it takes for one-half of the atoms in a sample to decay. (See Figure 18.6.) An isotope's half-life is independent of the amount of isotope present. It is not affected by temperature or pressure. Even if the radioactive element is part of a compound, the half-life remains unchanged.

Figure 18.6
The half-life of a radionuclide is the time taken for half of the original number of atoms to decay. During each succeeding half-life period, half of the remaining (undecayed) nuclei will decay. Thus, an initial mass of 120 g of radioactive iodine-131, having a half-life of 8 d, would dwindle to 60 g in 8 d (one half-life), 30 g in 16 d (two half-lives), 15 g in 24 d (three half-lives), etc.

18.4 Types of Radiation

As discussed in Chapter 2, radioactive substances have three "natural" ways of decaying – alpha, beta, and gamma radiation.

Alpha Particles

An unstable nucleus can approach the band of stability (Figure 18.4) by ejecting a helium nucleus. Recall that a helium nucleus is also called an **alpha particle**.

$$^{210}_{84}Po \quad \rightarrow \quad ^{206}_{82}Pb \quad + \quad ^{4}_{2}He$$

| a radioactive isotope | a stable isotope | an alpha particle |

Note that in the above **nuclear equation**, the mass number on the reactant side of the equation (210) is equal to the total mass number on the product side (206 + 4).

The atomic numbers are also equal on both sides of the equation (84 = 82 + 2). This is reflected by experimental findings that the total number of nucleons and the overall electrical charge are both conserved in a nuclear reaction.

Sample Problem 18.1

Radon-222 decays by emitting an alpha particle. What is the isotope produced when radon-222 decays?

Solution

$$^{222}_{86}Rn \rightarrow {}^{4}_{2}He + {}^{A}_{Z}X$$

(a) The total electrical charge is conserved in a nuclear reaction. Thus,

$$86 = 2 + Z$$
$$Z = 84$$

(b) The total mass number (number of nucleons) is conserved in a nuclear reaction. Thus,

$$222 = 4 + A$$
$$A = 218$$

Therefore, the isotope produced is $^{218}_{84}Po$.

Beta Particles

Another way a radioactive nucleus can approach the band of stability is to eject an electron from its nucleus. This electron is called a

beta particle. A beta particle is the same as the "normal" electrons surrounding the nucleus, but it is moving very fast as it leaves the nucleus. The loss of a negative particle from the nucleus effectively changes a neutron in the nucleus into another proton. Thus, the mass number of a beta-emitting nucleus remains the same, but the atomic number (number of positive charges in the nucleus) *goes up* by one.

$$^{14}_{6}\text{C} \rightarrow \, ^{14}_{7}\text{N} + \, ^{0}_{-1}\text{e}$$

| a radioactive | a stable | a beta particle |
| isotope | isotope | (an electron) |

Note that in this equation, the mass number of an electron is zero and its "atomic number" is just its electrical charge of -1.

Both alpha and beta decay produce new elements. The process of changing one element into a different one is called **transmutation**.

Gamma Rays

Like X rays, **gamma radiation** is a high energy form of light. (Gamma radiation is even more energetic than the more familiar X rays.) We already know that atoms can emit visible light when their orbital electrons gain energy and then fall back to a lower energy state. It is reasonable to hypothesize that a similar process goes on within the nucleus. Because the forces involved in the nucleus are so great, it is not surprising that the radiation emitted by an excited nucleus is so much more energetic than visible light.

As the following equation shows, a gamma ray–emitting nucleus has the same proton and neutron make-up before and after emitting the gamma ray. Transmutation does not take place during gamma decay.

$$^{60}_{27}\text{Co}^{m} \rightarrow \, ^{60}_{27}\text{Co} + \, ^{0}_{0}\gamma$$

m (metastable) means	ground	a gamma
that the nucleus is	nuclear	photon
in a high energy state	state	

The characteristics of the three types of radioactive decay are summarized in Table 18.2.

Table 18.2 Radiation Characteristics

Radiation Type	Mass Number	Charge	Identity
Alpha (α)	4	$+2$	Helium nucleus
Beta (β)	0	-1	Electron
Gamma (γ)	0	0	Very high energy light

Exercise

4. For each of the following, identify the unknown particle or isotope.

   (a) $^{234}_{91}\text{Pa} \rightarrow \, ^{0}_{-1}\text{e} \, + \, ?$

   (b) $^{210}_{84}\text{Po} \rightarrow \, ^{206}_{82}\text{Pb} \, + \, ?$

Radiation Penetrating Ability

The three types of radiation differ greatly in their ability to penetrate substances. Heavy, doubly charged alpha particles interact strongly with matter and do not travel very far. Singly charged beta particles travel a little farther through matter. Like the more familiar X rays, uncharged and virtually massless gamma rays can pass through relatively thick layers of material. (See Table 18.3.)

Table 18.3 Penetrating Ability of Different Types of Radiation

Radiation Type	Ability to Penetrate Matter
Alpha Particles	Stopped by 4 cm of air or a single sheet of paper.
Beta particles	Stopped by about 12 cm of air or a stack of paper several millimetres thick.
Gamma rays	Decreased in intensity by about 10% by 30 mm of lead. (For comparison, X rays are decreased by about 10% by 0.5 mm of lead.)

18.5 Uses of Radioactive Isotopes

More and more radioactive isotopes are being used in a variety of devices and processes. Every home or apartment should have a smoke/fire detector. Most of these detectors make use of a small amount of an alpha-emitting radioisotope for their operation. Surgical and medical supplies such as sutures and bandages are sterilized after packaging by exposing them to gamma rays. A new method of preserving food also involves irradiation with gamma rays. The radiation kills bacteria that would otherwise grow and cause food to rot. Radioisotopes are also used in a wide range of medical applications for diagnostic procedures and for destroying cancerous tissue. (See Table 2.3 in Chapter 2.)

Radioisotopes have many industrial and scientific applications. These include tracers, monitors, energy sources, imagers, and clocks (for dating materials). They are also used to cause mutations and for sterilization in the studies of plants and insects. (See Table 18.4 on page 484.)

THE OPERATION OF A FIRE DETECTOR

A home fire detector contains a small amount of the alpha emitting radioisotope americium-241. The alpha particles ionize the air between a pair of electrodes. This allows a small current to flow between them, as shown in the diagram.

A fire produces smoke particles and invisible combustion products. These substances attract ions. When they enter the space between the electrodes, they "mop up" the radiation-induced ions, interrupting the current flow. An electronic sensor reacts to the interrupted current flow by triggering an alarm.

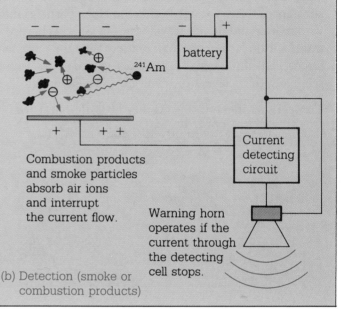

Operation of a fire detector

1. Americium-241 emits α particles which ionize air molecules.

detecting cell

2. Charged plates attract the ions and small current flows.

3. Warning horn is silent as long as a current flows.

(a) Normal (no fire present)

Combustion products and smoke particles absorb air ions and interrupt the current flow.

Warning horn operates if the current through the detecting cell stops.

(b) Detection (smoke or combustion products)

Table 18.4 Industrial and Scientific Uses of Radioisotopes

Use	Sample Applications
Tracers	**Wear on surfaces** Radioisotopes in surface materials collect in lubricating oil where the amount can be detected.
	Leaks Leaks in pipes and pipelines can be detected if radioisotopes flow through and collect at the leak points.
	Labels Chemical reactions can be studied by "labelling" reactants with radioactive isotopes. The destination of each isotope can be determined by using radiation detectors. Similarly, the paths followed by nutrients in living organisms can

		be traced by means of radioactive labelling. The movement of radioactively tagged insects or pollen can be followed.
Monitors	**Gauges** The thickness of thin sheets of materials produced by automated processes can be measured and regulated by measuring the radiation that gets through the sheet. **Detectors** The presence of smoke or combustion products in air can be detected by interrupting a radiation-induced electrical current. The resulting change in conductivity triggers an alarm.	
Energy sources	**Electricity** The radiation from radioisotopes produces heat which can be used to produce electricity. Such small, compact nuclear generators are used to power satellites, navigation devices (lighted buoys), and remote weather stations.	
Materials treating	**Sterilization** Gamma radiation can be used directly to sterilize medical equipment and preserve food. **Strengthening** Radiation causes an increase in strength of some plastics.	
Imagers	**Photography** Photographs of the inside of metallic parts can be made using gamma rays to locate hidden flaws inside the metal.	
Clocks	**Dating** The age of artifacts can be measured by detecting the amount of carbon-14 present. The age of rocks can be measured based on their content of radioisotopes.	
Genetic Agent	**Mutations** Mutations in plants and animals can be induced in order to study heredity and to create new varieties which may be useful in agriculture. **Insect control** Insects can be controlled by sterilizing males by radiation and releasing them. No offspring are produced by females mating with these sterilized males.	

18.6 Induced Nuclear Reactions

In section 18.4 you learned that unstable (radioactive) nuclei could become more stable by emitting radiation in the form of alpha or beta particles, or gamma rays. In this section, you will learn of two other methods by which nuclei can react. The first of these is a process called **nuclear fission**. In fission, a heavy nucleus splits into two fragments of roughly equivalent masses. The second process is **nuclear fusion**. In fusion two light nuclei are joined into a single nucleus. Both of these processes are usually accompanied by the release of extremely large amounts of energy. Nuclear fission is the process that takes place in nuclear power reactors and in atomic (fission) bombs. Nuclear fusion is the process that gives the sun its heat and light and powers hydrogen (fusion) bombs. Table 18.5 compares the energy of nuclear processes with the energies of the more familiar physical and chemical changes.

Table 18.5   Comparison of Physical, Chemical, and Nuclear Reactions

Reaction Class	Reaction Description		Approximate Energy Involved
	Taking Apart	Putting Together	
Physical	Vaporization	Condensation	50 kJ/mol
Chemical	Decomposition	Synthesis	500 kJ/mol
Nuclear	Fission	Fusion	10^{10} kJ/mol

18.7 Nuclear Reactions and Potential Energy

Why do some atoms undergo fission and others fusion? Why is energy released in both cases? One major tendency in nature is for things to "fall" to their lowest potential energy state. Atoms emit light energy as their electrons fall to their lowest allowed potential energy states. Atoms nearly always release heat and/or light as they join together to form compounds (thus lowering their potential energy). Similarly, when atomic nuclei react, they tend to shift to a lower potential energy. The shift in this case is extremely large.

The Mass-Energy Relationship

The mass of a nucleus is different from the total mass of the separate nucleons of which it is made. Consider the following example.

$$10\ {}^{1}_{1}\text{H} \qquad + \qquad 10\ {}^{1}_{0}\text{n} \qquad \rightarrow \qquad {}^{20}_{10}\text{Ne}$$

| 10 protons | 10 neutrons | neon nucleus |

10 protons each of mass 1.007825 u

10 neutrons each of mass 1.00865 u

neon nucleus

$m = 10 \times 1.007825\ \text{u}$
$\quad = 10.0783\ \text{u}$

$m = 10 \times 1.00865\ \text{u}$
$\quad = 10.0865\ \text{u}$

$m_{\text{nucleons}} = 20.1648\ \text{u}$

$m_{\text{Ne}} = 19.9924\ \text{u}$

$$m_{\text{difference}} = m_{\text{nucleons}} \quad - \quad m_{\text{Ne}}$$
$$= 20.1648\ \text{u} - 19.9924\ \text{u}$$
$$= 0.1724\ \text{u}$$

Thus, the total mass of 10 protons and 10 neutrons separately is 0.172 u more than the mass of 10 protons and 10 neutrons assembled in the neon nucleus. This mass difference is called the **mass defect**. Albert Einstein's famous relationship, $E = mc^2$, shows that mass is a form of potential energy. The mass lost (mass defect) in assembling separate nucleons into a nucleus is released as a specific amount of kinetic and radiant energy. Once a nucleus has formed and has released its energy of formation, it would require that same amount of energy (called the **binding energy**) to decompose it into nucleons again. Figure 18.7 shows a graph of the average mass per nucleon in an element's nucleus plotted against the atomic mass number of the element. Since mass is a form of potential energy, this is in effect a nuclear potential energy graph.

Figure 18.7 on page 488 shows that those elements lighter than 56 u tend to decrease their potential energy by undergoing fusion. Isotopes heavier than 56 u tend to decrease their potential energy by undergoing fission. Whether by fission or fusion, the tremendous amount of energy resulting from nuclear reactions is simply the conversion of this potential energy (mass) into kinetic energy, heat, and radiation. Since the potential energy drops so much more quickly when lighter particles undergo fusion than when heavier particles undergo fission (Figure 18.7), fusion reactions produce much more energy (per unit mass) than fission reactions do.

Figure 18.7
Nuclear potential energy measured as mass per nucleon. Both high and low atomic mass nuclei have large nuclear potential energies. The iron-56 nucleus has the lowest nuclear potential energy. Thus, the fusion of lighter nuclei and the fission of heavier ones both release kinetic and other forms of energy.

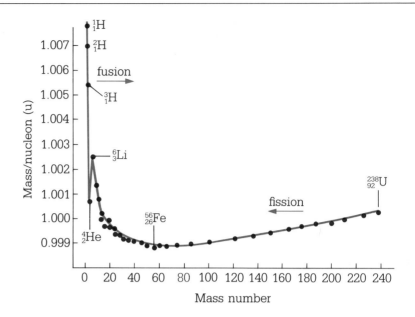

THE LAW OF CONSERVATION OF MASS

To understand the huge amounts of energy involved in nuclear changes, we have to consider the relationship between energy and mass. In ordinary chemical reactions, there is no detectable change in mass. Experimentally, we have found that the mass of the products in any chemical reaction is always equal to the mass of the reactants. This result is the basis for the Law of Conservation of Mass. But even in ordinary chemistry this law may not be totally correct. It is thought that very small (unmeasurable) amounts of mass do disappear in exothermic chemical reactions and very small extra amounts turn up in endothermic reactions. In nuclear reactions, these changes *are* significant, and the Law of Conservation of Mass has had to be modified.

Since mass and energy are interconvertible ($E = mc^2$), nuclear physicists use a new law — the Law of Conservation of Mass-Energy. It states that in any reaction the total quantity of mass-energy (mass + energy) remains constant.

18.8 Fission

Fission is not usuallly a spontaneous process like alpha, beta, and gamma decay. A nucleus has to be made unstable before it will split into two or more parts. The easiest way to cause fission is to destabilize the nucleus by adding a neutron. Only a few isotopes, called **fissionable isotopes**, can be destabilized in this fashion and produce more neutrons when they split. Only one isotope, uranium-235, occurs naturally. Uranium-235 makes up only 0.71% of uranium atoms. (Uranium-238 makes up almost all of the rest.) All other fissionable isotopes are human-made. Plutonium-239 and uranium-233, for example, can be produced in nuclear reactors from uranium-238 and thorium-232 respectively. Since they can be used to make fissionable isotopes, uranium-238 and thorium-232 are called **fertile isotopes**.

Because a neutron is electrically neutral, it can very easily penetrate the nucleus and be captured by it. When this happens to the nucleus of a fissionable isotope, a new highly unstable isotope is formed. For example,

$$\underset{\substack{\text{fissionable}\\\text{isotope}}}{^{235}_{92}\text{U}} \quad + \quad \underset{\substack{\text{slow}\\\text{neutron}}}{^{1}_{0}\text{n}} \quad \rightarrow \quad \underset{\substack{\text{highly}\\\text{unstable}\\\text{isotope}}}{^{236}_{92}\text{U}}$$

The uranium-236 isotope is so unstable that it has a lifetime of less than 1 ns (10^{-9} s) before it undergoes fission.

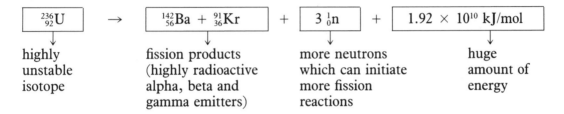

$$\boxed{^{236}_{92}\text{U}} \quad \rightarrow \quad \boxed{^{142}_{56}\text{Ba} + {}^{91}_{36}\text{Kr}} \quad + \quad \boxed{3\,{}^{1}_{0}\text{n}} \quad + \quad \boxed{1.92 \times 10^{10} \text{ kJ/mol}}$$

| highly unstable isotope | fission products (highly radioactive alpha, beta and gamma emitters) | more neutrons which can initiate more fission reactions | huge amount of energy |

The reaction shown here is just one of many possibilities. The uranium-236 nucleus can break apart in many other ways, producing different **fission products**.

The production of three neutrons for each fission that takes place can initiate two or more new fissions for each old one, depending on the number of neutrons that escape or react with non-fissionable nuclei. These two new fissions, in turn, can produce enough neutrons to initiate four more fissions, etc. When conditions are right for this to happen, a fission **chain reaction** takes place. (See Figure 18.8 on page 490.)

Fission Devices

If enough very pure fissionable material such as uranium-235 or plutonium-239 is brought together suddenly, a chain reaction will occur. In a fraction of a second, most of the nuclei will undergo fission, releasing a tremendous burst of energy. This is the reaction that occurs when an atomic bomb explodes. An atomic bomb is a device designed to assemble or compress enough fissionable material so that a chain reaction can occur. The minimum amount of fissionable material needed is called the **critical mass**. (See Figure 18.9 on page 490.)

Figure 18.8

If the concentration and amount of fissionable isotope is high enough, the neutrons released by one fission can initiate more than one fission in other nuclei. These in turn cause many more fissions. This increasing cascade of exploding nuclei is called a nuclear chain reaction.

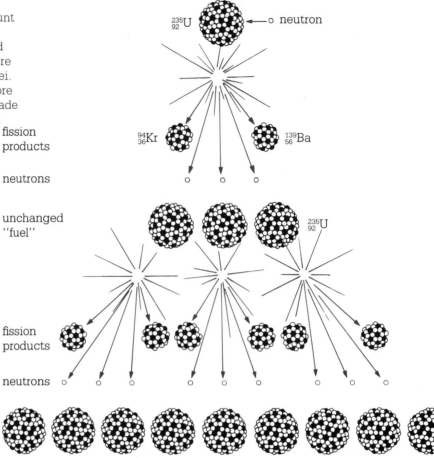

fission products

neutrons

unchanged "fuel"

fission products

neutrons

Figure 18.9

Obtaining a critical mass

(a) A small amount of fissionable material allows too many neutrons to escape without initiating further fissions. A chain reaction cannot occur. As the mass of fissionable material increases, the chance of these neutrons initiating further fissions increases. A critical mass is the minimum mass necessary for a chain reaction to occur.

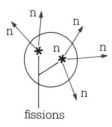

fissions

In a small amount of fissionable material, neutrons escape and a chain reaction does not start.

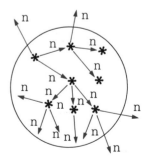

In a "critical mass" of fissionable material, neutrons do not escape as easily, and a chain reaction occurs.

(b) The proportion of neutrons absorbed can also be increased by compressing (via an inward directed symmetrical chemical explosion) a sub-critical mass of fissionable isotope. Compression pushes the nuclei closer together, increasing a neutron's chance of striking a nucleus. Atomic (fission) bombs obtain a critical mass in this way.

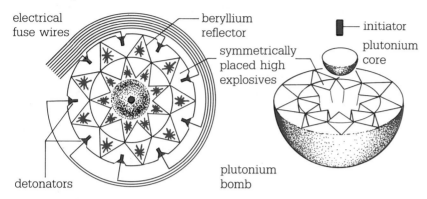

electrical fuse wires

beryllium reflector

symmetrically placed high explosives

detonators

plutonium bomb

initiator

plutonium core

plutonium bomb

SPECIAL TOPIC 18.3

THE MOUSETRAP "BOMB"

Imagine a field completely covered with thousands of mousetraps. Some of the traps are set; some are not. Some of the traps are loaded with as many as three ping pong balls; others have none. Now imagine that a single ping pong ball is thrown out onto the field. If it strikes a trap that is set and loaded, up to three balls will fly into the air. These three, in turn, may land on other traps.

If conditions are favourable — that is, if lots of the traps are set, and they have two or more balls on them — the air above the field might soon be filled with a "blizzard" of ping pong balls. In a few seconds it will all be over. The mousetrap "bomb" will have exploded. Under what conditions would the "bomb" not explode?

1. If the field of mousetraps is too small (a subcritical mass of mousetraps), then most of the balls fired into the air by a triggered trap might land outside the field and fail to set off more traps. A mousetrap "bomb" must have a minimum size — the critical mass.

2. If too many of the traps are not set, then the chances of the balls from one triggered trap setting off another will be low. The chain reaction of ping pong balls will die out. A mousetrap bomb must have a minimum percentage of set traps for it to be able to "explode."

Like a mousetrap "bomb", a fission bomb must have a very high proportion of "set traps" or fissionable atoms. It must, as well, have a large enough number of these atoms (a critical mass) so that the fission initiating neutrons do not escape before they are captured by another fissionable nucleus.

Old Age and the Bomb

Most of the world's stock of atomic bombs are made of plutonium-239, with some plutonium-241 present as an impurity. The radioactive decay of the plutonium-241 causes the metal core of the bomb to heat up and to accumulate impurities, mainly americium. Impurities change the crystal structure of the plutonium metal, causing it to swell. Hot, expanding metal in the core of a complex and precisely built bomb eventually damages it. Thus, the plutonium core of such bombs must be periodically removed, the metal reprocessed to remove impurities, and the bomb remanufactured. In effect, plutonium fission bombs slowly die of old age. What opportunities does this fact offer arms control negotiators?

Nuclear Power Reactors

The enormous energy available in fissionable nuclei has prompted a vast research and development effort to use uranium as a commercial energy source. Nuclear power reactors are the result of this effort. Nuclear power reactors use fissionable isotopes in a very dilute form so that an explosive critical mass cannot form. Canadian CANDU reactors use natural uranium containing only 0.71% of the fissionable isotope uranium-235. Other types of reactors use slightly enriched uranium (2%–4% uranium-235). By contrast, the uranium or plutonium used to make nuclear weapons must contain at least 93.5% of the fissionable isotope. A nuclear reactor cannot become an atomic bomb. A reactor can, however, destroy itself by overheating or, as in the accident at Chernobyl, USSR, in 1986, blow itself apart by chemical means.

Neutron Control and Nuclear Power Reactors

A nuclear power reactor operates by precisely controlling neutrons. At operating conditions, every fission will, on average, initiate just one new fission. If too many neutrons are lost or absorbed, the rate of fission decreases and the reactor loses power. If too many neutrons initiate new fissions, too much heat is produced and the reactor's cooling systems could be swamped. This would result in skyrocketing temperatures in the reactor's core. If unchecked, the reactor's fuel and supporting structure could get hot enough to melt (a reactor meltdown).

Neutrons are efficiently absorbed by fissionable nuclei if they are moving relatively slowly. Thus, in nuclear reactors, fast-moving neutrons produced by fission are first slowed down by a **moderator**. This is a substance that does not absorb neutrons very well, but slows them down on collision. Carbon and the hydrogen-2 isotope (deuterium or heavy hydrogen, used in the form of "heavy water") are such efficient moderators that natural uranium can be used as a fuel. (See Figure 18.10.) The hydrogen-1 isotope found in regular water absorbs neutrons more easily. This is why reactors using ordinary water as a moderator must use enriched uranium.

A nuclear reactor can be shut down by placing neutron-absorbing substances into the reactor core. **Control rods** of a neutron absorber,

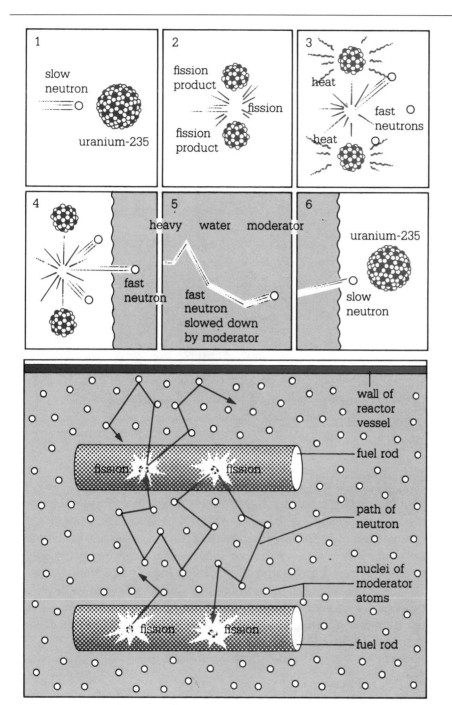

Figure 18.10
Nuclear fission in a reactor. Fast neutrons are not as efficiently absorbed by a fissionable nucleus as slow neutrons are. The fast neutrons produced by a fission must first be slowed down by a moderator. These slow neutrons are absorbed by other fissionable nuclei, which in turn undergo fission, maintaining a controlled chain reaction in the reactor.

such as cadmium, are moved into and out of the reactor core in order to decrease or increase the number of neutrons available to initiate fission reactions. Control rods are like the accelerators of a reactor – the farther they are pulled out, the more neutrons are available for fissions and the more energy the reactor produces, and vice versa. (See Figure 18.11.) In an emergency, fission in the reactor can be stopped by injecting a solution of a neutron absorbing element (for example, gadolinium) into the heavy water moderator.

Figure 18.11

A nuclear reactor is controlled by moving neutron absorbing rods, called control rods, into or out of the reactor's core. In an emergency, the fission reaction can also be interrupted by injecting a neutron-absorbing solution, called a reactor poison, into the heavy water moderator.

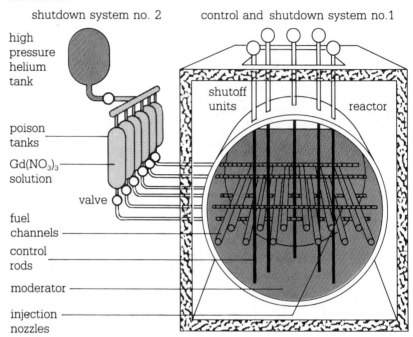

Nuclear Fuel

A nuclear reactor uses up and produces fissionable fuel at the same time. Some reactors, like the CANDU, produce almost as much fuel as they consume. **Breeder reactors** are those that produce more fuel than they use up. Research into breeder reactors is currently taking place in Europe and the USSR. One of the reactions that "breeds" more fissionable fuel involves uranium-238 (a fertile isotope) which makes up more than 99% of natural uranium. The uranium-238 nucleus absorbs a neutron and undergoes radioactive decay to produce a fissionable isotope.

$$^{238}_{92}U \quad + \quad ^{1}_{0}n \quad \rightarrow \quad ^{239}_{92}U$$

fertile isotope	reactor neutron	radioactive isotope (half-life 23.5 min)

$$^{239}_{92}U \quad \rightarrow \quad ^{239}_{93}Np \quad + \quad ^{0}_{-1}e$$

radioactive isotope (half-life 2.35 d)   beta particle

$$^{239}_{93}Np \quad \rightarrow \quad ^{239}_{94}Pu \quad + \quad ^{0}_{-1}e$$

fissionable isotope (half-life 24 400 a)   beta particle

The buildup of fission products forces reactor operators to replace the fuel long before all the fissionable uranium-235 and newly produced plutonium-239 have been used up. Some of these fission products are neutron absorbers. If left in place, they would capture enough of the reactor's neutrons to shut it down. "Spent" fuel can be chemically reprocessed to recover the fissionable plutonium-239 and to remove other fission products. The disposal of radioactive fission products in spent fuel is a major problem. Several disposal techniques have been developed, but none has been authorized for use. As a temporary measure, "spent" fuel rods are usually stored under water in large vaults at nuclear reactor sites.

Exercise

5. Reprocessing spent nuclear fuel yields significant quantities of fissionable plutonium-239. Other highly radioactive fission products are also obtained. Many nations that do not manufacture nuclear weapons, Canada included, have so far refused to build nuclear fuel reprocessing plants. Suggest at least two reasons for these nations making this decision.

Reactor Shutdown

Most of the heat output of reactors comes directly from fission. But an important additional amount is provided by the radioactive decay of fission products. This means that a reactor cannot be shut down instantly. The fission reaction itself can be quenched with neutron absorbers, but the radioactive decay of fission products continues to produce significant amounts of heat for a long time.

Imagine a new type of automobile powered by a nuclear reactor. When you turn off the key, this engine does not quit. It just drops to about one third of its normal power output, and slowly loses its remaining power thereafter. There is nothing you can do to stop it completely, except wait. You would want to be sure that when you parked such a car, its cooling system would continue to function. Otherwise, the idling engine would overheat, seize up, and damage itself seriously. Even with an ordinary gasoline engine, such events are expensive. If the engine contained radioactive fuel, there could be a deadly release of radioactive material, evacuation of the neighbourhood, and perhaps loss of life. That is why nuclear power reactors must have fail-proof cooling systems that can operate for months after shutdown. (See Figure 18.12.)

One of the most feared accidents at a nuclear plant is the loss of reactor coolant. Without coolant, the reactor heats up excessively. Vital components could melt or break. Spurts of cooling water hitting red hot material would suddenly turn into high pressure steam and could rupture pressure vessels and pipes. At very high temperatures, chemical reactions between steam and metal produce the explosive gas, hydrogen. If the hydrogen mixed with oxygen (air), chemical explosions could result that might breach containment structures. This could lead to a release of deadly radioactive materials to the atmosphere.

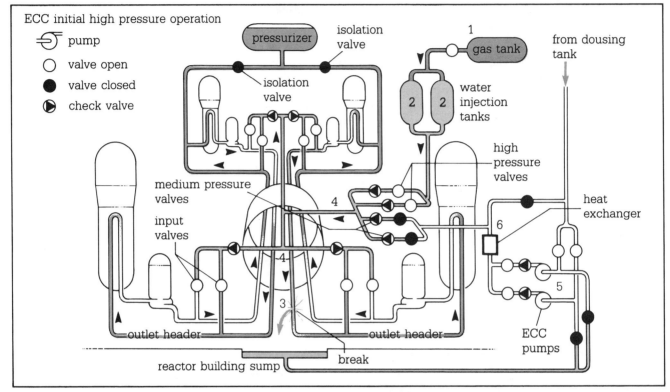

Figure 18.12
A CANDU nuclear power station emergency core cooling system. The emergency core cooling system consists of three separate subsystems. First, a high pressure gas tank (1) can force cold water from the water injection tanks (2) into the reactor core and boilers to initiate a crash cooling immediately after an accident such as a cooling pipe break (3). Next, a much larger amount of water in a dousing tank at the top of the reactor's containment building flows by gravity through the reactor core-cooling pipes (4). Finally, twin emergency core cooling pumps or ECC pumps (5) – supplied by a separate emergency power system – can maintain circulation of the water in the reactor's sump through a (cooling) heat exchanger (6) and the reactor's core.

18.9 Fusion

In fusion, lighter nuclei are merged into a larger one. This process requires an extremely high "ignition" temperature of about 10^8°C. Two nuclei must approach each other with enough kinetic energy to overcome their mutual electrostatic repulsion and get close enough for the strong nuclear force to bind them together. Temperatures of this magnitude are hard to achieve. They are found only in the core of stars like the sun, during a fission bomb explosion, and in a few fusion research reactors.

The fusion that requires the lowest ignition energy is the reaction of hydrogen-2 (deuterium) and hydrogen-3 (tritium). These isotopes have a nuclear charge of only $1+$ and repel each other much less than any of the nuclei of the higher elements.

$$\ce{^2_1H} + \ce{^3_1H} \rightarrow \ce{^4_2He} + \ce{^1_0n} + 1.7 \times 10^9 \text{ kJ}$$

deuterium   tritium   helium-4   neutron   energy

The Hydrogen Bomb

The hydrogen or thermonuclear bomb is ignited by the high energy gamma and X radiation produced by a fission bomb. These rays interact with the deuterium and tritium fusion fuel (in the form of solid lithium hydride, $\ce{^6_3Li^2_1H}$, and $\ce{^6_3Li^3_1H}$) to produce the intense temperatures needed for fusion to occur. (See Figure 18.13.) In addition, lithium-6 absorbs neutrons to generate more tritium fuel (and energy) as the fusion reaction proceeds.

$$\ce{^6_3Li} + \ce{^1_0n} \rightarrow \ce{^3_1H} + \ce{^4_2He} + 4.62 \times 10^8 \text{ kJ}$$

Tritium is a radioactive isotope manufactured only by nuclear reactions. It has a half-life of 12.3 a. Thus, over 8% of the tritium in nuclear weapons decays into stable helium-3 every year. This causes thermonuclear weapons to become disabled as they grow older – hydrogen bombs also die of old age! It has been estimated that if replacement of tritium in weapons were stopped, about half of the world's current stock of nuclear bombs would be disabled by old age within a decade. (See Figure 18.14.)

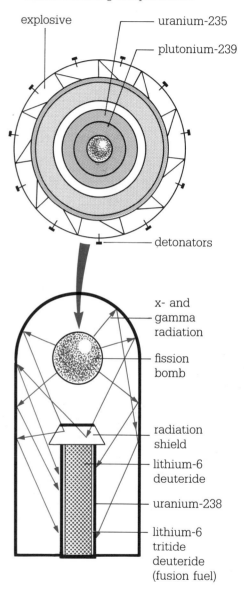

Figure 18.13
A thermonuclear (fusion) bomb is triggered by gamma and X radiation from a fission bomb. This radiation is absorbed by the lithium-6 deuteride and tritide, causing it to heat up to fusion-initiating temperatures.

explosive — uranium-235 — plutonium-239 — detonators — x- and gamma radiation — fission bomb — radiation shield — lithium-6 deuteride — uranium-238 — lithium-6 tritide deuteride (fusion fuel)

Figure 18.14
Fusion weapons become more and more ineffective as their radioactive tritium decays with time. If tritium is not replaced regularly, the weapons eventually become unusable.

Fusion Power Reactors

Fission reactions can be controlled and contained. However, for fusion reactions to occur, extremely high temperatures are needed just to initiate the process. No known materials can withstand such temperatures, so alternative containment procedures are being examined. Fusion research reactors *have* been able to initiate fusion reactions for very short periods of time. So far, however, no reactor has yet been able to reach the "break even" point where as much power is produced as it takes to run the reactor.

There are certain advantages to devising technology for controlled fusion reactions. First of all, few radioactive products are formed from the reaction. Also, our planet has an almost inexhaustible supply of deuterium in the form of water in the oceans. (Compare this to the limited supply of fossil fuels and uranium.) Since fusion fuel would be added in very small quantities as needed, there is very little possibility of the reaction getting out of control, and even less of significant radiation escaping to the environment.

With these advantages in mind, scientists, engineers, and technologists are developing two techniques which they hope will lead to a practical fusion power generator. These are magnetic confinement and laser implosion. (See Figures 18.15 and 18.16.)

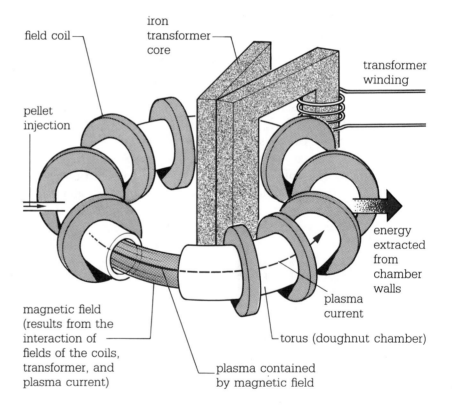

Figure 18.15

The design of the first Canadian magnetic confinement research fusion reactor, to be part of the fusion research facility at Varennes, Quebec. Powerful magnets are used to maintain an extremely hot ionized gas (called a plasma) of tritium and deuterium away from the walls of the reactor. If the density and temperature of this plasma can be maintained at a high enough level in this "magnetic bottle", controlled fusion reactions could occur. The heat produced in such a reactor would be removed by liquid lithium. The lithium would also absorb neutrons to produce more tritium fuel.

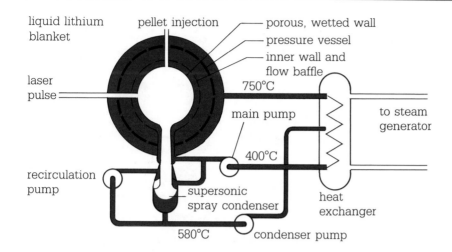

liquid lithium blanket
pellet injection
porous, wetted wall
pressure vessel
inner wall and flow baffle
750°C
laser pulse
main pump
to steam generator
400°C
recirculation pump
supersonic spray condenser
heat exchanger
580°C
condenser pump

Figure 18.16
A laser implosion fusion reactor. Powerful laser pulses can be used to compress and heat a small pellet of deuterium and tritium. If the temperature, pressure, and confinement time are large enough, the deuterium and tritium will undergo fusion—in effect, the pellet would explode like a miniature fusion bomb. A continuous series of these exploding pellets would provide the reactor's energy.

18.10 Radiation and People

Compare nuclear technology with the technology that harnesses combustion reactions, for example. Fires due to faulty equipment or human error cause serious damage, and cause many injuries and deaths every year. However, fire, in the form of controlled combustion reactions, is also responsible for heating many homes and buildings, running transportation vehicles, and powering factories. Similarly, nuclear technology brings with it the potential for both harm and good. Is the development and use of this technology worth its price?

We all live in an unavoidable bath of background radiation. This natural radioactivity comes from long lived radioactive isotopes like uranium-238 (half-life of 4.5×10^9 a), the radioisotopes they produce when they decay (for example, radium and radon), and cosmic radiation from space. This unavoidable radiation level is, of course, increased by the use of nuclear materials and devices such as x-ray machines and television sets. (See Table 18.6 and Figure 18.17.)

Figure 18.17
How radioactive substances can be transmitted to humans

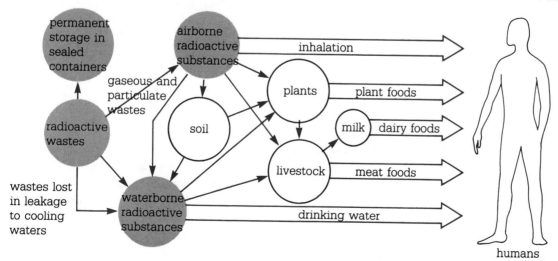

Table 18.6 Sources of Radiation Exposure

Source of Exposure	Exposure (mSv/a)★
Average cosmic radiation (increases by 0.03 mSv for every 150 m above sea level)	0.38
Average radiation from radioisotopes in the soil, in food and in the human body (due to potassium-40 and carbon-14)	0.24
Average radiation due to building materials and radon gas seepage into living space (based on 18 h/d in house)	0.18
Total average background radiation	0.80
A heavy smoker (due to polonium-210)	80
Medical diagnostic X rays (average)	0.70
A transatlantic air flight (more exposure to cosmic rays)	0.02 (per flight)
Living in the Vatican (more radioactivity in the stone)	8.0
Living in a home built since 1970 in Sweden (due to radon gas seepage from rock the house is built on)	48
Living in the vicinity of a CANDU nuclear station (under normal operating conditions i.e. no accidental radiation release)	0.02
Working in Canada's nuclear industry (average)	3.65
Watching televison for 4 h/d (X rays from the picture tube)	0.01

★ A Sievert (Sv) is the SI unit of biologically absorbed radiation energy. It accounts for the different amounts of damage caused by different types of radiation. One Sievert (1 Sv) of radiation, whatever its source, causes the same amount of biological damage. For example, 1 mSv corresponds to the biological damage caused by the absorption of about 5×10^{11} beta particles.

Biological Effects

The particles and rays from radioisotopes and cosmic radiation are capable of knocking electrons out of atoms, ionizing them. It is these chemically reactive ions that can cause biological damage. Radiation-induced damage can occur in all cell material. The most critical damage, however, occurs, in the cell's genetic material, its DNA. Most of the damage (well over 99%) is repaired by the cell's own biological repair mechanisms. Unrepaired damage may destroy the affected cell or change its genetic code. Altered codes may cause cancer in the individual or birth defects in offspring. The number of serious genetic disorders expected is about the same as the number of fatal cancers produced by a given exposure to ionizing radiation.

Estimating the effect of low levels of radiation on a population is notoriously difficult and controversial. Other variables, such as chemical pollutants for example, produce similar biological effects. This makes conclusions concerning radiation uncertain. Nevertheless, major scientific review committees have produced similar estimates of cancer risk for given doses of radiation. (See Table 18.7.) There is no known safe lower limit of exposure to radiation.

Table 18.7   Estimated Cancer Risks from Radiation

Exposure	Duration	Eventual Additional Cancer Deaths Per Million People*
1000 mSv	Over a short period of time	20 000
10 mSv	Over a short period of time	200
1.0 mSv/a (approximately natural background level)	Each Year	20/a
0.5 mSv/a	Each Year	1/a

* The cancer rate due to all causes is approximately 1500 per million people per year.

As with other technologies, nuclear technology is neither inherently good nor bad. Its beneficial application in energy production, medical procedures, and industrial applications is balanced against the serious problems associated with these uses. The danger of annihilation posed by nuclear weapons and the effects caused by radiation on living matter cannot be ignored.

Should society continue to depend on the benefits of nuclear technology? If so, what safeguards and regulations should govern it use? If not, what are the costs of eliminating its use?

Looking Back – Looking Ahead

Chemistry is unique in that it is the only science on which a whole industry is based. For example, there is no "biology industry", "physics industry", or "geology industry".

PUTTING CHEMISTRY TO WORK

DR. LINDA HEIER, M.D. / RADIOLOGIST

Radiology is the field of medicine that applies the latest advances in physics and chemistry to produce images of the internal workings of the body. Linda Heier is an assistant professor of radiology at a leading medical university.

"I have always been fond of machines and technology," Linda recalls. "Radiology, or diagnostic imaging as it is more often called, uses the most recent discoveries of science to diagnose and treat disease. It was ideal for me."

Linda's specialty is the central nervous system. "The brain is the most complex organ in the body — and the most difficult area in which to diagnose problems." A radiologist is actually a doctor's doctor. "I deal with other doctors during a diagnosis," Linda explains. "Most decisions regarding the cause of a patient's symptoms are made after a radiologist has taken a look inside (called imaging) using one or more methods." Linda performs two basic types of imaging.

"X rays are a form of imaging we refer to as 'non-invasive', meaning that we do not have to enter the body." Invasive techniques include, for example, injecting dye into blood vessels through fine tubes. These tubes can be moved throughout the body, allowing the radiologist to map each vessel — finding potential trouble spots in the blood supply system.

Some radiology treats the disease or condition. "My specialty," Linda explains, "is to treat abnormal blood vessels in the brain area by blocking them off using a form of glue or a remotely controlled balloon."

Linda especially enjoys the excitement of applying breakthroughs in science to medicine. "The most recent development," she explains, "has been the production of detailed three-dimensional images of the inside of a living body using magnetic resonance. The technique relies on radio signals from flipped hydrogen protons in a strong magnetic field. A computer generates a picture based on variations in signals from chemically different parts of the body." Magnetic resonance imaging does not use radiation or invasive techniques which could be stressful to a patient.

"It is the problem-solving that fascinates me," Linda confesses. "I'm a detective — with the added bonus that if I find the culprit in the 'crime', a person may have an added chance for life."

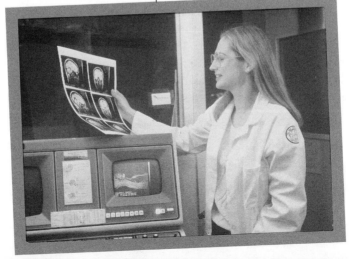

CHEMISTRY AND SOCIETY

At the end of Chapter 18, you read that the impact and influences of chemistry on society are probably greater than those of any other science. One way of examining the validity of this statement is to think about how chemistry is involved in the following photographs.

For each photograph, list as many principles and processes of chemistry as you can. Then consider the explicit and implicit impact of the principles and processes you listed.

Sulphur is one of the world's most valuable raw materials. Its principle use is in the manufacture of sulphuric acid, which is industry's most important chemical.

The dimensions of the Itaipu Dam on the Brazil — Paraguay border are 180 m high × 8 km long!

The chemical reactions responsible for producing iron and steel take place in a blast furnace. Once started, a typical blast furnace may operate non-stop for two years or more before it is worn out and must be rebuilt.

Lime is pumped into a lake to neutralize acidity caused by acid rain.

◁ Imperial Oil Limited's
Strathcona refinery

Nitric acid pouring from a
derailed tank car. The yellowish
brown colour is evidence that
the acid is already attacking
nearby metals. The acid was
neutralized using sodium
carbonate.
▽

◁ A model of a short section
of the DNA molecule

One method of fighting forest fires involves dropping chemicals from small airplanes to slow down the movement of the flames.

White clouds of aluminum oxide, produced by burning aluminum powder, billow from the solid booster rockets that lift the space shuttle Discovery.

When nitrogen dioxide reacts with sunlight, it produces the reddish brown colour characteristic of photochemical smog.

All sciences (and their associated technologies) have an impact on society to a greater or lesser extent. Chemistry, through the chemical industries, (arguably) influences our lives more than any other science.

In this textbook, many questions about chemistry and the work of chemists have been answered. At the same time, many more questions have been raised. We hope that you have a better understanding of what chemistry is and what chemists do. We also hope that what you have investigated in this chemistry course will help you become a more informed and more concerned member of society. The difficult decisions needed to maintain and improve our quality of life will be yours and your fellow citizens' to make in the years to come.

Vocabulary Checklist

You should now be able to define or explain the meaning of the following terms.

alpha particle	control rods	half-life	nucleon
band of stability	critical mass	mass defect	quark
beta particle	fertile isotope	moderator	radioactive
binding energy	fission products	nuclear equation	radioisotope
breeder reactor	fissionable isotope	nuclear fission	radionuclide
chain reaction	gamma radiation	nuclear fusion	transmutation

Chapter Objectives

You should now be able to do the following.

1. Appreciate some of the benefits and problems associated with nuclear technology.

2. Identify the force holding the atomic nucleus together as the strong nuclear force.

3. Identify the particles making up nucleons as quarks held together by the strong nuclear force.

4. Use the band of stability to predict which isotopes are stable.

5. Explain what is meant by half-life and determine how much of a given sample of a radioactive isotope would be left after a given number of half-lives.

6. List three kinds of natural radioactivity, their relative penetrating power and identify the particles associated with each.

7. Use nuclear equations to identify a missing reactant or product in a nuclear reaction.

8. List several uses of radioisotopes.

9. Distinguish between nuclear fission and fusion.

10. Identify lowered potential energy as the reason for fusion of light nuclei and fission of heavy nuclei, both releasing energy.

11. Specify the conditions necessary for fission and for fusion to take place.

12. Explain why a nuclear power reactor cannot explode like a fission bomb.

13. Discuss the operation of a nuclear reactor in terms of neutrons.

Understanding Chapter 18

REVIEW

1. Why does the nucleus of a carbon atom containing six mutually repelling positive protons (and six neutrons) *not* explode?

2. What keeps helium nuclei at room temperature from "condensing" into oxygen nuclei of lower potential energy? In other words, why does the following reaction *not* occur at room temperature?

$$4 \, {}^{4}_{2}\text{He} \rightarrow {}^{16}_{8}\text{O}$$

3. (a) What is a quark?
   (b) Sketch the internal structure of a proton and a neutron as viewed by the quark theory.

4. How is a radioactive isotope different from a stable isotope?

5. Copy the following information chart into your notebook. Fill in the missing information. *Do not write in your textbook.*

Type of Radiation	Particle Involved	Penetrating Power
	${}^{0}_{-1}\text{e}$	
Gamma rays		
		About 4 cm in air

6. List three uses for radioactive isotopes.

7. Explain the difference between nuclear fission and nuclear fusion.

8. Explain how energy can be released during both nuclear fission and nuclear fusion.

9. How can a nuclear fission reaction be induced?

10. What are the conditions necessary for a chain reaction nuclear fission to occur?

11. What is necessary for a nuclear fusion reaction between light nuclei to occur?

12. Why can't a nuclear power reactor blow up like a fission bomb?

13. The operation of a nuclear power reactor relies on the precise control of the neutrons in its core. Explain the function of (a) the moderator and (b) the control rods in this neutron control.

14. What is a breeder reactor?

15. A CANDU nuclear power reactor produces almost as much fissionable fuel as it uses. Why does the reactor's fuel need to be changed long before it is significantly depleted of fissionable material?

16. Explain why it is not possible to quickly shut off significant heat production in a fission reactor even though the fission process can be stopped in seconds.

APPLICATIONS AND PROBLEMS

17. Use the graph in Figure 18.4 to determine which of the following are unstable isotopes.

    (a) ${}^{206}_{82}\text{Pb}$ (b) ${}^{214}_{82}\text{Pb}$ (c) ${}^{18}_{8}\text{O}$ (d) ${}^{18}_{10}\text{Ne}$

18. Iodine-131, a radioactive fission product, has a half-life of 8 d. If a reactor core contained 1 kg of iodine-131 when it was shut down, how much iodine-131 would remain in the core after 40 d?

19. Balance the following nuclear equations and identify the missing particle or isotope in each case. Then list the type of nuclear reaction represented by each equation.

    (a) ${}^{235}_{92}\text{U} + {}^{1}_{0}\text{n} \rightarrow 3 \, {}^{1}_{0}\text{n} + {}^{131}_{53}\text{I} + ?$

    (b) ${}^{90}_{36}\text{Kr} \rightarrow {}^{90}_{37}\text{Rb} + ?$

    (c) ${}^{2}_{1}\text{H} + ? \rightarrow {}^{4}_{2}\text{He}$

    (d) $? + {}^{1}_{0}\text{n} \rightarrow {}^{85}_{36}\text{Kr} + {}^{153}_{58}\text{Ce} + 2 \, {}^{1}_{0}\text{n}$

CHALLENGE

20. A sample of rock was found to contain 0.01 mol of uranium-238 and 0.03 mol of lead-206. Lead-206 is the final stable isotope produced when uranium-238 and its subsequent radioactive products decay. Lead-206 is only produced by this radioactive decay. How long ago did the rock solidify? (Hint: The half-

life of uranium-238 is 4.5×10^9 a. Assume that no lead-206 was present in the rock at the time it solidified.)

21. Tritium, hydrogen-3, is slowly produced in the moderator of CANDU power reactors.

$$_1^2H + _0^1n \rightarrow _1^3H$$

Tritium is quite radioactive and accounts for some of the radioactivity released into the environment during normal operation of a reactor. Tritium is also a product in high demand by nations that manufacture nuclear weapons. Should Canadian power utilities sell tritium to such nations in order to reduce the overall cost of power? Justify your answer.

22. A population exposed to radiation develops more cancers and genetic defects than would normally be attributed to chance. Should the use of radiation-producing materials and processes be banned? Explain your answer.

23. The life of a star begins with the gradual accumulation, through gravity, of an enormous amount of matter, primarily hydrogen and some helium. As a star "ages", its composition changes. The proportion of hydrogen drops and that of helium rises. Explain how this composition change takes place.

24. Very massive stars (four or more times more massive than the sun) can undergo a very dramatic composition change. The proportion of hydrogen drops relatively quickly and the proportion of higher atomic mass elements rises in sequence (that is, from hydrogen to helium, to carbon, to oxygen, and so on.) This process speeds up and, in a spectacular stellar explosion (called a supernova), all of the rest of the elements up to and including uranium are produced and spewed into space. Explain how the higher potential energy elements beyond iron (see Figure 18.7) can be formed during such an explosion.

APPENDICES

Appendix A
The Metric System of Measurement

SI Base Units

Quantity	Unit	Symbol
Length	metre	m
Mass	kilogram	kg
Time	second	s
Electric current	ampere	A
Temperature	kelvin	K
Amount of substance	mole	mol
Luminous intensity	candela	cd

Metric Prefixes

Prefix	Symbol	Factor by Which the Base Unit is Multiplied	
exa	E	10^{18} =	1 000 000 000 000 000 000
peta	P	10^{15} =	1 000 000 000 000 000
tera	T	10^{12} =	1 000 000 000 000
giga	G	10^{9} =	1 000 000 000
mega	M	10^{6} =	1 000 000
kilo	k	10^{3} =	1 000
hecto	h	10^{2} =	100
deca	da	10^{1} =	10
		10^{0} =	1
deci	d	10^{-1} =	0.1
centi	c	10^{-2} =	0.01
milli	m	10^{-3} =	0.001
micro	μ	10^{-6} =	0.000 001
nano	n	10^{-9} =	0.000 000 001
pico	p	10^{-12} =	0.000 000 000 001
femto	f	10^{-15} =	0.000 000 000 000 001
atto	a	10^{-18} =	0.000 000 000 000 000 001

The Discovery of the Elements

The discovery of the elements paralleled the development of the periodic table classification system. With the ever increasing number of discovered elements, it became necessary to classify them. The ten or so elements known in ancient times were gradually augmented by new discoveries as chemists became increasingly successful at isolating and identifying the elements. The development of the spectroscope in the early 1800s led to a rash of discoveries. (See Figure B.1.)

The height of new element discovery coincided with the efforts of chemists, notably Dobereiner, Newlands, and Mendeleev, to classify them. Only 91 elements occur naturally on earth. The final filling in (technetium and promethium) and the extension of the elements past uranium in the periodic table had to await the discovery and development of nuclear reactions and new techniques.

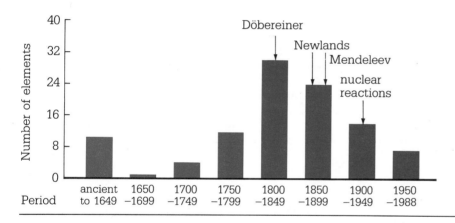

Figure B.1
The discovery of the elements. The height of new element discovery prompted chemists such as Döbereiner, Newlands, and Mendeleev to develop a system to organize and classify the elements. The discovery of new, human-made elements followed the discovery of nuclear reactions.

Table B1 The Elements, Their Dates of Discovery and The Derivations of Their Names and Symbols

Au. – Austrian	Fr. – French	I. – Italian	Sp. – Spanish	USA – United
GB – British	Ge. – German	Po. – Polish	Swe. – Swedish	States
Du. – Dutch	Hu. – Hungarian	R. – Russian	USSR – Soviet Union	

Element	Symbol	Atomic Number	Date of Discovery	Discoverer and Nationality	Name Derivation
Actinium	Ac	89	1899	A. Debierne (Fr.)	Greek: *aktis*, beam or ray
Aluminum	Al	13	1827	F. Woehler (Ge.)	Alum, the aluminum compound it was discovered in Latin: *alumen*, astringent taste

Element	Symbol	Atomic Number	Date of Discovery	Discoverer and Nationality	Name Derivation
Americium	Am	95	1944	A. Ghiorso R. James G. Seaborg S. Thompson (USA)	The Americas
Antimony	Sb	51	Ancient		Greek: *anti* plus *monos*, a metal not found alone
Argon	Ar	18	1894	Lord Raleigh William Ramsay (GB)	Greek: *argos*, inactive
Arsenic	As	33	1250	Albertus Magnus (Ge.)	Greek: *arsenikon*, yellow pigment
Astatine	At	85	1940	D. Corson K. Mackenzie E. Serge (USA)	Greek: *astatos*, unstable
Barium	Ba	56	1808	Humphrey Davy (GB)	Greek: *barys*, heavy
Berkelium	Bk	97	1950	G. Seaborg S. Thompson A. Ghiorso (USA)	Berkeley, California
Beryllium	Be	4	1828	F. Woehler A. Bussy (Ge., Fr.)	French, Latin: *beryl*, sweet
Bismuth	Bi	83	1753	C. Geoffroy (Fr.)	German: *weisse masse*, white mass
Boron	B	5	1808	Humphrey Davy J. Gay-Lussac L. Thenard (GB, Fr., Fr.)	Arabic: *buraq*, white
Bromine	Br	35	1826	A. Balard (Fr.)	Greek: *bromos*, stench
Cadmium	Cd	48	1817	F. Stromeyer (Ge.)	Latin: *cadmia*, calamine
Calcium	Ca	20	1808	Humphrey Davy (GB)	Latin: *calx*, lime
Californium	Cf	98	1950	G. Seaborg S. Thompson A. Ghiorso K. Street, Jr. (USA)	California

Element	Symbol	Atomic Number	Date of Discovery	Discoverer and Nationality	Name Derivation
Carbon	C	6	Ancient		Latin: *carbo*, charcoal
Cerium	Ce	58	1803	J. Berzelius W. Hisinger M. Klaproth (Swe., Swe., Ge.)	Asteroid Ceres
Cesium	Cs	55	1860	R. Bunsen G. Kirchhoff (Ge.)	Latin: *caesium*, blue for its spectral lines
Chlorine	Cl	17	1774	K. Scheele (Swe.)	Greek: *chloros*, light green
Chromium	Cr	24	1797	L. Vauquelin (Fr.)	Greek: *chroma*, colour
Cobalt	Co	27	1735	G. Brandt (Ge.)	German: *kobold*, goblin
Copper	Cu	29	Ancient		Latin: *cuprum*, copper
Curium	Cm	96	1944	G. Seaborg R. James A Ghiorso (USA)	Marie and Pierre Curie
Dysprosium	Dy	66	1886	L. de Boisbaudran (Fr.)	Greek: *dysprositos*, hard to get at
Einsteinium	Es	99	1952	A. Ghiorso (USA)	Albert Einstein
Element 104 (Kurchatovium proposed)	Ku	104	1964	Dubna Research Group (USSR)	– Igor Kurchatov
Element 105 (Hahnium proposed)	Ha	105	1967	Dubna Research Group (USSR)	– Otto Hahn
Element 106	–	106	1974	Dubna Research Group (USSR)	–
Element 107	–	107	1976	Dubna Research Group (USSR)	–
Erbium	Er	68	1843	C.G. Mosander (Swe.)	Ytterby, Sweden; from the rare earth deposits found there
Europium	Eu	63	1896	E. Demarcay (Fr.)	Europe
Fermium	Fm	100	1953	A. Ghiorso (USA)	Enrico Fermi
Fluorine	F	9	1886	H. Moissan (Fr.)	Latin: *fluere*, flow
Francium	Fr	87	1939	Marguerite Perey (Fr.)	France

Element	Symbol	Atomic Number	Date of Discovery	Discoverer and Nationality	Name Derivation
Gadolinium	Gd	64	1880	J. Marignac (Fr.)	Johan Gadolin, a Finnish chemist
Gallium	Ga	31	1875	L. de Boisbaudran (Fr.)	Latin: Gallia, France
Germanium	Ge	32	1886	Clemens Winkler (Ge.)	Latin: Germania, Germany
Gold	Au	79	Ancient		Latin: *aurum*, shining dawn
Hafnium	Hf	72	1923	D. Coster G. von Hevesey (Du., Hu.)	Latin: Hafnia, Copenhagen
Helium	He	2	1868 1895	P. Janssen (Fr.) (spectrum observed) W. Ramsay (GB) (isolated helium)	Greek: *helios*, sun
Holmium	Ho	67	1879	P. Cleve (Swe.)	Latin: Holmia, Stockholm
Hydrogen	H	1	1766	Henry Cavendish (GB)	Greek: *hydro*, water plus *genes*, forming
Indium	In	49	1863	F. Reich (Ge.)	Indigo spectral lines
Iodine	I	53	1811	B. Courtois (Fr.)	Greek: *iodes*, violet
Iridium	Ir	77	1803	S. Tennant (GB)	Latin: iris, rainbow
Iron	Fe	26	Ancient		Latin: *ferrum*, iron
Krypton	Kr	36	1898	William Ramsay M. Travers (GB)	Greek: *kryptos*, hidden
Lanthanum	La	57	1839	C. Mosander (Swe.)	Greek: *lanthanein*, concealed
Lawrencium	Lr	103	1961	A. Ghiorso T. Sikkeland A. Larsh R. Latimer (USA)	E.O. Lawrence, inventor of the cyclotron
Lead	Pb	82	Ancient		Latin: *plumbum*, lead
Lithium	Li	3	1817	A. Arfvedson (Swe.)	Greek: *lithos*, rocks
Lutetium	Lu	71	1907	G. Urbain K. von Welsbach (Fr., Au.)	Lutetia: ancient name for Paris

Element	Symbol	Atomic Number	Date of Discovery	Discoverer and Nationality	Name Derivation
Magnesium	Mg	12	1808	Humphrey Davy (GB)	Magnesia, Thessaly
Manganese	Mn	25	1774	J. Gahn	Latin: *magnes*, magnet
Mendelevium	Md	101	1955	A. Ghiorso G. Choppin G. Seaborg B. Harvey S. Thompson (USA)	Mendeleev
Mercury	Hg	80	Ancient		Latin: *hydrargyrum*, liquid silver
Molybdenum	Mo	42	1778	G. Scheele (Swe.)	Greek: *molybdos*, lead
Neodymium	Nd	60	1885	C. von Welsbach (Au.)	Greek: *neos*, new; *didymos*, twin
Neon	Ne	10	1898	William Ramsay M. Travers (GB)	Greek: neos, new
Neptunium	Np	93	1940	E. McMillan P. Abelson (USA)	Planet Neptune
Nickel	Ni	28	1751	A. Cronstedt (Swe.)	Swedish: *koppernickel*, false copper
Niobium	Nb	41	1801	Charles Hatchett (GB)	Greek: Niobe, daughter of Tantalus
Nitrogen	N	7	1772	D. Rutherford (GB)	Latin: *nitrum*, native soda
Nobelium	No	102	1958	A. Ghiorso T. Sikkeland J. Walton G. Seaborg (USA)	Alfred Nobel
Osmium	Os	76	1803	S. Tennant (GB)	Greek: *osme*, odour
Oxygen	O	8	1774	J. Priestley C. Scheele (GB, Swe.)	Greek: *oxys*, acid; *genes*, forming
Palladium	Pd	46	1803	W. Wollaston (GB)	Asteroid Pallas
Phosphorus	P	15	1669	H. Brandt (Ge.)	Greek: *phosphoros*, light bearing
Platinum	Pt	78	1735	A. de Ulloa (Sp.)	Spanish: *platina*, silver

Element	Symbol	Atomic Number	Date of Discovery	Discoverer and Nationality	Name Derivation
Plutonium	Pu	94	1940	G. Seaborg E. McMillan J. Kennedy A. Wahl (USA)	Planet Pluto
Polonium	Po	84	1898	Marie Curie (Po.)	Poland
Potassium	K	19	1807	Humphrey Davy (GB)	Latin: *kalium*, potash
Praseodymium	Pr	59	1885	A. von Welsbach (Au.)	Greek: *prasios*, green, *didymos*, twin
Promethium	Pm	61	1945	J. Marinsky L. Glendenin C. Coryell (USA)	Greek mythology, Prometheus
Protactinium	Pa	91	1917	O. Hahn L. Meitner (Ge., Au.)	Greek: *protos*, first; *actinium*, disintegrates into
Radium	Ra	88	1898	Marie & Pierre Curie (Fr., Po.)	Latin: radius, ray
Radon	Rn	86	1900	F. Dorn (Ge.)	Derived from radium, with suffix "on" added.
Rhenium	Re	75	1925	W. Noddak I. Tacke Otto Berg (Ge.)	Latin: Rhenus, Rhine
Rhodium	Rh	45	1804	W. Wollaston (GB)	Greek: *rhodon*, rose
Rubidium	Rb	37	1861	R. Bunsen G. Kirchhoff (Ge.)	Latin: rubidius, dark red spectral lines
Ruthenium	Ru	44	1844	K. Klaus (R.)	Latin: Ruthenia, Russia
Samarium	Sm	62	1879	L. de Boisbaudran (Fr.)	Samarskite, after Samarski, a Russian engineer
Scandium	Se	21	1876	L. Nilson (Swe.)	Scandinavia
Selenium	Se	34	1817	J. Berzelius (Swe.)	Greek: *selene*, moon
Silicon	Si	14	1824	J. Berzelius (Swe.)	Latin: *silex silcis*, flint
Silver	Ag	47	Ancient		Latin: *argentum*, silver
Sodium	Na	11	1807	Humphrey Davy (GB)	English: soda

Element	Symbol	Atomic Number	Date of Discovery	Discoverer and Nationality	Name Derivation
Strontium	Sr	38	1808	Humphrey Davy (GB)	Strontian, town in Scotland
Sulphur	S	16	Ancient		Latin: *sulphurium*
Tantalum	Ta	73	1802	A. Ekeberg (Swe.)	Greek mythology, Tantalus
Technetium	Tc	43	1937	C. Perrier E. Serge (I.)	Greek: *technetos*, artificial
Tellurium	Te	52	1782	F. Muller (Au.)	Latin: *tellus*, earth
Terbium	Tb	65	1843	C. Mosander (Swe.)	Ytterby, Sweden; from the rare earth deposits found there.
Thallium	Tl	81	1861	William Crookes (GB)	Greek: *thalos*, a budding twig
Thorium	Th	90	1828	J. Berzelius (Swe.)	Mineral theorite derived from Thor
Thulium	Tm	69	1879	P. Cleve (Swe.)	Thule, early name for Scandinavia
Tin	Sn	50	Ancient		Anglo-Saxon: tin
Titanium	Ti	22	1791	W. Gregor (GB)	Greek: giants, the Titans, from mythology
Tungsten	W	74	1783	J. de Elhuyar F. de Elhuyar (Sp.)	Swedish: *tung sten*, heavy stone
Uranium	U	92	1789	M. Klaproth (Ge.)	Planet Uranus
Vanadium	V	23	1801	A. del Rio (Sp.)	Norse: Vanadis, goddess of love and beauty
Xenon	Xe	54	1898	William Ramsay M. Travers (GB)	Greek: *xenos*, stranger
Ytterbium	Yb	70	1907	G. Urbain (Fr.)	Ytterby, Sweden; from the rare earth deposits found there.
Yttrium	Y	39	1843	C. Mosander (Swe.)	Ytterby, Sweden; from the rare earth deposits found there.
Zinc	Zn	30	1746	A. Marggraf (Ge.)	German: *zink*, obscure origin
Zirconium	Zr	40	1789	M. Klaproth (Ge.)	Zircon, in which it was found

An Extended Atomic Model – Quantum Theory

The first ionization energy versus atomic number graph discussed in Chapter 4 (Figure 4.21) has a number of unexplained inconsistencies in it. The first ionization energy tends to increase as the atomic number increases across a row (same outer electron energy level) of the periodic table. (See Figure C.1). There are times, however, when the first ionization energy actually *decreases* when the nuclear positive charge increases – for example, going from beryllium to boron and going from nitrogen to oxygen.

Figure C.1

First ionization energy v. atomic number

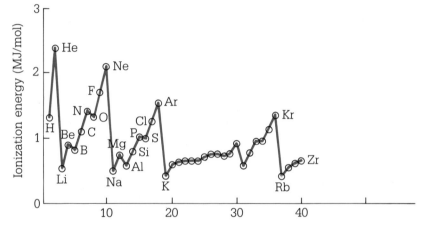

The model we have developed (Chapter 4) cannot be used to explain these discrepancies. Therefore, either a modification of this model or the devolopment of a new model is required.

Quantum Theory

The Bohr model of the atom provides the essential basis for a modification. Bohr had proposed that the energy absorbed by or produced from atoms can only be found as definite packages of energy or **quanta.** The atomic model conceives of matter as being made up of particles called atoms. Similarly, **quantum theory** conceives of energy as "particles" called quanta. The observation of line spectra of the elements and their interpretation as quanta of energy was essential in the development of the quantum theory and its mathematical application called **quantum mechanics.**

Particle Waves

In addition to this evidence, the French scientist Louis de Broglie suggested in 1924 that moving particles of matter actually have some wave-like characteristics. According to de Broglie, a moving electron, for example, should have some wave characteristics such as a definite wave length and frequency of vibration. Experiments with beams of moving particles have confirmed de Broglie's hypothesis. Moving particles actually do have wave-like properties. The smaller and the faster moving the particles are, the more pronounced these properties are.

Quantum Mechanics

In 1926, Erwin Schrödinger combined the quantum and particle wave theories in a mathematical equation that could predict and explain many of the previously unexplained atomic features – features such as the ionization energy inconsistencies. Schrödinger's quantum mechanical calculations produced results that were stated only as probabilities. Quantum mechanical calculations cannot predict actual paths or trajectories for moving electrons (because of their wave nature); they can only predict the chance of an electron showing up in a particular region of space around the nucleus. For example, Schrödinger's calculations for the location of the electron around a proton in the hydrogen atom produced a probability graph in three dimensions that showed the chance of finding the electron in any one location. This probability graph for finding the electron in a hydrogen atom is called an **orbital.** (See Figure C.2) The darker regions of the graph represent a higher probability of finding the electron; the lighter regions represent a lower probability. Note that closer to the nucleus the probability of finding the electron is greater.

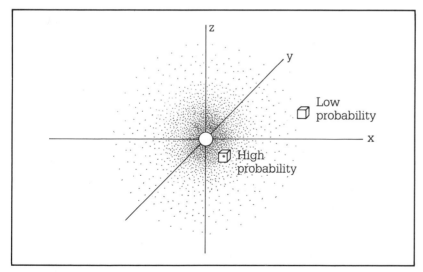

Figure C.2
Probability of finding the electron in the hydrogen atom. Schrödinger's probability calculations for finding an electron in the first Bohr energy level of the hydrogen atom may be plotted as a three-dimensional graph called an orbital. The one shown here is called the "1s" orbital. (The electron is in the first Bohr energy level and its probability graph is spherical.)

QUANTUM THEORY AND PROBABILITY*

Quantum theory gives a description of the energy levels and locations of the electrons around the nucleus in an atom, but the description of the location is in the form of *probabilities* rather than exact locations. To visualize a probability pattern, imagine two skywriting aircraft over a remote desert on a calm day. All day long the pilots circle and dive, moving in a disorganized way but always keeping an eye on a television transmitter atop a rocky outcropping. The top of the tower can be thought of as the nucleus.

Late in the afternoon, a search plane is sent out to locate them. What the search plane discovers is shown in the diagram. The smokey contrails left by the skywriters form a sphere of smoke several miles in diameter. This sphere is centred on the tower. The smoke is densest near the tower and becomes less and less dense with increasing radius. A few contrails can be seen at very great distances, indicating occasional side trips far away from the tower.

To the search pilot, the smoke cloud is helpful but not specific. The skywriters are still in the air, but where? The answer given by the smoke cloud is that the planes are *probably* somewhere close to the tower. One or both, however, might be a great distance away.

If the electrons in a helium atom were to emit a thin stream of smoke as they move around the nucleus, they would leave a pattern similar to the cloud around the television tower. This imaginary smoke cloud around an atom is a probability pattern. Quantum mechanics gives

useful information about the probability of finding an electron at a location near the nucleus.

For an atom with many electrons, quantum mechanics predicts that there will be a pair of electrons relatively close to the nucleus, four pairs in a somewhat higher energy level farther away from the nucleus, and four pairs in a higher level still further from the nucleus. From element 19 onward, the arrangements become more complicated.

*Reprinted from Richardson, Blizzard, Humphreys, *Structure and Change: An Introduction to Chemical Reactions* (John Wiley & Sons: Toronto, 1981).

As shown in Figure C.2, the electron occupies a rather "fuzzy" spherical region of space around the nucleus within which it has a high probability of being found. An orbital could be considered a time exposure photograph of the electron that records the electron locations over a long period of time.

The Pauli Exclusion Principle

The magnetic properties of atoms and ions contributed to quantum theory. These magnetic properties were accounted for by the hypothesis that electrons spin on their own axis as well as moving around the nucleus. Since a moving electric charge generates a magnetic field, the magnetic properties of elements such as iron and cobalt could be explained.

Why do the elements have different magnetic properties? This was explained by proposing that electrons can spin in only two opposite directions. Thus if two electrons spinning in opposite directions were close to each other, their magnetic fields would be opposite and a magnetic attraction between them would partially oppose their electrostatic repulsion. Their magnetic fields would also cancel, and no external magnetic effects would be noticed. In 1925, the German physicist, Wolfgang Pauli, incorporated this idea into quantum theory by stating his exclusion principle: Only two electrons of opposite spin can occupy any given orbital. Three electrons (three negative charges) in the same orbital (region of space) result in too much electrostatic repulsion and one of the electrons would be ejected.

Using Quantum Theory

When the Schrödinger wave equation is solved for atoms with many electrons, it is found that the first Bohr energy level consists of one orbital which is essentially spherical in shape. This is designated as the "1s" orbital: "1" for the (first) energy level and "s" for its (spherical) shape. In the second Bohr energy level it is found that there are four separate orbitals. One of these is a "2s" orbital of the same shape as the "1s" orbital but larger in radius and higher in energy. The other three orbitals are a set of "2p" ("pointed") orbitals all with the same energy, which is a little higher than that of the "2s" orbital. (See Figures C.3 and C.4.)

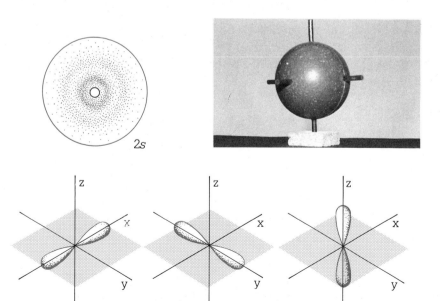

Figure C.3
The orbitals in the second Bohr energy level. Quantum mechanical calculations for the second Bohr energy level predict the existence of four orbitals. The "2s" orbital is spherical, and the three "2p" orbitals are oriented along the x, y, and z axes.

517

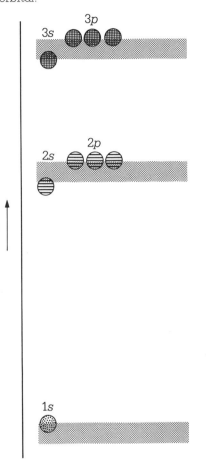

All of the three "*p*" orbitals illustrated in Figure C.3 have the same shape but are oriented at right angles to each other. The designations $2p_x$, $2p_y$, and $2p_z$ are simply used for convenience to indicate the three directions in space in which the orbitals point. In all other respects these orbitals are identical.

Electron Configurations

Across a row in the periodic table, an element differs from the preceding adjacent element by one proton (and some neutrons) in the nucleus and one electron in an orbital outside the nucleus. In order to have the minimum potential energy, electrons will fill orbitals in order of increasing orbital energy (from the lower energy orbital to the next higher energy orbital, etc.). Thus, the electron arrangement for a boron atom, $_5$B, would be

$$1s^2, 2s^2, 2p_x^1$$

Such a list of occupied orbitals is called an **electron configuration.** The superscript after each orbital indicates the number of electrons in the orbital.

The next element, $_6$C, presents a problem – does the sixth electron go into the $2p_x^1$ orbital, filling it up to become $2p_x^2$ or does it go into the empty $2p_y^0$ or $2p_z^0$ orbital? Since electrons repel each other and the "*p*" orbitals all have the same energy level, the new electron will attain a lower potential energy by going into an empty orbital, rather than a half-filled "*p*" orbital. Thus, the expected electron configuration for $_6$C, is

$$1s^2, 2s^2, 2p_x^1, 2p_y^1$$

Explaining the Ionization Energy Graph

These quantum mechanical results can be used to explain the slight inconsistencies noted earlier in the ionization energy versus atomic number graph in Chapter 4, Figure 4.21. As shown in Figure C.5,

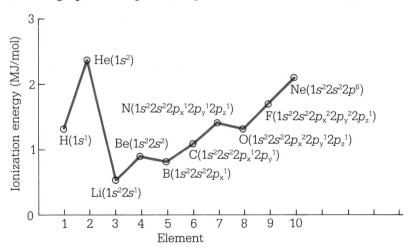

the energy needed to remove one of the electrons from the "2s" orbital of a beryllium atom is greater than that needed to remove the electron from the "2p" orbital of a boron atom, even though boron has an extra proton in its nucleus. The "p" orbital must, therefore, be at a higher energy level and it would require less energy to remove its electron. Thus, the results of quantum theory can be used to explain some of the inconsistencies that our earlier model could not.

Exercise

1. As shown in Figure C.5, the first ionization energy for oxygen is slightly lower than that for nitrogen, although oxygen has a larger positive charge in its nucleus. Use the quantum mechanical model idea of orbitals to explain why this occurs.

Ionization energy data is only one example of experimental support for the quantum model of the atom. This much more complex model can be used to explain many of the characteristics our simpler model cannot. Some of the more puzzling properties of elements – for example, the excellent electrical conductivity of the non-metal carbon or the behaviour of semi-conductors in transistors and "chips" – cannot be adequately explained without using quantum theory and quantum mechanics.

Appendix D
Answers to Selected Problems

Answers for all numerical problems found in Exercises throughout the textbook may be found here.

Chapter 2

1. % of hydrogen = 11.1%; % of oxygen = 88.9%
2. % of carbon = 42.1%; % of hydrogen = 6.44%; % of oxygen = 51.5%
3. (a) 9.8×10^{14} g/cm^3 (b) 9.8×10^{14} times
4. 9.8×10^{16} g or 9.8×10^{13} kg
6. 6.94 u
7. 24.31 u

Chapter 6

1. (a) 23.0 g   (b) 27.0 g   (c) 28.1 g
   (d) 39.9 g   (e) 207.2 g   (f) 126.9 g
2. (a) 44.0 g   (b) 80.1 g   (c) 78.0 g
   (d) 32.0 g   (e) 98.1 g   (f) 180.0 g
   (g) 342.0 g   (h) 227.0 g
3. (a) 153.3 g   (b) 111.1 g   (c) 110.3 g
   (d) 142.1 g   (e) 294.2 g   (f) 58.3 g
   (g) 310.3 g   (h) 96.0 g
4. (a) 1.20×10^{24} (b) 2.53×10^{23} (c) 1.51×10^{21} (d) 3.31×10^{24}
5. (a) 6.02×10^{24} (b) 1.20×10^{21} (c) 7.37×10^{23} (d) 5.12×10^{23}
6. (a) 3.90×10^{24} H atoms; 1.95×10^{24} S atoms; 7.80×10^{24} O atoms
   (b) 7.46×10^{22} C atoms; 1.49×10^{23} H atoms; 7.46×10^{22} O atoms
   (c) 4.82×10^{26} C atoms; 1.08×10^{27} H atoms
7. (a) 1.65×10^{15} a; 1.91×10^{10} a
   (b) $\$2.4 \times 10^{14}$
   (c) 1.91×10^{7} a
   (d) 6×10^{25} apples
8. (a) 1.25 mol   (b) 0.417 mol   (c) 7.36 mol
   (d) 0.42 mol   (e) 3.88 mol   (f) 1.46 mol
9. (a) 49.6 g   (b) 15 g   (c) 4340 g
   (d) 100 g   (e) 0.1 g   (f) 0.40 g
   (g) 1040 g   (h) 7.0×10^{-3} g
10. (a) 75.0% carbon; 25.0% hydrogen
    (b) 81.8% carbon; 18.2% hydrogen
    (c) 27.4% sodium; 1.2% hydrogen; 14.3% carbon; 57.1% oxygen

11. FeO has 77.7% Fe and 22.3% O; Fe_2O_3 has 69.9% Fe and 30.1% O
12. NH_4NO_3 has 35.0% N; $(NH_4)_2SO_4$ has 21.2% N
13. (a) CH_2 (b) Al_2O_3 (c) PbN_2O_6
14. C_3H_2Cl
15. C_3H_8O
16. C_6H_6
17. Si_2F_6
18. $C_6H_{12}O_6$

Special Topic 6.1
1. 6.12×10^3 or 6120 cm
2. 2.25×10^3 or 2250 s
3. 8.5×10^3 or 8500 sheets
4. 7.14 h
5. 2.83×10^3 or 2830 g

Special Topic 6.2
(a) 5   (b) 1   (c) 5   (d) 7   (e) 3   (f) 4   (g) 6   (h) 2
(a) 7300   (b) 35.90   (c) 860   (d) 0.2   (e) 472

Chapter 7

2. (a) $MgBr_2$ (e) Li_3N (i) MgH_2
    (b) K_2O (f) $NaCl$ (j) $SrCl_2$
    (c) BeF_2 (g) Al_2O_3
    (d) CaS (h) BaI_2

3. (a) magnesium sulphide   (e) aluminum bromide
    (b) sodium oxide   (f) lithium phosphide
    (c) calcium chloride   (g) aluminum sulphide
    (d) potassium nitride   (h) barium fluoride

4. (a) iron(II) chloride   (e) sulphur(IV) oxide
    (b) iron(III) chloride   (f) sulphur(VI) oxide
    (c) tin(II) fluoride   (g) copper(I) sulphide
    (d) tin(IV) fluoride   (h) copper(II) sulphide

5. (a) $NiCl_2$ (b) As_2S_3 (c) SF_4 (d) FeP (e) Co_3N_2 (f) PbO

6. (a) nitrogen dioxide   (c) diphosphorus pentoxide
    (b) phosphorus trichloride   (d) diarsenic trisulphide

7. (a) $SiCl_4$ (b) NI_3 (c) N_2O_3 (d) UF_6

8. (a) plumbous chloride   (c) ferrous oxide
    (b) plumbic chloride   (d) ferric oxide

9. (a) lead(II) chloride   (c) iron(II) oxide
    (b) lead(IV) chloride   (d) iron(III) oxide

10. The actual valence represented by "ous" or "ic" varies from element to element. For example, "ous" may mean 1, 2, or 3.

11. (a) carbon dioxide   (f) aluminum oxide
    (b) lithium sulphide   (g) phosphorus(III) hydride or
    (c) nickel(II) chloride   phosphorus trihydride
    (d) dinitrogen oxide or   (h) beryllium chloride
       nitrogen(I) oxide   (i) lead(II) oxide
    (e) sulphur(IV) bromide or   (j) calcium bromide
       sulphur tetrabromide

12. (a) K_2S (e) FeO   (i) $SnCl_2$
     (b) $CuBr_2$ (f) Mg_3N_2 (j) CaF_2
     (c) N_2O_5 (g) FeP
     (d) SF_4 (h) Al_4C_3

13. (a) barium cyanide   (c) strontium hydroxide
     (b) ammonium bromide

14. (a) LiOH   (b) $Al(CN)_3$ (c) $(NH_4)_3P$

15. (a) Na_2CO_3 (e) $Ca_3(PO_4)_2$
     (b) $Al(NO_3)_3$ (f) $KMnO4$
     (c) $CuSO_4$ (g) Na_2CrO_4
     (d) $Mg(ClO_3)_2$ (h) $(CH_3COO)_2Ca$

16. (a) chloric acid   (e) aluminum sulphate
     (b) calcium chromate   (f) barium nitrate
     (c) iron(III) phosphate   (g) potassium dichromate
     (d) magnesium nitrate   (h) ammonium carbonate

17. (a) K_2SO_3 (c) $LiClO_2$ (e) $Mg(ClO_2)_2$
     (b) HNO_2 (d) $BaSO_3$ (f) NaClO

18. (a) sodium sulphite
     (b) chlorous acid
     (c) potassium chlorite
     (d) iron(II) nitrite
     (e) calcium hypochlorite
     (f) sulphurous acid

19. (a) $KHSO_4$ (c) $Fe(H_2PO_4)_2$ (e) $NaHSO_3$
     (b) $Ca(HCO_3)_2$ (d) $MgHPO_4$

20. (a) lithium hydrogen sulphite
     (b) barium hydrogen sulphate
     (c) ammonium hydrogen carbonate
     (d) sodium hydrogen phosphate

21. (a) $SnCl_2 \cdot 2H_2O$ (c) $Na_2SO_4 \cdot 10H_2O$
     (b) $CH_3COONa \cdot 3H_2O$

22. (a) calcium chloride dihydrate
     (b) zinc sulphate heptahydrate
     (c) nickel(II) chloride hexahydrate

23. (a) aluminum iodide   (g) barium carbonate
     (b) sodium oxide   (h) hydrogen fluoride
     (c) calcium nitrate   (hydrofluoric acid)
     (d) cobalt(II) nitrate   (i) sulphur dioxide
     (e) lithium sulphite   (sulphur(IV) oxide)
     (f) water   (j) aluminum sulphate

Chapter 8

1. Answers will vary, depending on the reactions that have been carried out.

2. (a) $Ca + H_2O \longrightarrow Ca(OH)_2 + H_2$
     (b) $C_3H_8 + O_2 \longrightarrow CO_2 + H_2O$

3. (a) $2Na + Cl_2 \longrightarrow 2NaCl$
     (b) $2H_2 + O_2 \longrightarrow 2H_2O$
     (c) $4K + O_2 \longrightarrow 2K_2O$
     (d) $H_2 + Cl_2 \longrightarrow 2HCl$
     (e) $N_2 + 3H_2 \longrightarrow 2NH_3$

(f) $2CO + O_2 \longrightarrow 2CO_2$

4. (a) $2Na + 2H_2O \longrightarrow 2NaOH + H_2$

(b) $CaCO_3 + 2HCl \longrightarrow CaCl_2 + H_2O + CO_2$

(c) $Mg + 2HCl \longrightarrow MgCl_2 + H_2$

(d) $2KClO_3 \longrightarrow 2KCl + 3O_2$

(e) $C_3H_8 + 5O_2 \longrightarrow 3CO_2 + 4H_2O$

(f) $2Al + 3H_2SO_4 \longrightarrow Al_2(SO_4)_3 + 3H_2$

(g) $2Fe_2O_3 + 3C \longrightarrow 4Fe + 3CO_2$

(h) $4NH_3 + 5O_2 \longrightarrow 4NO + 6H_2O$

(i) $2Fe(OH)_3 + 3H_2SO_4 \longrightarrow Fe_2(SO_4)_3 + 6H_2O$

(j) $CaC_2 + 2H_2O \longrightarrow Ca(OH)_2 + C_2H_2$

(k) $CH_3OH + O_2 \longrightarrow CO + 2H_2O$

(l) $5FeCl_2 + KMnO_4 + 8HCl \longrightarrow 5FeCl_3 + KCl + MnCl_2 + 4H_2O$

(m) $3Cu + 8HNO_3 \longrightarrow 3Cu(NO_3)_2 + 2NO + 4H_2O$

5. SO_2: sulphur (IV) oxide (sulphur dioxide)
SO_3: sulphur (VI) oxide (sulphur trioxide)
H_2O: water
H_2SO_4: sulphuric acid (hydrogen sulphate)

6. (a) $4Li + O_2 \longrightarrow 2Li_2O$

(b) $P_4 + 3O_2 \longrightarrow 2P_2O_3$

(c) $3Mg + N_2 \longrightarrow Mg_3N_2$

(d) $K_2O + H_2O \longrightarrow 2KOH$

(e) $SO_2 + H_2O \longrightarrow H_2SO_3$

7. (a) $2Ag_2O \longrightarrow 4Ag + O_2$

(b) $CuCO_3 \longrightarrow CuO + CO_2$

(c) $2KHCO_3 \longrightarrow K_2CO_3 + CO_2 + H_2O$

8. (a) $Zn + Pb(NO_3)_2 \longrightarrow Pb + Zn(NO_3)_2$

(b) $Ag + CuSO \longrightarrow$ NO REACTION

(c) $2Al + 3NiCl_2 \longrightarrow 3Ni + 2AlCl_3$

(d) $Mg + H_2SO_4 \longrightarrow MgSO_4 + H_2$

9. (a) $Cl_2 + 2KI \longrightarrow I_2 + 2KCl$

(b) $I_2 + NH_4Cl \longrightarrow$ NO REACTION

(c) $Br_2 + 2NaI \longrightarrow I_2 + 2NaBr$

10. (a) $Na_2CO_3 + CuSO_4 \longrightarrow CuCO_3 + Na_2SO_4$ (precipitate $CuCO_3$)

(b) $(NH_4)_2SO_4 + CaCl_2 \longrightarrow CaSO_4 + 2NH_4Cl$ (precipitate $CaSO_4$)

(c) $2KOH + Co(NO_3)_2 \longrightarrow Co(OH)_2 + 2KNO_3$ (precipitate $Co(OH_2)$)

(d) $Na_2S + Pb(NO_3)_2 \longrightarrow PbS + 2NaNO_3$ (precipitate PbS)

11. $Mg(OH)_2 + 2HCl \longrightarrow MgCl_2 + 2H_2O$

12. (a) $Cu(OH)_2 + 2HNO_3 \longrightarrow Cu(NO_3)_2 + H_2O$

(b) $3LiOH + H_3PO_4 \longrightarrow Li_3PO_4 + 3H_2O$

(c) $2Fe(OH)_3 + 3H_2SO_3 \longrightarrow Fe_2(SO_3)_3 + 6H_2O$

(d) $Ca(OH)_2 + 2HBr \longrightarrow CaBr_2 + 2H_2O$

Salt produced: (a) copper(II) nitrate
(b) lithium phosphate
(c) iron(III) sulphite
(d) calcium bromide

13. (a) exothermic (b) endothermic (c) exothermic
(d) exothermic (e) endothermic

14. (a) $2CO + O_2 \longrightarrow 2CO_2 + energy$

(b) $2KClO_3 + energy \longrightarrow 2KCl + 3O_2$

Chapter 9

1. (a) 2.5 mol nitrogen; (b) 7.5 mol hydrogen
2. (a) (i) 15 mol; (ii) 100 mol; (iii) 1.0 mol
   (b) 10.0 mol propane
   (c) 40.00 mol propane and 200.0 mol oxygen used 120.0 mol carbon dioxide and 160.0 mol water produced
3. (a) 62.5 mol oxygen; (b) 0.80 mol octane;
   (c) 2.5 mol water and 0.28 mol octane
4. 8.16×10^5 kg oxygen
5. (a) 2504 g (2.504 kg or 2.504×10^3 g) oxygen
   (b) 558 g ethanol
6. (a) $Ca(OH)_2 + Na_2CO_3 \longrightarrow 2NaOH + CaCO_3$
   (b) 243 g NaOH
7. 703 g iron(III) oxide; 564 g sulphur dioxide
8. (a) sodium is the limiting reactant; (b) 31.8 g NaCl
9. 4.68 g $MgCl_2$
10. (a) 2.58 kg oxygen is the excess reactant; (b) 2.58 kg
11. (a) 3438 g; (b) 87.3%
12. (a) 1765 g ammonium nitrate; (b) 1562 g ammonium nitrate

Chapter 10

6. A very hot day: 40°C = 313 K; coldest day on record: −88°C = 185 K; oil temperature for cooking french fries: 162°C = 435 K

Chapter 11

1. 26.4°C
2. 1.7 L
3. 60 L
4. 3.8×10^3 m³
5. 4.2 L
6 3.75 L
7. 348 L
8. 99.8 L
9. 17.5 mL
10. 1.76×10^3 kPa
11. 2.3 m³
12. 1.46×10^4 kPa
13. 74 balloons
14. 1.2×10^2 min or 2 h
17. 78 kPa
18. 21 kPa
19. 57.9 g/mol
20 (a) 2.23 mol; (b) 1.34×10^{24} oxygen molecules
21. 18 L
22. 37.2 L

Chapter 12

1. 2.96 L
2. 88°C
3. 102 g
4. (a) 255 L   (b)255 L
5. 2.19 g
6. 4.37 L
7. (a) 7.5 L;   (b) 15 L
8. (a) 25 L;   (b) 119 L
9. 12.8 L

Chapter 14

2.   41.6 g
3.   21%
4.   425 g
5.   28.6%
6.   12 L
7.   30 g
8.   0.26 L
9.   15 pg (picograms)
10. (a) 27.1 g   (b) 33.3 g   (c) 13.8 g   (d) 14 g   (e) 15.4 g
11. (a) 20.8 mL   (b) 22.7 mL   (c) 1.67 mL
12. 11.6 mol/L
13. 1.38 mol/L
14. 1.05 L
15. 0.756 mol/L
16. 122 g/mol
17. 4.84 mol/L (new); 1.68 mol/L (old)

Chapter 15

3. (a) $26 423   (b) $33 433   (c) $13 598 527
    (d) (i) 51.93%   (ii) 48.07%

Glossary

A

absolute zero   A temperature of zero on the Kelvin scale. All molecular motion ceases at this temperature.

acid   A substance which produces hydrogen ions (H^+) when dissolved in water. A proton donor.

addition reaction   A reaction in which atoms are added to a molecule containing a double or triple bond.

alchemy   A type of pre-science developed by the Egyptians during the 3rd century B.C. Alchemists tried to change base metals into gold, and sought a potion that would prolong life.

alcohol   An organic compound containing at least one hydroxyl (–OH) group.

aldehyde   An organic compound containing a carbonyl group to which a hydrogen atom is attached (–CHO).

aliphatic   A term used to describe hydrocarbons consisting of chains or non-aromatic rings of carbon atoms.

alkali metals   The elements in the lithium group of the periodic table: lithium, sodium, potassium, rubidium, cesium, and francium.

alkaline earth metals   The elements in the beryllium group of the periodic table: beryllium, magnesium, calcium, strontium, barium, and radium.

alkanes   Hydrocarbons in which carbon atoms are bonded to each other by single bonds to form a straight or branched chain.

alkenes   Hydrocarbons in which at least one carbon-carbon double bond exists.

alkylation   The process of joining together smaller organic molecules to form larger molecules.

alkynes   Hydrocarbons in which at least one carbon-carbon triple bond exists.

alloy   A mixture formed by two or more metals.

alpha particles   Positively charged particles which move at high speed. They are emitted from some radioactive materials. Identical to the nucleus of a helium atom.

amorphous solid   A solid with no regular, crystalline structure.

anhydrous   Describes a substance from which water of crystallization has been removed.

anions   Negatively charged ions.

anode   A positively charged electrode in an electrolysis cell or discharge tube.

aqueous ions   Ions surrounded by water molecules in solution.

aromatic   Describes compounds based on the benzene ring structure.

atom   The smallest particle of an element.

atomic core   The nucleus and all inner electron energy levels of an atom.

atomic mass   The mass of an atom expressed in atomic mass units.

atomic mass unit   A unit used to measure the masses of atoms; equal to $1/12$ of the mass of a carbon-12 atom.

atomic number   The number of protons in the nucleus of an atom of an element. Each element has a different atomic number.

average atomic mass   The weighted average of the masses of the isotopes in a sample of an element.

Avogadro's constant   The number of particles in a mole, 6.02×10^{23}.

Avogadro's law   Equal volumes of all gases at the same temperature and pressure contain the same number of molecules.

B

balanced equation   An equation with the same number of atoms of each element on each side of the equation.

band of stability   The region on a graph of number of neutrons versus number of protons for different nuclei which includes all of the stable nuclei.

barometer   A device used to measure atmospheric pressure.

base   A substance which produces hydroxide ions (OH^-) when dissolved in water. A proton donor.

beta particle   An electron emitted from an atom during radioactive decay.

binary compounds   Compounds composed of only two elements.

binding energy   The amount of energy required to decompose an atomic nucleus into its separate protons and neutrons.

boiling point   The temperature at which the vapour pressure of a liquid is equal to the atmospheric pressure.

bond   The force of attraction holding atoms together in a compound.

bond length The inter-atomic distance where the forces of attraction are balanced by the forces of repulsion.

bonding capacity The number of electrons lost, gained, or shared by an atom when it bonds chemically.

Boyle's law The volume of a fixed mass of gas varies inversely as the pressure when the temperature is constant.

breeder reactor A nuclear reactor that produces more fissionable material than it uses during its operation.

brittle Describes the property of breaking or shattering when a substance is subjected to mechanical stress.

burette A graduated glass tube with a stopcock, used to deliver accurately measured volumes of liquid.

C

carbon-dating A technique used to estimate the age of old objects that were once alive. It is based on the observation that the ratio of carbon-14 to carbon-12 in an object decreases after the death of the organism.

carboxylic acids A group of organic compounds containing the carboxyl group (–COOH).

catalyst A substance that speeds up the rate of a chemical reaction without undergoing permanent change itself.

cathode The negatively charged electrode in an electrolysis cell or discharge tube.

cathode rays A stream of negatively charged particles found emanating from the cathode.

catenation The ability of carbon atoms to covalently bond to other carbon atoms to form long chains of atoms.

cation A positively charged ion.

chain reaction A reaction which is self-sustaining. In a nuclear chain reaction, the fission of a nucleus caused by a neutron produces more neutrons which can cause more fission reactions, and so on.

Charles' law The volume of a fixed mass of gas varies directly as the absolute temperature of the gas when the pressure is held constant.

chemical bond The force of attraction holding atoms together in a compound.

chemical change A change in which at least one new substance with different properties from the starting materials is formed.

chemical family A group of elements found in a vertical column in the periodic table.

chemical equation A summary of a chemical reaction using chemical formulas.

chemical formula A shorthand representation of a substance, showing the number of atoms of each element in a formula unit of the substance.

chemical property A property of a substance describing how it reacts when it is involved in a chemical change.

chemical reaction The same as a chemical change.

chemical symbol A representation of an element by one or two letters.

chemistry The study of the composition, properties, and structure of matter, and the changes that matter undergoes.

classification A technique of grouping like substances or objects according to similarities.

coefficient A number placed in front of a formula in a balanced equation. The number multiplies all atoms in the formula that follows.

control rods Rods made of a neutron-absorbing material, which can be inserted into a nuclear reactor to control the number of fission reactions occurring.

combining capacity See bonding capacity.

combustion The reaction of a substance with oxygen. Heat and light are produced during the reaction.

compound A pure substance formed by the chemical combination of two or more elements.

concentrated solution A solution containing a large amount of solute relative to solvent.

continuous spectrum The pattern of colours obtained when white light is passed through a prism.

covalence The bonding capacity of an element when it bonds covalently.

covalent bond A chemical bond in which one or more pairs of electrons are shared by two atoms.

cracking The technique of breaking down relatively large organic molecules into simpler molecules.

criss-cross rule A procedure used to predict the formula of a compound. The valences or combining capacities of the combining atoms or ions are criss-crossed to give the number of atoms or ions in the formula.

crystal A solid consisting of atoms, ions, or molecules arranged in a definite, repeating pattern.

D

Dalton's law of partial pressures The total pressure of a mixture of gases is equal to the sum of the partial pressures of the components of the mixture.

decomposition A chemical reaction in which a compound is broken down into two or more simpler substances.

diatomic molecule A molecule consisting of two atoms.

diffusion The movement of gas molecules from an area of high concentration to an area of low concentration.

dilute solution A solution containing a small amount of solute relative to solvent.

dipole The partial separation of positive and negative charge in a molecule.

dipole-dipole force The force of attraction between polar molecules.

discharge tube A glass tube with electrodes, containing a gas at a very low pressure; a high electrical voltage can be applied to the gas.

dissociation The separation of ions when an ionic solid is dissolved in water.

distillation The evaporation of a liquid followed by condensation of the vapour.

double bond A covalent bond in which two pairs of electrons are shared by the bonding atoms.

double replacement reaction A chemical reaction in which two elements replace each other in compounds.

ductile Describes the property of a metal that enables it to be drawn out into a wire.

E

electrodes Electrically conducting rods or plates placed in a solution or tube.

electrolysis The chemical reaction caused by the passage of electricity through a substance.

electrolytes Solutes that dissolve in water to form solutions that conduct electricity.

electron A negatively charged particle found in atoms.

electron affinity The energy released or absorbed when a neutral atom gains an electron to form a negative ion.

electron bombardment Fast moving electrons striking materials or atoms.

electron shells Another term for the energy levels in which electrons in atoms are found.

electronegativity A measure of the attraction that an atom has for electrons in a bond.

electrovalence The bonding capacity of an element when it bonds ionically.

element A pure substance that cannot be chemically broken down into simpler substances.

empirical Pertaining to experiments, observations, or experiences.

empirical formula The formula that gives the simplest whole number ratio of atoms in a compound.

empirical knowledge Knowledge obtained by observations and experiments.

end point The point during a titration when the indicator changes colour.

endothermic reaction A chemical reaction that absorbs energy from its surroundings.

energy levels Permitted amounts of energy that electrons in an atom may possess. In any atom, only certain energy levels are allowed.

enzyme A biological catalyst.

esters A group of organic compounds formed by the reaction of an alcohol with a carboxylic acid.

ethers A group of organic compounds consisting of two alkyl groups separated by and bonded to an oxygen atom.

excess reactant A reactant that is present in a quantity greater than that required to react completely with another reactant.

excited state An atom in which the electrons are not in their lowest possible energy levels.

exothermic reaction A chemical reaction tht releases energy to its surroundings.

experimental yield The actual amount of product obtained in a chemical reaction.

extrapolating Extending a graph beyond the region for which data exist.

F

feedstock The raw materials used in a refinery or petrochemical plant.

fertile isotope An isotope that can be converted into a fissionable isotope by nuclear reactions.

first ionization energy The energy needed to remove the least tightly held electron from an atom.

fission The process in which the nucleus of an atom is split into smaller nuclei by neutron bombardment; large amounts of energy are usually released in the process.

fission products The smaller, usually highly radioactive isotopes that result from the fission of a larger isotope.

fissionable isotope An isotope that will undergo fission when it absorbs a neutron, releasing energy and more neutrons.

flame test A procedure in which a compound of a metal is placed in a flame; the colour of the flame is used to identify the presence of certain metals.

force A push or pull.

formula A shorthand representation of a substance, indicating the number of atoms of each element in the formula unit of the substance.

formula mass The sum of the atomic masses of the atoms in a formula unit.

formula unit The simplest whole number ratio of atoms or ions of the elements in a compound.

fossil fuels Fuels formed from matter that was once living: coal, oil, gas, and peat.

fractional distillation The process in which a mixture of liquids is separated into its components because of differences in boiling points.

functional group A group of atoms found in a series of compounds; responsible for the properties of the compounds.

fusion A nuclear reaction in which small nuclei combine to produce a larger nucleus; large amounts of energy are usually released during this process.

G

galvanizing The procedure in which an iron object is coated with a layer of zinc.

gamma rays High energy electromagnetic radiation given off by some radioactive materials.

gas A physical state of matter which expands to fill the available space, and which is easily compressed.

Gay-Lussac's law of combining gas volumes When gases react chemically, their combining volumes, measured at the same temperature and pressure, are in simple whole number ratios.

Graham's law of diffusion The rate of diffusion of a gas is inversely proportional to the square root of its molecular mass.

ground state Describes an atom in which the electrons are in their lowest possible energy levels.

groups Vertical columns of elements in the periodic table.

H

heat of fusion The energy needed to melt a given amount of a solid substance.

heat of vaporization The energy needed to vaporize a given amount of a liquid substance.

half-life The time it takes for one half of an amount of a radioactive sample to decay.

halogens The elements in the fluorine group of the periodic table: fluorine, chlorine, bromine, iodine, and astatine.

heterogeneous Describes a sample of matter composed of two or more components which are not uniformly distributed; consisting of two or more phases.

homogeneous Describes a sample of matter that is uniform throughout; consisting of one phase.

homologous series A series of organic compounds with the same general formula; each member of the series differs from adjacent members by a constant number of atoms.

hydrate A compound which contains water molecules as part of its crystal structure.

hydrated ions See aqueous ions.

hydration The attraction between water molecules and dissolved ions.

hydrocarbons Compounds composed of hydrogen and carbon.

hydrogen bond The attraction between a positive hydrogen atom of one molecule and a highly electronegative atom (F, O, or N) in another molecule.

hydrogenation The addition of hydrogen atoms to a molecule at the sites of double or triple bonds between carbon atoms.

hydronium ion The H_3O^{3+} ion, which represents a hydrated proton.

hydroxyl group The functional group $-OH$ responsible for the properties of alcohols.

hypothesis A tentative explanation for an observation or pattern.

I

ideal gas A hypothetical gas which obeys the ideal gas law under all conditions of temperature and pressure. The molecules of an ideal gas are assumed to have no volume and to have no attractive forces between them.

ideal gas constant A constant in the ideal gas law equation, represented by R and equal to $8.31 \text{ L·kPa·mol}^{-1}.\text{K}^{-1}$

ideal gas law A mathematical equation relating the number of moles of an ideal gas to pressure, volume, and temperature: $PV = nRT$

immiscible Describes substances that do not dissolve in each other.

indicator A compound that detects the presence of acids and bases by changing to different colours.

inorganic compound A compound which does not contain carbon, with the exception of carbonates, cyanides, carbon monoxide, and carbon dioxide.

intuition The ability to perceive "truth" without apparent reasoning or analysis.

inverse variation The value of the dependent variable decreases by the same factor by which the independent variable increases.

ionic bond The electrostatic attraction between positive and negative ions.

ionic equation Chemical equation that specifies the individual ions that are present in and react in a solution.

ionization energy The energy needed to remove an electron from an isolated, gaseous atom.

ionization reaction A reaction of neutral molecules with a solvent (usually water) that produces ions.

ions Atoms or groups of atoms that are positively or negatively charged due to a loss or gain of electrons.

isomers Compounds with same molecular formula but different structures.

isotopes Atoms of an element with the same number of protons but a different number of neutrons.

K

kelvin The unit of temperature on the Kelvin scale.

Kelvin scale The SI scale for the measurement of temperature; zero kelvins (0 K) is equal to $-273.16°C$.

ketones A group of organic compounds containing the carbonyl group (C=O) between two alkyl groups.

kinetic energy The energy an object has due to its motion.

kinetic molecular theory The theory that explains the behaviour of matter in terms of the motion of particles.

L

law A statement of a regularity or pattern observed in a large number of experiments.

law of conservation of energy Energy is neither created nor destroyed in a physical or chemical change.

law of conservation of mass When a physical or chemical reaction takes place, the mass of the products is equal to the mass of the reactants.

law of constant composition The elements that make up a compound are always present in the same proportions by mass.

Lewis structure A representation of a compound, showing all atoms present, along with the valence electrons.

Lewis symbol A representation of an atom, showing all atoms present along with the valence electrons.

limiting reactant The reactant in a chemical reaction that is completely used up; the reactant that limits the amount of product obtained from the reaction.

linear relationship When the data points from an experiment are plotted, a straight line graph results; there is direct variation between the quantities under consideration.

liquid The physical state of matter with no definite shape, but with a definite volume.

M

malleable Describes substances which can be hammered into thin sheets.

mass The amount of matter in an object.

mass defect The difference between the mass of a nucleus and the sum of the masses of the protons and neutrons that make up the nucleus.

mass-mass problems Problems involving a balanced equation in which the mass of one substance is given and the mass of another substance has to be found.

mass number The total number of protons and neutrons in the nucleus of an atom.

mass spectrometer A device used to measure the masses of atoms or molecules, based on their deflection in magnetic and electrical fields.

mass spectrum The pattern of different masses and their relative abundances obtained when a

sample of material is analyzed using a mass spectrometer.

matter Anything that has mass and takes up space.

mechanical mixture A heterogeneous mixture of two or more substances.

melting point The temperature at which a solid changes to a liquid.

metalloids Elements with some metallic properties and some non-metallic properties.

metals Elements with certain physical properties – shiny, ductile, conduct heat and electricity; elements that lose electrons when they react.

miscible Describes two substances that dissolve in each other in all proportions.

mixture A physical intermingling of two or more substances that retain their properties.

model A physical or mental representation of a system that cannot be observed directly.

moderator A substance that slows down fast neutrons in a nuclear reactor without absorbing them.

molar mass The mass in grams of 1 mol of atoms or molecules.

molar concentration The number of moles of solute contained in 1 L of solution.

molar volume The volume occupied by 1 mol of a gas at standard temperature and pressure; equal to 22.4 L.

mole An amount of matter that contains 6.02×10^{23} particles.

mole ratio The ratio of the number of moles of one substance in a chemical reaction to the number of moles of another substance, as expressed in the balanced equation for the reaction.

molecular formula A formula that indicates the actual number of atoms of each element in a molecule.

molecular mass The sum of the atomic masses of the atoms in a molecule.

molecule A particle consisting of two or more atoms held together by covalent bonds.

monomer A substance whose molecules can be linked together in a chain to form a polymer.

N

natural resources Materials in the air, earth, or oceans that we make use of to produce energy or useful substances.

net ionic equation An equation for a reaction that shows only the substances or ions that take part in the reaction; spectator ions are omitted.

neutron A neutral particle found in the nucleus of the atom.

neutralization A chemical reaction involving the combination of an acid and a base to produce a salt and water.

neutralization point The point in a titration at which the number of moles of hydroxide ion is equal to the number of moles of hydrogen ion.

noble gases The elements in the helium group 18 of the periodic table; helium, neon, argon, krypton, xenon, and radon.

non-electrolyte A substance that dissolves in water without forming ions.

non-metals Elements that generally do not display the properties of metals; characterized by relatively high ionization energies; found in the upper right corner of the periodic table.

non-polar bond A covalent bond in which electrons are equally shared by the two atoms.

normal boiling point The temperature at which the vapour pressure of a liquid is equal to standard atmospheric pressure, 101.3 kPa.

nuclear equation An equation that describes the isotopes and radiation particles involved in a nuclear reaction.

nuclear fission See fission.

nuclear fusion See fusion.

nuclear force The force holding the particles in the nucleus together.

nuclear reactor A device that uses a controlled nuclear reaction to produce heat which is used to generate electricity.

nucleons The protons and neutrons in the nucleus of the atom.

nucleus A tiny, dense, positively charged region at the centre of the atom; contains protons and neutrons.

O

octet rule A generalization that atoms react in such a way that they acquire eight electrons in their valence level.

orbit According to the Bohr model of the atom, the circular path followed by an electron as it revolves around the nucleus.

orbital A region of space around the nucleus where there is a high probability of finding an electron.

organic chemistry The study of the compounds of carbon.

organic compounds Compounds containing carbon with the exceptions of carbonates, cyanides, carbon dioxide, and carbon monoxide.

oxyacid An acid containing oxygen as well as hydrogen and another element.

P

partial pressure The pressure exerted by one component in a mixture of gases.

pascal The SI unit for pressure; 1 Pa (one pascal) is equivalent to a force of 1 N (one newton) acting on an area of 1 m^2.

percentage composition The composition of a compound expressed in terms of the percentage by mass of each element in the compound.

percentage yield The ratio of the actual yield in a reaction to the theoretical yield expressed as a percentage.

periodic law When the elements are arranged in the order of increasing atomic number, there is a periodic repetition of elements with similar properties.

periodic table An arrangement of the elements in order of atomic number. Elements with similar properties and similar electron arrangements are in the same column.

periods The horizontal rows of elements in the periodic table.

pH scale A number scale used to represent the acidity of a substance.

phase A physically distinct region of matter, either solid, liquid, or gas.

photoionization The removal of electrons from isolated atoms by means of light energy.

physical change A change in which the chemical composition of the starting materials does not change.

physical property A property of a substance that can be observed without changing the chemical composition of the substance.

polar covalent bond A covalent bond between two atoms in which the shared electrons are more strongly attracted to one atom than the other.

polar molecule A molecule with a slight positive charge at one end and a slight negative charge at the other end.

polyatomic ion A group of atoms carrying an overall charge; the atoms are bonded to each other by covalent bonds.

polymer A compound formed by the linking together of a large number of simple molecules.

potential energy The energy an object possesses due to its position relative to a body that either attracts or repels the object.

precipitate A solid formed from a reaction that takes place in solution.

pressure Force per unit area.

products The substances formed in a chemical reaction.

proton A positively charged particle found in the nucleus of the atom.

pure substance A subtance composed of only one type of atom or molecule.

Q

qualitative property A property of matter that can be described without the use a measurement.

quantitative property A property of matter that requires the use of a measurement.

quantum mechanics A theory of the structure of atoms based on discrete packages of energy (quanta), and on the wave properties of electrons.

quark A subnucleon particle; quarks make up protons and neutrons.

R

radioactivity The spontaneous disintegration of atomic nuclei involving the emission of particles or energy.

radioisotopes An isotope which emits radioactivity.

radionuclide A radioactive isotope.

raw materials The starting materials in a chemical reaction; usually applied to an industrial reaction.

reactants The starting materials in a chemical reaction.

reserves The estimated amounts of natural resources available for our use in the future.

resonance hybrid A Lewis structure for a molecule or ion, which is an average of two or more equally valid structures.

reversible reaction A chemical reaction in which the products can react to form the original reactants.

S

salt A compound formed from the reaction of an acid and a base.

saturated compound An organic compound that contains no double or triple bonds.

saturated solution A solution that cannot dissolve any more solute at the same temperature.

scientific literature The journals and reference books that describe and summarize scientific experiments, experimental results, and theories.

scientific notation A method of expressing numbers in the form $a.bc \times 10^n$.

seed crystal A crystal that provides a nucleus around which a dissolved solute can crystallize.

semi-metal A material with some metallic properties and some non-metallic properties.

shared pair Two electrons shared by two atoms bonded together.

shielding effect The decrease in the force of attraction of the nucleus for valence electrons, caused by the presence of electrons in inner energy levels.

significant digits Those digits in a measurement known to be certain plus one uncertain or estimated digit.

single replacement reaction A chemical reaction in which one element replaces another element in a compound.

solid The physical state of matter which has both a fixed shape and a fixed volume.

solubility The amount of a solute that will dissolve in a known amount of solvent at a certain temperature.

solute The substance that dissolves in the solvent to form a solution; usually the substance present in the lesser amount in a solution.

solution A homogeneous mixture of two or more substances.

solvent The substance that dissolves the solute to form a solution; usually the substance present in the greater amount in a solution.

spectator ion An ion present in a chemical reaction but not involved in the reaction.

standard boiling point See normal boiling point.

standard temperature and pressure or STP $0°C$ (-273.16 K) and 101.3 kPa.

stock solution A solution of known concentration for laboratory use.

stoichiometry Calculations involving the amounts of reactants and/or products in a chemical reaction.

strong (acid, base, or electrolyte) A substance that completely ionizes or dissociates in water.

sublimation The change of state from solid to gas, or from gas to solid.

substitution reaction A reaction involving an organic compound in which a hydrogen atom is replaced by another atom or group of atoms.

supersaturated solution A solution that contains more dissolved solute than can normally be dissolved at a given temperature.

synthesis A chemical reaction in which two or more elements or compounds combine to form a single compound.

T

technology The application of scientific knowledge and principles to produce something society considers useful.

temperature A measure of the average kinetic energy of a sample of molecules.

theoretical yield The amount of product that should be produced from a known mass of reactant in a chemical reaction, according to the balanced equation for the reaction.

theory A hypothesis that has been tested by experiment, and has been shown to be consistent with many observations.

thermal pollution The increase in temperature of a body of water caused by the release of waste heat from a chemical or nuclear reaction.

titration The laboratory procedure in which the volume of one solution required to react with a known volume of another solution, or mass of another substance, is determined.

transmutation The nuclear process by which an isotope of one element changes into an isotope of another element.

triple bond A covalent bond between two atoms in which three pairs of electrons are shared by the atoms.

U

unsaturated compound An organic compound containing at least one double or triple bond.

unsaturated solution A solution that is capable of dissolving more solute at the same temperature.

V

valence The combining capacity of an element; the number of bonds an atom forms in a compound.

valence electrons The electrons in the highest energy level of an atom.

valence shell (or level) The outermost energy level containing electrons in an atom.

valence shell electron pair repulsion theory A theory to predict the shapes of molecules; it is based on the assumption that pairs of electrons around an atom mutually repel each other, and position themselves around the atom to be as far away as possible from each other.

vapour pressure The pressure exerted by a vapour in equilibrium with its liquid or solid.

volumetric flask A glass flask (used to make solutions) which has a narrow neck on which is etched a line; when the flask is filled to this line, the volume of liquid in the flask is known exactly.

W

water of crystallization or hydration Water molecules that form part of the structure of a crystalline substance.

wave mechanics See quantum mechanics.

weak (acid, base, or electrolyte) A substance that ionizes or dissociates into ions only slightly in water.

Credits

INDEX

A

Absolute zero, 268
Acid-base indicators, 347
Acid-base strength, measuring, 348
Acid-base-titrations, 374–79
Acids, 197, 347, 348–50
Acid salt, 200–201
Activity series, 216–17, 218
Adamek, Dr. Edward, 468
Air, 300
Alchemy, 32–33
Alcohols, 455–58
   naming, 455–56
   physical properties, 456–57
   uses of, 457–58
Aldehydes, 464–67
Alden, Margaret, 306
Aliphatic hydrocarbons, 440
Alkali metals, 90
Alkaline earth metals, 92
Alkanes, 440, 441–42
   physical properties, 444–45
   structure and bonding of, 443–44
   substitution reactions, 445–46
Alkenes, 440, 446–47
   addition reactions, 449
   chemical and physical properties,
      448
   polymerization, 450
Alkyne, 441, 450, 451
Alpha particles, 49, 481
   scattering experiment, Rutherford's,
      44–45
Aluminum, properties and uses,
   75–76
Amorphous solids, 9
Analytical chemist, 169
Anhydrous substance, 203
Anions, 343
Anode, 40, 343
Aqueous ions, 338, 340
Aristotle, 32
Aromatic hydrocarbons, 452–55

Atmospheric modeller, 267
Atom
   excited state, 61
   ground state, 61
   nuclear, 44–45
   nucleus of, 45
   Thomson model, 44
Atomic core, 120
Atomic mass
   average, 57
   determining, 55–57
   unit, 55–57
Atomic number, defined, 46
Atomic radius graph, 115, 120
Atomic theory, Dalton's, 36–38
Atoms, 32, 125–26
   combining capacities, 188
   and electrostatic force, 116
Avogadro, Amadeo, 166
Avogadro's constant, 166
Avogadro's hypothesis, 298–99
Avogadro's law, 298, 299, 321

B

Balanced equations, 25–26, 208,
   230–32
Band of stability, 479
Bartlett, Neil, 108
Bases, 347, 348–50
Becquerel, Henri, 49
Benzene, 428–32, 453–55
   structure of, 452
Beta particles, 49, 481
Binary acids, 196–97
Binary compounds, 190
Binding energy, 487
Bohr atom, 60–63
Bohr diagram, helium atom, 110
Bohr, Niels, 60–61
Boiling point, standard, 264
Boiling water until it freezes, 265

Bond(s)
   double, 145
   in-between, 149–54
   length, 143
   spectrum, 154
   triple, 145
   type, predicting, 147–48
Bonding model, 135–38
Boyle, Robert, 33, 34
Boyle's law, 291, 292
Breeder reactors, 494
Brown, Mary, 242
Burette, 374, 376

C

Canada Customs tariff administrator,
   404
Carbon, 439
Carbon dating, 52
Carboxylic acids, 459–61
Catalyst, 213
Catenation, 146
Cathode, 343
Cathode rays, 40
Cations, 343
Cavendish, Henry, 71
Celsius temperature scale, 269
Chadwick, Sir James, 43
Chain reaction, 489
Charles' law, 285–86, 287–88
Chemical-based industries, 409–12
Chemical bonds, 130
Chemical change, 10–11
Chemical equations, 25–26, 207–11
   balanced, 208
   balancing by inspection, 209–10
   balancing complex, 209–11
   balancing simple, 208–209
   for aqueous reactions, 351
   skeleton, 208
   word, 207